Time Domain Electromagnetics

ACADEMIC PRESS SERIES IN ENGINEERING

Series Editor
J. David Irwin
Auburn University

Designed to bring together interdependent topics in electrical engineering, mechanical engineering, computer engineering, and manufacturing, the Academic Press Series in Engineering provides state-of-the-art handbooks, textbooks, and professional reference books for researchers, students, and engineers. This series provides readers with a comprehensive group of books essential for success in modern industry. Particular emphasis is given to the applications of cutting-edge research. Engineers, researchers, and students alike will find the Academic Press Series in Engineering to be an indispensable part of their design toolkit.

Published books in the series:

Industrial Controls and Manufacturing, 1999, E. Kamen
DSP Integrated Circuits, 1999, L. Wanhammar
Single and Multi-Chip Microcontroller Interfacing, 1999, G. J. Lipovski
Control in Robotics and Automation: Sensor-Based Integration, 1999, B. K. Ghosh, N. Xi, T. J. Tarn
Soft Computing and Intelligent Systems, 1999, N. K. Sinha, M. M. Gupta
Introduction to Microcontrollers, 1999, G. J. Lipovski

Time Domain Electromagnetics

Edited by

S. M. RAO
Department of Electrical Engineering
Auburn University
Auburn, AL

ACADEMIC PRESS
SAN DIEGO / SAN FRANCISCO / NEW YORK / BOSTON / LONDON / SYDNEY / TOKYO

This book is printed on acid-free paper.

Figures and select material in Chapter 1: E. K. Miller, "Time Domain Modelling in Electromagnetics," *Journal of Elecromagnetic Waves and Applications,* Vol. 8, No. 9/10 (1994): 1125–1172.
Chapter 5: D. A. Vechinski, S. M. Rao, and T. K. Sarkar, "Transient Scattering from Three-Dimensional Arbitrarily Shaped Dielectric Bodies," *Journal of the Optical Society of America A,* Vol. 11 (April 1994) 1458–1470.
Pages 269–273: F. J. German, "General Electromagnetic Scattering Analysis by TLM Method," *Electronic Letters,* Vol. 30, No. 9 (28 April 1994), published by Michael Faraday House.
Chapter 9: P. Bonnet, X. Ferrieres, F. Isaac, F. Paladian, J. Grando, J. C. Alliot, and J. Fontaine, "Numerical Modeling of Scattering Problems Using a Time Domain Finite Volume Method," *Journal of Electromagnetic Waves and Applications,* Vol. 11, No. 8 (1997): 1165–1189.

ACADEMIC PRESS
A Harcourt Science and Technology Company
525 B Street, Suite 1900, San Diego, CA 92101-4495
http://www.apnet.com

Academic Press
24-28 Oval Road, London NW1 7DX
http://www.hbuk.co.uk/ap/

Library of Congress Cataloging-in-Publication Data
Rao, S. M. (Rao, Sadasiva Madiraju), 1953–
Time domain electromagnetics / S.M. Rao.
p. cm. – (Academic Press series in engineering)
Includes bibliographical references and index.
ISBN 0-12-580190-4
1. Electromagnetism–Mathematics. 2. Time domain analysis. 3. Integral equations–Numerical solutions. 4. Differential equations–Numerical solutions. I. Title. II. Series.
QC760.R53 1999
537–dc21 98-52663
CIP

Printed in the United States of America

99 00 01 02 03 COB 9 8 7 6 5 4 3 2 1

Contents

Preface

In recent times, we have seen increased interest in the direct time domain methods to calculate electromagnetic scattering/interaction phenomenon. This may be due to the surge in activities in the areas of EMP, short-pulse radar, or other related applications. It may also be due to the fact that the time domain methods have several advantages over conventional frequency domain methods. For example, time domain methods work better for wideband signature studies, are better suited for parallel processing, and provide better visual representations for understanding the field interactions.

Although time domain methods have been available to the user for more than a decade, there is no single textbook dedicated to this subject for the interested student or practicing engineer. Most of the material is scattered in research papers. It is true that some related material has appeared as chapters in some books. Unfortunately, these books typically have very little to do with time domain studies and such treatment is often piecemeal. Thus, we believe that a textbook devoted entirely to time domain methods is long overdue, and we desire to fill this gap.

It is our intention to develop this book as a kind of textbook useful for teaching and for self-learning. For this reason, the emphasis is more on describing the already existing techniques (published in the literature) in a detailed manner rather than giving a piecemeal description of all the current and possible future techniques. However, we assume the reader has some familiarity with Maxwell's equations and basic mathematics. Furthermore, the book may be broadly divided into two parts. In the first part, we deal with the solution of integral equations. These methods have not been given the deserved attention until now, and this effort is intended to promote them or at least to put them on equal footing with the differential equation-based techniques. The second part examines the differential equation methods.

Finally, we note with profound regret that, during the course of development of this book, one of the contributors, Prof. K. R. Umashankar, passed away. We respectfully acknowledge his contributions to this book and hope that his efforts will not go in vain.

ACKNOWLEDGMENTS

I would like to express my appreciation to all the authors who responded to my requests, sometimes unreasonable, with enthusiasm and in a timely manner so that the project could be completed in the proposed time frame. Further, I would like to extend my gratitude to Douglas Vechniski and Surendra Singh for reading the entire manuscript and suggesting ways to improve the reading, removing inconsistencies, and checking the errors. Finally, I express my gratitude to my wife, Kalyani, and kids, Yeshaswi and Siri, for their understanding, love, and patience.

S. M. Rao
Auburn, Alabama
April, 1999

Contributors

Pierre Bonnet, Electromagnetics Department, ONERA, P.O. Box 72, F-92332, Châtillon Cedex, France.

Antonije R. Djordjevic, Department of Electrical Engineering, University of Belgrade, Belgrade 11120, Yugoslavia.

Xavier Ferrieres, Electromagnetics Department, ONERA, P.O. Box 72, F-92332, Châtillon Cedex, France.

Frederick J. German, Raytheon Systems Company, McKinney, TX 75070-2899, USA.

Griffin K. Gothard, Harris Corporation/GASD, Melbourne, FL 32902, USA.

Patricia Klotz, Numerical Modelisation Department, ONERA, P.O. Box 72, F-92332, Châtillon Cedex, France.

Bas L. Michielsen, Electromagnetics Department, ONERA, P.O. Box 72, F-92332, Châtillon Cedex, France.

Edmund K. Miller, 3225 Calle Celestial, Santa Fe, NM 87501-9613, USA.

Sadasiva M. Rao, Department of Electrical Engineering, Auburn University, Auburn, AL 36849, USA.

Jean L. Roumiguières, LASMEA, University "Blaise Pascal" of Clermont-Ferrand, 24, Avenue des Landais, 63177 Aubière Cedex, France.

Tanmoy Roy, Sun Microsystems Inc., 901 San Antonio Road, MS MPK15-103, Palo Alto, CA 94303, USA.

M. Salazar-Palma, Universidad Politécnica de Madrid, ESTI de Telecomunicación, Ciudad Universitaria, 28040 Madrid, Spain.

Tapan K. Sarkar, Department of EE & CS, Syracuse University, Syracuse, NY 13244, USA.

Korada R. Umashankar, Department of EE & CS, University of Illinois at Chicago, Chicago, IL 60607-7053, USA.

Douglas, A. Vechinski, Nichols Research Corporation, 1130 Eglin Pkwy, Suite A, Shalimar, FL 32579, USA.

CHAPTER 1

Introduction

E. K. MILLER

Although the Maxwell curl equations are usually first encountered in the time domain (TD), i.e., with time as an explicit, independent variable, until relatively recently most electromagnetic instruction and research has taken place in the frequency domain (FD) where time-harmonic behavior is assumed. A principle reason for favoring the FD over the TD in the precomputer era had been that a FD approach was generally more tractable analytically. Furthermore, the experimental hardware available for making measurements in past years was largely confined to the FD.

The inferior position of TD electromagnetics (EM) began to change with the arrival of the digital computer, which has profoundly affected what can be done not only numerically (or computationally) but also experimentally. Since the beginning of what has come to be called computational electromagnetics (CEM) in the early 1960s, there has been a steady growth in TD modeling. This growth, which began slowly at first, was primarily confined to integral equation (IE) treatments, but it has become almost explosive during the past 10 years as TD differential equation (DE) modeling has attracted wide attention. This chapter summarizes the status of direct TD (as opposed to Fourier-transformed FD results) modeling and highlights some of the current research areas.

The remainder of this chapter surveys the previous aspects in more detail. In Section 1.1, some important aspects related to radiation phenomena using time domain snapshots are discussed. Some general EM modeling choices, followed by the reasons why TD modeling in particular might be advantageous and a brief account of the evolution of TDEM modeling, are considered in Section 1.2. Basic steps in developing a TD model and its subsequent application are discussed in Section 1.3. In Sections 1.4 and 1.5, some specific issues related to TDIE and TDDE modeling, including spatial meshes, closure conditions, obtaining far fields, and alternate formulations are considered. In Section 1.6, topics common to time domain modeling such as increasing late-time stability, extracting resonances, signal processing of TD results, total/scattered field formulations, and handling dispersion, nonlinearities, and time variations are summarized. An extensive reference list is also included at the end of the chapter.

1.1 AN INITIAL EXPLORATION OF TIME DOMAIN PHENOMENA

Time domain modeling in electromagnetics has been of interest since the advent of Maxwell's equations. Despite the fact that, historically, most analysis and experimentation was performed in the frequency domain, EM fields are dynamic phenomena, and even FD results contain an explicit time-harmonic variation. Unfortunately, it is rare that FD solutions are examined as a function of time by the simple expedient of determining the real components of the fields as the time phasor rotates. Observing the time behavior of FD fields could add greatly to our physical understanding as is demonstrably the case when a TD result is available. Although the transient response of an object can be obtained directly in the TD, or from transformed FD data, the emphasis here is on the former, so the results presented are called time domain, rather than transient responses.

1.1.1 The Infinite-Length Wire Antenna

Consider, for example, the feedpoint current and broadside radiated field of an infinite, circular, perfect electric conducting (PEC) cylinder excited by a Gaussian voltage pulse $V(t) = V_o e^{-a^2t^2}$ as presented in Fig. 1.1, computed from a TD IE model [1]. (Since the cylinder diameter is small compared with the wavelength, it is appropriate to refer to the geometry as a "wire" structure as is done here). Initially, the current and radiated field follow the rise in voltage. However, after the voltage peak is reached they begin to fall slightly faster in value and then exhibit a negative undershoot which lasts well beyond the time at which the voltage becomes negligible. As time continues to progress, the current and charge decay back toward zero as the two halves of the antenna return to a neutral state.

These effects are more clearly demonstrated in Fig. 1.2. Here, we observe that the feedpoint current and broadside radiated field are essentially identical in their time variations. Also, note that the feedpoint current and voltage appear proportional until the voltage peak is reached, after which the current decreases somewhat faster and exhibits an overshoot.

This simple computer experiment displays some very fundamental physics. It is relevant at this point in discussing TD radiation to include the Lienard–Wichert potentials [2],

$$\boldsymbol{E} = \frac{e}{4\pi\epsilon_0}\left[\frac{1}{s^3}\left(\boldsymbol{r} - \frac{r\boldsymbol{u}}{c}\right)\left(1 - \frac{u^2}{c^2}\right) + \frac{1}{c^2 s^3}\left\{\boldsymbol{r} \times \left[\left(\boldsymbol{r} - \frac{r\boldsymbol{u}}{c}\right) \times \frac{d\boldsymbol{u}}{dt'}\right]\right\}\right], \tag{1.1.1}$$

where $s = r - (\boldsymbol{u} \cdot \boldsymbol{r})/c$, $\boldsymbol{u}$ is the charge velocity, and $t' = t - r/c$ is the retarded time with c

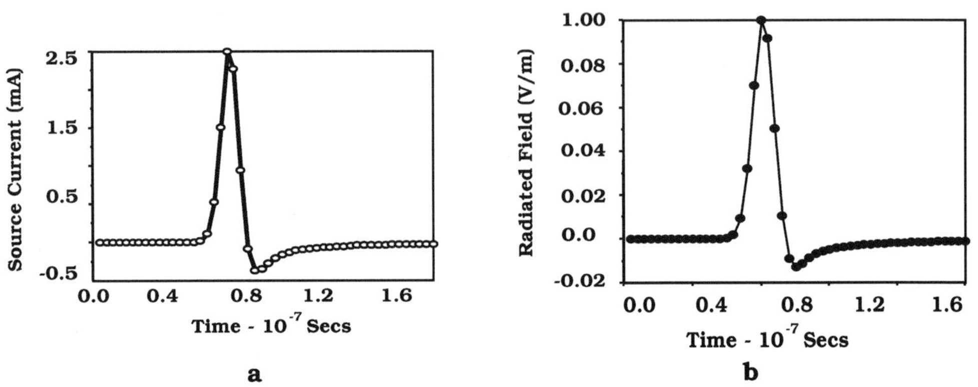

FIGURE 1.1
Exciting an infinite, circular wire by the Gaussian voltage pulse. (a) The time domain feedpoint current and (b) the negative of the broadside radiated field.

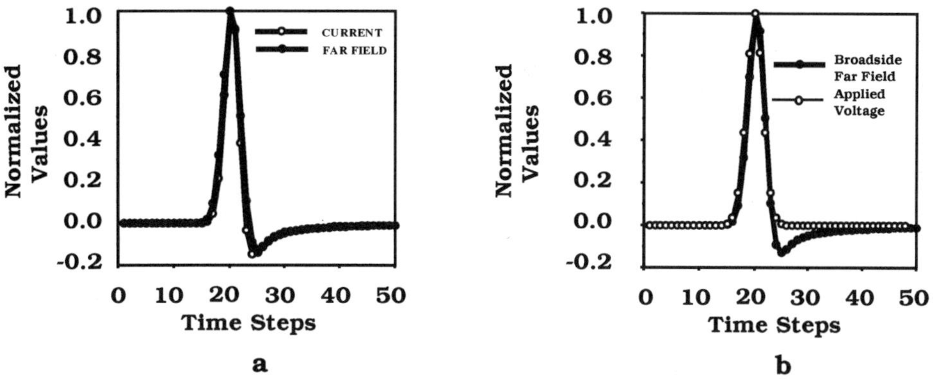

FIGURE 1.2
Comparison of (a) the feedpoint current with the negative of the broadside radiated field and (b) the exciting voltage with the feedpoint current.

representing the speed of light. The magnetic field is given by

$$\boldsymbol{B} = \frac{1}{rc}(\boldsymbol{r} \times \boldsymbol{E}). \tag{1.1.2}$$

These equations show explicitly that the only source of radiated EM fields is accelerated charge as the $\boldsymbol{du}/dt'$ term produces a $1/r$ field. Therefore, it is insightful to consider where charge acceleration is occurring as might be deduced from the previous results.

Beginning with the initial "turning on" of the exciting voltage, charge on the antenna, originally at rest and in equilibrium, is set into motion by the electric field that results, as shown in Fig. 1.3. The positive charge moves to the right and the negative charge to the left; both cause positive currents but their pulses travel in opposite directions. Although the numerical model used in obtaining the results of Fig. 1.1 is limited to finite-length wires, until end reflections occur the behavior is identical to that of an infinite wire, as is also the case here. As the voltage increases, proportionately more charge is set into motion with a proportionate increase in the radiated field. This process continues until part of the outward-propagating current and charge are reflected back toward the feedpoint, when the current and radiated field no longer follow the excitation voltage.

The feedpoint current grows with the increasing excitation voltage, and since the conduction current is approximately $I = Qv \approx Qc$ the feedpoint charge density follows this same buildup. This increasing charge density continues to undergo the same effective acceleration since its

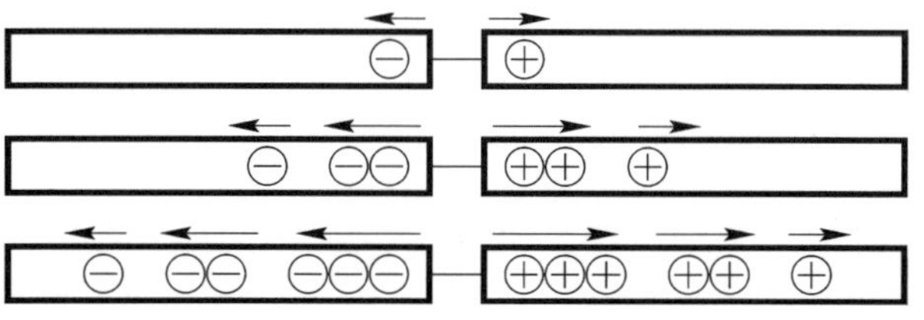

FIGURE 1.3
A qualitative picture of the charge and current caused by a Gaussian voltage pulse applied to a wire antenna.

velocity changes from zero to near c as it is "pushed" out onto the antenna. The feedpoint charge acceleration is accompanied by a radiation field that builds up with the same time variation. Although we may visualize the charge that leaves the feedpoint as continuing to flow along the antenna's surface, in reality the charge motion is more like a domino effect; the charges that comprise the current flow do not move very far before their motion is transmitted to neighboring charges that continue the current flow.

This behavior continues until about the time when the voltage peaks, after which the current decays more rapidly, reaching a negative peak of approximately 20% of the positive peak value and decaying back toward zero over a substantially longer period of time. While the reason for the more rapid current decay may not be obvious, the cause of the negative undershoot is more so. It can only happen because some of the charge flowing away from the source region is reflected back and whose reversal in direction results in a sign change of its current. This reversal of charge flow also represents an acceleration, although of opposite sign when compared with the charge flow caused by the original excitation. Consequently, we should also expect a sign change in the radiated field, which indeed occurs.

1.1.2 The Finite-Length Wire Antenna

The previously discussed effects are shown more clearly in Fig. 1.4, in which snapshots of the current and charge on a finite-length wire are shown at several instants of time. The effect of the current reflection is seen as a decrease in the amplitude of its pulse as it propagates down the wire accompanied by a negative trailing part which is clearly associated with the negative current at the feedpoint. The cause of the current reflection is evidently due to a spatially varying wave impedance down the wire. This conclusion is consistent with the space-dependent wave impedance for an infinite antenna of radius a in a medium with wave impedance η_0 and wavenumber k_0 analyzed in the frequency domain as given by [3]

$$Y(z) \approx 4\pi/[\eta_0 \ln(2iz/k_0a^2\Gamma)], \tag{1.1.3}$$

resulting in an "impedance-reflection" effect.

The feedpoint current and broadside radiated field for the finite-length wire antenna are presented in Fig. 1.5. It should be noted that the current and field are identical to that of an infinite-length wire (see Figs. 1.1 and 1.2) until end reflections affect them. Peaks in these quantities thereafter alternate since the radiated field maxima occur upon end reflection of the charge pulses, whereas the current peaks occur as the current pulses arrive back at the antenna's center.

Since there would be no radiated field without current and charge flowing on an object such as a wire, it seems reasonable to ask whether these sources might provide some indication of how quickly an impulsively excited wire radiates the energy stored in the near field that they produce. One way to obtain this stored energy would be to integrate the square of the electric and magnetic fields in its vicinity. Another, less computationally involved, way would be to integrate the square of the current and charge over the object as a function of time, an approach demonstrated here. The quantities evaluated are

$$W_I(t) = \int_{C(r)} I^2(z,t)\,dz \quad \text{and} \quad W_Q(t) = C^2 \int_{C(r)} Q^2(z,t)\,dz, \tag{1.1.4}$$

where $C(r)$ defines the wire contour in space, here a straight line along the z-axis.

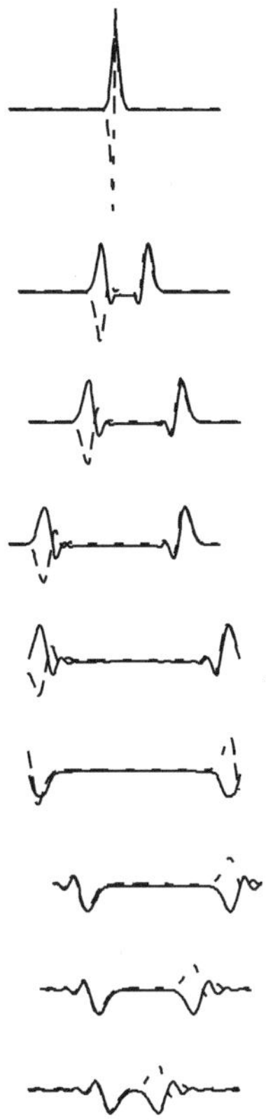

FIGURE 1.4
Snapshots of the current (solid line) and charge (dashed line) on a wire antenna excited at its center by a Gaussian voltage pulse.

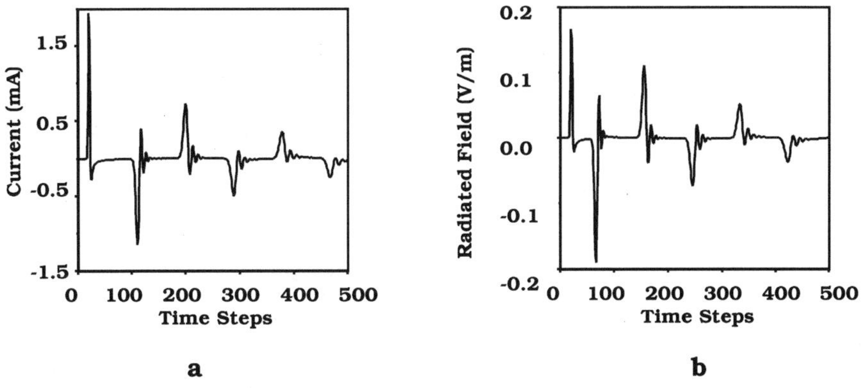

FIGURE 1.5
The feedpoint current (a) and broadside radiated field (b) of a 1-m dipole excited at its center by a Gaussian voltage pulse.

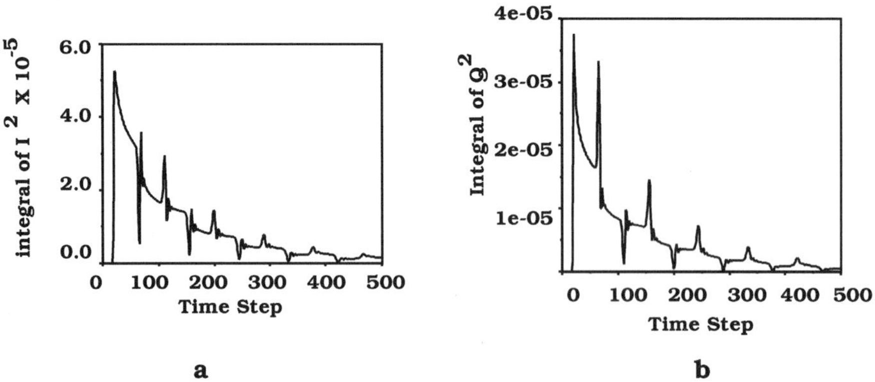

FIGURE 1.6
Plots of the $W_I(t)$ (a) and $W_Q(t)/c^2$ (b) for a Gaussian pulse excited, finite-length wire antenna.

These quantities are plotted in Fig. 1.6 for the antenna, whose source current and radiated field were shown in Fig. 1.4. These plots clearly demonstrate the interchange of energy between kinetic energy of charge motion (current), or the magnetic field, and the potential energy of charge separation, or the electric field, along the antenna. As the oppositely signed charge waves meet at the antenna's center, as occurs after each end reflection, they exactly cancel, upon which W_Q becomes zero while W_I peaks. The converse behavior occurs during end reflection in which the charge piles up and the current goes to zero. The decay of both quantities between these two limits demonstrates a continuous radiation process, whereas the abrupt change in their sum on end reflection illustrates a more abrupt radiation process.

The feedpoint current and broadside radiated field that result when the same antenna is excited by a Gaussian voltage step (an integral of a Gaussian pulse) are shown in Fig. 1.7. After the current reaches a maximum value, it declines monotonically in time due to partial reflection of the outward-propagating current wave which in turn produces a similarly shaped broadside radiation field. As time approaches infinity, the current and radiation field will both decay to zero,

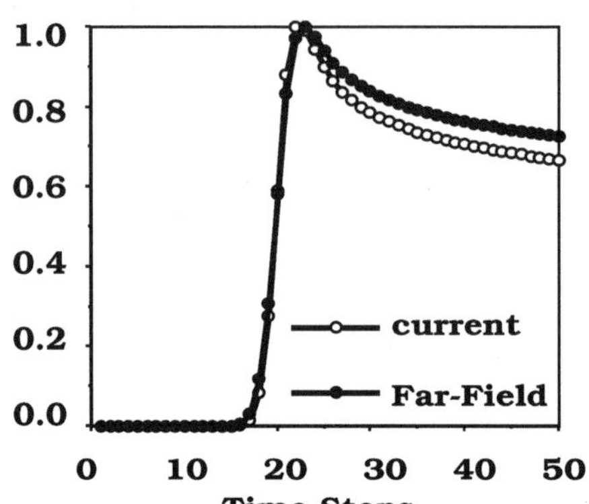

FIGURE 1.7
(a) Feedpoint current and (b) broadside radiated field for a finite-length antenna excited by a Gaussian-step (integral of Gaussian pulse) voltage pulse.

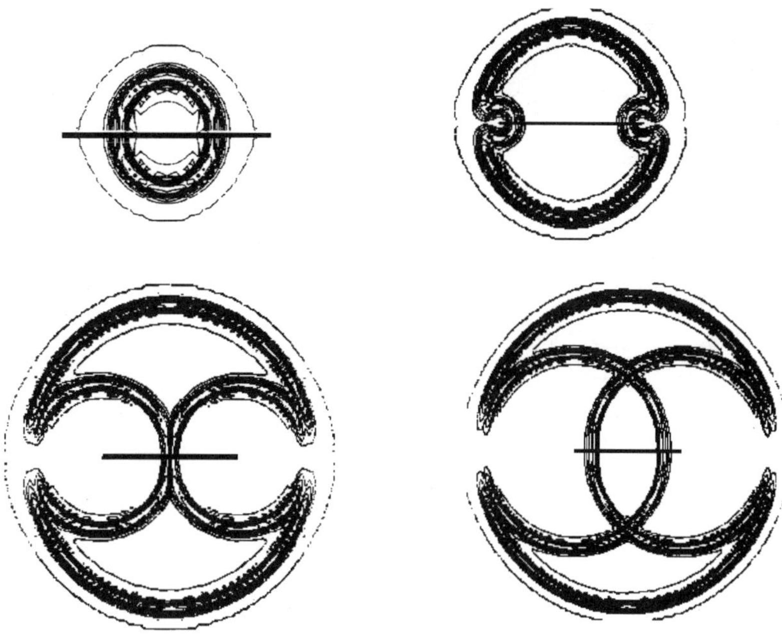

FIGURE 1.8
Space-contour plots of the near electric field around an impulse excited dipole antenna.

whereas the two halves of the antenna will have a static charge distribution because the exciting field maintains a constant voltage difference between them.

Examining the near fields of an impulsively excited antenna can be especially illuminating regarding the question of where radiation originates, as shown in Fig. 1.8 in which several snapshots of the near-electric-field contours around the impulsively excited, finite-length dipole are presented. These plots are two-dimensional slices of three-dimensional space. These plots clearly illustrate the dominant radiation mechanisms. The larger "bubble" grows at approximately the speed of light, being centered at the feedpoint and caused by the *turn-on* of the exciting field. The smaller bubbles grow at a similar rate but are centered about the antenna ends due to the acceleration caused by end reflection of the current/charge waves.

The fact that this near-field behavior is indicative of the origin of radiation is confirmed by the two plots of Fig. 1.9. These plots exhibit a high degree of similarity, showing that the near-field bubbles represent the beginning stage of far-field radiation pulses. Following this process further would reveal additional end-centered field bubbles of decreasing amplitude as the stored energy is radiated away from the antenna.

1.1.3 The Finite-Length Wire Scatterer

It is useful to examine results similar to those discussed previously for a Gaussian, plane wave pulse, incident from broadside upon a finite-length wire. The center current and broadside scattered field for this problem are shown in Fig. 1.10. These plots display radiation physics in a different way than for the antenna case but are similarly revealing. During the first two cycles, the current displays a distinct radiation-decay signature due to impedance reflection, but as the higher frequency components radiate away more quickly the scattering current becomes similar to that

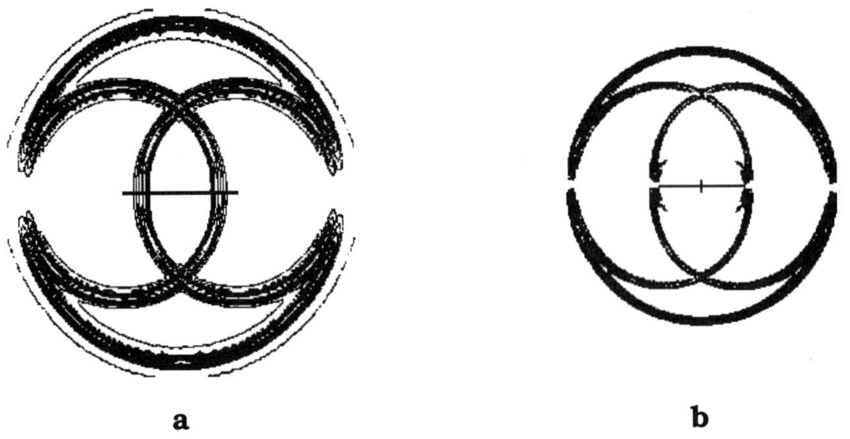

FIGURE 1.9
Space contour plots of (a) the near electric field and the far radiated field (b) for a Gaussian pulse excited finite-length wire.

seen in the antenna case of Fig. 1.5, as does the late-time scattered field. It should be noticed that the current builds up to a maximum with the arrival of the incident wave, after which it displays an exponential decay until it changes sign quite abruptly, followed by periodic sign reversals and a decaying amplitude. The scattered field exhibits a "specular" flash due to the initial excitation followed by a behavior quite similar to that of the current.

Plots of the spatial current at several instants of time are presented in Fig. 1.11. Here the early time current distribution is seen to be uniform over the center of the wire which, after rising to a maximum, slowly decreases due to the same reflection phenomenon as already seen in the source current for the antenna problem. Eventually, the end-reflected current drives the current negative along most of the length of the wire. The specific relative influence of these two reflection effects depends on the wire length, the time variation of the incident pulse, and the wire radius which determines the impedance reflection effect. Note that the early time uniform current demonstrates the fact that the current at a given point is determined solely by the local incident field and the current arriving there from the nearby parts of the wire. For a straight wire, this

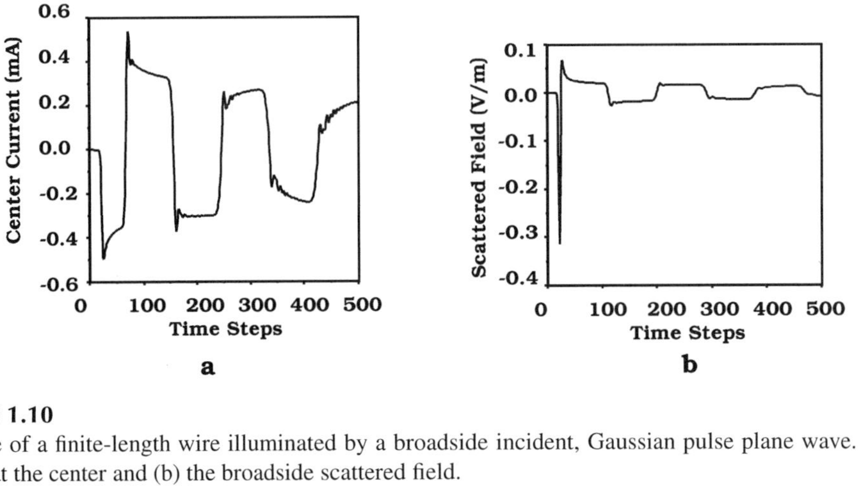

FIGURE 1.10
The case of a finite-length wire illuminated by a broadside incident, Gaussian pulse plane wave. (a) The current at the center and (b) the broadside scattered field.

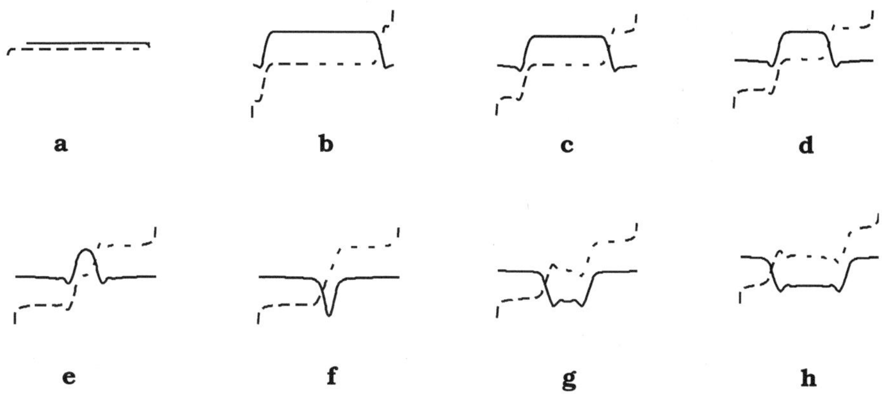

FIGURE 1.11
Snapshots of the current excited by a broadside-incident, Gaussian plane wave pulse on a finite-length wire.

early current can be approximated by a simple time integral of the incident field as discussed in Section 1.6.

Finally, W_I and W_Q are plotted in Fig. 1.12 for the scattering case. The alternating maxima of these two quantities show that the energy is oscillating respectively between being primarily magnetic and primarily electric in nature. To the extent that W_I and W_Q actually represent the energy stored in an object's near fields, their sum must monotonically decrease in time due to energy loss as a result of radiation. This possibility evidently needs further exploration since, among other questions, it is not clear that the proportionality constant used in computing W_Q in Eq. (1.1.4) should be exactly c^2 everywhere over an object's surface. For example, since the charge must be slowing as it nears the end of a wire, using Qc in that region as the measure of electric energy may overestimate its true value.

1.1.4 Late-Time Radiation from an Impulsively Excited Perfect Conductor

In this section, how a PEC can radiate away the energy it "collects" from an incident field, such as the one considered for the scattering case is discussed. By definition, the total tangential electric

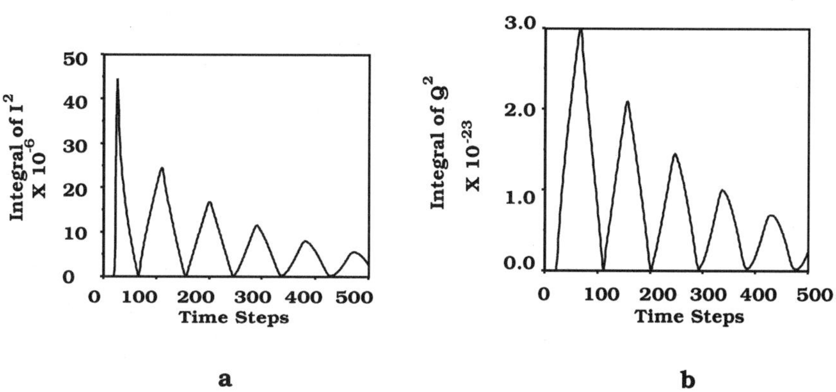

FIGURE 1.12
Plots of $W_I(t)$ (a) and $W_Q(t)/c^2$ (b) for a straight wire excited by a broadside-incident, Gaussian plane wave pulse.

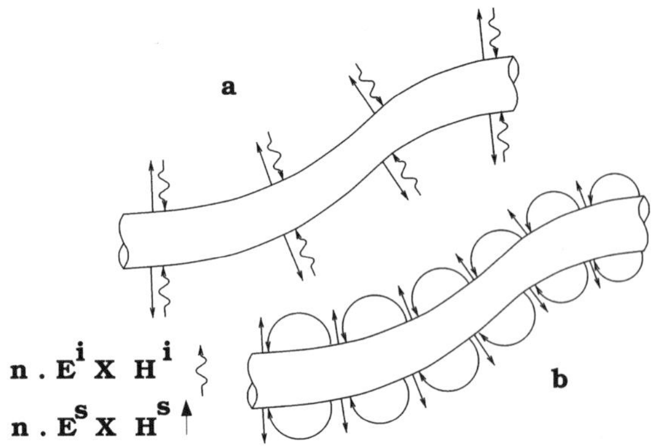

FIGURE 1.13
These diagrams conceptually depict how a perfectly conducting object can (a) collect and (b) radiate EM energy even though the component of the Poynting vector normal to the surface is identically zero.

field at the surface of a PEC is zero. During the time the incident field is present on such an object, the boundary condition is $[\boldsymbol{E}^{\mathrm{i}} + \boldsymbol{E}^{\mathrm{s}}]_{\tan} = 0$, where the first term accounts for energy collection from the incident field and the latter is the "response" field which accounts for both near-field stored energy and far-field radiated energy. However, after the excitation is no longer significant, then $\boldsymbol{E}^{\mathrm{s}}_{\mathbf{tan}} = 0$. This implies that the normal component of the Poynting vector over the object's surface is zero everywhere. This situation is illustrated conceptually in Fig. 1.13.

However, if the Poynting vector's normal component is everywhere zero, how can the object radiate away the remaining collected energy? The answer is contained in the previous observation that $\boldsymbol{E}^{\mathrm{res}}_{\mathrm{tan}}$ accounts for both near-field and far-field energy. However, rather than balancing the effect of the incident field at late time, these two components of $\boldsymbol{E}^{\mathrm{res}}_{\mathrm{tan}}$ now balance each other, as illustrated in Fig. 1.13b. Since far-field energy is by definition outward flowing, this implies that as the stored near-field energy collapses back onto the object, representing an inward energy flow; this is balanced by an equal outward flow, some of which remains in the near field and some of which is converted into far-field energy. Thus, through the tangential field boundary condition, a PEC converts transient near-field, stored energy to far-field radiated energy.

1.1.5 Some Special Capabilities of Time Domain Models

It is worth mentioning several of the advantages in performing time domain modeling. First, wideband data are made available from one model computation as opposed to the frequency domain approach, in which many frequency samples are required to obtain the equivalent data. Second, it provides a more straightforward approach in modeling impedance nonlinearities in the time domain. Third, time domain models can handle time variations of load impedances. For example, the use of a time domain model for nonlinear loading is demonstrated in Fig. 1.14 in which the input current, broadside radiated field, and the radiated field spectrum are presented [4]. The radiated field is seen to have two opposite-sign pulses, caused by the initial turn on of the drive voltage, and the stopping of the current-charge pulses as their outward propagation is stopped, in accord with the Lienard–Wichert potentials. Such loading might be used for pulse shaping.

Time varying problems are also well suited to time domain modeling. For example, when a dipole is illuminated by a time-harmonic plane wave incident from broadside while its center load varies sinusoidally in value, we obtain the results of Fig. 1.15. Here, a 16 MHz, broadside-incident, plane wave illuminating a half-wave dipole having a center load whose resistance varies

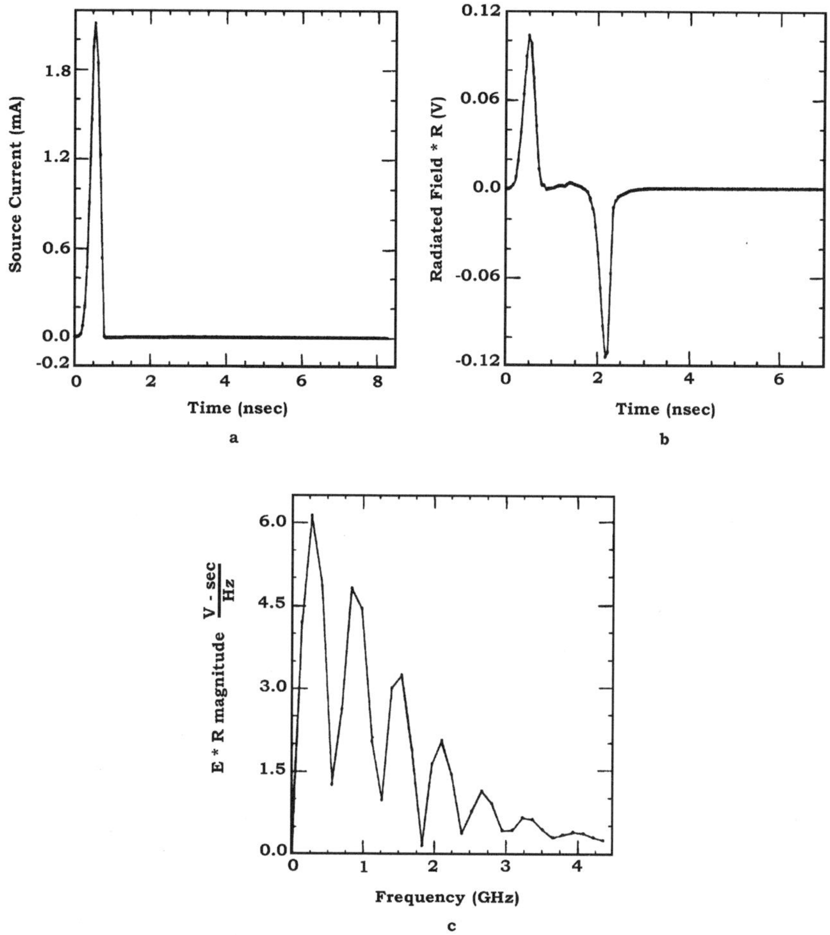

FIGURE 1.14
A dipole antenna continuously loaded with diodes. (a) The feedpoint current, (b) the broadside radiated field, and (c) the spectrum of the radiated field.

sinusoidally at 4 MHz is shown. Interaction of the incident field and the time-varying load causes intermodulation that produces upper and lower sidebands in the scattered field, a phenomenon that can significantly modify its radar cross section as the resulting frequency spectrum demonstrates. It may be noted that dynamically varying the reflectivity of a scatterer can change the scattered field spectrum from what it would be otherwise.

1.2 MODELING CHOICES IN CEM

In discussing CEM in general and TD modeling in particular, it is appropriate to consider two basic questions:

1. What alternative modeling approaches are available for CEM?
2. What are the advantages of TD models relative to the other possibilities?

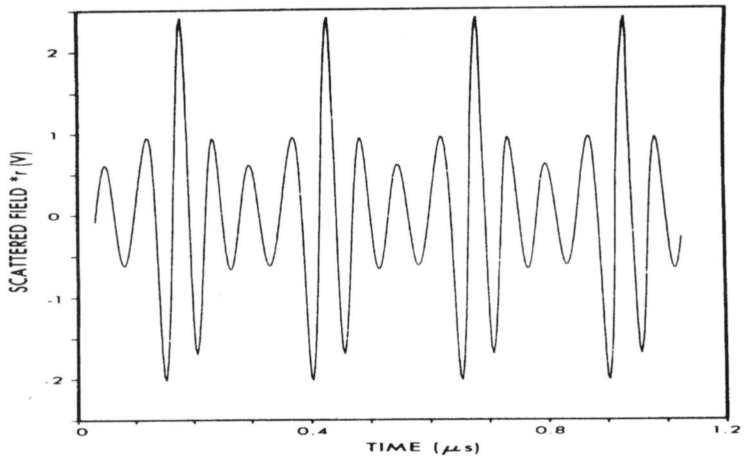

a

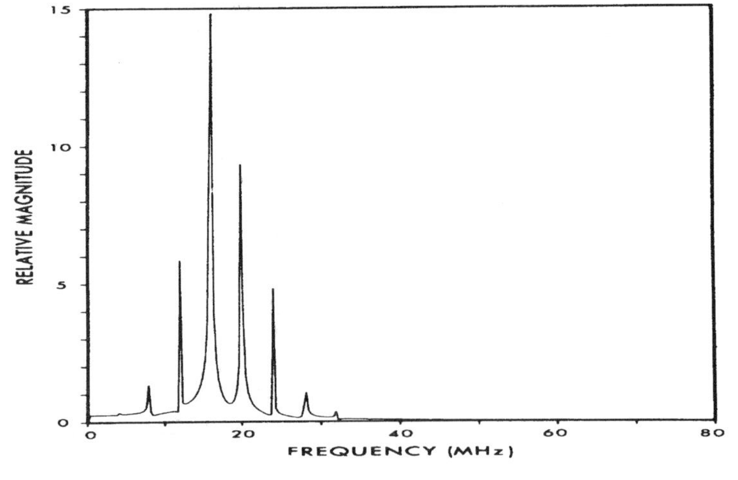

b

FIGURE 1.15
(a) Broadside scattered field and (b) the spectrum of a time-dependent loaded scatterer.

To answer both questions, we observe that there are four major, first principles, models in CEM, given by,

1. Time domain differential equation (TDDE) models, the use of which has increased tremendously over the past several years, primarily as a result of much larger and faster computers.
2. Time domain integral equation (TDIE) models, although available for more than 30 years, have gained increased attention in the past decade. The recent advances in this area make these methods very attractive for a large variety of applications.

3. Frequency domain integral equation (FDIE) models remain the most widely studied and used models; they were the first to receive detailed development.
4. Frequency domain differential equation (FDDE) models, whose use has also increased considerably in recent years, although most work to date has emphasized low-frequency applications.

1.2.1 Why Model in the Time Domain?

Besides physical interpretability, as demonstrated previously, there are two basic reasons for modeling in the time domain which provide a distinct advantage in most applications in which transient results are available:

1. **Computational efficiency:** For certain problems and/or approaches, fewer arithmetic operations are required when performed in the time domain. For example, in applications in which the early time peak response of an object to an impulsive field is sought, a TD model offers an intrinsically more efficient approach compared to a FD model, which requires frequency samples across a broad bandwidth followed by a Fourier (or other) transform to obtain the desired result. When seeking broadband information, a TD model is also a more natural choice because it provides a transient response whose bandwidth is limited only by the frequency content of the excitation and the time and space sampling used in developing the model. In addition, TD models may offer a naturally better match to massively parallel computer architectures than do FD models.
2. **Problem requirements:** Problems that involve nonlinear media or components can usually be modeled in a more straightforward and efficient manner in TD, as can problems involving time-varying media and components. An additional benefit of TD modeling is that time gating can be used in modeling, as in measurements, to remove the effects of unwanted reflections or to simulate larger objects. An example of the latter application is that of replacing an infinite cylindrical antenna model with a three-dimensional (3D) wire model whose behavior at a midpoint feed at early times, prior to end reflections, will be identical to that of an infinite structure [5]. Finally, body resonances, or singularity expansion method (SEM) poles, may be computed more directly from a TD model.

1.2.2 Evolution of Time Domain Modeling

Development of computer-era TD CEM models might be traced to physical optics work [6–8], in which the relationship between an object's ramp response and its cross-sectional area along the propagation direction of an incident plane wave was derived. Representative examples of the growing variety of TD research that followed include the original TDDE approach by Yee [9] which forms the basis of the widely used finite-difference time domain (FDTD) model. An extensive survey of the applications of this method is available [10]. A related application of a TDIE to acoustics was presented by Mitzner [11]. This work was closely followed by TDIE EM applications [12–15]. An alternate implementation of TDDE models was shortly thereafter initiated as the transmission-line method (TLM) by Johns and Beurle [16]. Recently, TD versions of the method of lines (TDML) and the geometrical theory of diffraction (TDGTD) were presented by Nam *et al.* [17] and Veruttipong [18], respectively. It seems likely that TD versions of other modeling approaches can also be expected to be developed.

Accompanying this initial research into TD CEM models was continuing work of a more analytical nature, including a series of papers in the early 1960s, one of which was a study by Brundell [19] on transient current waves propagating azimuthally around an infinite circular cylinder. Related papers by Wu [20] and Einarsson [21] investigated the impulse response of an infinite dipole antenna. Another fundamental analytical study of antennas excited by impulsive sources was presented by Franceschetti and Pappas [22]. Tijhuis *et al.* [23] reexamined a classical problem, the transient response of a thin, straight wire.

An increasing amount of TDIE modeling has followed. For example, Miller *et al.* [24] emphasized wire applications of the electric field IE (EFIE) which is further developed together with surface modeling using the magnetic field IE (MFIE) [25]. Other examples of developing TD models include Lui and Mei [26], Bennett [27, 28], Bennett and Mieras [29, 30], Gomez *et al.* [31], Marx [32], Bretones *et al.* [33], Gomez *et al.* [34, 35], Rao and Wilton [36], Vechinski *et al.* [37, 38], and Walker *et al.* [39, 40]. Application examples have grown commensurately, as demonstrated by some nonlinear modeling [41, 42], and as illustrated by using the time-gating feature of TD modeling for simulating infinite structures with a 3D wire model [5]. Selective overviews of this early TD research are given by Bennett and Ross [43], Miller and Landt [4], and Miller [44, 45].

1.2.3 Some General References

Although the literature devoted to TD EM is rapidly expanding, there are few books devoted to the topic. Two edited books are by Felsen [46] and Miller [47]. The former covers a variety of topics in TD modeling and analysis, whereas the latter systematically addresses the topic of TD measurements in electromagnetics together with an associated discussion of modeling and signal processing applications. Also, books by Kunz and Luebbers [48] and Taflove [49] are devoted exclusively to the FDTD formulation, whereas the TLM is the topic of a book by Christopoulos [50]. Recent edited books devoted to a related topic, ultra-wideband EM, include Noel [51], Bertoni *et al.* [52], and Taylor [53], whereas Lamensdorf and Susman [54] presented work on pulsed antennas. Periodic publications in which TD articles are routinely published include the following:

- *IEEE Transactions on Antennas and Propagation*
- *IEEE Antennas and Propagation Magazine*
- *IEEE Transactions on Microwave Theory and Techniques*
- *IEEE Transactions on Electromagnetic Compatibility*
- *The IEEE Proceedings*
- *Journal of Electromagnetic Waves and Applications*
- *International Journal of Numerical Modeling*
- *Journal of Computational Physics*
- *The Journal and Proceedings of the Applied Computational Electromagnetics Society*
- *Radio Science*
- *Electromagnetics*
- *The Journal of the Acoustical Society of America*

Also, many special issues of these journals have been devoted in whole, or in part, to various aspects of TD modeling. The following is a sampling of such publications:

- *IEEE Proceedings*, vol. 77, No. 5, May 1989
- *IEEE Proceedings*, vol. 79, No. 10, October 1991

- *IEEE Proceedings*, vol. 80, No. 1, January 1992
- *IEEE Transactions on Antennas and Propagation*, vol. 37, No. 5, May 1989
- *International Journal of Numerical Modeling*, vol. 2, No. 4, 1989.
- W. R. Stone (ed.), *Radar Cross Sections of Complex Objects*, New York: IEEE Press, 1990.
- *Computer Physics Communications*, vol. 68, Nos. 1–3, 1991
- *International Journal of Numerical Modeling*, vol. 5, No. 3, August 1992
- *International Journal of Numerical Modeling*, vol. 6, No. 1, February 1993
- *International Journal of Numerical Modeling*, vol. 7, No. 2, April 1994
- *Journal of the Optical Society of America*, vol. 11, April 1994

1.3 GENERAL ASPECTS OF TIME DOMAIN MODELING

The formulation and numerical development of a TD model in general involves many basic steps whether a DE or an IE approach is being followed. In the following, some of these considerations are discussed.

1.3.1 Model Development

For any numerical solution, it is necessary to develop the required equations and solve them on a computer. The equations thus developed must include the physics of the problem as well as the geometrical features. The following four steps are carried out in EM time domain problems:

1. Develop time-dependent integral equations using potential theory along with appropriate boundary conditions (see Section 1.4) or, alternatively, begin with the time-dependent Maxwell curl equations or their equivalent (see Section 1.5).
2. Sample these equations in space and time utilizing an appropriate geometrical space grid and suitable basis and testing functions. Note that, depending on the choice of formulation, the space grid may cover the structure and/or the surrounding space.
3. Develop a set of simultaneous equations relating known and unknown quantities. Generally, the known and unknown quantities are the excitation field or its derivatives and the radiated/scattered field or induced current and charge, respectively.
4. Generate a computer solution of this system in space and time as an initial-value problem.

1.3.2 Explicit vs Implicit Solution

Note that TD models can be either "explicit" ($\Delta t \leq R_{\min}/c$, where Δt is the time step, $R_{\min}$ is the minimum spatial sampling interval, and c is the speed of light) or "implicit" ($\Delta t > R_{\min}/c$). In the former case, spatially adjacent field and source samples do not interact within the same time step (due to the causal nature of EM fields because c is finite) and so the system of equations which they must satisfy can be solved algebraically, i.e., a matrix does not need to be solved for either a DE or IE model. In an implicit solution, same-time-step interactions are permitted, although because of the finite speed of light, the number of these interactions is limited, so advancing the solution from time step i to time step $i + 1$ requires the solution of, at most, a sparse rather than a full matrix.

1.3.3 Excitation Requirements

In TD problems, normally a "numerically gentle" turning on of the excitation is required, whether it be an incident plane wave in a scattering computation or an applied voltage for a radiation problem. This is necessary to avoid introducing excitation frequency components that exceed the highest frequency for which the model is valid. For impulsive, or step-function sources, for example, the initial excitation value would be required to be less than some fraction of its peak, and the rise time would need to extend over some minimum number of time steps. It is convenient in TD modeling to use a Gaussian-pulse time variation, given by

$$g(t) \propto e^{-a^2t^2},$$

whose corresponding frequency coverage is given by

$$G(w) \propto \left(\frac{1}{a}\right) e^{\frac{-\pi^2\omega^2}{a^2}}.$$

Thus, by varying the single parameter a, coverage of a wide frequency range can be assured.

An exception to this general rule, as discussed later, is provided by the case in which a time-harmonic excitation is simultaneously turned on everywhere in the region being modeled and the goal is to advance the solution until a steady state is reached.

Although obtaining wideband information from a single computation is one of the major advantages of using a TD model, time-harmonic, continuous-wave (CW) excitation is sometimes employed instead. When this is done for radar scattering, it may not be necessary to propagate the incident plane wave entirely across the object and continue the solution until the scattered field reaches a steady state, analogous to time stepping for impulsive excitation until the response decays to zero. Instead, the incident field can be turned on over the whole object simultaneously and then the time stepping can be continued until the response has reached a steady state. For large enough objects the steady-state response in such cases might be attained in a time interval shorter than the propagation, or transit, time across the body. In essence, this means that entire-body interactions for large objects may not be needed to obtain their CW behavior with acceptable accuracy. Consequently, a substantial reduction in computer time can be realized, possibly reducing the frequency dependence from $\approx f^5$ to $\approx f^2$ for 3D problems, as discussed in Section 1.4.7. Walker [40] describes other approaches intended to reduce the frequency dependence of TDIE models.

1.3.4 TD Solution

As an initial-value problem, a TD computation begins with specified values of the unknowns. Most often for a TDIE model, all current samples $I_{i;j}$ and charge samples $Q_{i;j}$ would be assumed to be zero prior to beginning the computation at $t=0$ (the first subscript refers to the space index and the second to the time index). Alternatively, the initial values might be nonzero when relaxation phenomena, such as the time required for a static charge distribution on a PEC to return to neutrality after closing a switch, are of interest.

The termination point of a TD computation depends on the problem requirements. In an application such as electromagnetic pulse interactions, in which the peak current is needed, the time-stepping solution might be stopped when the first current maximum has been reached. If the total energy collected by an object such as a straight wire is needed instead, the computation could be terminated when the integral $\int\int E^{\mathrm{i}}_{\mathrm{tan}}(x,t)I(x,t)\,dx\,dt$ stabilizes, where $E^{\mathrm{i}}_{\mathrm{tan}}(x,t)$ is the tangential component of the incident field at location x and time t. This integral provides a measure of how much energy has been collected by the object. If the goal is to obtain a wideband frequency response using a Fourier transform, the computation would normally proceed until the waveforms of interest have suitably converged. For an application in which body resonances (SEM poles) are to be estimated from time waveforms, a time interval of two to four times L/c, where L is the object's largest dimension, would be needed [4].

As mentioned earlier, space and time sampling determine the equivalent bandwidth of a TD model. The spatial sampling density, or conversely, the sampling interval, Δx, is driven by the maximum frequency, f_{max}, or minimum wavelength, λ_{min}, for which it is desired that the model produce reasonably valid results. Analogous to FD modeling, it has been found [25] that $\Delta x < \lambda_{\text{min}}/n$ is needed, where $n \approx 6$–10 for wires and around 4 for surfaces.

The space–time sampling employed in developing a numerical model establishes only the maximum frequency for which valid results might be obtained from that model. It should be noted that, in reality, it is the spectral content of the excitation that determines whether the intended frequency range is actually covered.

1.4 TIME DOMAIN INTEGRAL EQUATION MODELING

A TDIE is based on the simple scalar Green's function:

$$G(\boldsymbol{r}, t; \boldsymbol{r}', \tau) = \frac{\delta\left(t - \tau - \frac{R}{c}\right)}{R},$$

where $c = 1/\sqrt{\mu\epsilon}$ represents the medium wave speed, $R = |\boldsymbol{r} - \boldsymbol{r}'|$, and $\boldsymbol{r}'$ and $\boldsymbol{r}$ denote source and observation points at time instants τ and t, respectively [27]. A detailed analysis of the application of TDIE to various geometries is presented in Chapters 2–5. In the following, some general issues related to TDIE are discussed.

Many questions arise in implementing a TDIE model, including the following:

1. Since TDIE solutions are based on the well-known method of moments, what kinds of basis and testing functions are appropriate for the problems of interest?
2. Should a single-field representation be employed, i.e., either an EFIE or MFIE alone, or is a combined-field form needed?
3. What kinds of element shapes should be used, e.g., triangular or quadrilateral?
4. Will faceted representations of surfaces provide the required modeling fidelity or are curvilinear elements needed?

These questions are considered in the following sections.

1.4.1 Some Representative TDIEs

It is useful to summarize some of the TDIEs that have been employed. For a smooth, closed surface, the MFIE is quite useful and is given by

$$\boldsymbol{J}(\boldsymbol{r}, t) = 2\boldsymbol{a}_n \times \boldsymbol{H}^{\text{inc}}(\boldsymbol{r}, t) + \frac{1}{2\pi} \int_S \boldsymbol{a}_n \times \left\{ \left[\frac{1}{R} + \frac{1}{c} \frac{\partial}{\partial \tau} \right] \boldsymbol{J}(\boldsymbol{r}', \tau) \times \frac{\boldsymbol{a}_R}{R^2} \right\} ds', \qquad (1.4.1)$$

where $\boldsymbol{J}(\boldsymbol{r}, t)$ is induced current density on S and $\boldsymbol{H}^{\text{inc}}$ is the incident field. Because of the low-order singularity contained in the kernel function, delta function basis and testing functions have been found to yield acceptable results [13, 27]. This is also the case for FD MFIEs [25]. When using delta function basis and testing, the self-term integral contribution of Eq. (1.4.1) is usually ignored and the nonself terms involve only the patch areas but not their shapes or possible curvatures. In effect, although the surface current is distributed over a given patch, its contribution to the field is approximated as a current moment concentrated at a single point, for example, at the patch's geometric center.

For open surfaces, a TDIE model based on the EFIE is needed, a form of which is given by [27]

$$-\epsilon\left\{\frac{\partial \boldsymbol{E}^{\text{inc}}(\boldsymbol{r}, t)}{\partial t}\right\}_{\text{tan}} = \left\{\left[\nabla(\nabla\cdot) - \frac{\partial^2}{c^2\partial t^2}\right] \times \left[\int_S \frac{\boldsymbol{J}(\boldsymbol{r}', \tau)}{4\pi R}\, ds'\right]\right\}_{\text{tan}}. \tag{1.4.2}$$

A specialized version of TDIE can be developed for wires, using a thin-wire approximation, to yield various forms, one version of which is [24]

$$\begin{aligned}\boldsymbol{a}_s \cdot \boldsymbol{E}^{\text{inc}}(s, t) = \frac{\mu}{4\pi} \int_l \Bigg\{ & \left[\frac{\boldsymbol{a}_s \cdot \boldsymbol{a}_s'}{R} \frac{\partial I(\boldsymbol{a}_s', \tau)}{\partial \tau}\right. \\ & + \left[\frac{c\boldsymbol{a}_s \cdot \boldsymbol{a}_R}{R^2}\right]\left[\frac{\partial I(\boldsymbol{a}_s', \tau)}{\partial s'} - \frac{cQ(\boldsymbol{a}_s', \tau)}{R}\right]\Bigg]\Bigg\} ds',\end{aligned} \tag{1.4.3}$$

where

$$Q(\boldsymbol{a}_s', \tau) = -\int_{-\infty}^{\tau} \frac{\partial I(\boldsymbol{a}_s', \tau)}{\partial s'}\, dt', \tag{1.4.4}$$

l denotes the parameter along the wire geometry, and $\boldsymbol{a}_s'$ and $\boldsymbol{a}_s$ are tangent vectors to the wire at source and observation points s' and s, respectively.

TDIEs have been developed and applied to a variety of objects, such as those depicted in Fig. 1.16. Particular algorithms include the TD EFIE for wires [14, 24, 55], a TD version of the Hallen IE for wires [26], the TD MFIE for 2D and 3D PEC bodies [13], the TD EFIE for thin plates and shells [28], the TD EFIE for triangular faceted PEC bodies [36], the TD CFIE for penetrable bodies [30, 38], and the TD EFIE for hybrid objects involving wires and fins [28].

1.4.2 A Prototype TDIE Model

A "prototype" TDIE can be written in the form

$$g(x, t) = \int_{C(\boldsymbol{r})} f(x', t) K(x, x')\, dx', \tag{1.4.5}$$

where $g(x, t)$ is the excitation or forcing function, $f(x, t)$ is the unknown response to be obtained, $K(x, x')$ is the IE kernel function, and $C(\boldsymbol{r})$ denotes the object geometry. The observation and source coordinates are x and x', respectively. Use of a space–time pulse approximation (i.e., constants over each space segment and time step), point sampling, and an explicit solution treatment produces, with N the number of space samples, a discretized form of the IE

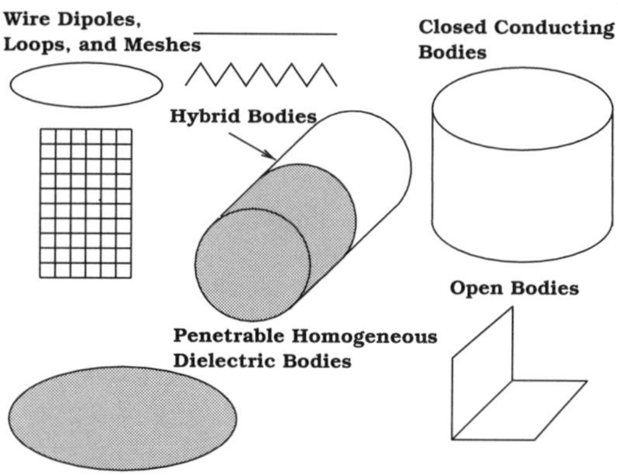

FIGURE 1.16
Representative objects for which TDIE models have been employed.

given by

$$g_{i;m} = \sum_{i'=1}^{N} A_{i';m'} Z_{i,i'} \tag{1.4.6}$$

$$Z_{i,i'} = \int_{\Delta C_i'} K(x_i, x_{i'})\, dx_{i'}, \tag{1.4.7}$$

where the sampled quantities are

$$g_{i;m} = g(x_i, t_m); \qquad A_{i';m'} = f(x_{i'}, t_{m'}); \qquad m' = m - |i - i'|, \tag{1.4.8}$$

leading to a solution that can be written as

$$A_{i;m} = \frac{1}{Z_{i,i}}\left[g_{i;m} - \left(\sum_{i'=1}^{N} A_{i';m'} Z_{i,i'}\right)_{(i)}\right], \tag{1.4.9}$$

where the subscript i indicates that the term $i' = i$ is omitted from the summation [24].

1.4.3 Alternate Forms for a TDIE Solution

The generic TDIE can be written symbolically for a more general geometry than a straight wire as follows:

$$\sum_{k=1}^{N} Z_{i,k} I_{k;m-f(i,k)} = V_{i;m}, \tag{1.4.10}$$

for $i = 1, 2, \cdots, N$ and $m = 1, 2, \cdots, N_t$, where N and N_t are the number of space samples and time steps, respectively. The quantity $f(i, k)$ is a "time-lag," geometry-dependent index that takes into account the propagation time between observation sample i and source sample k. The explicit forms of the interaction coefficients, $Z_{i,k}$, and the excitation, $V_{i;m}$, depend on the choice of basis and testing functions. These might be as simple as pulses and delta functions, respectively, or of higher order to provide a smoother current and field representation, the details of which are absorbed into the $Z_{i,k}$ coefficients. Once the current samples, $I_{i;m-f(i,k)}$, have been found from solving Eq. (1.4.10), the space and time-dependent solution current can be obtained from the basis function expansion used in developing Eq. (1.4.10).

Various alternate solution forms are available for this equation. In the most straightforward version using an explicit solution, the present current sample at space point i and time step m, $I_{i;m}$, is written in terms of past current samples as

$$I_{i;m} = \frac{1}{Z_{i,i}}\left[V_{i;m} - \sum_{k_1} Z^{(1)}_{i,k_1} I_{k_1;m-1} - \sum_{k_2} Z^{(2)}_{i,k_2} I_{k_2;m-2} - \cdots - \sum_{k_{n-1}} Z^{(n-1)}_{i,k_{n-1}} I_{k_{n-1};m-n+1} - \sum_{k_n} Z^{(n)}_{i,k_n} I_{k_n;m-n}\right], \quad (1.4.11)$$

where the superscript on each impedance matrix is an integer that denotes the number of time lags between i and current samples located in space at index k_m. Each matrix would have only those nonzero terms that describe interaction with space point i of the specified time lag.

If an implicit solution were to be developed instead, the current might be expressed as

$$I_{i;m} = \sum_{k_0} Y^{(0)}_{i,k_0} V_{i;m} - \sum_{k_1} M^{I(1)}_{i,k_1} I_{k_1;m-1} - \sum_{k_2} M^{I(2)}_{i,k_2} I_{k_2;m-2} - \cdots - \sum_{k_{n-1}} M^{I(n-1)}_{i,k_{n-1}} I_{k_{n-1};m-n+1} - \sum_{k_n} M^{I(n)}_{i,k_n} I_{k_n;m-n}, \quad (1.4.12)$$

where the leading Y matrix accounts for same-time-step interactions and the other terms account for the indicated number of time lags, but with the M matrices given by the time-lag impedance matrices multiplied by Y. Note that changing from an explicit to an implicit procedure changes the details of the time-stepping solution but not the form of the solution. The superscript $I(m)$ is used to denote that the M matrices provide a solution in terms of past currents.

Another alternate form of the implicit solution can be written in terms of the excitation only. Note that a more fundamental expression for the currents induced on an object by an impulsive field involves past values of the excitation only rather than an intermediate expression involving past currents. Formally, this form of the response becomes

$$I_{i;m} = \sum_{k_0} Y^{(0)}_{i,k_0} V_{i;m} - \sum_{k_1} M^{V(1)}_{i,k_1} V_{k_1;m-1} - \sum_{k_2} M^{V(2)}_{i,k_2} V_{k_2;m-2} - \cdots - \sum_{k_{n-1}} M^{V(n-1)}_{i,k_{n-1}} V_{k_{n-1};m-n+1} - \sum_{k_n} M^{V(n)}_{i,k_n} V_{k_n;m-n}, \quad (1.4.13)$$

where the superscript $V(m)$ denotes that the M matrix is associated with the excitation. Note that deriving the M^V matrices would involve recursive application of Eq. (1.4.13).

1.4.4 Excitation of a TDIE Model

The excitation term in the TDIE model is discussed in this section. Although any temporal waveform may be used, a plane wave impulse is perhaps the most useful. With the impulse response known, the response due to any other plane wave incident from the same direction and arbitrary time variation may be obtained by a convolution operation. From the viewpoint of limited computer time and storage, an ideal impulse function may not be implemented since its frequency spectrum extends from zero to infinity with a constant amplitude. We instead use an approximate smoothed impulse which has a Gaussian temporal variation. This type of pulse is effectively time and band limited and is well suited to numerical computation. It has a rapid decay to a negligible value in both the time and frequency domains. The incident pulse used in this work for the TDIE solutions, discussed in Chapters 2–5, is the Gaussian plane wave defined as

$$\boldsymbol{E}^{\mathrm{i}}(\boldsymbol{r}, t) = \boldsymbol{E}_{\mathrm{o}} \frac{4}{T\sqrt{\pi}} \mathrm{e}^{-\gamma^2}, \tag{1.4.14}$$

with

$$\gamma = \frac{4}{T}(ct - ct_{\mathrm{o}} - \boldsymbol{r} \cdot \boldsymbol{a}_k), \tag{1.4.15}$$

where $\boldsymbol{a}_k$ is the unit vector in the direction of propagation of the incident wave, T is the pulse width of the Gaussian impulse, $\boldsymbol{E}_{\mathrm{o}} \cdot \boldsymbol{a}_k = 0$, $\boldsymbol{r}$ is a position vector relative to the origin, c is the velocity of propagation in the external medium, and t_{o} is a time delay which represents the time at which the pulse peaks at the origin. The time delay is introduced to ensure a smooth rise of the incident field from a zero value. The pulse width T is defined such that for $ct - ct_{\mathrm{o}} - \boldsymbol{r} \cdot \boldsymbol{a}_k = \pm T/2$; the exponential has fallen to about 2% of its peak value. The Fourier transform of Eq. (1.4.14) referenced to the origin is

$$\mathcal{F}(\boldsymbol{E}^{\mathrm{i}}) = \frac{\boldsymbol{E}_{\mathrm{o}}}{c} \mathrm{e}^{-\left(\frac{\pi T f}{4c}\right)^2} \mathrm{e}^{-j2\pi f t_{\mathrm{o}}}. \tag{1.4.16}$$

The temporal and spectrum plots of the two pulses used in this work are shown in Fig. 1.17. Here, we define a light meter (LM) as the length of time that it takes for the electromagnetic wave to travel 1 m in a free space medium.

1.4.5 Physical Implication of a TDIE Explicit Model

From a physical viewpoint, it is worthwhile to note that a good approximation for the current induced by a time-varying incident field on a long, straight wire of length L and radius a is

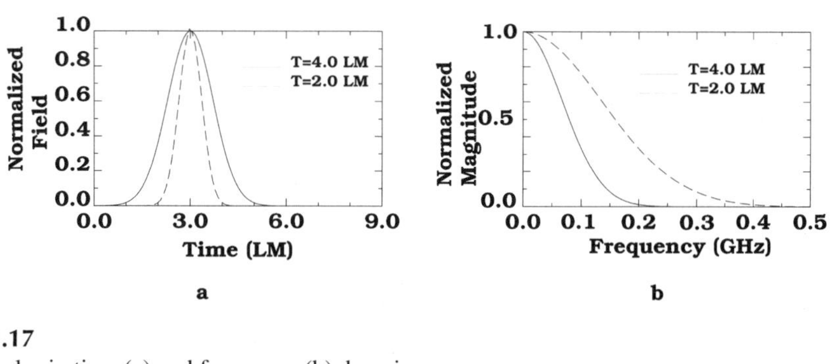

FIGURE 1.17
Gaussian pulse in time (a) and frequency (b) domains.

given by

$$I^{\text{total}}(\boldsymbol{r}, t) \approx \frac{\boldsymbol{a}_s \cdot \boldsymbol{E}^{\text{i}}(\boldsymbol{r}, t)}{R_{\text{surge}}} + I^{\text{prop}}(\boldsymbol{r}, t)$$
$$\approx \frac{1}{R_{\text{surge}}} \int_L \boldsymbol{a}'_s \cdot \boldsymbol{E}^{\text{i}}(\boldsymbol{r}', t')\, dl'$$
$$\approx \frac{1}{R_{\text{surge}}} \{\boldsymbol{a}_s \cdot [\boldsymbol{E}^{\text{i}}(\boldsymbol{r}, t) + \boldsymbol{E}^{\text{scat}}(\boldsymbol{r}, t)]\}, \tag{1.4.17}$$

where R_{surge} represents the local response [5], $\boldsymbol{a}_s$ and $\boldsymbol{a}'_s$ represent unit vectors tangent to the wire at $\boldsymbol{r}$ and $\boldsymbol{r}'$, respectively, and $t' = t - |\boldsymbol{r} - \boldsymbol{r}'|/c$ is the retarded time. In other words, the current at a given space location and observation time can be alternately approximated by

- the sum of a term induced by the present incident field and a propagated term from other parts of the wire; or
- a time integration of the retarded excitation over the wire; or
- the total local field;

with the field contribution divided by R_{surge} in every case. These various expressions emphasize the "local" nature of the response by identifying the directly induced and interaction terms. When the wire is loaded with resistance R_{Load} at the observation point, the current can be approximated alternatively by

$$I^{\text{total}}(\boldsymbol{r}, t) \approx \frac{R_{\text{o}}}{R_{\text{o}} + R_{\text{Load}}} \frac{c}{R_{\text{surge}}} \int_0^T \boldsymbol{a}'_s \cdot \boldsymbol{E}^{\text{i}}(\boldsymbol{r}', t')\, dt', \tag{1.4.18}$$

where the integration is with respect to the retarded time and $R_{\text{o}} = 60(\Omega - 3.39)$ is related to the wire "fatness" factor $\Omega = 2\ln(L/a)$ [56].

1.4.6 A Near-Neighbor TD Approximation

An ongoing need in CEM is that of reducing model complexity, i.e., decreasing the number of arithmetic operations needed to solve a given problem to a desired accuracy. Two readily stated, but not so readily realized, ways of accomplishing this goal can be identified for IEs, whether FD or TD. One is to develop formulations whose system matrices are easier to compute, thus reducing matrix fill time. Another is to use formulations and/or numerical implementations whose system matrices are easier to solve, thus reducing the solution time. If, for example, a system matrix can be developed by judicious choice of basis and testing functions so that most interaction coefficients are negligible and can be dropped from the computation, significant time savings might result in both matrix fill and matrix solution times. An example of this is the "impedance matrix localization" procedure [57]. Although it is not entirely accurate to do so, such procedures might be described as "near-neighbor" approximations (NNA) because their simplest implementations might involve dropping any interactions from an IE model whose magnitudes are below some specified level or which are separated by more than some minimum distance. If implemented in this simple fashion, the NNA is quite ad hoc and of unpredictable performance. However, the potential advantage of any TD NNA is possibly to reduce both the number of spatial interactions and the number of time steps that are needed for acceptably accurate and converged results to be obtained. When using a TD model with impulsive excitation, the time window, or number of required time steps, is related to, and is usually a multiple of, the object's transit time. This is required for the currents to decay back to zero and for charge neutrality to be restored. Hence, for this case, NNA may be applied only after a careful and thorough analysis of the problem.

However, if time-harmonic excitation is used with a TD model, the number of required time steps may be determined by how far apart on the object being modeled significant physical interactions occur. Suppose the number of numerical interactions included at each time step is allowed to expand in proportion to the speed of light as the time-stepping solution continues to take into account the outward expanding fields produced by the sources on each subdomain. Then, only after a time of $\approx L/c$ had passed would all N^2 interactions be taking place at each time step. However, if the solution were to stabilize acceptably before this time, the operation count could then be reduced from being proportional to f^5, which a 3D TDIE model normally requires, to possibly as little as f^3, assuming that a limited number of near-neighbor interactions are needed. The asymptotic limit of f^5 would be reached only when mutual interactions over the entire object are needed.

1.5 TIME DOMAIN DIFFERENTIAL EQUATION MODELING

As already noted, a TDDE model begins with the time-dependent Maxwell curl equations with possible additional analytical manipulation, such as developing variational expressions. Many questions arise with respect to such issues as the following:

- Are the first-order curl equations to be used, in which a two-field representation involving both $\boldsymbol{E}$ and $\boldsymbol{H}$ are developed, or will a second-order wave equation description in only one of these field variables be used?
- If a two-field representation is used, will the components be evaluated at common spatial nodes or be staggered in space, as is done in the FDTD approach?
- Are auxiliary field quantities, such as scalar and vector potentials, needed in the formulation or will auxiliary numerical conditions need to be imposed?
- Is a regular mesh or a more general "unstructured" mesh to be employed?
- What order and what kinds of basis and testing functions will be used to represent the spatial and temporal variation of the field samples and to satisfy the defining equations?

These and other related questions are considered in the following sections along with a comparison of the TDIE and TDDE models.

1.5.1 Space–Time Sampling of TDDE

The essence of DE modeling is approximating space derivatives with weighted samples and, similarly, sampling and satisfying the applicable DEs. As noted by Botha and Pinder [58], finite-difference approximations and their higher order variations differ not in respect to the underlying philosophy of representing differential operators by sampled differences but only in terms of how the quantities of interest are assumed to vary spatially and how the equations that describe their behavior are to be satisfied.

For example, for the 2D wave equation in Cartesian space,

$$(\nabla^2 + k^2) f(x, y; k) = 0, \tag{1.5.1}$$

using linear basis functions for the field, and point sampling of the equation, leads to

$$(4 - \Delta^2 k^2) f_{i,j} = [f_{i+1,j} + f_{i-1,j} + f_{i,j+1} + f_{i,j-1}], \tag{1.5.2}$$

where i denotes the space point $x_i = i\Delta$ and j denotes the space point $y_j = j\Delta$, with Δ being the sample spacing in x and y as shown in Fig. 1.18. This discretized form of the wave equation may be interpreted as a spatial "filter" which relates sampled values of f at i, j to those in the immediate surroundings. If k were made zero, to recover the Laplace equation, the discretized equation is seen to approximate the field at a given point, or node, from an average of the four nearby fields. For nonzero k, the right-hand divisor is less than 4, so the averaging is weighted differently, as determined by the wavenumber and sampling interval. The spatial arrangement of the samples and their weights can be represented diagrammatically in a form sometimes referred to as a "computational molecule" [59]. Various other descriptors that might be used in referring to the samples and their locations are grid, mesh, lattice, cell, stencil, or tessellation, depending on the context and application discipline. The points at which fields or equations are evaluated are usually called nodes. As seen in Fig. 1.18, the computational molecule for the Laplace equation may be illustrated by weighting the field samples at their locations by the associated circled values to evaluate the field sample at the central node.

Consider the time-dependent, 2D wave equation,

$$\left(\nabla^2 - \frac{1}{c^2}\frac{\partial^2}{\partial t^2}\right) f(x, y; t) = 0. \tag{1.5.3}$$

After straightforward space–time sampling, and adding the subscript m for the time index,

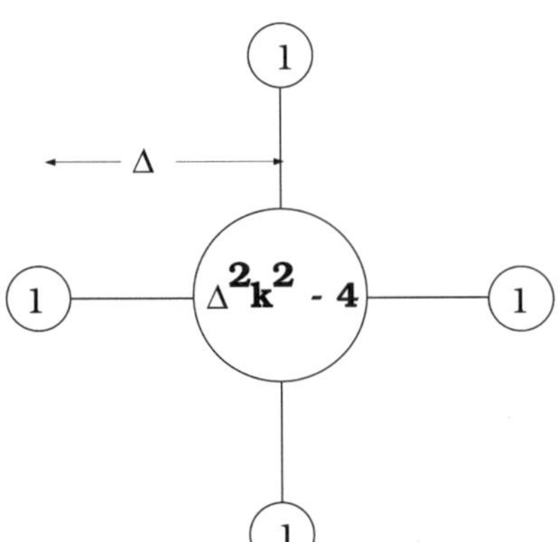

FIGURE 1.18
Representative computational molecule for the 2D Laplace equation.

leads to

$$f_{i,j;m+1} = \left(\frac{c\Delta t}{\Delta}\right)^2 [f_{i+1,j;m} + f_{i-1,j;m} + f_{i,j+1;m} + f_{i,j-1;m} - 4f_{i,j;m}] + 2f_{i,j;m} - f_{i,j;m-1}, \tag{1.5.4}$$

where a stable solution is ensured if $c\Delta t/\Delta \le 1/\sqrt{2}$ [58]. Thus, it is possible to compute the field value at an arbitrary point at time step $m + 1$ from previous time samples at the same point and the nearest spatial samples. An alternative to this particular computational molecule is the Yee lattice, which is based on the Maxwell curl equations rather than on the wave equation as discussed in the next section. Note that a wave equation approach can be based on using only one field quantity, such as $\boldsymbol{E}$ or $\boldsymbol{H}$ alone. However, such an approach involves second-order space and time derivatives of that quantity, whereas the use of both fields, as the Yee lattice entails, results in only first-order derivatives. A similar observation can be made by eliminating one of the field quantities from the curl equations, which then yields mixed second-order derivatives in space and time together.

1.5.2 Some Spatial-Mesh Alternatives

The Yee Lattice An especially attractive feature of the original FDTD formulation [9] is the simplicity of the regular, Cartesian mesh on which it is based. The components of the electric field are centered along, and parallel to, the edges of one set of cubes (assuming equally spaced samples in each direction), whereas the magnetic fields are centered along, and parallel to, the edges of another set of cubes which are offset from the first set by half the edge width in each direction. (A detailed analysis of the FDTD method with several application examples is presented in Chapter 6.) This arrangement is called the Yee lattice. An important feature of this representation is that the $\boldsymbol{E}$ and $\boldsymbol{H}$ fields need not be computed at the same space points and, therefore, need not be computed at the same instants of time because of causality. Consequently, upon updating the $\boldsymbol{E}$ fields at time step m, the $\boldsymbol{H}$ fields, being a half-mesh spacing distant, are updated at time step $m + 1/2$. This staggered mesh permits the electric and magnetic fields to be alternately computed in a "leapfrogging" manner so that simultaneous solution of coupled field components is never required. Representative Yee lattice equations in 3D [i, j, and k signify a point at $x_i = i\Delta$, $y_j = j\Delta$, and $z_k = k\Delta$ and similarly $i + 0.5$, j, k is at $x_i = (i + 0.5)\Delta$, $y_j = j\Delta$, $z_k = k\Delta$, etc.] can be written, with x, y, and z denoting the Cartesian field components, as

$$H_{z;i,j,k-0.5;m+0.5} = H_{z;i,j,k-0.5;m-0.5} - \frac{\Delta t}{\mu\Delta}[(E_{x;i,j+0.5,k-0.5;m} - E_{x;i,j-0.5,k-0.5;m}) - (E_{y;i+0.5,j,k-0.5;m} - E_{y;i-0.5,j,k-0.5;m})] \tag{1.5.5}$$

and

$$E_{z;i-0.5,j-0.5,k;m+1} = E_{z;i-0.5,j-0.5,k;m} - \frac{\Delta t}{\epsilon\Delta}[(H_{x;i-0.5,j,k;m+0.5} - H_{x;i-0.5,j-1,k;m+0.5}) - (H_{y;i,j-0.5,k;m+0.5} - H_{y;i-1,j-0.5,k;m+0.5})]. \tag{1.5.6}$$

The $\boldsymbol{H}$ fields are seen to be evaluated temporally at alternate half-time steps from the electric fields and spatially at alternate half mesh spacings.

Relationships like those discussed previously can be derived either directly from the Maxwell curl equations or from their integral forms as Ampere's and Faraday's laws, given by

$$\int_S \frac{\partial \boldsymbol{D}}{\partial t} \cdot \boldsymbol{a}_n \, dS = \oint_C \boldsymbol{H} \cdot \boldsymbol{dl} \tag{1.5.7}$$

$$\int_S \frac{\partial \boldsymbol{B}}{\partial t} \cdot \boldsymbol{a}_n \, dS = -\oint_C \boldsymbol{E} \cdot \boldsymbol{dl}, \tag{1.5.8}$$

where S is an area, C is the boundary of that area, and $\boldsymbol{a}_n$ is a unit normal to the area. An advantage of the integral forms is the relative ease with which they can be used to develop computational molecules for irregular meshes, compared with the differential equations, thereby leading to better spatial resolution and resulting in finite-area and finite-volume solution procedures [60, 61]. One disadvantage of the Yee lattice is that the separate field components are not available over the same set of points in space, which complicates defining object geometries and obtaining the far field. For example, Trueman *et al.* [62, 63] carefully compared FDTD results and measurements for the resonance structure of a long rod and found that the effective physical size of an object being modeled is about one-fourth of a cell size larger than intended by their numerical model. By appropriately reducing the model size using the "quarter-cell margin" correction, they obtained much improved agreement. Another disadvantage of the regular lattice is that a smoothly curved surface can only be approximated in a stair-step fashion. This stair-step approximation can lead to anomalous results unless the mesh size is made much less than a wavelength. Thus, for example, while sampling intervals of $\lambda/6$ or so can be used in IE models, a stair-step FDTD model might require mesh sizes as small as $\lambda/120$ to obtain 1 dB RCS [Rader Cross Section] accuracy in a curved boundary (an ogive), compared with a piecewise linear, boundary-conforming mesh that yields the same accuracy with a $\lambda/15$ mesh [64]. The use of a such a small mesh can greatly increase the overall FDTD computer time because the simplest Cartesian mesh is uniform everywhere throughout the solution space, and, being an explicit algorithm, it also requires a smaller time step consistent with the mesh size.

Boundary-Conforming Meshes As a means of better representing curved surfaces while avoiding the excessive computational cost imposed by small mesh size, work has also been done on boundary-conforming meshes. In this approach, a regular mesh is employed everywhere away from the object's surface. The mesh cells that would otherwise intersect the surface are instead deformed to provide a piecewise linear (or higher order) approximation to the surface [59]. This approach retains the simplicity of a regular mesh while more accurately representing a curved boundary. It also provides a logical transition from a regular mesh to a body-fitted, nonorthogonal or irregular mesh.

An FDTD implementation of boundary-conforming meshes is described by Yee *et al.* [65]. In their treatment, the locally conformal mesh immediately adjacent to the object being modeled extends outward to overlap with the external Cartesian mesh over an interval of about three zones. This approach preserves the computational simplicity of the Cartesian mesh over most of the solution space while providing a much more accurate representation of a curved boundary. Their test cases for an infinite, circular cylinder demonstrate a substantial improvement in accuracy over a stair-step approximation when using comparable mesh sizes. However, the added complexity imposed by such approaches has evidently inhibited their acceptance in FDTD models.

Body-Fitted, Nonorthogonal (Irregular) Meshes A logical extension, beyond deforming a regular mesh where it intersects with a curved boundary, is to deform the mesh throughout the solution space. The use of numerically defined, generalized coordinates for handling curved or irregular objects is described by Fusco [66], who extended the earlier work of Holland [67]. He presents results for scattering from circular and square cylinders that exhibit excellent agreement with independent data. The mesh employed was produced by a mesh-generating program originally written for hydrodynamic applications. Another hydrodynamics-based meshing approach is described by Mohammedian *et al.* [68]. They employ the method of characteristics and present a multizone scheme for meshing each of the various parts of the problem space in a different way to conform to the local geometry.

Additional Aspects of Meshes and Sampling Note that DE meshes provide a link between the equations being solved and the spatial sampling of the unknowns needed to satisfy them. How variations of the unknowns are represented spatially and temporally and how the sample points are spatially distributed are key to determining the various properties of the solution process. For example, the differential operations in Maxwell's curl equations are approximated in the FDTD approach by using all space samples from one time step earlier than that of the updated samples being computed from the time derivatives. Were this not the case, the model would become implicit because the updated samples would then be related to their adjacent neighbors. It should also be mentioned that TDIE models can become implicit even though explicit time stepping is utilized when continuity conditions result in higher order, spatially adjacent current bases being related within the same time step.

A characteristic of spatial meshes is that their discretization can cause frequency dispersion and direction-dependent propagation. Another consequence of the discretization is that the time step Δt cannot be arbitrarily close to Δ/c for an explicit model; instead, it must be maintained below this limit, with the specific difference depending on the spatial dimensionality of the mesh. Thus, Δt must not exceed the Courant stability criterion, given by

$$\begin{aligned} c\Delta t_{\max} &= \frac{1}{\sqrt{\frac{1}{\Delta x^2} + \frac{1}{\Delta y^2} + \frac{1}{\Delta z^2}}} \\ &= \frac{\Delta}{\sqrt{3}} \\ &= 0.577\Delta \end{aligned} \tag{1.5.9}$$

for equal sampling spacing in all three dimensions. If it exceeds this limit, a diverging solution results.

The maximum mesh size, $h_{\max}$, must also be small enough that mesh-induced anisotropy and dispersion effects are below an acceptable level. For the 1D case, a numerical dispersion relation can be derived which is given by [69, 70]

$$\sin^2\left(\frac{\pi v_{\mathrm{p}}\Delta t}{\lambda}\right) = \left(\frac{c\Delta t}{\Delta x}\right)^2 \sin^2\left[\frac{\pi\Delta x}{\lambda}\right], \tag{1.5.10}$$

where v_{p} is the phase speed. The 2D dispersion relation is

$$\sin^2\left(\frac{\omega\Delta t}{2}\right) = \left(\frac{c\Delta t}{\Delta x}\right)^2 \sin^2\left[\frac{k_x\Delta x}{2}\right] + \left(\frac{c\Delta t}{\Delta y}\right)^2 \sin^2\left[\frac{k_y\Delta y}{2}\right]. \tag{1.5.11}$$

However, computational experience has shown that using $\Delta_{\max} \leq \lambda/10$ produces good results for many applications [71].

Most of the TDDE models being used employ staggered meshes similar in concept to the Yee lattice, except for variations arising from the local mesh description itself, i.e., each of the field samples is located at a different point in space. It is reasonable to consider the ramifications of computing all field values at the same space point. For 3D problems, for example, six simultaneous field components would be needed at each time step. Consequently, a 6×6 matrix would need to be inverted at each field node before beginning the time stepping. Because the matrix coefficients would be identical for all cells of equal size, such a computation would impose no significant cost, enabling all six field components to be updated at each node for the price of a single matrix multiply. For irregular meshes, one matrix inversion could be required for each node in the mesh, and then time stepping would proceed in the usual way.

An interesting aspect of TDDE models of PEC objects is that although the fields do not need to be computed inside such bodies, it can sometimes be more efficient, computationally, to do so anyway because of the overhead otherwise involved. A common strategy is to update all fields in the mesh in a first pass before the boundary conditions at conducting surfaces are implemented in a second pass. The latter would typically involve only a small number of additional calculations compared with the first pass and would thus represent a relatively modest additional computation. Furthermore, many more nodes would generally be inside the PEC body ($\approx V/\Delta^3$), which is more than the number of surface nodes, $\approx S/\Delta^2$, where V and S represent the volume and surface area, respectively. Therefore, for some problems, implementing the logic required to omit these nodes from the computation might consume more computer time than simply computing, and then discarding, the internal fields.

Depending on the specific procedures used in developing a numerical model, various kinds of spurious responses can occur because of discretization and numerical roundoff. A simple example is provided by numerically solving a problem in which evanescent fields occur, such as a plasma below cutoff (where the plasma frequency exceeds the wave frequency). A plane wave analytical solution for this problem has two exponentially growing fields in the $+x$ and $-x$ directions. One of these fields would normally be omitted because of boundary conditions. However, a numerical solution obtained from the integration of the wave equation in the $+x$ direction would inevitably be dominated by the solution that grows with increasing x, even though the other decaying solution is dictated by the boundary condition. Similarly, a numerical model based on a derivation that employs the operator identity $\nabla \cdot \nabla \times \boldsymbol{A} = 0$ may anomalously produce a solution that does not satisfy this requirement. It has been observed that node-oriented basis functions or elements are subject to this problem, but edge-oriented basis functions are not [72, 73]. Note that spurious responses of various kinds appear in all DE models, for example [74], and elimination of these spurious solutions is currently being studied.

1.5.3 Mesh Closure Conditions

An IE formulation automatically provides a match to outward-propagating fields for modeling exterior problems because a radiation condition is encompassed by the Green's function upon which such a model is based. Thus, the IE model naturally includes near-field and evanescent-wave interactions while also accounting for the far radiation fields, with no additional analytical or numerical manipulation.

A DE model, however, produces purely outward-propagating fields only at some distance far enough from a body that there are no significant near-fields present. However, extending field sampling out to the far-field region can require a much larger mesh than might be desirable. Furthermore, even if such a large mesh was acceptable, some means of closing the mesh is still needed. Otherwise, the mesh would extend to infinity if knowledge of the far-field behavior somehow not incorporated into the model. This situation arises because computing a field equation at a given point in space from a differential relationship requires other field samples from the adjacent nodes in the mesh. This leapfrogging effect requires field samples on a mesh expanding ever outward from the object being modeled, an impractical requirement from a numerical viewpoint. The only alternative is to truncate, or close, the mesh by introducing some auxiliary relationships so field samples that are one mesh interval outward from the outermost field equation node are not needed. From the viewpoint of computational efficiency and of minimizing the overall operation count, it is desirable to close the mesh as close as possible to the object being modeled.

Closing a DE mesh requires that the information which would otherwise come from sampling the next set of nodes outward from the body be replaced by other analytical information. Mesh closure conditions can be described as being "local" or "global." In the local closure condition (LCC), an outward-propagating nature is assumed for the fields at the closure boundary, which permits the next-layer field samples that would otherwise be needed to be approximated in terms of already sampled nearby fields and appropriate analytical expressions. Because no "artificial" field reflections should occur at this nonphysical closure boundary, closure conditions are often described as "absorbing boundary conditions." It should be realized, however, that if the closure boundary is so close to an object that it intercepts near-field energy, then numerically "absorbing" such fields must alter the object's electromagnetic response because this absorbed energy is lost to further interactions with other portions of the object.

Local Closure Conditions Simple closure conditions illustrate conceptually what is involved in mesh closure. For the 1D wave equation in the TD, the typical computational molecule is given by

$$f_{i;m+1} = \left(\frac{c\Delta t}{\Delta}\right)^2 [f_{i+1;m} - 2f_{i;m} + f_{i-1;m}] + 2f_{i;m} - f_{i;m-1}. \tag{1.5.12}$$

If the space sampling is to be stopped at some maximum location x_b, then the samples at x_i, $i > b$, need to be avoided. By assuming that the field behavior has "settled down" to an outward-propagating wave at this point and confining the numerical solution to a region $i < b$, we could use the following as a closure condition:

$$f(x_{i+1}; t_m) \approx f(x_i; t_{m-1}) \tag{1.5.13}$$

so that the computational molecule on the closure boundary becomes

$$f_{i;m+1} = \left(\frac{c\Delta t}{\Delta}\right)^2 [f_{i;m-1} - 2f_{i;m} + f_{i-1;m}] + 2f_{i;m} - f_{i;m-1}, \tag{1.5.14}$$

yielding the desired result, i.e., space sampling can be stopped at x_i. Comparable closure conditions for 2D (ρ, ϕ) and 3D (r, θ, ϕ) problems are

$$f(\rho_{i+1}, \phi_j; t_m) \approx \sqrt{\frac{\rho_i}{\rho_{i+1}}}\, f(\rho_i, \phi_j; t_{m-1}) \tag{1.5.15}$$

and

$$f(r_{i+1}, \theta_k, \phi_j; t_m) \approx \left(\frac{r_i}{r_{i+1}}\right) f(r_i, \theta_k, \phi_j; t_{m-1}), \tag{1.5.16}$$

assuming that the fields are propagating radially outward and the geometric multiplier accounts for field spreading. Since these conditions are not likely to be met unless the closure boundary is very far from the object being modeled, it is necessary to develop LCCs that apply for nonradial propagation as well as for broad frequency bands.

One of the first, more general, LCCs for exterior DE models was developed by Lindman [75], who described his approach as a "free-space" boundary condition. Other LCCs were derived by Engquist and Majda [76], Bayliss and Turkel [77], and Mur [78]. The details of the LCCs vary with respect to the number of near-field samples that are used in a series expansion of the local fields [79]. A LCC can also be described as a "one-way wave operator" [80]. A limiting form of an LCC is reported by Arendt *et al.* [81] as the "on-surface" radiation condition whose application has been found to be best suited to convex bodies. Still another LCC, the *concurrent complementary operators method*, has been developed [82, 83]. This involves the cancellation of the first-order reflection that can otherwise arise by averaging two independent solutions of a problem that are designed to generate errors equal in magnitude but 180° out of phase.

Another class of LCC is, in finite-element terminology, the so-called "infinite elements." In this approach, a problem is separated into an interior region where standard elements are employed and an exterior region where the elements extend to infinity and have infinite area. The infinite elements incorporate an explicit decay law to describe the exterior fields of the form $e^{-\alpha\rho}$ or $e^{-\alpha x-\beta y}$, for example, when modeling waveguide problems having evanescent fields [84, 85]. Other applications would be better modeled by infinite elements whose geometric attenuation is matched to the problem's needs.

The empty-space volume between the object being modeled and the closure surface is known as the "white space" region. Trueman *et al.* [62, 63] explored this problem systematically for various shapes of long cylinders. They found that the cell size and the size of the whitespace are somewhat interrelated, and that approximately 20 cells, each of size $\lambda/10$, in white space are needed for a 100-cell rod to produce accurate results. Furthermore, when the rod length is increased to 200 cells, 20 cells, with $\lambda/20$ cell size, will also produce comparably good results.

Global Closure Conditions Global closure conditions (GCCs) take a different, more computationally demanding, approach to the problem. Recognizing that LCCs are approximate at best, requiring some compromise between modeling efficiency (moving the closure boundary toward the object) and accuracy (moving the boundary farther from the object), the GCC provides a rigorous treatment from the start. It involves developing a source-integral expression for the fields outside the closure boundary, where the unknowns are tangential field samples on that boundary. The extra equations needed to obtain these field samples are obtained by transforming this source integral into an integral equation [86]. Using the same technique, a TD version of an IE GCC condition could also be derived from the time-dependent Stratton–Chu integrals [2].

An IE GCC, in contrast to a LCC, has the disadvantage of generating a full, rather than sparse, matrix for the outermost set of sampling nodes. On the other hand, the IE can be applied arbitrarily close to the object being modeled. The computational trade-offs in its application are similar to those encountered in LCCs. An implicit FDTD implementation of an IE GCC is described by Barkeshli *et al.* [87] and Ziolkowski *et al.* [88].

Another GCC is derived from using a modal expansion, rather than an integral equation, for the fields on the closure surface. This method, however, is not as general as using an IE because it is most suitable for application to a constant-coordinate surface, such as a sphere in a spherical coordinate system. This approach was used in the "unimoment" method developed by Morgan and Mei [89].

The Measured Equation of Invariance Closure Condition The measured equation of invariance (MEI) approach is a hybrid GCC/LCC for exterior problems that combines the advantages of each closure condition [90]. It exploits the fact that spatial field samples developed from integrating specified source distributions using an appropriate Green's function must satisfy not only the radiation condition but also whatever DE is applicable to that problem, for example, the wave equation for EM applications.

For problems in which a field-sampling region has no sources, it follows that an appropriate linear combination of the field samples must add to zero, i.e.,

$$\sum_{i=1}^{N} a_i f_i^{(j)} = 0, \qquad i = 1, 2, \cdots, N, \tag{1.5.17}$$

where $f_i^{(j)}$ denotes the field value at node i due to source distribution j. There is no loss of generality in making this equation inhomogeneous by setting one of the a_i's to unity and solving for the remaining $N-1$ a_i's in terms of field samples at $N-1$ of the N nodes in Eq. (1.5.17). These field samples are obtained by integrating $N-1$ linearly independent source distributions over the surface of the object being modeled, resulting in an $(N-1)^{\text{th}}$ order linear system whose solution yields the remaining a_i's. Thus, by choosing one field sample location on the closure boundary, an equation is obtained as a weighted sum of the neighboring field samples in the mesh, yielding a sparse closure condition but one which satisfies the radiation condition because the field samples in it are related by a Green's function. In effect, Eq. (1.5.17) provides an alternative to using another spatial sample of the defining DE, thereby stopping the outward progression of field sampling layers. An example of using the MEI closure approach for a 2D static problem is described by Gothard *et al.* [91], in which the approach is discussed in more detail. Although the MEI approach has been applied to FD problems only, its possible use for TD remains to be developed.

1.5.4 Handling Small Features in DE Models

An attractive feature of FDTD-type models is the simplicity of the mesh employed. However, when using explicit techniques and the same mesh size everywhere in space, the number of nodes, or spatial unknowns, and the number of time steps are driven by the size of the smallest feature to be resolved in the model. Alternate schemes can be developed to circumvent the increased computational burden that this approach can impose. One alternative is to use variable time and space sampling in different regions of the solution space, permitting each of the regions to be modeled explicitly while avoiding the oversampling imposed when using uniform space and time intervals everywhere. This approach requires imposition of appropriate continuity conditions at common boundaries between the regions. Another approach is to use an implicit procedure, which permits larger time steps than would be capable with an explicit technique, in regions in which finer spatial sampling is needed.

Other less numerically complicated procedures are also feasible for some aspects of treating features smaller than the spatial mesh size in DE models. An approach based on Babinet's principle for handling narrow apertures and slots in PEC bodies is described by Demarest [92]. Babinet's principle permits the meshes on either side of a narrow slot in a PEC body to be decoupled so that the mesh size does not need to match the slot width. Taflove *et al.* [93] developed a "thin-wire" (and slot) approach for FDTD modeling based on the integral form of Maxwell's equations. They obtained accurate results for conductor sizes as small as 1/3000 of the mesh size. Riley and Turner [94] developed an iterative, "feedback" procedure for modeling thin slots. Their treatment incorporates the dual of the Pocklington IE whose solution is used as a magnetic current element in the appropriate curl-E equations in an FDTD code. A model for a wire antenna, using FDTD, was described by Tirkas and Balanis [95], who also employed the integral form of Maxwell's equations for the antenna description.

1.5.5 Obtaining Far Fields from DE Models

Evaluating far fields when using an IE model is quite straightforward. For PEC bodies, the far field is normally obtained using the Stratton–Chu integrals [2], which involve integrating the electric surface current, $\boldsymbol{a}_n \times \boldsymbol{H}_{\text{total}}$, over the surface, where $\boldsymbol{H}_{\text{total}}$ is the total magnetic field. The far fields for penetrable bodies are obtained from the same integral expressions using the tangential components of both the surface $\boldsymbol{E}$ and $\boldsymbol{H}$ fields.

A DE-based model, by contrast, usually requires more "postprocessing" to obtain the far fields because, in models such as FDTD, the magnetic and electric fields are not evaluated at the same points in space. Thus, while a PEC body may be defined by the vanishing of $\boldsymbol{E}$-field components tangential to its surface, the magnetic fields are located one-half mesh size away from it. Obtaining the surface current requires that the off-surface magnetic fields be extrapolated to the body's surface.

Alternatively, the $\boldsymbol{E}$ and $\boldsymbol{H}$ fields might be integrated over a mathematical surface enclosing the body for which the closure boundary would be a natural choice. However, if a staggered mesh were used in the model, such as in FDTD, it would be necessary to transfer one of these field components to a surface on which the other component is known in order to perform the integration. Some specific examples of approaches taken to obtain far fields from such models are described by Luebbers *et al.* [69].

1.5.6 Variations of TDDE Models

Other variations of TDDE models have also been explored. Perhaps the most popular is the TLM developed by Johns and Beurle [18], who took a conceptually different approach from that of the straightforward differencing of the Maxwell curl equations. They introduced the idea of a network of fictitious transmission lines in space whose interconnections at common nodes led to reflected and transmitted fields (voltages and currents) along the line sections that met. (A detailed analysis of TLM including several application examples are presented in Chapter 7.) The TLM approach can be interpreted as an implementation of Huygen's principle, in which fields spreading in space can be developed as a series of secondary sources. It can be shown that TLM and FDTD reduce to equivalent representations in some simplifying cases, but TLM can impose a larger storage and operation-count costs. This apparently occurs because TLM carries more information in the solution about oppositely propagating waves along the transmission line mesh as contrasted with the summed fields provided by the FDTD model. The equivalence between the TLM and a modified FDTD approach was shown by Chen *et al.* [96]. They showed this equivalence by using a mesh in which the field components were all defined at the same cell-centered node and decomposed the fields into components traveling toward and away from the nodes.

A comparison of TLM and FDTD leads to the following general observations:

- TLM can be described as embodying the method of characteristics in which the partial fields of oppositely propagating waves are separately developed, whereas FDTD yields the summed field.
- The fields in TLM are usually determined at common points in space, using a symmetrical, condensed-node formulation, leading to a requirement for

solving simultaneously for the field components there (through the scattering matrix), but an expanded-node formulation which separates the fields similar to the Yee lattice has also been used.

- Because the $\boldsymbol{E}$ and $\boldsymbol{H}$ fields are determined at common points, defining a surface in TLM is less ambiguous than it is for the FDTD.
- Another consequence of this added information in the TLM solution is that knowledge of wave propagation in opposite directions along each leg of a transmission line of known impedance makes it possible to solve directly for the electrical properties required to provide a reflectionless boundary [97].

As with the FDTD approach, TLM models can be based on scattered field or total field formulations. Because TLM models involve space and time discretization, they are also subject to problems of mesh dispersion and anisotropy. When "graded" meshes are used, in which the spatial samples are of different sizes in the mesh, anisotropic propagation properties can be amplified. As is the case with FDTD modeling, TLM applications have expanded substantially during the past few years; however, apparently because TLM is less intuitively structured than FDTD, it is not as widely used.

Recently, two other TDDE methods have been proposed: the time domain finite element (TDFE) method and finite volume time domain method (FVTD). The TDFE follows closely the similar and more popular technique in FD (a detailed explanation of this method and representative examples are presented in Chapter 8). The FVTD is based on the integral form of Maxwell's equations and employs an unstructured body-fitted mesh for accurate modeling of the problem (a detailed explanation of this method and representative examples are presented in Chapter 9).

Another TDDE approach is the method of lines in the TD or the TDML. A model based on TDML is described by Nam *et al.* [17]. They used it for application to microstrip lines. This approach reduces the dimensionality of the discretized numerical solution by representing the field variation in one transverse dimension as an analytical function. For example, the model can be discretely sampled in x and z, with the behavior along the y-axis direction expanded in a set of modes. This approach might be described as employing a mixed basis set that is a subdomain basis in two dimensions and an entire domain basis in the third.

1.5.7 Comparison of TDDE and TDIE Models

The two primary choices for TD modeling are those based on DEs and IEs, although in principle, GTD [18] and mode-based TD models might also be developed. Whatever the details of the specific approach, the method of moments provides a way of solving IEs and DEs, involving approximating integrals as finite sums and derivatives as finite differences in the generic forms

$$\int f\,dx \approx \sum f_i \Delta x \quad \text{and} \quad \frac{\partial}{\partial x} \approx \frac{f_i - f_{i-1}}{\Delta x}, \tag{1.5.18}$$

leading, after some additional manipulation, to a linear system of equations or "system" matrix. The process of discretizing and quantifying DEs numerically is known by various names, including finite-difference, finite-area (or volume), and finite-element procedures. The term finite-element is usually, but not necessarily, associated with a variational formulation, whereas use of a designation other than finite difference usually refers to the use of more general basis and testing functions. A numerical model based on an IE is also called a "boundary element" method in structural dynamics and acoustics.

It may be noted that in most cases, computer modeling involves replacing an infinite domain, first principles, analytical description of a problem by a finite domain, discretized, numerical one. The numerical model is finite in nature because only a limited number of unknowns of limited precision can be used in the solution process. An analytical model, however, entails, symbolically at least, an infinite dimensionality such as that exhibited by a series expansion for a sphere. However, it is worth noting that, from a practical viewpoint, the analytical model is finite in nature because the process of quantitatively evaluating any analytical model is automatically subject to limited precision and accuracy. This means that any observable of interest will exhibit no quantitative change over some specified dynamic range after an appropriate number of terms have been summed in its series solution. This will be the case whenever we deal with *numerical* answers as opposed to *analytical* solutions.

The basic difference between IE and DE models is the form of the field propagator used to develop a relationship between the fields and the sources that produce them, as illustrated in Fig. 1.19. Note that

- IEs use Green's function, which implies (i) the field at R_o involves integration over S and (ii) the solution space has dimensionality of S (i.e., unknowns are confined to surface).
- DEs use curl equations which implies (i) the field at R_o involves only adjacent field values and (ii) the solution space has the dimensionality of V because unknowns occur throughout the volume. Furthermore, for open-region problems, the volume space V extends far beyond the original surface and, theoretically, the infinite volume.

Some other essential differences between DE and IE models are as follows:

1. Reducing mathematical operators to a numerical form can be intrinsically less robust numerically for differential operators than for integral operators because numerical errors are additive, whether due to the subtraction of finite differencing or the addition of numerical quadrature.
2. The DE form produces a system matrix that has only a few nonzero, and simpler, coefficients per row because of the local, rather than the global, interactions. However, the size of

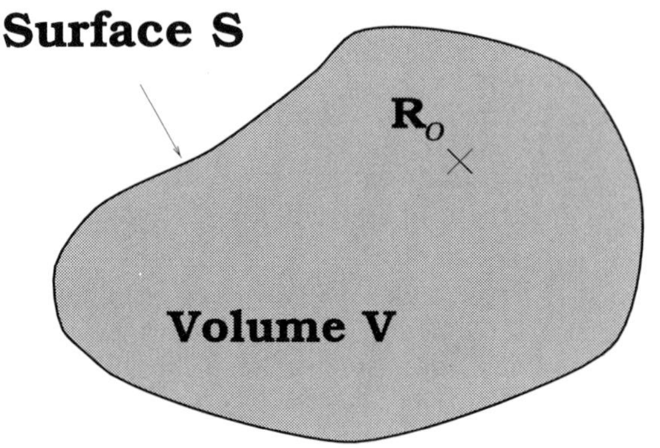

FIGURE 1.19
Representative example of an interior problem illustrating the difference between DE and IE sampling requirements for homogeneous regions.

the matrix for a DE solution, at least for open-region problems, is much larger than the corresponding IE solution since the solution space extends far beyond the object's surface. However, recent developments in the mesh closure conditions may alleviate this problem.

3. Generally, the IE solutions tend to be more accurate for similar mesh densities. Even for closed-region problems, the DE solutions typically require a much higher discretization than those derived from IE methods. Thus, the matrix sizes are generally larger in DE solutions.
4. In terms of medium nonlinearity, inhomogeneity, and time variation, the DE form is simpler for more general problems than is the corresponding IE form because of the need for Green's function for an IE model.

Note that if Δt were made progressively larger in an IE formulation, the number of samples interacting within a time step would increase monotonically. In the limit $\Delta t \to \infty$, the implicit scheme becomes equivalent to a FDIE model. On the other hand, an implicit TDDE approach, even in the limit $\Delta t \to \infty$, generates at most a sparse matrix because of the local nature of the curl operator, as contrasted with the global nature of an integral operator, assuming that subdomain bases and testing functions are used for the IE model.

1.6 SPECIFIC ISSUES RELATED TO TIME DOMAIN MODELING

As a rapidly evolving subdiscipline of CEM, TD modeling in general and TDDE modeling in particular involve a number of active research topics directed toward increasing its capabilities, including

1. The kinds of meshes used for DE models
2. The specific ways by which TDDE models can be implemented
3. Mesh closure conditions
4. Developing mesh descriptions for handling features smaller than the average mesh size
5. Handling frequency dispersion
6. Obtaining the far fields from the TDDE solution

Similarly, TDIE topics of interest include enhancing the basic numerical implementation, developing more efficient procedures for modeling larger problems, and hybridizing IE and DE models.

As discussed later, many problems peculiar to TD modeling require consideration. These problems include improving the late-time stability of TD models, which can sometimes produce diverging, nonphysical solutions beyond some observation time. The fact that the late-time behavior of impulsively excited objects is composed of characteristic resonance frequencies, or poles, has also motivated much research in the SEM. Extracting useful physical knowledge from the time domain waveforms associated with TD models has also been addressed as a general problem in signal processing. Whether a total or scattered field formulation is used remains an important modeling choice in DE models, as does the handling of frequency-dispersive, time-varying, and nonlinear components and/or media in all TD models.

1.6.1 Increasing the Stability of the Time-Stepping Solution

TD solutions have sometimes been found to diverge after a sufficient number of time steps have been computed, evidently because of the accumulation of numerical "noise" in the solution. This noise can have its source in numerical roundoff errors or from analytical and numerical approximations made in developing the computer model. In either case, an interpretation of the divergent result is that such approximations introduce low-amplitude, right-half-plane, nonphysical poles into the model. An example of the latter problem is using segment lengths shorter than the wire diameter, thereby violating the thin-wire approximation [4]. Such poles can come to dominate the overall solution after enough time has passed or when the correct response has decayed enough. In either case, a sharply diverging numerical solution can result.

Various techniques for solving the late-time divergence problem have been reported [55, 98–101]. Tijhuis [101] investigated using an improved time-interpolation scheme to increase the accuracy of time derivatives. Rao *et al.* [55] used a conjugate gradient technique to control error accumulation over time. Smith [100] describes a procedure that exploits the fact that late-time instabilities, generally being of a high-frequency nature relative to the correct response, can be filtered from the solution by an averaging technique. Note that late-time instabilities occur for objects such as thin PEC plates that do not exhibit internal resonances, and so they represent a distinctly different kind of numerical anomaly. However, these oscillations can be effectively suppressed by following the averaging scheme which is simple, accurate, and involves a negligible amount of extra computation.

Let $I_{m,n}$ be the current coefficient at the mth patch at a time instant $t = n\Delta t$. In the current stabilization scheme, $I_{m,n+1}$ is calculated using explicit methods, and the averaged value, $\widetilde{I}_{m,n}$, is approximated as

$$\widetilde{I}_{m,n} = \frac{1}{4}(\widetilde{I}_{m,n-1} + 2I_{m,n} + I_{m,n+1}). \tag{1.6.1}$$

1.6.2 Exploiting EM Singularities

Motivation for pursuing electromagnetic resonances can be found throughout the electromagnetics literature [102]. However, it was not until computer techniques became more developed that SEM resonances could be exploited practically. Baum [103] popularized this idea in the 1970s. However, given the nature of Maxwell's equations, other kinds of physical "resonances" can also be identified and can be made similarly accessible from various observed fields. These resonances include source locations determined from far-field patterns, plane wave arrival angles from measurements made along a line or over a plane, and stratified media inversion from frequency-dependent reflected fields [104].

The usual SEM poles can provide information about the size and shape of a PEC object [86]. The oscillatory component of the pole is more sensitive to size because of its dependence on resonance path lengths on the object. The lossy pole component is more sensitive to shape because it is related to radiative damping whose relative importance is determined by shape-induced charge acceleration at curves, bends, and edges. These relationships suggest that SEM poles might provide a mechanism for target recognition because the poles apparently may be unique to a given object. However, the problem of observing poles in noisy data has inhibited their practical application for target recognition [105, 106].

1.6.3 Signal Processing as a Part of TD Modeling

Time domain EM fields can be analyzed for many applications using some of the powerful tools that have been developed in the signal processing community for time-series analysis. Representative examples of some of this work are described by Candy *et al.* [105] and Dudley and Goodman [106]. A particular example of using an earlier signal processing technique for

EM application is Prony's method, developed originally in 1795 [107, 108] for obtaining the coefficients of the exponential series

$$f_n = f(n\Delta t) = \sum R_\alpha e^{s_\alpha n\Delta t}; \qquad n = 1, 2, \cdots, N_{\mathrm{p}}, \tag{1.6.2}$$

which is sampled at a sequence of evenly spaced observation times. Prony's method was the basis for much of the early work in extracting SEM poles from time domain signals [109], but it is recognized, in its original form, to be sensitive to noise. Thus, much recent work in SEM pole estimation has involved other processing techniques [105, 106].

Two interesting possibilities of using other signal processing techniques for specific EM applications are described by Cordaro and Davis [110], Wills [111], Dubard *et al.* [112], and Bi *et al.* [113]. Cordaro and Davis [110] show that it is not necessary to have EM observables available to obtain SEM poles. Instead, by constructing a state-transition matrix from the discretized TDIE for an object, they demonstrate that the object poles are the eigenvalues of that matrix and are thus available directly from the defining equations. This approach is inherently more accurate than estimating poles from TD responses by yielding poles several layers away from the $j\omega$ axis (a capability not generally provided by signal processing). In addition, the eigenvalue computation gives the poles directly, in contrast to their evaluation from a FD matrix which requires a search of the complex frequency plane. Beginning with the source-free version of Eq. (1.4.11), a state vector can be defined as

$$[X_{i;m}]^T = [I_{i;m}, I_{i;m-1}, \cdots, I_{i;m-M}]^T, \tag{1.6.3}$$

where M is the number of time steps required for one transit time across the object and the superscript T denotes a transpose. Then the state matrix can be written as

$$[\Phi] = \begin{bmatrix} Z^{(1)}_{i,k_1} & Z^{(2)}_{i,k_2} & Z^{(3)}_{i,k_3} & \cdots & Z^{(M)}_{i,k_M} \\ I & 0 & 0 & \cdots & 0 \\ 0 & I & 0 & \cdots & 0 \\ 0 & 0 & I & \cdots & 0 \\ \vdots & \vdots & \vdots & \vdots & \vdots \\ 0 & 0 & 0 & I & 0 \end{bmatrix}, \tag{1.6.4}$$

leading to the state equation

$$[X_{i;m+1}] = [\Phi_{i,j}][X_{j,m}] \quad \text{for } m \geq 1, \tag{1.6.5}$$

where the eigenvalues of $[\Phi]$ are now given by $\exp(s_\alpha \Delta t)$. Note that the rank of $[\Phi]$ is given by the total number of space samples, X_s, multiplied by M.

Wills [111] attacked the problem of reducing the number of time steps needed to develop a complete TD response using TLM modeling. In many cases, for high-Q objects in particular, the number of time steps required to obtain the entire response, i.e., for steady state to be reached, can be excessive. However, the late-time response, defined as starting when the object is no longer responding to any excitation, can be represented as a constant coefficient, linear system. By obtaining the coefficients of this system, it is possible to extrapolate the response to steady state, thus avoiding the expense of the late-time calculation. The time savings can exceed a factor of 10, depending on the specific problem. Dubard *et al.* [112] and Bi *et al.* [113] describe similar techniques for early termination of a time-stepping solution.

1.6.4 Total-Field and Scattered-Field Formulations

IE models are usually formulated in terms of the scattered fields caused by some primary excitation because the source integrals employed typically involve only the induced, secondary sources caused by interaction between the excitation and object or region being modeled. The incident field might be included in a different way, for example, by integrating its tangential components over some mathematical surface enclosing the region of interest. However, this step would increase the computation cost while possibly also introducing some inaccuracy in specifying the incident field. Thus, it is the scattered fields, and the secondary sources that cause them, that are the usual unknowns in IE models.

DE models, on the other hand, employ both scattered-field formulation (SFF) and total-field formulation (TFF) with equal facility. The primary field (e.g., an incident plane wave) can be propagated from its boundary values at the closure surface through a spatial mesh, or it can be specified over the surface of an object being modeled within that mesh. The former approach is a TFF, whereas the latter is a SFF. They differ with respect to whether the numerically solved fields include the incident field.

Whether to use a SFF or a TFF in a DE model depends on the specific application and the accuracy needed. If the problem is to find the "leakage" fields inside a metallic envelope having several small apertures, a TFF might be more accurate because the leakage fields are expected to be small relative to the incident field, and the total field in the envelope is the unknown being directly computed. If, however, these small leakage fields are to be obtained instead from an SFF, their computation requires summing the incident and scattered fields, requiring near cancellation of these nearly equal and opposite-signed fields within the envelope. This implies a higher accuracy requirement on the SFF solution than for the TFF due to the errors that arise when subtracting two nearly equal numbers.

Both the SFF and TFF formulation can provide a separate accuracy check for a TD model that is based on how well the incident and scattered fields cancel in certain regions and times. For example, for a TDIE model of a PEC 3D body illuminated by a Gaussian-pulse plane wave, the surface current on the shadow side directly opposite the specular point is known to be zero until enough time has passed for creeping waves to propagate around the surface of the body. The equivalence principle formulation on which an IE for such a problem is based has the incident and scattered fields propagating through empty space. Both components arrive at this shadow point, due to a shorter propagation path, before the creeping wave can do so. During this early time, they cancel exactly analytically. Their failure to cancel numerically is a measure of solution inaccuracy. A similar check is provided by a TDDE model.

1.6.5 Handling Frequency Dispersion and Loading in TD Models

Although a TD model is intrinsically better suited than a FD approach for handling medium and/or boundary-condition nonlinearities and time variations, the converse is true when frequency-dependent phenomena are encountered. This is because frequency dependencies in a TD computation require the equivalent of a convolution integral to be evaluated at each time step. The need for this arises because the various resolvable frequency components of a wideband field or source are affected differently in proportion to, for example, the medium frequency dependency that describes a given problem. Evaluating a convolution integral at each time step for each space sample, however, could impose a substantial computational and storage cost. Fortunately, this degree of rigor is not usually required, and various simplifying methods have been developed that permit efficient and accurate TD modeling of dispersion.

One TD method for modeling dispersion is based on approximating the frequency dependence of dispersive media or component by a pole series. Since the time response associated with a pole series can be expressed as a series of damped exponentials, a time convolution can be avoided. This is the kind of relationship involved in the signal processing procedure called Prony's method, which was originally developed to estimate the parameters of time waveforms comprising damped exponentials. The medium displacement currents that result from this kind of frequency dependency are called "Prony currents" [114]. Other approaches for treating frequency dependence in a TD model are described by German and Gothard [115]. They outline a convolution integral treatment and test its application to media that have Debye and Lorentz frequency dependencies. For the former, the permittivity is given by

$$\epsilon(\omega) = \epsilon_\infty + \frac{\epsilon_{dc} - \epsilon_\infty}{1 + j\omega t_o}, \tag{1.6.6}$$

where ϵ_{dc} is the zero-frequency permittivity, ϵ_∞ is its high-frequency limit, and t_o is the relaxation time for the medium. Similarly, the Lorentz permittivity is given by

$$\epsilon(\omega) = \epsilon_\infty - \frac{\omega_0^2(\epsilon_{dc} - \epsilon_\infty)}{\omega^2 + 2j\omega\delta - \omega_0^2}, \tag{1.6.7}$$

where ω_0 is the medium resonance frequency and δ is its damping coefficient. A DE-based approach was also reported by Nickisch and Franke [116] for modeling pulse propagation in the ionospheric plasma by using the constitutive partial differential equations that define the permittivity tensor.

Impedance boundary conditions and loading are often of interest in modeling. One application, that of synthesizing an absorptive coating on a 2D target to minimize a broadband response, is described by Strickel and Taflove [117]. An approach for handling 2D, thin, resistive sheets is described by Wu and Han [118]. The use of the surface impedance concept for modeling lossy objects is also discussed by Maloney and Smith [119] and Beggs *et al.* [120] as a means of avoiding the volume sampling of a penetrable object that would otherwise be needed.

1.6.6 Handling Medium and Component Nonlinearities or Time Variations in TD Models

A definite advantage of TD models is the capability of handling medium and component nonlinearities or time variations. Although such phenomena can be included in both TDDE and TDIE formulations, their degree of applicability is generally better suited to DE models. One reason is because of the local nature of DE models. Another is because a Green's function is not available for nonlinear or time-varying media so that accounting for the global interactions of an IE models is more difficult.

However, nonlinear and time-varying loads on objects can be quite readily treated using TDIE models. For example, if an object being modeled using the TDIE EFIE were to have L noninteracting, nonlinear loads, at each of the N_t time steps used to compute the time domain response, generally L separate transcendental equations of the form

$$I_{i;m}^L = \frac{I_{i;m}^U}{1 + Y_{i,i} Z_i^L\left(I_{i;m}^L\right)}; \qquad i = 1, 2, \cdots, L; \quad m = 1, 2, \cdots, N_t \tag{1.6.8}$$

would have to be solved, where the superscripts L and U on the currents denote loaded and unloaded quantities, respectively, and Z_i^L is the nonlinear load at space sample i. If up to L loads, instead, interacted within a time step (due to using an implicit solution or because of basis function overlap), then an equation of the form

$$\left[\delta_{i,k} + Y_{i,k} Z_k^L \left(I_{k;m}^L\right)\right] I_{k;m}^L = I_{i;m}^U \qquad i, k = 1, 2, \cdots, L \tag{1.6.9}$$

would need simultaneous solution at each time step.

Similar relationships apply if the loads are permitted to vary in time (independent of the exciting field), and are,

$$I_{i;m}^L = \frac{I_{i;m}^U}{1 + Y_{i,i} Z_i^L(t_m)}; \qquad i = 1, 2, \cdots, L; \quad m = 1, 2, \cdots, N_t \tag{1.6.10}$$

for non-interacting loads, and

$$\left[\delta_{i,k} + Y_{i,k} Z_k^L(t_m)\right] I_{k;m}^L = I_{i;m}^U; \qquad i, k = 1, 2, \cdots, L \tag{1.6.11}$$

for interacting loads, respectively.

A time-varying load might be used to dynamically alter the radiating or scattering properties of an object [5]. Such a load is easier to model than a nonlinear load because the load values themselves are independently established by a given load's specified time variation. An especially simple case of nonlinear loading is a single, current direction-dependent load, such as an idealized diode whose forward and reverse biased impedances are different constants. In this case, the load value is easily determined because, by itself, the load cannot change the direction of the current. In general, however, handling nonlinear loads will require iteration at each time step to satisfy the V–I curve of each load.

1.6.7 Hybrid TD Models

A hybrid model in CEM is usually based on a formulation that involves two kinds of field propagators or (in the case of IE models) one that combines the MFIE and EFIE. Hybrid models are of interest for the possibility they offer of reducing the overall operation count by using the field propagators to which each is best suited for the various parts of a problem. A simple FD example would be a wire antenna interacting with a large, but finite, metal sheet acting as a ground plane for which a combination of the EFIE and GTD is well-suited [121]. Another kind of FD hybrid model employs a special Green's function for part of the problem to limit the number of unknowns to a smaller region, for example, a monopole antenna on a sphere for which a spherical expansion of a point current source permits the unknowns to be restricted to the wire [122].

One of the first hybrid TD models, reported by Taflove and Umashankar [123], was hybridized with respect to the field propagator and used both FD and TD models. They employed a hybrid IE–DE model; the TDDE was used to model the interior geometry of a complex envelope illuminated through an aperture by an external field, and the FDIE was used to model the exterior envelope response. Time-harmonic excitation and a total-field formulation were used, providing a solution that had a 60-dB dynamic range. Barkeshli *et al.* [87] modeled a similar problem entirely in the TD using a TDIE for the exterior and a TDDE for the interior, developing the solution using an implicit formulation and preconditioned iteration.

1.6.8 The Concept of Pseudo-Time in Iterative FD Solutions

Iteration of a FDDE or FDIE can be likened, in some respects, to providing a pseudo-time-like solution, in which each iteration step takes the place of a time step in a time-dependent solution procedure. In either case, the system matrix represents a set of constant interaction coefficients whose values are determined by Maxwell's equations and by the problem's physical characteristics. The sequence of iterates can be viewed as a generalized signal, the appropriate processing of which might make it feasible to estimate a converged result without needing to time step (iterate) the solution to convergence. The potential utility of this viewpoint arises from the observation that, when beginning an iterative solution with an initial guess, the error fields caused by the boundary conditions not being satisfied result in an error current being added to the correct solution. The iteration process is intended to drive the error current to zero analogous to the way that a TD response goes to zero at late times. A current distribution that produces errors in the boundary field on the object being modeled might also be viewed instead as the correct current for a modified excitation that consists of the difference between the desired incident field and the error. In other words, any mathematically possible current distribution is the correct solution for some particular excitation; the problem is finding that distribution for the excitation of interest.

A particular context for exploring pseudo-time was reported by Ling [124]. Using a total-field formulation, he represents the potential function for a 2D problem as

$$F(r, a; t) = e^{i(kx-\omega t)} + \frac{f(r, a; t)e^{i(kr-\omega t)}}{\sqrt{kr}}, \tag{1.6.12}$$

where $f(r, a; t)$ is the scattered field to be found. Substituting $F(r, a; t)$ into the time-dependent wave equation yields

$$\frac{1}{c^2}\frac{\partial^2 f}{\partial t^2} - 2ik\frac{1}{c}\frac{\partial f}{\partial t} = \frac{\partial^2 f}{\partial r^2} + 2ik\frac{\partial f}{\partial r} + \frac{1}{r^2}\frac{\partial^2 f}{\partial a^2} + \frac{1}{4}\frac{f}{r^2}, \tag{1.6.13}$$

which is solved as an initial-value problem by propagating the solution in time, either until the transients die out or until

$$\frac{\partial f}{\partial t} = 0 \quad \text{as } t \to \infty. \tag{1.6.14}$$

This procedure represents an intermediate modeling approach between an FD formulation and a TD formulation that uses time-harmonic excitation. The non-time-harmonic time variation used by Ling is a transient equivalent to that produced in an iterative solution of the FD model. In either case, one might say that a time-harmonic field exists in space, and at some instant of time an object is turned on, resulting in transients whose dying out leads to the steady-state solution that is being sought.

1.6.9 Exploiting Symmetries in TD Modeling

Problem symmetries can profoundly reduce computing requirements in CEM applications [125]. Symmetries can arise in the geometry of an object being modeled and can occur in various forms, including rotation (continuous as in a circular loop or discrete as in a polygon), translation (continuous such as in a finite, elliptical cylinder with translation symmetry along the cylinder axis or discrete as in a periodic structure), and reflection (two parallel wires). Such symmetries reduce both the fill time of IE models and the subsequent solution time. A simple example is a plane wave axially incident on a body of rotation, where the currents vary azimuthally as $\sin(\phi)$. This known angular variation reduces the number of unknowns from being proportional to the surface area to depending only on the object's generating arc length [126, 127]. Although problem symmetries have been quite widely exploited in FD models, more can be done to use symmetries for reducing solution time in TD modeling.

1.7 CONCLUDING REMARKS

This chapter has presented a wide ranging discussion of time domain modeling in electromagnetics. Although several varieties of TD models are possible, attention has been focused on DE and IE types. Among the points made here are the following:

1. TD models are competitive with their FD counterparts in total operation count and may provide more efficient results for some applications.
2. Among the reasons for choosing a TD model are its capabilities for modeling both the nonlinearities and time-varying physical properties of a problem.
3. A TD model offers ready access to object resonances either through signal processing of a computed time domain response or directly from the eigenvalues of the TDIE model when represented in state-matrix form.
4. TDDE models are more general, with respect to handling medium inhomogeneities and other complicating features, compared to their TDIE counterparts.
5. TDIE models can offer a more straightforward treatment of exterior problems and their far fields as well as small geometrical features such as wires.

In conclusion, as computer storage and speed have increased, the scope of problems amenable to TD modeling has dramatically expanded. The fastest growing area of CEM modeling is now in TD models, both DE and IE solutions.

ACKNOWLEDGMENTS

I value the discussions I had with the following colleagues concerning many specific points in this chapter: Professor A. C. Cangellaris, Professor K. R. Demarest, Dr. F. J. German, Professor L. S. Riggs, and Professor C. W. Trueman. I also greatly appreciate the care with which Professor S. M. Rao, the editor of this book, has reorganized some of the material contained herein and corrected various inconsistencies.

BIBLIOGRAPHY

[1] J. A. Landt, E. K. Miller, and M. Van Blaricum, *WT-MBA/LLL1B: A Computer Program for the Time Domain Electromagnetic Response of Thin-Wire Structures*, Lawrence Livermore Laboratory Report No. UCRL-51585, 1974.
[2] J. A. Stratton, *Electromagnetic Theory*, McGraw-Hill, New York, 1941.
[3] J. B. Anderson, "Admittance of Infinite and Finite Cylindrical Metallic Antenna," *Radio Sci.*, vol. 3, pp. 607–621, 1968.
[4] E. K. Miller and J. A. Landt, "Direct Time Domain Techniques for Transient Radiation and Scattering from Wires," *Proc. IEEE*, vol. 68, pp. 1396–1423, 1980.

[5] J. A. Landt and E. K. Miller, "Transient Response of the Infinite Cylindrical Antenna and Scatterer," *IEEE Trans. Antennas Propagat.*, vol. AP-24, p. 246, 1976.
[6] E. M. Kennaugh and R. Cosgriff, "The Use of Impulse Response in Electromagnetic Scattering Problems," *1958 IRE Natl. Conv. Rec.*, pt. 1, pp. 72–77, 1958.
[7] E. M. Kennaugh and D. L. Moffatt, "On the Axial Echo Area of the Cone Sphere Shape," *Proc. IRE [Correspondence]*, vol. 50, pp. 199, 1962.
[8] E. M. Kennaugh and D. L. Moffatt, "Transient and Impulse Response Approximations," *Proc. IEEE*, vol. 53, pp. 893–901, 1965.
[9] K. S. Yee, "Numerical Solution of Initial Boundary Value Problems Involving Maxwell's Equations in Isotropic Media," *IEEE Trans. Antennas Propagat.*, vol. 14, pp. 302–307, 1966.
[10] K. L. Shlager and J. B. Schneider, "A Selective Survey of the Finite-Difference Time Domain Literature," *IEEE Antennas Propagat. Mag.*, vol. 37, pp. 39–57, 1995.
[11] K. M. Mitzner, "Numerical Solution for Transient Scattering from a Hard Surface of Arbitrary Shape-Retarded Potential Technique," *J. Acoust. Soc. Am.*, vol. 42, pp. 391–397, 1967.
[12] C. L. Bennett and W. L. Weeks, "Electromagnetic Pulse Response of Cylindrical Scatterers," in *1968 IEEE G-AP International Symposium*, Northeastern University, Boston, pp. 176–183, 1968.
[13] C. L. Bennett and W. L. Weeks, "Transient Scattering from Conducting Cylinders," *IEEE Trans. Antennas Propagat.*, vol. 18, pp. 627–633, 1970.
[14] E. P. Sayre and R. F. Harrington, "Transient Response of Straight Wire Scatterers and Antennas," *IEEE G-AP International Symposium*, Northeastern University, Boston, pp. 160–164, 1968.
[15] E. P. Sayre and R. F. Harrington, "Time Domain Radiation and Scattering by Thin Wires," *Appl. Sci. Res.*, vol. 26, pp. 413–444, 1972.
[16] P. B. Johns and R. L. Beurle, "Numerical Solution of Two-Dimensional Scattering Problems Using a Transmission-Line Matrix," *Proc. IEEE*, vol. 118, Pt. H, pp. 1203–1208, 1971.
[17] S. Nam, H. Ling, and T. Itoh, "Characterization of Uniform Microstrip Line and Its Discontinuities Using the Time Domain Method of Lines," *IEEE Trans. Microwave Theory Tech.*, vol. 37, pp. 2051–2057, 1989.
[18] T. W. Veruttipong, "Time Domain Version of the Uniform GTD," *IEEE Trans. Antennas Propagat.*, vol. 38, pp. 1757–1764, 1990.
[19] P. O. Brundell, *Transient Electromagnetic Waves around a Cylindrical Transmitting Antenna*, Erricsson Tech., vol. 16, pp. 137–162, 1960.
[20] T. T. Wu, "Transient Response of a Dipole Antenna," *J. Math. Phys.*, vol. 2, pp. 982–984, 1961.
[21] O. Einarsson, "The Step-Voltage Current Response of an Infinite Conducting Cylinder," *Trans. R. Inst. Technol.*, Stockholm, Sweden, pp. 191, 1962.
[22] G. Franceschetti and C. H. Pappas, "Pulsed Antennas," *IEEE Trans. Antennas Propagat.*, vol. 22, pp. 651–661, 1974.
[23] A. G. Tijhuis, P. Zhongqiu, and A. R. Bretones, "Transient Excitation of a Straight Thin-Wire Segment: A New Look at an Old Problem," *IEEE Trans. Antennas Propagat.*, vol. 40, pp. 1132–1146, 1992.
[24] E. K. Miller, A. J. Poggio, and G. J. Burke, "An Integro-Differential Equation Technique for the Time Domain Analysis of Thin-Wire Structures, Part I—The Numerical Method," Part II—Numerical Results," *J. Comput. Phys.*, vol. 12, 1973.
[25] A. J. Poggio and E. K. Miller, "Integral Equation Solutions of Three-Dimensional Scattering Problems," in *Computer Techniques for Electromagnetics*, R. Mittra, ed., Pergamon, New York, 1973.
[26] T. K. Lui and K. K. Mei, "A Time Domain Integral Equation Solution for Linear Antennas and Scatterers," *Radio Sci.*, vol. 8, pp. 797–804, 1973.
[27] C. L. Bennett, "The Numerical Solution of Transient Electromagnetic Scattering Problems," in *Electromagnetic Scattering*, P. G. E. Uslenghi, ed., Academic Press, New York, pp. 393–428, 1978.
[28] C. L. Bennett, "Time Domain Solutions via Integral Equations–Surfaces and Composite Bodies," in *Theoretical Methods for Determining the Interaction of Electromagnetic Waves with Structures*, Sijthoff and Noordhoff, Rockville, MD, pp. 255–275, 1981.
[29] C. L. Bennett and H. Mieras, "Time Domain Scattering from Open Thin Conducting Surfaces," *Radio Sci.*, vol. 16, pp. 1231–1239, 1981.

[30] H. Mieras and C. L. Bennett, "Space–Time Integral Equation Approach to Dielectric Targets," *IEEE Trans. Antennas Propagat.*, vol. 30, pp. 2–9, 1982.
[31] R. Gomez, J. A. Morente, and A. Salinas, "Time Domain Analysis of an Array of Straight-Wire Coupled Antennas," *IEEE Electronic Lett.*, vol. 22, pp. 316–318, 1986.
[32] E. Marx,"Electromagnetic Pulse Scattered by a Sphere," *IEEE Trans. Antennas Propagat.*, vol. 35, pp. 412–417, 1987.
[33] A. R. Bretones, A. Salinas, R. Gomez-Martin, and A. Perez, "The Comparison of a Time Domain Numerical Code (DOTIG1) with Several Frequency-Domain Codes Applied to the Case of Scattering from a Thin-Wire Cross," *ACES J.*, pp. 121–129, 1989.
[34] R. Gomez, A. Salinas, A. R. Bretones, J. Fornieles, and M. Martin, "Time Domain Integral Equations for EMP Analysis," *Int. J. Numerical Modeling*, vol. 4, pp. 153–162, 1991.
[35] R. Gomez, A. Salinas, and A. R. Bretones, "Time Domain Integral Equation Methods for Transient Analysis," *IEEE AP-S Mag.*, vol. 34, pp. 15–22, 1992.
[36] S. M. Rao and D. R. Wilton, "Transient Scattering by Conducting Surfaces of Arbitrary Shape," *IEEE Trans. Antennas Propagat.*, vol. 39, pp. 56–61, 1991.
[37] D. A. Vechinski and S. M. Rao, "A Stable Procedure to Calculate the Transient Scattering by Conducting Surfaces of Arbitrary Shape," *IEEE Trans. Antennas Propagat.*, vol. 40, pp. 661–665, 1992.
[38] D. A. Vechinski, S. M. Rao, and T. K. Sarkar, "Transient Scattering from Three-Dimensional Arbitrarily Shaped Dielectric Bodies," *J. Optical Soc. Am.*, vol. 11, pp. 1458–1470, 1994.
[39] S. P. Walker, M. J. Bluck, M. D. Pocock, C. Y. Leung, and S. J. Dodson, "Curvilinear, Isoparametric Modeling for RCS Prediction Using Time Domain Integral Equations," in *12th Annual Review of Progress in Applied Computational Electromagnetics*, vol. 1, pp. 196–204, 1996.
[40] S. P. Walker, "Developments in Time Domain-Equation Modeling at Imperial College," *IEEE Antennas Propagat. Mag.*, vol. 39, pp. 7–19, 1997.
[41] J. A. Landt, E. K. Miller, and F. J. Deadrick, "Time Domain Modeling of Nonlinear Loads," *IEEE Trans. Antennas Propagat.*, vol. 31, pp. 121–126, 1983.
[42] T. K. Sarkar and D. D. Weiner, "Scattering Analysis of Nonlinearly Loaded Antennas," *IEEE Trans. Antennas Propagat.*, vol. 24, pp. 125–131, 1976.
[43] C. L. Bennett and G. F. Ross, "Time Domain Electromagnetics and Its Applications," *Proc. IEEE*, vol. 66, pp. 299–318, 1978.
[44] E. K. Miller, "An Overview of Time Domain Integral-Equation Models in Electromagnetics," *J. Electromagnetic Waves Appl.*, vol. 1, pp. 269–293, 1987.
[45] E. K. Miller, "Time Domain Modeling in Electromagnetics," *J. Electromagnetic Waves Appl.*, vol. 8, pp. 1125–1172, 1994.
[46] L. Felsen, *Transient Electromagnetics*, Springer-Verlag, New York, 1976.
[47] E. K. Miller, *Time Domain Measurements in Electromagnetics*, Van Nostrand–Reinhold, New York, 1986.
[48] K. S. Kunz and R. J. Luebbers, *The Finite Difference Time Domain Method in Electromagnetics*, CRC Press, Boca Raton, FL, 1993.
[49] A. Taflove, *Computational Electromagnetics: The Finite-Difference Time Domain Method*, Artech House, Boston, 1995.
[50] C. Christopoulos, *Transmission-Line Modeling (TLM) Method*, IEEE Press, Piscataway, NJ, 1995.
[51] B. Noel, "Ultra-Wideband Radar," in *Proceedings of the First Los Alamos Symposium*, CRC Press, Boca Raton, FL, 1991.
[52] H. L. Bertoni, L. Carin, and L. B. Felsen, *Ultra-Wideband, Short-Pulse Electromagnetics*, Plenum, New York, 1993.
[53] J. D. Taylor, *Introduction to Ultra-Wideband Radar Systems*, CRC Press, Boca Raton, FL, 1995.
[54] D. Lamensdorf and L. Susman, "Baseband-Pulse-Antenna Techniques," *IEEE Antennas Propagat. Mag.*, vol. 36, pp. 20–30, 1994.
[55] S. M. Rao, T. K. Sarkar, and S. A. Dianat, "A Novel Technique to the Solution of Transient Electromagnetic Scattering from Thin Wires," *IEEE Trans. Antennas Propagat.*, vol. 34, pp. 630–634, 1986.

[56] R. M. Bevensee, H. S. Cabayan, F. J. Deadrick, R. W. Egbert, L. C. Martin, E. K. Miller, J. T. Okada, A. J. Poggio, and J. L. Willows, *External Coupling of EMP to Generic System Structures*, Lawrence Livermore National Laboratory, Rep. No. M–090, 1978.
[57] F. X. Canning, "The Impedance Matrix Localization (IML) Method for Moment–Method Calculations," *IEEE Antennas Propagat. Mag.*, vol. 32, No. 5, pp. 18–30, 1990.
[58] J. F. Botha and G. F. Pinder, *Fundamental Concepts in the Numerical Solution of Differential Equations*, Wiley, New York, 1983.
[59] B. W. Arden and K. N. Astill, *Numerical Algorithms: Origins and Applications*, Addison-Wesley, Reading, MA, 1970.
[60] D. J. Riley and C. D. Turner, "VOLMAX: A Solid Model-Based, Transient Volumetric Maxwell Solver Using Hybrid Grids," *IEEE Antennas Propagat. Mag.*, vol. 39, pp. 20–33, 1997.
[61] J. S. Shang, "Characteristic-Based Algorithms for Solving the Maxwell Equations in the Time Domain," *IEEE Antennas Propagat. Mag.*, vol. 37, pp. 15–25, 1995.
[62] C. W. Trueman, S. J. Kubina, R. J. Luebbers, K. S. Kunz, S. R. Mishra, and C. Larose, "RCS of Cubes, Strips, Rods and Cylinders by FDTD," in *Proceedings of the 8th Annual Review of Progress in Computational Electromagnetics*, Naval Postgraduate School, Monterey, CA, pp. 487–494, 1992.
[63] C. W. Trueman, S. J. Kubina, S. R. Mishra, and C. Larose, "RCS of Four Fuselage-like Scatterers at HF Frequencies," *IEEE Trans. Antennas Propagat.*, vol. 40, pp. 236–240, 1992.
[64] R. Holland, V. P. Cable, and L. Wilson, "A 2D Finite-Volume Time Domain Technique for RCS Evaluation," in *Proceedings of the 7th Annual Review of Progress in Applied Computational Electromagnetics*, Naval Postgraduate School, Monterey, CA, pp. 667–681, 1991.
[65] K. S. Yee, J. S. Chen, and A. H. Chang, "Conformal Finite-Difference Time Domain (FDTD) with Overlapping Grids," *IEEE Trans. Antennas Propagat.*, vol. 40, pp. 1068–1075, 1992.
[66] M. Fusco, "FDTD Algorithm in Curvilinear Coordinates," *IEEE Trans. Antennas Propagat.*, vol. 38, pp. 76–89, 1990.
[67] R. Holland, "Finite Difference Solutions of Maxwell's Equations in Generalized Nonorthogonal Coordinates," *IEEE Trans. Nucl. Sci.*, vol. 30, pp. 4689–4691, 1983.
[68] A. H. Mohammedian, V. Shankar, and W. F. Hall, "Computation of Electromagnetic Scattering and Radiation Using a Time Domain Finite-Volume Discretization Procedure," *Comp. Phys. Commun.*, vol. 68, pp. 175–196, 1991.
[69] R. J. Luebbers, K. Kunz, M. Schneider, and F. Hunsberger, "A Finite Difference Time Domain Near Zone to Far Zone Transformation," *IEEE Trans. Antennas Propagat.*, vol. 39, pp. 429–433, 1991.
[70] R. Mittra, "The Finite Difference Time Domain (FDTD) Method," in *Short-Course Notes for Computational Methods in Electromagnetics*, Southeastern Center for Electrical Engineering Education, Kissimmee, FL, 1992.
[71] A. Taflove and K. Umashankar, "The Finite-Difference Time Domain Method for Numerical Modeling of Electromagnetic Wave Interactions," *Electromagnetics*, vol. 10, 1991.
[72] J. M. Jin and J. L. Volakis, "A Finite Element-Boundary Integral Formulation for Scattering by Three-Dimensional Cavity-Backed Apertures," *IEEE Trans. Antennas Propagat.*, vol. 39, pp. 97–108, 1991.
[73] J. M. Jin and J. L. Volakis, "Electromagnetic Scattering by and Transmission Through a Three-Dimensional Slot in a Thick Conducting Plane," *IEEE Trans. Antennas Propagat.*, vol. 39, pp. 543–550, 1991.
[74] J. Nielsen, "Spurious Modes of the TLM-Condensed Node Formulation," *IEEE Microwave & Guided Wave Lett.*, vol. 1, pp. 201–203, 1991.
[75] E. Lindman, "Free-Space Boundary Conditions for the Time-Dependent Wave Equation," *J. Comput. Phys.*, vol. 18. pp. 66–78, 1975.
[76] B. Engquist and A. Majda, "Absorbing Boundary Conditions for Numerical Computation of Waves," *Math. Comput.*, vol. 31, pp. 639–651, 1977.
[77] A. Bayliss and E. Turkel, "Radiation Boundary Conditions for Wavelike Equations," *Comm. Pure Appl. Math.*, vol. 33, pp. 707–725, 1980.
[78] G. Mur, "Absorbing Boundary Conditions for the Finite-Difference Approximation of the Time Domain Electromagnetic-Field Equations," *IEEE Trans. Electromagnetic Compat.*, vol. 23, pp. 377–382, 1981.

[79] C. H. Wilcox, "An Expansion Theorem for Electromagnetic Fields," *Commun. Pure Appl. Math.*, vol. 9, pp. 115–132, 1956.
[80] D. R. Wilton and W. Richards, "Application of Near-Field Radiation Conditions to the Solution of Electromagnetic Scattering Problems," *URSI Radio Science Symposium*, Syracuse, NY, 1988.
[81] S. Arendt, K. Umashankar, A. Taflove, and G. Kreigsmann, "Extension of On-Surface Radiation Condition Theory to Scattering by Two-Dimensional Homogeneous Dielectric Objects," *IEEE Trans. Antennas Propagat.*, vol. 38, pp. 1551–1558, 1990.
[82] O. M. Ramahi, "Concurrent Implementation of the Complementary Operators Method in 2-D Space," *IEEE Microwave Guided Wave Lett.*, vol. 7, pp. 165–167, 1997.
[83] O. M. Ramahi and J. B. Schneider, "Comparative Study of the PML and C-COM Mesh-Truncation Techniques," *IEEE Microwave Guided Wave Lett.*, vol. 8, pp. 55–57, 1998.
[84] K. Hayata, M. Eguchi, and M. Koshiba, "Self-Consistent Finite/Infinite Element Scheme for Unbounded Guided Wave Problems," *IEEE MTT-S Trans.*, vol. 36, pp. 614–616, 1988.
[85] J. M. Jin, *The Finite Element Method in Electromagnetics*, Wiley, New York, 1993.
[86] E. K. Miller, "Model-Based Parameter-Estimation Techniques in Electromagnetics," in *Electromagnetic Modeling and Measurements for Analysis and Synthesis Problems*, B. de Neumann, ed., Kluwer, Dordrecht, pp. 205–256, 1991.
[87] S. Barkeshli, H. A. Sabbagh, D. J. Radecki, and M. Melton, "A Novel Implicit Time-Domain Boundary-Integral/Finite-Element Algorithm for Computing Transient Electromagnetic Field Coupling to a Metallic Enclosure," *IEEE Trans. Antennas Propagat.*, vol. AP-40, pp. 1155–1164, 1992.
[88] R. W. Ziolkowski, N. K. Madsen, and R. C. Carpenter, "Three-Dimensional Computer Modeling of Electromagnetic Fields: A Global Lookback Lattice Transaction Scheme," *J. Comp. Phys.*, vol. 50, pp. 360–408, 1983.
[89] M. A. Morgan and K. K. Mei, "Finite-Element Computation of Scattering by Inhomogeneous Penetrable Bodies of Revolution," *IEEE AP-S Trans. Antennas Propagat.*, vol. 27, pp. 202–214, 1979.
[90] K. K. Mei, R. Pous, Z. Chen, Y. W. Liu, and M. D. Prouty, "Measured Equation of Invariance: A New Concept in Field Computation," *IEEE Trans. Antennas Propagat.*, vol. 42, pp. 320–328, 1994.
[91] G. K. Gothard, S. M. Rao, T. K. Sarkar, and M. Salazar Palma, "Finite Element Solution of Open Region Electrostatic Problems Incorporating the Measured Equation of Invariance," *IEEE Microwave Guided Wave Lett.*, vol. 5, pp. 252–254, 1995.
[92] K. Demarest, "A Finite-Difference Time Domain Technique for Modeling Narrow Apertures in Conducting Scatterers," *IEEE Trans. Antennas Propagat.*, vol. 35, pp. 826–831, 1987.
[93] A. Taflove, K. Umashankar, B. Beker, F. Harfoush, and K. S. Yee, "Detailed FDTD Analysis of Electromagnetic Fields Penetrating Narrow Slots and Lapped Joints in Thick Conducting Screens," *IEEE Trans. Antennas Propagat.*, vol. 36, pp. 247–257, 1988.
[94] D. J. Riley and C. D. Turner, "Hybrid Thin-Slot Algorithm for the Analysis of Narrow Apertures in Finite-Difference Time Domain Calculations," *IEEE Trans. Antennas Propagat.*, vol. 38, pp. 1943–1950, 1990.
[95] P. A. Tirkas and C. A. Balanis, "Finite-Difference Time Domain Method Antenna Radiation," *IEEE Trans. Antennas Propagat.*, vol. 40, pp. 334–340, 1992.
[96] Z. Chen, M. M. Ney, and W. J. R. Hoefer, "A New Finite-Difference Time Domain Formulation and Its Equivalence with the TLM Symmetrical Condensed Node," *IEEE MTT-S Trans.*, vol. 39, pp. 2160–2169, 1991.
[97] F. J. German, G. K. Gothard, L. S. Riggs, and P. M. Goggans, "The Calculation of Radar Cross-Section (RCS) Using the TLM Method," *Int. J. Numerical Modeling*, vol. 2, pp. 267–278, 1989.
[98] B. P. Rynne and P. D. Smith, "Stability of Time-Marching Algorithms for the Electric-Field Integral Equation," *J. Electromagnetic Waves Appl.*, vol. 4, pp. 1181–1205, 1990.
[99] B. P. Rynne, "Time Domain Scattering from Arbitrary Surfaces Using the Electric-Field Integral Equation," *J. Electromagnetic Waves Appl.*, vol. 5, pp. 93–112, 1991.
[100] P. D. Smith "Instabilities in Time Marching Methods for Scattering: Cause and Rectification," *Electromagnetics*, vol. 10, pp. 439–451, 1990.
[101] A. G. Tijhuis, "Toward a Stable Marching-on-in Time Method for Two-Dimensional Transient Electromagnetic Scattering Problems," *Radio Sci.*, vol. 19, pp. 1311–1317, 1984.

[102] S. A. Schelkunoff, *Advanced Antenna Theory*, Wiley, New York, 1952.
[103] C. E. Baum, "Emerging Technology for Transient and Broadband Analysis and Synthesis of Antennas and Scatterers," *Proc. IEEE*, vol. 64, pp. 1598–1616, 1976.
[104] E. K. Miller, "Natural-Mode Methods in Frequency and Time Domain Analysis," in *Theoretical Methods for Determining the Interaction of Electromagnetic Waves with Structures*, Sijthoff & Noordhoff, Rockville, MD, pp. 173–212, 1981.
[105] J. V. Candy, G. A. Clark, and D. M. Goodman, "Transient Electromagnetic Signal Processing: An Overview of Techniques," in *Time Domain Measurements in Electromagnetics*, E. K. Miller, ed., Van Nostrand–Reinhold, New York, 1986.
[106] D. G. Dudley and D. M. Goodman, "Transient Identification and Object Classification," in *Time Domain Measurements in Electromagnetics*, E. K. Miller, ed., Van Nostrand–Reinhold, New York, 1986.
[107] F. Hildebrand, *Introduction to Numerical Analysis*, McGraw-Hill, New York, 1956.
[108] R. Prony "Essai Experimental et Analytique sur les Lois de la Dilatabilite de Fluides Elastiques et sur Celles de la Force Expansive de la Varpeur de L'alkoal, a Differentes Temperatures," *J. l'Ecole Polytech.*, vol. 2, pp. 24–76, 1795.
[109] A. J. Poggio, M. L. VanBlaricum, E. K. Miller, and R. Mittra, "Evaluation of a Processing Technique for Transient Data," *IEEE AP-S Trans. Antennas Propagat.*, vol. 26, pp. 165–173, 1978.
[110] J. T. Cordaro and W. A. Davis, "Time Domain Techniques in the Singularity Expansion Method," *IEEE AP-S Trans.*, vol. 29, pp. 534–538, 1981.
[111] J. D. Wills, "Spectral Estimation for the Transmission Line Matrix Method," *IEEE Trans. Microwave Theory Tech.*, vol. 38, pp. 448–451, 1990.
[112] J. L. Dubard, D. Ponpei, J. Le Roux, and A. Papiernik, "Characterization of Microstrip Antennas Using the TLM Simulation Associated with a Prony–Pisarenko Method," *Int. J. Numerical Modeling*, vol. 3, pp. 269–285, 1990.
[113] Z. Bi, Y. Shen, K. Wu, and J. Litva, "Fast Finite-Difference Time Domain Analysis of Resonators Using Digital Filtering and Spectrum Estimation Techniques," *IEEE MTT-S Trans.*, vol. 40, pp. 1611–1619, 1992.
[114] R. Holland, "Time Domain Treatment of Maxwell's Equations in Frequency-Dependent Media," in *Proceedings of the 2nd Annual Review of Progress in Applied Computational Electromagnetics*, Naval Postgraduate School, Monterey, CA, 1986.
[115] F. J. German and G. K. Gothard, "The Analysis of Pulse Propagation in Linear Dispersive Media with Absorption by the Finite-Difference Time Domain Method," in *Proceedings of the 8th Annual Review of Progress in Applied Computational Electromagnetics*, Naval Postgraduate School, Monterey, CA, pp. 495–514, 1992.
[116] L. J. Nickisch and P. M. Franke, "Finite Difference Time Domain Solution of the Maxwell Equations for the Dispersive Ionosphere," in *Proceedings of the 8th Annual Review of Progress in Applied Computational Electromagnetics*, Naval Postgraduate School, Monterey, CA, pp. 515–522, 1992.
[117] M. A. Strickel and A. Taflove, "Time Domain Synthesis of Broadband Absorptive Coatings for Two-Dimensional Conducting Targets," *IEEE Trans. Antennas Propagat.*, vol. 38, pp. 1084–1091, 1990.
[118] L. K. Wu and L. T. Han, "Implementation and Application of Resistive Sheet Boundary Condition in the Finite-Difference Time Domain Method," *IEEE Trans. Antennas Propagat.*, vol. 40, pp. 628–633, 1992.
[119] J. G. Maloney and G. S. Smith, "The Use of Surface Impedance Concepts in the Finite-Difference Time Domain Method," *IEEE Trans. Antennas Propagat.*, vol. 40, pp. 38–48, 1992.
[120] J. H. Beggs, R. J. Luebbers, K. S. Yee, and K. S. Kunz, "Finite Difference Time Domain Implementation of Surface Impedance Boundary Conditions," *IEEE Trans. Antennas Propagat.*, vol. 40, pp. 49–56, 1992.
[121] G. A. Thiele and T. H. Newhouse, "A Hybrid Technique for Combining Moment Methods with the Geometrical Theory of Diffraction," *IEEE Trans. Antennas Propagat.*, vol. 17, pp. 62–69, 1969.
[122] F. M. Tesche and A. R. Neureuther, "Radiation Patterns for Two Monopoles on a Perfectly Conducting Sphere," *IEEE Trans. Antennas Propagat.*, vol. 18, pp. 692–694, 1970.
[123] A. Taflove and K. Umashankar, "A Hybrid Moment Method/Finite Difference Time Domain Approach to Electromagnetic Coupling and Aperture Penetration into Complex Bodies," in *Applications*

of the Method of Moments to Electromagnetic Fields, B. J. Strait, ed., St. Cloud, FL: SCEEE Press, pp. 361–426, 1980.

[124] T. T. Ling, "A Time-Dependent Method for the Numerical Solution of Wave Equations in Electromagnetic Scattering Problems," *Comp. Phys. Commun.*, vol. 68, pp. 213–223, 1991.

[125] E. K. Miller, "A Selective Survey of Computational Electromagnetics," *IEEE Antennas Propagat.*, vol. 36, pp. 1281–1305, 1988.

[126] C. L. Bennett, "Time Domain Solutions of Rotationally Symmetric Scattering Problems," Proceedings of 1971 Fall URSI Meeting, UCLA, Los Angeles, p. 81.

[127] E. Y. Sun and W. V. T. Rusch, "EFIE Time Domain Scattering from Bodies of Revolution," *IEEE AP-S Symposium*, University of Western Ontario, London, Ontario, Canada, pp. 926–929, 1991.

CHAPTER 2

Wire Structures: TDIE Solution

S. M. RAO
Department of Electrical Engineering, Auburn University

T. K. SARKAR
Department of Electrical Engineering and Computer Science, Syracuse University

Many practical geometries can be modeled as wire-grid structures to accurately compute the radiation pattern or radar cross section. These models usually use thin wires for modeling purposes which result in evaluating simple one-dimensional integrals in the numerical solution procedure.[1] Thus, the algorithm developed using a thin-wire approximation is very efficient from a computational point of view. There are several computer codes available based on wire-modeling techniques, both in the frequency and in the time domain [1–4]. All these codes are very popular and currently used in a variety of situations to model complex bodies.

In order to develop an efficient, general-purpose, wire-grid modeling algorithm, whether in the time domain or in the frequency domain, the following basic steps are required:

- Use an efficient numerical procedure to analyze the arbitrary wire problem.
- Extend the numerical procedure to handle multiple wires, wire junctions, and wires with different radii.
- Add pre- and postprocessors such as an efficient geometry generation package, graphical interface to display model and results, and other user-defined parameters.

In this chapter, we provide detailed mathematical equations and various numerical methods to analyze arbitrary wire scatterers in the time domain. The wire scatterers are assumed to be perfectly conducting, finite-length, circular cylinders which may be either thin-wall tubes or solid wires. In this situation, the induced current varies primarily along the length of the wire. Any variation in the circumferential direction is ignored. It is hoped that the material presented in this chapter provides enough insight for the reader to develop algorithms for the development of a full-blown thin-wire code if desired.

[1] We define a thin wire as a cylindrical tube whose length to radius ratio (ℓ/a) is large ($\approx$100).

First, we develop the thin-wire analysis with a comprehensive treatment of a single, straight, thin wire to provide the basic background necessary to handle arbitrary wires. For this case, we develop the thin-wire theory, starting with Maxwell's equations and utilizing potential theory. The well-known equivalence principle [5] is used to obtain the necessary integral equations. We discuss various methods and formulations to solve the straight, thin-wire problem. We note advantages and disadvantages of these methods.

Next, we discuss the analysis of an arbitrary wire. Here, we provide a detailed treatment of the different numerical methods used to obtain an accurate and stable solution. Lastly, we include a short discussion on multiple-wire and wire junction problems. We hope that this discussion helps the reader to develop a thin-wire time domain code to solve complex problems.

2.1 BASIC ANALYSIS

In this section, we consider some preliminary steps involved in developing expressions for the electric and magnetic fields in the time domain in terms of electromagnetic field sources and potentials. These expressions are later utilized in a variety of radiation and scattering problems in the following sections and chapters. We begin the analysis by writing Maxwell's equations in the time domain, given by

$$\nabla \times \boldsymbol{E}(\boldsymbol{r}, t) = -\mu \frac{\partial \boldsymbol{H}(\boldsymbol{r}, t)}{\partial t} \tag{2.1.1}$$

$$\nabla \times \boldsymbol{H}(\boldsymbol{r}, t) = \epsilon \frac{\partial \boldsymbol{E}(\boldsymbol{r}, t)}{\partial t} + \boldsymbol{J}^{\mathrm{i}}(\boldsymbol{r}, t) \tag{2.1.2}$$

$$\nabla \cdot \boldsymbol{E}(\boldsymbol{r}, t) = \frac{q_v^{\mathrm{i}}(\boldsymbol{r}, t)}{\epsilon} \tag{2.1.3}$$

$$\nabla \cdot \boldsymbol{H}(\boldsymbol{r}, t) = 0, \tag{2.1.4}$$

where $\boldsymbol{J}^{\mathrm{i}}(\boldsymbol{r}, t)$ and $q_v^{\mathrm{i}}(\boldsymbol{r}, t)$ represent the impressed source current density and source charge density, respectively, and are related by the continuity equation, given by

$$\nabla \cdot \boldsymbol{J}^{\mathrm{i}}(\boldsymbol{r}, t) = -\frac{\partial q_v^{\mathrm{i}}(\boldsymbol{r}, t)}{\partial t}. \tag{2.1.5}$$

Using potential theory, it is convenient to represent the electric and magnetic fields in terms of intermediate quantities given by vector potential $\boldsymbol{A}(\boldsymbol{r}, t)$ and scalar potential $\Phi(\boldsymbol{r}, t)$. Thus, we define

$$\boldsymbol{H} \triangleq \frac{1}{\mu} \nabla \times \boldsymbol{A}. \tag{2.1.6}$$

Note that $\boldsymbol{A}(\boldsymbol{r}, t)$ is a purely mathematical quantity with no physical significance. By substituting Eq. (2.1.6) into Eq. (2.1.1), we have

$$\nabla \times \boldsymbol{E}(\boldsymbol{r}, t) = -\nabla \times \frac{\partial \boldsymbol{A}(\boldsymbol{r}, t)}{\partial t} \Rightarrow \nabla \times \left[\boldsymbol{E}(\boldsymbol{r}, t) + \frac{\partial \boldsymbol{A}(\boldsymbol{r}, t)}{\partial t}\right] = 0. \tag{2.1.7}$$

In Eq. (2.1.7), the curl of a vector, $\boldsymbol{E}(\boldsymbol{r}, t) + (\partial \boldsymbol{A}(\boldsymbol{r}, t))/\partial t$, equals zero. We know from classical vector calculus that when the curl of a vector is zero, the vector can then be equated to the gradient of a scalar function. Thus, we have

$$\boldsymbol{E}(\boldsymbol{r},t)+\frac{\partial \boldsymbol{A}(\boldsymbol{r},t)}{\partial t}=-\nabla\Phi(\boldsymbol{r},t), \tag{2.1.8}$$

where Φ is a scalar function, also known as the scalar potential. Note that the negative sign on the gradient of the scalar potential in Eq. (2.1.8) is introduced only for mathematical convenience. From Eq. (2.1.8), we have

$$\boldsymbol{E}(\boldsymbol{r},t)=-\frac{\partial \boldsymbol{A}(\boldsymbol{r},t)}{\partial t}-\nabla\Phi(\boldsymbol{r},t). \tag{2.1.9}$$

Next, we substitute Eqs. (2.1.6) and (2.1.9) into Eq. (2.1.2) to obtain

$$\nabla\times\nabla\times\boldsymbol{A}=-\mu\epsilon\frac{\partial}{\partial t}\left[\frac{\partial \boldsymbol{A}}{\partial t}+\nabla\Phi\right]+\mu\boldsymbol{J}^{\mathrm{i}}. \tag{2.1.10}$$

Now, using the vector identity $\nabla\times\nabla\times\boldsymbol{A}=\nabla(\nabla\cdot\boldsymbol{A})-\nabla^2\boldsymbol{A}$ and defining the wave speed $c=1/\sqrt{\mu\epsilon}$, we can rewrite Eq. (2.1.10) as

$$\nabla^2\boldsymbol{A}-\frac{1}{c^2}\frac{\partial^2\boldsymbol{A}}{\partial t^2}=\nabla\left[\nabla\cdot\boldsymbol{A}+\frac{1}{c^2}\frac{\partial\Phi}{\partial t}\right]-\mu\boldsymbol{J}^{\mathrm{i}}. \tag{2.1.11}$$

Using the Lorentz gauge condition,

$$\nabla\cdot\boldsymbol{A}=-\frac{1}{c^2}\frac{\partial\Phi}{\partial t}, \tag{2.1.12}$$

we have the required differential equation

$$\nabla^2\boldsymbol{A}-\frac{1}{c^2}\frac{\partial^2\boldsymbol{A}}{\partial t^2}=-\mu\boldsymbol{J}^{\mathrm{i}}. \tag{2.1.13}$$

Furthermore, taking the divergence of $\boldsymbol{E}(\boldsymbol{r},t)$ in Eq. (2.1.9), utilizing Eqs. (2.1.3) and (2.1.12), and noting that $\nabla\cdot\nabla\Phi=\nabla^2\Phi$, we have another differential equation for the scalar potential, given by

$$\nabla^2\Phi-\frac{1}{c^2}\frac{\partial^2\Phi}{\partial t^2}=-\frac{q_v^{\mathrm{i}}}{\epsilon}. \tag{2.1.14}$$

Notice that Eqs. (2.1.13) and (2.1.14) have the same mathematical form with well-known solutions, given by

$$\boldsymbol{A}(\boldsymbol{r},t)=\mu\int_v\frac{\boldsymbol{J}^{\mathrm{i}}(\boldsymbol{r}',t-R/c)}{4\pi R}\,dv' \tag{2.1.15}$$

and

$$\Phi(\boldsymbol{r},t)=\frac{1}{\epsilon}\int_v\frac{q_v^{\mathrm{i}}(\boldsymbol{r}',t-R/c)}{4\pi R}\,dv', \tag{2.1.16}$$

where $R = |\boldsymbol{r} - \boldsymbol{r}'|$. In Eqs. (2.1.15) and (2.1.16), $\boldsymbol{r}$ and $\boldsymbol{r}'$ represent the location of the observation and source point, respectively. Once we obtain the vector and scalar potentials, it is straightforward to calculate $\boldsymbol{H}(\boldsymbol{r}, t)$ and $\boldsymbol{E}(\boldsymbol{r}, t)$ via Eqs. (2.1.6) and (2.1.9), respectively. It is also possible to derive an alternate expression for the electric field by differentiating both sides of Eq. (2.1.9) with respect to time and utilizing Eq. (2.1.12) to obtain

$$\frac{\partial \boldsymbol{E}}{\partial t} = -\frac{\partial^2 \boldsymbol{A}}{\partial t^2} + c^2 \nabla(\nabla \cdot \boldsymbol{A}), \tag{2.1.17}$$

which is in terms of the vector potential only.

2.2 ANALYSIS OF A STRAIGHT WIRE

In this section, integral equations are derived for the current induced on a thin, finite-length perfect electrical conducting cylindrical tube excited by an incident electromagnetic field. The incident electric field, $\boldsymbol{E}^{\mathrm{i}}(\boldsymbol{r}, t)$ is assumed to be invariant around the cylinder circumference and polarized along the length of the cylinder.

Consider a thin wire of length $2h$ m and radius a m, located symmetrically along the z-axis as shown in the Fig. 2.1. The incident electric field induces current $I(z, t)$, which is a function of z and t only because of the thin-wire approximation. Using the reduced kernel approximation [6], the vector potential $\boldsymbol{A}(x, y, z, t) = A_z(x, y, z, t)\boldsymbol{a}_z$ is given by

$$A_z(x, y, z, t) = \mu \int_{z'=-h}^{h} \frac{I(z', t - R/c)}{4\pi R}\, dz', \tag{2.2.1}$$

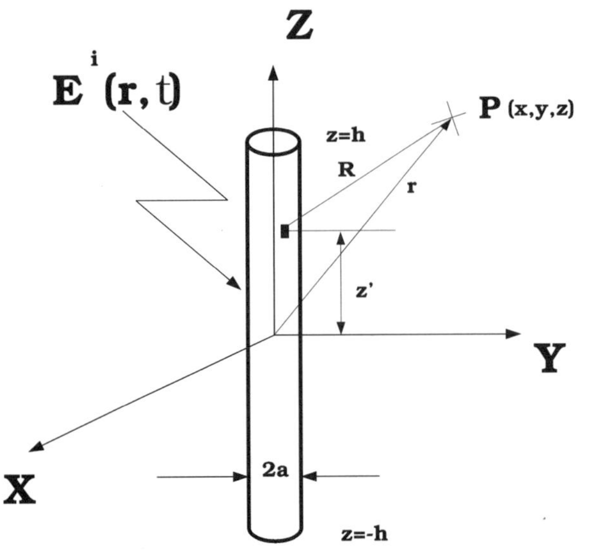

FIGURE 2.1
Finite-length wire excited by an incident electromagnetic plane wave.

where

$$R = \sqrt{[x^2 + y^2 + (z - z')^2 + a^2]}, \tag{2.2.2}$$

and a is the radius of the wire.

The total electric field is the sum of the incident and scattered fields. Next, we apply the boundary condition for the total electric field, which implies that the z component of the total electric field must vanish on the conducting surface. Thus, we derive the electric field integral equation for a straight wire scatterer using Eq. (2.1.17), setting $x = y = 0$ in Eq. (2.2.2), and noting that $\nabla(\nabla \cdot \boldsymbol{A}) = (\partial^2 A_z)/(\partial z^2)$, given by

$$\frac{\partial^2 A_z}{\partial z^2} - \frac{1}{c^2}\frac{\partial^2 A_z}{\partial t^2} = -\frac{1}{c^2}\frac{\partial E_z^{\mathrm{i}}}{\partial t} \quad \text{for } z \in (-h, h), \tag{2.2.3}$$

where E_z^{i} is the z component of the incident electric field, $\boldsymbol{E}^{\mathrm{i}}$, and A_z is given by

$$A_z(z, t) = \mu \int_{z'=-h}^{h} \frac{I\left(z', t - \frac{|z-z'|}{c}\right)}{4\pi\sqrt{|z - z'|^2 + a^2}}\, dz'. \tag{2.2.4}$$

In the operator notation, Eq. (2.2.3) can be written as

$$\boldsymbol{L}(t_n)\boldsymbol{I}(t_n) = \boldsymbol{V}(t_n), \tag{2.2.5}$$

where $\boldsymbol{L}(t_n)$ is an integrodifferential operator, $\boldsymbol{I}(t_n)$ is the induced current vector, and $\boldsymbol{V}(t_n)$ represents the excitation vector obtained from the time derivative of the incident field, evaluated at $t = t_n$. Furthermore, we note that Eq. (2.2.3), along with Eq. (2.2.4), is a second-order integrodifferential, hyperbolic equation with initial condition $I(z, 0) = 0$ and boundary conditions $I(\pm h, t) = 0$. The numerical solution of the hyperbolic differential equation is well-known in mathematical literature [7]. In the following sections, we present two numerical procedures to solve Eq. (2.2.5) using the method of moments [8] and the conjugate gradient method [9].

2.2.1 Method of Moments Solution

For the method of moments solution, we divide the wire axis into $N+1$ equal subdomains of width Δz, with N match points, $z_1, z_2, \cdots, z_N$, as shown in Fig. 2.2. Note that the current is zero at both ends of the wire which implies that $I = 0$ for $z = z_0 = z_{N+1}$ at all times. Furthermore, we divide the time axis into small time intervals given by Δt and denote $t_n = n\Delta t$ for $n = 0, 1, 2, \cdots, \infty$.

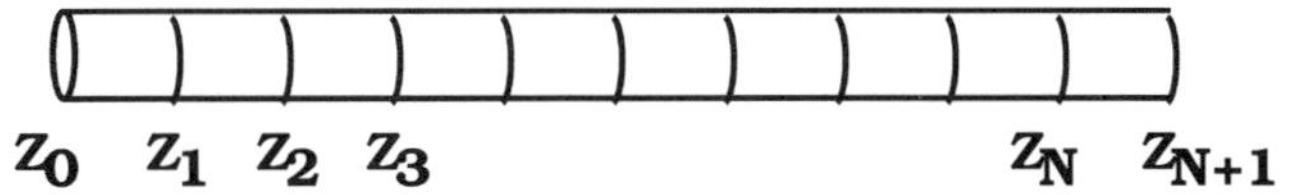

FIGURE 2.2
Wire divided into subdomains.

First, a set of basis functions for expansion purposes is defined, given by

$$f_m(z) \equiv \begin{cases} 1 & z_m - \frac{\Delta z}{2} \le z \le z_m + \frac{\Delta z}{2} \\ 0 & \text{otherwise} \end{cases}. \tag{2.2.6}$$

These expansion functions are just the standard pulse functions on a linear segment. Using these expansion functions, we approximate the current I as

$$I \approx \sum_{k=1}^{N} I_k(t) f_k(z), \tag{2.2.7}$$

where $I_k(t)$ is the unknown coefficient at the kth zone. Using Eq. (2.2.7) and denoting $A_{m,n} = A_z(z_m, t_n)$, we may rewrite Eq. (2.2.4) as

$$\begin{aligned} A_{m,n} &= \mu \int_{z'=-h}^{h} \frac{I\left(z', t_n - \frac{|z_m - z'|}{c}\right)}{4\pi\sqrt{|z_m - z'|^2 + a^2}}\, dz' \approx \mu \int_{z'=-h}^{h} \frac{\sum_{k=1}^{N} I_k\left(t_n - \frac{|z_m - z'|}{c}\right) f_k(z')}{4\pi\sqrt{|z_m - z'|^2 + a^2}}\, dz' \\ &= \sum_{k=1}^{N} I_k\left(t_n - \frac{|z_m - z_k|}{c}\right) \kappa_{m,k}, \end{aligned} \tag{2.2.8}$$

where

$$\begin{aligned} \kappa_{m,k} &= \mu \int_{z'=z_k-\Delta z/2}^{z_k+\Delta z/2} \frac{dz'}{4\pi\sqrt{|z_m - z'|^2 + a^2}} \\ &= \frac{\mu}{4\pi} \left\{ \ln\left[z_m - z_k + \frac{\Delta z}{2} + \sqrt{\left(z_m - z_k + \frac{\Delta z}{2}\right)^2 + a^2} \right] \right. \\ &\quad \left. - \ln\left[z_m - z_k - \frac{\Delta z}{2} + \sqrt{\left(z_m - z_k - \frac{\Delta z}{2}\right)^2 + a^2} \right] \right\}. \end{aligned} \tag{2.2.9}$$

Here we observe that we can also write Eq. (2.2.8) as

$$A_{m,n} = I_{m,n}\kappa_{m,m} + \not{A}_{m,n}, \tag{2.2.10}$$

where

$$\not{A}_{m,n} = \sum_{\substack{k=1 \\ k \ne m}}^{N} I_k\left(t_n - \frac{|z_m - z_k|}{c}\right) \kappa_{m,k}. \tag{2.2.11}$$

Note that the bar on $\not{A}_{m,n}$ explicitly denotes that the $k = m$ term is omitted from the summation. We also observe that the current I_k in Eq. (2.2.11) is computed at time instants earlier than t_n because $|z_m - z_k| \ne 0$. This fact can be exploited to our advantage as follows: By selecting $c\Delta t \le \Delta z$ we note that, even if $|z_m - z_k| = \Delta z$, the current I_k in the summation of Eq. (2.2.11) is known since it occurs at time instant $t \le t_{n-1}$. Here, we are implicitly assuming that when the currents at time instant $t = t_n$ are computed, the currents at earlier time instants $t = t_j$, $j = 1, \cdots, n-1$, are known. Thus, the only unknown in Eq. (2.2.10) is the term $I_{m,n}$.

Next, we approximate the derivatives in Eq. (2.2.3) by central difference approximation to obtain

$$\frac{A_{m+1,n} - 2A_{m,n} + A_{m-1,n}}{\Delta z^2} - \frac{A_{m,n+1} - 2A_{m,n} + A_{m,n-1}}{(c\Delta t)^2} = -F_{m,n}, \tag{2.2.12}$$

where

$$F_{m,n} = \frac{1}{c^2}\frac{\partial E_z^{\mathrm{i}}(z_m, t_n)}{\partial t}. \tag{2.2.13}$$

Also, we note that the derivative on the incident field may be performed analytically if a closed-form expression is available for the time variation as in the case of the Gaussian impulse plane wave described in Chapter 1, Section 1.4.4. Rearranging the terms in Eq. (2.2.12), we have

$$\begin{aligned} A_{m,n+1} = {} & 2A_{m,n} - A_{m,n-1} + (c\Delta t)^2 F_{m,n} \\ & + \left[\frac{c\Delta t}{\Delta z}\right]^2 [A_{m+1,n} - 2A_{m,n} + A_{m-1,n}]. \end{aligned} \tag{2.2.14}$$

Finally, if we replace $n \Longrightarrow n-1$ and using Eq. (2.2.10), we can write Eq. (2.2.14) as

$$\begin{aligned} I_{m,n}\kappa_{m,m} = {} & -A_{m,n} + 2A_{m,n-1} - A_{m,n-2} + (c\Delta t)^2 F_{m,n-1} \\ & + \left[\frac{c\Delta t}{\Delta z}\right]^2 [A_{m+1,n-1} - 2A_{m,n-1} + A_{m-1,n-1}]. \end{aligned} \tag{2.2.15}$$

By examining Eq. (2.2.15), it is evident that the left-hand side only involves the term at $t = t_n$, whereas the right-hand side contains the terms for $t \le t_{n-1}$. The algorithm may be started by assuming $I_{m,0} = I_{m,1} = 0$ and calculating $I_{m,2}$ using Eq. (2.2.15). Once we obtain $I_{m,2}$, coupled with the knowledge of $I_{m,0}$ and $I_{m,1}$, we proceed to calculate $I_{m,3}$ again using Eq. (2.2.15). This procedure can be continued to calculate currents at successive time instants $t_4, t_5, \cdots$ until the transient currents die down. For an incident Gaussian plane wave pulse, described in Chapter 1, Section 1.4.4, this usually happens after several pulse widths.

The method described so far, popularly known as the marching on in time (MOT) method, is quite efficient for computing the transient response from a thin wire. An important advantage of this method which has been frequently stressed [10–13] is the fact that no matrix inversion is required as long as we choose $\Delta t \le (\Delta z)/c$. In the following section we present an alternate method, based on the well-known CG algorithm [14], which relaxes this restriction.

2.2.2 Conjugate Gradient Method Solution

For the conjugate gradient (CG) method solution, referring to Eq. (2.2.5), we define an error functional given by

$$E(t_n) = \int_{-h}^{h} |\boldsymbol{L}(t_n)\boldsymbol{I}(t_n) - \boldsymbol{V}(t_n)|^2 \, dz \tag{2.2.16}$$

at time instant $t = t_n$. The method starts with an initial guess $\boldsymbol{I}_o(t_n)$ and generates the vectors

$$\boldsymbol{R}_o = \boldsymbol{V} - \boldsymbol{L}\boldsymbol{I}_o \tag{2.2.17}$$

$$\boldsymbol{P}_o = \boldsymbol{L}^*\boldsymbol{R}_o, \tag{2.2.18}$$

where $\boldsymbol{L}^*$ represents the adjoint operator of $\boldsymbol{L}$ which, in this case, is the same as $\boldsymbol{L}$. Furthermore, $\boldsymbol{I}$, $\boldsymbol{P}$, and $\boldsymbol{R}$ represent column vectors of dimension N and the subscripts on these vectors represent the iteration number. The conjugate gradient method then develops

$$\boldsymbol{I}_{k+1} = \boldsymbol{I}_k + \alpha_k \boldsymbol{P}_k \tag{2.2.19}$$

$$\boldsymbol{R}_{k+1} = \boldsymbol{R}_k - \alpha_k \boldsymbol{L}\boldsymbol{P}_k \tag{2.2.20}$$

$$\boldsymbol{P}_{k+1} = \boldsymbol{P}_k + \beta_{k+1}\boldsymbol{L}^*\boldsymbol{R}_{k+1} \tag{2.2.21}$$

$$\alpha_k = \frac{\langle \boldsymbol{L}\boldsymbol{P}_k, \boldsymbol{R}_k\rangle}{\langle \boldsymbol{L}\boldsymbol{P}_k, \boldsymbol{L}\boldsymbol{P}_k\rangle} \tag{2.2.22}$$

$$\beta_k = \frac{\langle \boldsymbol{L}\boldsymbol{R}_k, \boldsymbol{L}\boldsymbol{R}_k\rangle}{\langle \boldsymbol{L}\boldsymbol{R}_{k-1}, \boldsymbol{L}\boldsymbol{R}_{k-1}\rangle}, \tag{2.2.23}$$

where $\langle \boldsymbol{A}, \boldsymbol{B}\rangle$ represents the usual vector dot product.

The solution procedure starts by computing $A_z(z_m, t_n) = A_{m,n}$, using Eq. (2.2.8), at $t = 0,\ \Delta t,\ 2\Delta t$ with the assumption that $I(z) = 0$ at $t = 0$ and Δt, and with an initial guess equal to zero at $t = 2\Delta t$. Then, the double derivatives in Eq. (2.2.3) are approximated by the following finite difference procedure which provides higher order approximation for the derivatives [7], given by

$$\frac{\partial^2 A_z(z_m, t_n)}{\partial z^2} \approx \frac{A_{m+1,n+1} - 2A_{m,n+1} + A_{m-1,n+1}}{4(\Delta z)^2} + \frac{A_{m+1,n} - 2A_{m,n} + A_{m-1,n}}{2(\Delta z)^2} + \frac{A_{m+1,n-1} - 2A_{m,n-1} + A_{m-1,n-1}}{4(\Delta z)^2}, \tag{2.2.24}$$

and

$$\frac{\partial^2 A_z(z_m, t_n)}{\partial t^2} \approx \frac{A_{m+1,n+1} - 2A_{m+1,n} + A_{m+1,n-1}}{4(\Delta t)^2} + \frac{A_{m,n+1} - 2A_{m,n} + A_{m,n-1}}{2(\Delta t)^2} + \frac{A_{m-1,n+1} - 2A_{m-1,n} + A_{m-1,n-1}}{4(\Delta t)^2}. \tag{2.2.25}$$

Once the double-derivative operation is performed, we evaluate the residual vector $\boldsymbol{R}_o$ at $t = 2\Delta t$ by calculating $\boldsymbol{V} - \boldsymbol{L}\boldsymbol{I}_o$. Equations (2.2.19)–(2.2.23) are successively applied until a desired error criterion is achieved. This completes the evaluation of the current distribution on the wire at $t = 2\Delta t$. The current at later instants is obtained by repeating the solution procedure outlined here for each time step.

The main advantage of the CG method described in this section is the independence in choosing the time step which is not coupled to the spatial increment Δz. For this procedure, it is quite possible to choose a larger time step than $(\Delta z)/c$. Furthermore, since the error is minimized at each time step to a desired value, accumulation of error is minimized in the late time. In the following section, we present a numerical example to illustrate this fact.

Lastly, there exists yet another method based on the application of the CG method to solve the wire problem [15]. In this work, the initial value problem is formulated as a boundary value problem with $z \in (-h,\ h)$ and $t \in (0, T_{\max})$, where $T_{\max}$ is the upper value of the time scale beyond which the response is assumed to be zero. After dividing the time interval $(0, T_{\max})$ into N_t equal time steps, the error is minimized to a desired value on a rectangular grid $N \times N_t$ using the CG algorithm. Unfortunately, this method is expensive in terms of computer time and storage.

2.2.3 Numerical Example

As an example, consider a perfectly conducting straight wire of length 2.0 m and radius 0.01 m, as shown in Fig. 2.1, illuminated by a Gaussian electromagnetic plane wave pulse given by Eq. (1.4.14). For this case, $\boldsymbol{E}_{\mathrm{o}} = 120\pi \boldsymbol{a}_z$, $ct_{\mathrm{o}} = 3$ LM, $T = 2$ LM, and $\boldsymbol{a}_k = -\boldsymbol{a}_x$. The wire is divided into 10 equal subdomains so that each subdomain is 0.2 m long. Because of this discretization, the time step in the method of moments (MoM) solution is restricted to less than or equal to 0.2 LM. However, for the CG scheme, this restriction does not apply.

Figure 2.3 shows the transient current induced at the center of the wire scatterer as a function of time. The results are compared with an inverse discrete Fourier transform (IDFT) solution, obtained by solving the same problem in the frequency domain at 128 frequencies for the interval 2–256 MHz and transforming the results into time domain using the discrete inverse Fourier transform method. For the conjugate gradient method, the iterations were continued until the error was $<10^{-6}$ at each instant. It is quite obvious from the figure that both methods, i.e., the MoM procedure and CG method, compare well with the IDFT solution.

2.3 ANALYSIS OF AN ARBITRARY WIRE

In this section, we discuss the mathematical steps to analyze an arbitrary wire excited by a transient electromagnetic plane wave. It is obvious that any wire-grid model used to approximate a complex body consists of multiple arbitrary wires coupled with wire junctions. Thus, the analysis of an arbitrary wire forms an important step in the development of a thin-wire code to calculate the transient response from complex geometrical shapes.

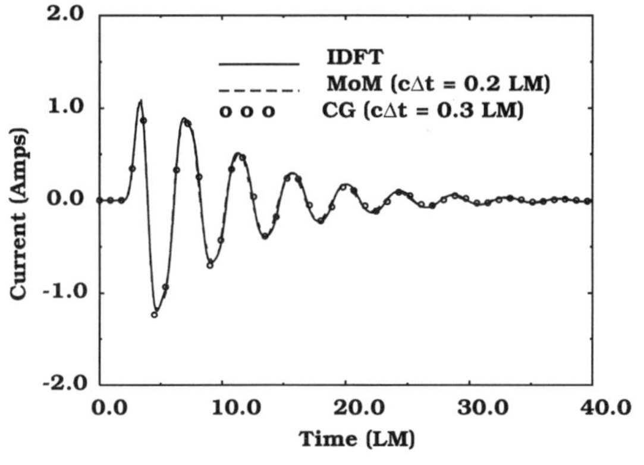

FIGURE 2.3
Current induced at the center of a conducting wire scatterer ($2h = 2.0$ m, $a = 0.01$ m) excited by a normal incident Gaussian plane wave.

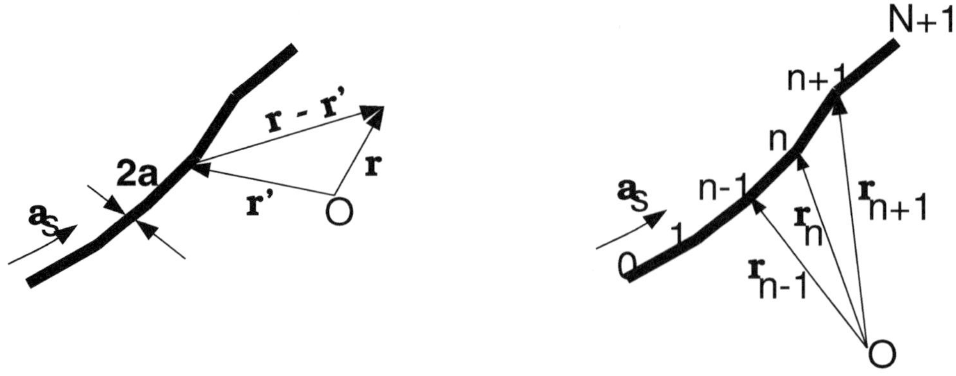

FIGURE 2.4
Arbitrary wire with segmentation scheme.

Let S denote an open or closed perfectly conducting arbitrarily oriented wire modeled by a series of straight wire segments, as shown in Fig. 2.4. An electric field $\boldsymbol{E}^{\mathrm{i}}(\boldsymbol{r}, t)$, defined in the absence of the scatterer, is incident on and induces a surface current $I(\boldsymbol{r}, t)$ on S. Following the mathematical steps outlined in Section 2.1, the scattered electric field $\boldsymbol{E}^{\mathrm{s}}(\boldsymbol{r}, t)$ computed from the surface current is given by

$$\boldsymbol{E}^{\mathrm{s}}(\boldsymbol{r}, t) = -\frac{\partial \boldsymbol{A}(\boldsymbol{r}, t)}{\partial t} - \nabla \Phi(\boldsymbol{r}, t), \tag{2.3.1}$$

where the magnetic vector potential and electric scalar potentials are given by

$$\boldsymbol{A}(\boldsymbol{r}, t) = \mu \int_l \frac{\boldsymbol{a}_s' I(\boldsymbol{r}', t - R/c)}{4\pi R}\, dl' \tag{2.3.2}$$

and

$$\Phi(\boldsymbol{r}, t) = \frac{1}{\epsilon} \int_l \frac{q_l(\boldsymbol{r}', t - R/c)}{4\pi R}\, dl'. \tag{2.3.3}$$

In Eqs. (2.3.2) and (2.3.3), $R = \sqrt{|\boldsymbol{r} - \boldsymbol{r}'|^2 + a^2}$, μ and ϵ denote the permeability and permittivity of the surrounding medium, respectively, $\boldsymbol{r}$ and $\boldsymbol{r}'$ are the locations of the observation and source points on the wire, a is the radius, and c is the velocity of the electromagnetic wave. Also, l represents the parameter along the length of the wire. The linear charge density q_l is related to the induced current I by

$$\frac{\partial q_l}{\partial t} = -\frac{\partial I}{\partial l}. \tag{2.3.4}$$

Differentiating Eq. (2.3.1) with respect to time and using Eq. (2.3.4), we obtain the following expression for the time derivative of the scalar potential as

$$\Psi \triangleq \frac{\partial \Phi}{\partial t} = \frac{-1}{\epsilon} \int_l \frac{\partial I(\boldsymbol{r}', t - R/c)/\partial l}{4\pi R}\, dl'. \tag{2.3.5}$$

The integrodifferential equation for I can be derived using the boundary condition $(\boldsymbol{E}^{\mathrm{i}}+\boldsymbol{E}^{\mathrm{s}})_{\tan} = 0$ on S as

$$\left[\frac{\partial \boldsymbol{A}}{\partial t} + \nabla\Phi\right]_{\tan} = \boldsymbol{E}^{\mathrm{i}}_{\tan}. \tag{2.3.6}$$

The charge density appearing in the scalar potential of Eq. (2.3.6) may be eliminated by differentiating Eq. (2.3.6) with respect to time and using Eq. (2.3.5). Thus, the electric field integral equation for an arbitrary wire in time domain is given by

$$\left[\frac{\partial^2 \boldsymbol{A}}{\partial t^2} + \nabla\Psi\right]_{\tan} = \left[\frac{\partial \boldsymbol{E}^{\mathrm{i}}}{\partial t}\right]_{\tan}, \tag{2.3.7}$$

which needs to be solved for the unknown current $I(\boldsymbol{r}, t)$.

For numerical purposes, the wire is divided into N linear segments and the position vectors $\boldsymbol{r}_n$, $n = 0, 1, \cdots, N+1$, defined with respect to the global coordinate origin ○, locate the endpoints of the linear segments along the wire axis. Notice that we can assign different radii for each segment which allows us to model wire structures with nonuniform radii. Furthermore, the tangential unit vector at $\boldsymbol{r} = \boldsymbol{r}_m$ along the wire segment may be denoted as $\boldsymbol{a}_{sm}$, given by

$$\boldsymbol{a}_{sm} = \frac{\boldsymbol{r}_{m+1/2} - \boldsymbol{r}_{m-1/2}}{|\boldsymbol{r}_{m+1/2} - \boldsymbol{r}_{m-1/2}|}. \tag{2.3.8}$$

Next, we discuss the numerical solution procedure to solve Eq. (2.3.7) using the MoM.

2.3.1 Moment Method Solution

The first step in the MoM solution procedure is to define a set of basis functions to approximate the unknown function $I(\boldsymbol{r}, t)$. Here, we define the basis functions as

$$f_m(\boldsymbol{r}) \equiv \begin{cases} 1 & \boldsymbol{r} \in \boldsymbol{r}_{m-1/2} \text{ to } \boldsymbol{r}_{m+1/2} \\ 0 & \text{otherwise.} \end{cases} \tag{2.3.9}$$

These expansion functions are the standard pulse functions on a linear segment. Furthermore, the same expansion functions are also used as testing functions, and the inner product is defined as

$$\langle \boldsymbol{a}, \boldsymbol{b} \rangle = \int_l \boldsymbol{a} \cdot \boldsymbol{b}\, dl'. \tag{2.3.10}$$

Next, we divide the time axis into small time intervals given by Δt and denote $t_n = n\Delta t$ for $n = 0, 1, 2, \cdots, \infty$. Applying the testing procedure to Eq. (2.3.7) gives

$$\left\langle f_m \boldsymbol{a}_{sm}, \left[\frac{\partial^2 \boldsymbol{A}}{\partial t^2} + \nabla\Psi\right]\right\rangle = \left\langle f_m \boldsymbol{a}_{sm}, \frac{\partial \boldsymbol{E}^{\mathrm{i}}}{\partial t}\right\rangle. \tag{2.3.11}$$

Approximating the time derivatives of the potential function by central difference, we get

$$\langle f_m \boldsymbol{a}_{sm}, \boldsymbol{L}[I] \rangle = \left\langle f_m \boldsymbol{a}_{sm}, \frac{\partial \boldsymbol{E}^{\mathrm{i}}}{\partial t} \right\rangle, \tag{2.3.12}$$

where

$$\boldsymbol{L}[I] = \frac{\boldsymbol{A}(\boldsymbol{r}, t_{n+1}) - 2\boldsymbol{A}(\boldsymbol{r}, t_n) + \boldsymbol{A}(\boldsymbol{r}, t_{n-1})}{\Delta t^2} + \nabla \Psi(\boldsymbol{r}, t_n). \tag{2.3.13}$$

The testing of the vector potential may be approximated by using a 1-point integration. This is done by evaluating $\boldsymbol{A}$ at $\boldsymbol{r}_m$ and multiplying by the length of the subdomain, $\Delta l_m = |\boldsymbol{r}_{m+1/2} - \boldsymbol{r}_{m-1/2}|$. Thus,

$$\langle f_m \boldsymbol{a}_{sm}, \boldsymbol{A}(\boldsymbol{r}, t_n) \rangle \approx \boldsymbol{A}(\boldsymbol{r}_m, t_n) \cdot \Delta l_m \boldsymbol{a}_{sm}. \tag{2.3.14}$$

The testing of the incident field is performed in a manner similar to that of the vector potential. We therefore make the approximation

$$\left\langle f_m \boldsymbol{a}_{sm}, \frac{\partial \boldsymbol{E}^{\mathrm{i}}(\boldsymbol{r}, t_n)}{\partial t} \right\rangle \approx \frac{\partial \boldsymbol{E}^{\mathrm{i}}(\boldsymbol{r}_m, t_n)}{\partial t} \cdot \Delta l_m \boldsymbol{a}_{sm}. \tag{2.3.15}$$

We now evaluate $\nabla \Psi(\boldsymbol{r}, t_n)$. By using the fact that the line integral of the gradient of a potential function is the function evaluated at its endpoints, we can rewrite the testing of the gradient term as

$$\begin{aligned} \langle f_m \boldsymbol{a}_{sm}, \nabla \Psi(\boldsymbol{r}, t_n) \rangle &= \int_l \nabla \Psi(\boldsymbol{r}, t_n) \cdot f_m \boldsymbol{a}_s \, dl' \\ &\approx \Psi(\boldsymbol{r}_{m+1/2}, t_n) - \Psi(\boldsymbol{r}_{m-1/2}, t_n). \end{aligned} \tag{2.3.16}$$

Thus, using Eqs. (2.3.14)–(2.3.16), Eq. (2.3.12) may be written as

$$\begin{aligned} &\left[\frac{\boldsymbol{A}(\boldsymbol{r}, t_{n+1}) - 2\boldsymbol{A}(\boldsymbol{r}, t_n) + \boldsymbol{A}(\boldsymbol{r}, t_{n-1})}{\Delta t^2} \right] \cdot \Delta l_m \boldsymbol{a}_{sm} + \Psi(\boldsymbol{r}_{m+1/2}, t_n) - \Psi(\boldsymbol{r}_{m-1/2}, t_n) \\ &\quad = \frac{\partial \boldsymbol{E}^{\mathrm{i}}(\boldsymbol{r}_m, t_n)}{\partial t} \cdot \Delta l_m \boldsymbol{a}_{sm}, \end{aligned} \tag{2.3.17}$$

for $m = 1, 2, \cdots, N$.

Next, we consider the expansion procedure. Let the surface current be approximated as

$$I(\boldsymbol{r}, t) = \sum_{k=1}^{N} I_k(t) f_k(\boldsymbol{r}), \tag{2.3.18}$$

where $I_k(t)$ is the unknown coefficient $\boldsymbol{r} = \boldsymbol{r}_k$, $k = 1, 2, \cdots, N$. Substituting Eq. (2.3.18) into Eq. (2.3.2) results in

$$\boldsymbol{A}(\boldsymbol{r}, t_n) = \mu \int_l \frac{\boldsymbol{a}_s' \sum_{k=1}^{N} I_k(t_n - R/c) f_k(\boldsymbol{r}')}{4\pi R} \, dl' \approx \mu \sum_{k=1}^{N} I_k(t_n - R_{mk}/c) \kappa_{m,k} \, \boldsymbol{a}_{sk}, \tag{2.3.19}$$

where

$$\kappa_{m,k} = \int_l \frac{f_k\, dl'}{4\pi R_m}, \tag{2.3.20}$$

$$R_m = \sqrt{|\boldsymbol{r}_m - \boldsymbol{r}'|^2 + a^2}, \tag{2.3.21}$$

$$R_{mk} = |\boldsymbol{r}_m - \boldsymbol{r}_k|. \tag{2.3.22}$$

Notice that, in obtaining Eq. (2.3.19), the current is assumed to be constant with time within a subdomain so that the integral may be simplified. The calculation of the $1/R$ integral in Eq. (2.3.20) is trivial and is given by

$$\kappa_{m,k} = \begin{cases} \dfrac{\Delta l_k}{4\pi\sqrt{|\boldsymbol{r}_m - \boldsymbol{r}_k|^2 + a^2}} & \text{for } k \neq m \\ \kappa_{m,m} & \text{for } k = m. \end{cases} \tag{2.3.23}$$

where

$$\kappa_{m,m} = \frac{1}{4\pi}\left[\ln\left(s_2 + \sqrt{s_2^2 + a^2}\right) - \ln\left(s_1 + \sqrt{s_1^2 + a^2}\right)\right], \tag{2.3.24}$$

$$s_1 = |\boldsymbol{r}_m - \boldsymbol{r}_{m-1/2}|, \tag{2.3.25}$$

$$s_2 = |\boldsymbol{r}_m - \boldsymbol{r}_{m+1/2}|. \tag{2.3.26}$$

Extracting the self term, Eq. (2.3.19) may be rewritten as

$$\boldsymbol{A} = \kappa_{m,m} I_m(t_n)\boldsymbol{a}_{sm} + \bar{\boldsymbol{A}}(\boldsymbol{r}, t_n), \tag{2.3.27}$$

where $\bar{\boldsymbol{A}}$ represents the rest of the terms in the summation. Note that the bar on $\bar{\boldsymbol{A}}$ explicitly denotes that the $k = m$ term is omitted from the summation. We also observe that $\bar{\boldsymbol{A}}$ in Eq. (2.3.27) is computed at time instants earlier than t_n because $R_{mk} \neq 0$. This fact can be exploited to our advantage just as in the case of a straight wire discussed in Section 2.2.

We now evaluate $\Psi(\boldsymbol{r}_m, t_n)$ given by Eq. (2.3.5). Using Eq. (2.3.18), we have

$$\Psi(\boldsymbol{r}_m, t_n) = \frac{-1}{\epsilon} \sum_{k=1}^{N} \int_l \frac{\partial\left[I_k(t_n - R_m/c) f_k\right]/\partial l'}{4\pi R_m}\, dl', \tag{2.3.28}$$

where $R_m = |\boldsymbol{r}_m - \boldsymbol{r}'|$. Since pulse basis functions are used, the derivative on the basis function f_k results in two delta functions; one at $r_{k+1/2}$ and one at $r_{k-1/2}$. We can "spread" the effect of these delta functions across the contour from $\boldsymbol{r}_{k-1}$ to $\boldsymbol{r}_{k+1}$ as shown in Fig. 2.5. This essentially amounts to approximating the derivative by a finite difference. We can now express Eq. (2.3.28) as

$$\Psi(\boldsymbol{r}_m, t_n) = \sum_{k=1}^{N} \Psi^+(\boldsymbol{r}_m, t_n) - \Psi^-(\boldsymbol{r}_m, t_n), \tag{2.3.29}$$

where

$$\Psi^+(\boldsymbol{r}_m, t_n) = \frac{-1}{4\pi\epsilon} \frac{I_k(t_n - R_{mk}^+/c)}{\Delta l_k^+} \int_{\boldsymbol{r}_k}^{\boldsymbol{r}_{k+1}} \frac{dl'}{\sqrt{|\boldsymbol{r} - \boldsymbol{r}'|^2 + a^2}}, \tag{2.3.30}$$

$$\Psi^-(\boldsymbol{r}_m, t_n) = \frac{-1}{4\pi\epsilon} \frac{I_k(t_n - R_{mk}^-/c)}{\Delta l_k^-} \int_{\boldsymbol{r}_{k-1}}^{\boldsymbol{r}_k} \frac{dl'}{\sqrt{|\boldsymbol{r} - \boldsymbol{r}'|^2 + a^2}}, \tag{2.3.31}$$

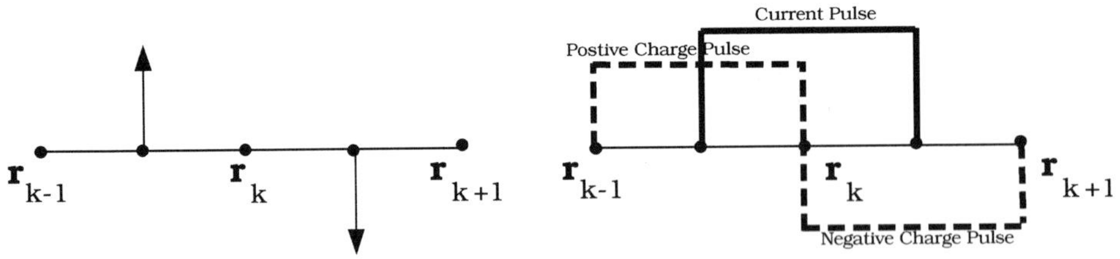

FIGURE 2.5
Approximating delta functions by pulse functions.

with

$$R_{mk}^{\pm} = \sqrt{|\boldsymbol{r}_m - \boldsymbol{r}_{k\pm 1/2}|^2 + a^2}, \tag{2.3.32}$$

$$\Delta l_k^- = |\boldsymbol{r}_k - \boldsymbol{r}_{k-1}|, \tag{2.3.33}$$

$$\Delta l_k^+ = |\boldsymbol{r}_{k+1} - \boldsymbol{r}_k|. \tag{2.3.34}$$

This completes the valuation of potential terms. Then, using Eq. (2.3.27) and replacing $n \Rightarrow n-1$, we can rewrite Eq. (2.3.17) as

$$\begin{aligned} \kappa_{m,m} I_m(t_n) \Delta l_m = {} & \frac{\partial \boldsymbol{E}^{\text{i}}(\boldsymbol{r}_m, t_{n-1})}{\partial t} \cdot \Delta l_m \boldsymbol{a}_{sm} \\ & - \left[\frac{\boldsymbol{A}(\boldsymbol{r}, t_n) - 2\boldsymbol{A}(\boldsymbol{r}, t_{n-1}) + \boldsymbol{A}(\boldsymbol{r}, t_{n-2})}{\Delta t^2} \right] \cdot \Delta l_m \boldsymbol{a}_{sm} \\ & - \Psi(\boldsymbol{r}_{m+1/2}, t_{n-1}) + \Psi(\boldsymbol{r}_{m-1/2}, t_{n-1}) \end{aligned} \tag{2.3.35}$$

for $m = 1, 2, \cdots, N$.

Finally, we select $c\Delta t \leq R_{\min}$, where $R_{\min}$ is the minimum of all the distances between any two distinct match points $\boldsymbol{r}_m$ and $\boldsymbol{r}_k$. Then, notice that the left-hand side of Eq. (2.3.35) involves only terms at $t = t_n$, whereas the right-hand side contains the terms retarded in time by at least one time step and hence assumed to be known. Therefore, the currents may be obtained by the MOT procedure. The numerical procedure starts at $n = 2$, assuming the current to be zero at $n = 0$ and $n = 1$, and then marches in time for $n = 3, 4, 5, \cdots$.

2.3.2 Conjugate Gradient Method

To apply the CG method, once again we use the current expansion functions defined in Eq. (2.3.9) and develop Eqs. (2.3.19) and (2.3.29) for vector and scalar potentials, respectively. However, the finite-difference approximation for the derivatives is modified as described in Section 2.2.2.

The second derivative in time for the vector potential term in Eq. (2.3.7) is modified as

$$\begin{aligned} \frac{\partial^2 \boldsymbol{A}(\boldsymbol{r}_m, t_n)}{\partial t^2} \approx {} & \frac{\boldsymbol{A}(\boldsymbol{r}_{m+1}, t_{n+1}) - 2\boldsymbol{A}(\boldsymbol{r}_{m+1}, t_n) + \boldsymbol{A}(\boldsymbol{r}_{m+1}, t_{n-1})}{4(\Delta t)^2} \\ & + \frac{\boldsymbol{A}(\boldsymbol{r}_m, t_{n+1}) - 2\boldsymbol{A}(\boldsymbol{r}_m, t_n) + \boldsymbol{A}(\boldsymbol{r}_m, t_{n-1})}{2(\Delta t)^2} \\ & + \frac{\boldsymbol{A}(\boldsymbol{r}_{m-1}, t_{n+1}) - 2\boldsymbol{A}(\boldsymbol{r}_{m-1}, t_n) + \boldsymbol{A}(\boldsymbol{r}_{m-1}, t_{n-1})}{4(\Delta t)^2}, \end{aligned} \tag{2.3.36}$$

whereas the gradient term is modified as

$$\nabla\Psi(\boldsymbol{r}_m, t_n) \approx \frac{\Psi(\boldsymbol{r}_{m+1/2}, t_n) - \Psi(\boldsymbol{r}_{m-1/2}, t_n)}{2\Delta l_m} + \frac{\Psi(\boldsymbol{r}_{m+1/2}, t_{n-1}) - \Psi(\boldsymbol{r}_{m-1/2}, t_{n-1})}{2\Delta l_m}. \tag{2.3.37}$$

Furthermore, the time derivative of the scalar potential in Eq. (2.3.5) is approximated as

$$\Psi(\boldsymbol{r}_m, t_n) \approx \frac{1}{2}\left[\frac{\partial\Phi(\boldsymbol{r}_m, t_n)}{\partial t} + \frac{\partial\Phi(\boldsymbol{r}_m, t_{n-1})}{\partial t}\right]. \tag{2.3.38}$$

Finally, using Eqs. (2.3.19), (2.3.29), and (2.3.36)–(2.3.38), Eq. (2.3.7) may be written in the operator form as

$$\boldsymbol{LI} = \boldsymbol{V} \tag{2.3.39}$$

and solved using the conjugate gradient method described in Section 2.2.2.

2.3.3 Numerical Examples

Consider a perfectly conducting 90° bent wire, lying in the xy plane and along the x and y axes, illuminated by a normally incident Gaussian plane wave pulse given by Eq. (1.4.14). Here, $\boldsymbol{E}_o = 120\pi\boldsymbol{a}_x$, $ct_o = 3$ LM, $T = 2$ LM, and $\boldsymbol{a}_k = -\boldsymbol{a}_z$. The two segments of wire are of 1-m length and the radius is 0.01-m. Each section of the wire is divided into five segments which implies $R_{\min} = 0.1414$ m. Figure 2.6 shows the current induced at the bend as a function of time for both MoM and CG solution schemes. For the MoM and CG solutions, the time steps are 0.1414 and 0.2 LM, respectively. Notice that the CG method allows a larger time step than the MoM solution, which is constrained by the minimum distance $R_{\min}$. The results are compared with the IDFT solution. The frequency domain solution is obtained by using the MoM with pulse basis functions and triangle-function testing. It is evident from Fig. 2.6 that all three results compare very well with each other.

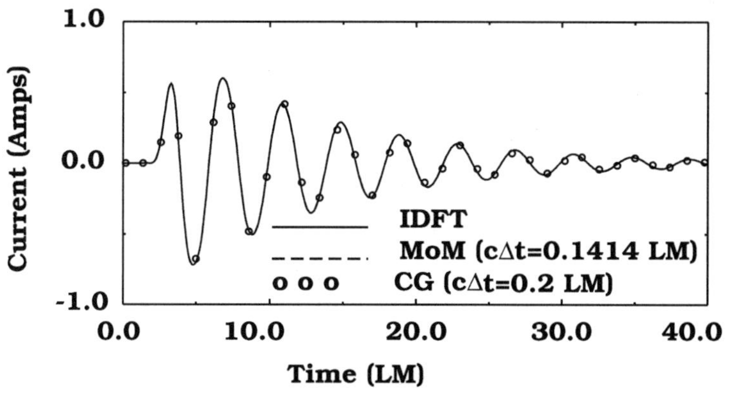

FIGURE 2.6
Current induced at the corner of a conducting bent-wire scatterer lying in the xy plane. Each section of the wire is 1 m long with 0.01-m radius.

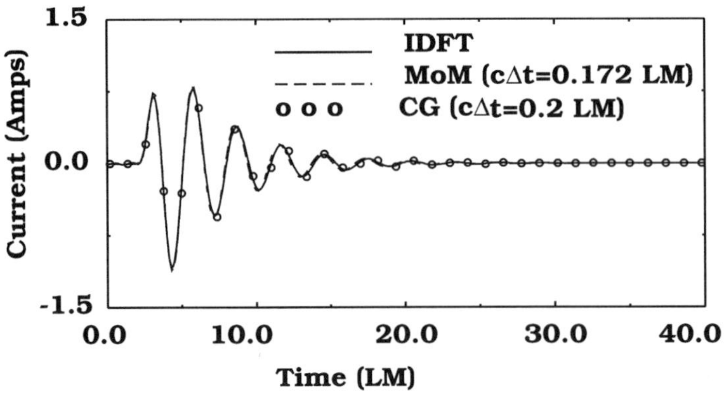

FIGURE 2.7
Transient current induced on a circular loop scatterer (radius 0.5 m) lying in the xy plane.

Next, we consider a circular loop with a 0.5-m radius, placed in the xy plane with the center coinciding with the origin of the coordinate axes. The circular loop is assumed to be constructed by a wire tube with a radius of 0.01 m. The whole structure is illuminated by a normally incident Gaussian plane wave pulse of the previous example. For the numerical solution, we divide the circular loop into 16 linear segments. Figure 2.7 shows the induced transient current as a function of time at a point on the loop coincident with the y-axis. Once again, we present the results obtained by both MoM and CG solutions and compare them with those of the IDFT method. Again, all three results agree very well with each other.

As a last example, we present the transient current induced at the discontinuity of a stepped-radius wire with the ratio of radii 1:10 in Fig. 2.8. The length of each segment is 1 m and the radii are 0.001 and 0.01 m, respectively. The incident pulse is normal to and polarized along the wire axis. Each section of the stepped-radius wire is divided into five segments for numerical purposes. For this division, note that a subdomain is equally divided into half subdomain lengths at the stepped-wire junction. While evaluating Eq. (2.3.20), care must be exercised to include

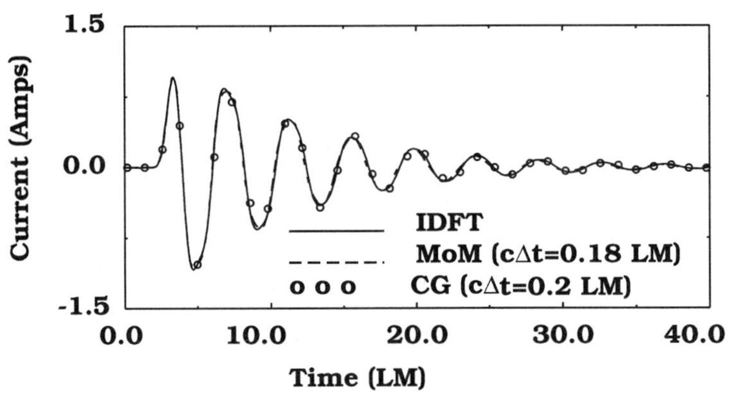

FIGURE 2.8
Transient current induced on a stepped-radius wire scatterer lying in the x-axis. Each section of the wire is 1 m long. The radius of the first section is 0.001 m and that of the second section is 0.01 m.

the different wire radii for each section. Once again, all three results, i.e., IDFT, MoM and CG solutions, compare well with each other.

2.4 IMPLICIT SOLUTION SCHEME

In the previous section, we presented the MoM solution procedure to analyze an arbitrary wire scatterer. This method is also popularly known as an explicit method. As already noted, this method is simple and efficient for the solution of wire problems. It requires minimum storage and no matrix inversion is necessary to obtain the transient induced current. Although this procedure is very suitable for a variety of wire modeling problems, its utility is somewhat limited by the duration of the time step one can choose. Recall that, for the explicit (MoM) scheme, the time step $c\Delta t$ must be smaller than $R_{\min}$, which can be excessively restrictive for certain applications. Although the CG method described in Section 2.3.2 overcomes this restriction, the method can be inefficient for larger time steps because of the requirement of a large number of iterations for convergence. In this section, we present another method, known as an implicit method, which sometimes allows us to use a much larger time step, at least for low-Q structures, and at the same time is quite efficient. Furthermore, the implicit scheme reduces to the explicit scheme if the sampling in time is properly chosen. We can describe the general outline of the implicit scheme as follows: Starting from Maxwell's equations and using standard analysis, the time domain integral equation for a general scattering object may be written as

$$\boldsymbol{L}\boldsymbol{J}\left(\boldsymbol{r}', t - \frac{|\boldsymbol{r} - \boldsymbol{r}'|}{c}\right) = \boldsymbol{E}^{\mathrm{i}}_{\tan}(\boldsymbol{r}, t), \tag{2.4.1}$$

where $\boldsymbol{L}$ represents the integrodifferential operator, $\boldsymbol{J}$ represents the induced current, and $\boldsymbol{E}^{\mathrm{i}}_{\tan}$ represents the tangential component of the incident field. Also, $\boldsymbol{r}, \boldsymbol{r}'$, and c represent the location of the observation point, the location of the source point, and the wave speed of the electromagnetic wave, respectively. For a numerical solution, we divide the space into a suitable grid and the time axis into equal intervals Δt. Then, at $\boldsymbol{r} = \boldsymbol{r}_m$ and $t = t_n$, Eq. (2.4.1) may be written as

$$\boldsymbol{L}\boldsymbol{J}\left(\boldsymbol{r}', t_n - \frac{|\boldsymbol{r}_m - \boldsymbol{r}'|}{c}\right) = \boldsymbol{E}^{\mathrm{i}}_{\tan}(\boldsymbol{r}_m, t_n), \tag{2.4.2}$$

which can be rewritten in the following form:

$$\boldsymbol{L}_1\boldsymbol{J}\left(\boldsymbol{r}', t_n - \frac{|\boldsymbol{r}_m - \boldsymbol{r}'|}{c}\right) + \boldsymbol{L}_2\boldsymbol{J}\left(\boldsymbol{r}', t_n - \frac{|\boldsymbol{r}_m - \boldsymbol{r}'|}{c}\right) = \boldsymbol{E}^{\mathrm{i}}_{\tan}(\boldsymbol{r}_m, t_n). \tag{2.4.3}$$

Notice that, while writing Eq. (2.4.3), we have replaced the operator $\boldsymbol{L}$ in Eq. (2.4.2) by two operators, $\boldsymbol{L}_1$ and $\boldsymbol{L}_2$, where

$$\boldsymbol{L} = \begin{cases} \boldsymbol{L}_1 & \text{if } \frac{|\boldsymbol{r}_m - \boldsymbol{r}'|}{c} < \Delta t \\ \boldsymbol{L}_2 & \text{if } \frac{|\boldsymbol{r}_m - \boldsymbol{r}'|}{c} \geq \Delta t. \end{cases} \tag{2.4.4}$$

We also note that in Eq. (2.4.4), the currents in the operator $\boldsymbol{L}_2$ are retarded in time, by at least Δt, and hence are known.

For the explicit scheme, we choose $\Delta t \leq R_{\min}/c$, where $R_{\min}$ represents the minimum of all the distances between any two observation points in the grid scheme. Then, if a proper numerical scheme is chosen, it is possible to transform $\boldsymbol{L}_1$ into a diagonal matrix. However, for the implicit scheme, we choose $\Delta t > R_{\min}/c$. Thus, we eliminate the restriction on the choice of the time step with respect to $R_{\min}$. Then, it is possible to transform the operator $\boldsymbol{L}_1$ into a sparse matrix, and we obtain $\boldsymbol{J}$ by inverting the sparse matrix and solving Eq. (2.4.3). It is obvious from this discussion that with a proper choice of Δt, one can alternate between explicit and implicit schemes.

2.4.1 Application to Arbitrary Wire

Consider an arbitrary wire. On the conductor, the total tangential electric field on the wire surface S is zero, i.e.,

$$(\boldsymbol{E}^{\mathrm{i}} + \boldsymbol{E}^{\mathrm{s}})_{\tan} = 0 \quad \text{for } \boldsymbol{r} \in S, \tag{2.4.5}$$

where $\boldsymbol{E}^{\mathrm{i}}$, $\boldsymbol{E}^{\mathrm{s}}$, and $\boldsymbol{r}$ represent the incident electric field, scattered electric field, and the field point, respectively. In addition, we know that

$$\boldsymbol{E}^{\mathrm{s}} = -\frac{\partial \boldsymbol{A}}{\partial t} - \nabla\Phi, \tag{2.4.6}$$

where the magnetic vector potential, $\boldsymbol{A}(\boldsymbol{r}, t)$, and the electric scalar potential, $\Phi(\boldsymbol{r}, t)$, are given by Eqs. (2.3.2) and (2.3.3), respectively.

Utilizing Eq. (2.4.6), Eq. (2.4.5) may be rewritten as

$$\boldsymbol{E}^{\mathrm{i}} \cdot \boldsymbol{a}_s = \left[\frac{\partial \boldsymbol{A}}{\partial t} + \nabla\Phi\right] \cdot \boldsymbol{a}_s. \tag{2.4.7}$$

We use the backward difference to numerically evaluate the time derivative of the vector potential. Hence, with the sampling time interval Δt, we have

$$\left[\frac{\boldsymbol{A}(\boldsymbol{r}, t_n) - \boldsymbol{A}(\boldsymbol{r}, t_{n-1})}{\Delta t} + \nabla\Phi(\boldsymbol{r}, t_n)\right] \cdot \boldsymbol{a}_s = \boldsymbol{E}^{\mathrm{i}}(\boldsymbol{r}, t_n) \cdot \boldsymbol{a}_s. \tag{2.4.8}$$

If we focus our attention on the time instant $t = t_n$, then $\boldsymbol{A}(\boldsymbol{r}, t_{n-1})$ is the retarded potential and considered to be known. Therefore, Eq. (2.4.8) can be rewritten as

$$[\boldsymbol{A}(\boldsymbol{r}, t_n) + (\Delta t)\nabla\Phi(\boldsymbol{r}, t_n)] \cdot \boldsymbol{a}_s = (\Delta t)\boldsymbol{E}^{\mathrm{i}}(\boldsymbol{r}, t_n) \cdot \boldsymbol{a}_s + \boldsymbol{A}(\boldsymbol{r}, t_{n-1}) \cdot \boldsymbol{a}_s. \tag{2.4.9}$$

This is the equation that needs to be solved for current $I(\boldsymbol{r}, t)$ or the charge $q_l(\boldsymbol{r}, t)$.

2.4.2 Numerical Implementation

For the numerical solution, we first divide the time axis into subintervals, each of duration Δt. The choice of Δt, in the implicit solution, is dictated purely by pulse width of the incident pulse. For a good numerical solution, we suggest that Δt be chosen approximately equal to one-tenth of the pulse width.

Next, we use the same expansion procedure described in Section 2.3.2 and obtain Eq. (2.3.19) for the vector potential. However, since the time step Δt is chosen independently and not related to the spatial discretization, Eq. (2.3.27) may be rewritten as

$$\boldsymbol{A}(\boldsymbol{r}_m, t_n) = \boldsymbol{A}_1(\boldsymbol{r}_m, t_n) + \boldsymbol{A}_2(\boldsymbol{r}_m, t_n), \tag{2.4.10}$$

where

$$\boldsymbol{A}_{1,2}(\boldsymbol{r}_m, t_n) \approx \mu \sum_{k=1}^{N} I_k(t_R)\kappa_{m,k}\,\boldsymbol{a}_{sk}, \tag{2.4.11}$$

and $t_R = t_n - R_{mk}/c$. For $\boldsymbol{A}_1(\boldsymbol{r}_m, t_n)$, we have $t_{n-1} < t_R \leq t_n$. Thus, the current values at $t = t_R$ are unknown. However, for $\boldsymbol{A}_2(\boldsymbol{r}_m, t_n)$, the retarded time t_R is less than t_{n-1}, and therefore the current values at these instants are known or can be calculated in an easy manner.

Now, consider the scalar potential term given by Eq. (2.3.3). Substituting the continuity equation (given by Eq. 2.3.4) into Eq. (2.3.3), we get

$$\Phi(\boldsymbol{r}_m, t_n) = \frac{-1}{4\pi\epsilon}\int_l \int_{t'=0}^{t_n - R/c} \frac{\partial I(\boldsymbol{r}', t')}{\partial l}\frac{1}{R}\,dt'\,dl', \tag{2.4.12}$$

where $R = \sqrt{|\boldsymbol{r}_m - \boldsymbol{r}'|^2 + a^2}$. Using Eq. (2.3.9) for the expansion of I in Eq. (2.4.12) results in

$$\Phi(\boldsymbol{r}_m, t_n) = \frac{-1}{4\pi\epsilon}\sum_{k=1}^{N}\int_{t'=0}^{t_n - R_{mk}/c} I_k(t')\,dt' \int_l \frac{\partial f_k}{\partial l}\frac{1}{R}\,dl'. \tag{2.4.13}$$

Approximating the spatial derivative, $\partial f_k/\partial l$ in Eq. (2.4.13), by finite difference, we have

$$\Phi(\boldsymbol{r}_m, t_n) = \sum_{k=1}^{N} \Phi_k^+(\boldsymbol{r}_m, t_n) - \Phi_k^-(\boldsymbol{r}_m, t_n), \tag{2.4.14}$$

where

$$\Phi_k^+ = \frac{-1}{4\pi\epsilon\Delta l_k^+}\int_{t'=0}^{t_R^+} I_k(t')\,dt' \int_{\boldsymbol{r}_k}^{\boldsymbol{r}_{k+1}} \frac{dl'}{\sqrt{|\boldsymbol{r}_m - \boldsymbol{r}'|^2 + a^2}}, \tag{2.4.15}$$

$$\Phi_k^- = \frac{-1}{4\pi\epsilon\Delta l_k^-}\int_{t'=0}^{t_R^-} I_k(t')\,dt' \int_{\boldsymbol{r}_{k-1}}^{\boldsymbol{r}_k} \frac{dl'}{\sqrt{|\boldsymbol{r}_m - \boldsymbol{r}'|^2 + a^2}}, \tag{2.4.16}$$

with

$$t_R^{\pm} = t_n - \frac{R_{mk}^{\pm}}{c}, \tag{2.4.17}$$

$$R_{mk}^{\pm} = \sqrt{|\boldsymbol{r}_m - \boldsymbol{r}_{k\pm 1/2}|^2 + a^2}, \tag{2.4.18}$$

$$\Delta l_k^- = |\boldsymbol{r}_k - \boldsymbol{r}_{k-1}|, \tag{2.4.19}$$

$$\Delta l_k^+ = |\boldsymbol{r}_{k+1} - \boldsymbol{r}_k|. \tag{2.4.20}$$

Next, we note that, as in the case of vector potential, the scalar potential term may also be split into two parts:

$$\Phi(\boldsymbol{r}_m, t_n) = \Phi_1(\boldsymbol{r}_m, t_n) + \Phi_2(\boldsymbol{r}_m, t_n), \tag{2.4.21}$$

where $\Phi_1(\boldsymbol{r}_m, t_n)$ and $\Phi_2(\boldsymbol{r}_m, t_n)$ represent the terms involving unknown and known current values, respectively, depending on the value of the retarded time. Thus, we can write Eq. (2.4.9) in a matrix form as

$$[\alpha][\boldsymbol{I}(t_n)] = [\boldsymbol{F}(t_n)] + [\beta][\boldsymbol{I}(t_R)]. \tag{2.4.22}$$

Referring to Eq. (2.4.22), we make the following observations.

- The elements of the $[\alpha]$ matrix are formed by the potential terms when the retarded time t_R satisfies the condition $t_{n-1} < t_R \leq t_n$.
- In the current formulation both the vector potential and the scalar potentials contribute to the $[\alpha]$ matrix.
- The $[\alpha]$ matrix is a sparse matrix, and its sparsity depends on the choice of Δt.
- The matrix elements, α_{mn}, are not functions of time and hence need to be computed only once at the first time step.
- The elements of the $[\boldsymbol{F}]$ matrix are obtained from incident field calculations as follows:

  The testing of the incident field is performed in a manner similar to that of the vector potential. We therefore make the approximation

$$\left\langle f_m \boldsymbol{a}_{sm}, \boldsymbol{E}^{\mathrm{i}}(\boldsymbol{r}, t_n)\right\rangle \approx \boldsymbol{E}^{\mathrm{i}}(\boldsymbol{r}_m, t_n) \cdot \Delta l_m \boldsymbol{a}_{sm}. \tag{2.4.23}$$

- The product term $[\beta][\boldsymbol{I}(t_R)]$ can be easily computed since $t_R \leq t_{n-1}$ and hence the currents at these instants are known.

Lastly, by solving Eq. (2.4.22) at each time step, the time domain current induced on the scatterer may be obtained iteratively.

2.4.3 Numerical Examples

In this section, we present numerical results for several wire geometries obtained by employing the implicit scheme. For every case, the wire geometry is excited by a Gaussian plane wave pulse, given by

$$\boldsymbol{E}^{\mathrm{i}}(\boldsymbol{r}, t) = \boldsymbol{E}_{\mathrm{o}} \frac{4}{T\sqrt{\pi}} e^{-\left[\frac{4}{T}(ct - ct_{\mathrm{o}} + \boldsymbol{r}\cdot\boldsymbol{a}_z)\right]^2}, \tag{2.4.24}$$

where $-\boldsymbol{a}_z$, T, and $\boldsymbol{E}_{\mathrm{o}}$ represent the unit vector along the direction of propagation of the incident wave, the pulse width of the Gaussian pulse, and the amplitude of the incident wave, respectively. Also, we choose $\boldsymbol{E}_{\mathrm{o}} = 377\ \boldsymbol{a}_x$ V/m, $T = 2$ LM, and $ct_{\mathrm{o}} = 3$ LM for the numerical examples presented in this section. The results are compared with data obtained in the frequency domain and transformed into the time domain using an IDFT and also with the explicit method discussed in the

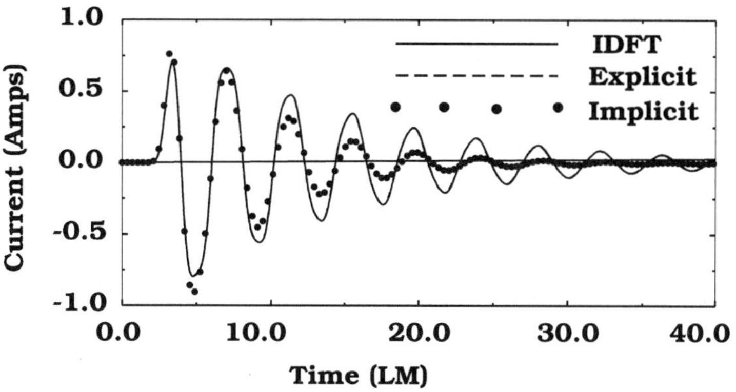

FIGURE 2.9
Transient current induced on a straight-wire scatterer lying along the x-axis. Length and radii of the wire are 2 and 10^{-6} m, respectively.

previous section. The IDFT solution is obtained by solving the problem at 128 frequency samples between 0 and 0.5 GHz and taking the inverse transform. We also note that the wire structures are high-Q structures; hence, we need to use a small time step in order to get an accurate solution.

As a first example, consider a straight wire of $l = 2$ m and radius $a = 10^{-6}$ m, placed along the x-axis and illuminated from the broadside direction. We note that for a straight-wire geometry, the induced current oscillates several times between positive and negative peaks along the time axis before becoming negligible. Thus, in this case we choose Δt to be very small, equal to 0.05 LM. For the explicit and IDFT solutions, we divide the wire into 40 segments. However, for the implicit solution, we divide the wire into 10 equal segments which makes the solution quite efficient in terms of computer resource requirements. The current induced at the center of the wire as a function of time is presented in Fig. 2.9. We note that the implicit solution compares very well with IDFT and the explicit solution in the early times but decays much faster for late times. Thus, the implicit solution scheme has a higher damping constant than does the explicit solution. This is an important point to note since the implicit solution generates a stable result even at very late times, whereas the explicit solution tends to break down, as will be shown for two- and three-dimensional geometries in Chapters 3–5.

Next, we consider an L-shaped wire with 1-m sides and radius $a = 10^{-6}$ m, located along the x and y axes. Again, we choose a time step of 0.05 LM for explicit and implicit solutions. Each section of the wire is divided into 10 and 5 subsections for explicit and implicit solution schemes, respectively. The current at the apex of the bend of the wire is shown in Fig. 2.10. Again, we observe similar behavior in this case as that of the straight-wire example.

As a third example, consider a circular loop of radius $R = 0.5$ m modeled by a wire of 10^{-6} m in radius. The loop is centered at the origin, placed in the xy plane, and illuminated by the Gaussian plane wave pulse described in Eq. (2.4.24). For the explicit and implicit schemes, the loop is divided into 36 and 12 subsections, respectively. The time step is again 0.05 LM. The current at the intersection of the y-axis and the circular loop is shown in Fig. 2.11. The comparison in earlier times is excellent, whereas the implicit solution decays much faster at late times.

Finally, we consider a stepped-radius wire structure with two different radii of 10^{-6} and 10^{-3} m. Each section is 1 m long and illuminated by a Gaussian pulse from the broadside direction. Each section is divided into 10 and 5 segments for the explicit and implicit schemes, respectively. The

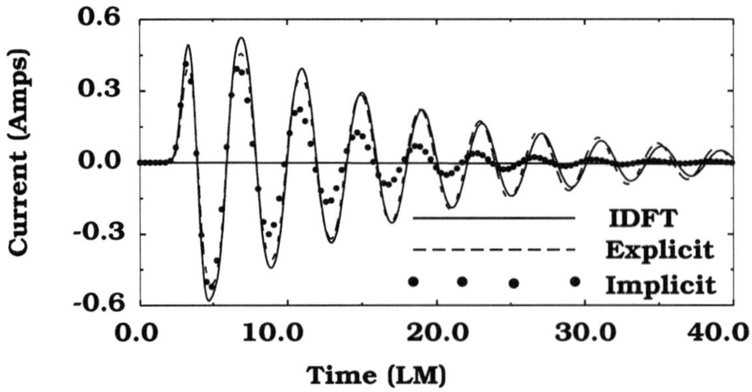

FIGURE 2.10
Transient current induced on a bent-wire scatterer lying along the x and y axes. Length and radii of each section of the wire are 1 and 10^{-6} m, respectively.

time step is 0.05 LM. The current at the junction of the two wires is given in Fig. 2.12. Again, we observe similar behavior as that in the previous examples.

2.5 ANALYSIS OF MULTIPLE WIRES AND WIRE JUNCTIONS

In this section, we briefly outline the numerical procedure for analyzing multiple wires and wire junctions—an essential step in the development of a user-oriented algorithm for handling complex bodies modeled with thin wires. The treatment of multiple wires is obvious since the mathematical steps presented in Sections 2.3 and 2.4 are also applicable to multiple wires. However, the wire junction problem may be handled as follows: First, we define the wire junction as a point at which more than two wires join. For example, consider a junction of three wires at a point as shown in Fig. 2.13a. The wires may have different radii but are simply shown as thick lines in the figure. We arbitrarily select one wire in the junction as a lead wire and refer to the rest of the wires as junction wires. We note that for the example shown in Fig. 2.13a, there is one lead wire and

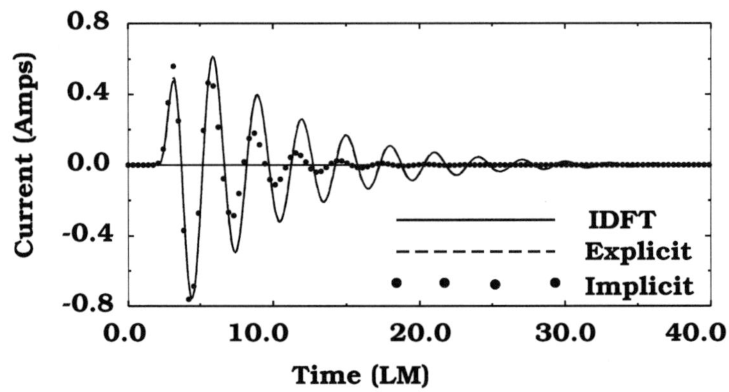

FIGURE 2.11
Transient current induced on a circular loop scatterer (radius 0.5 m) lying in the xy plane.

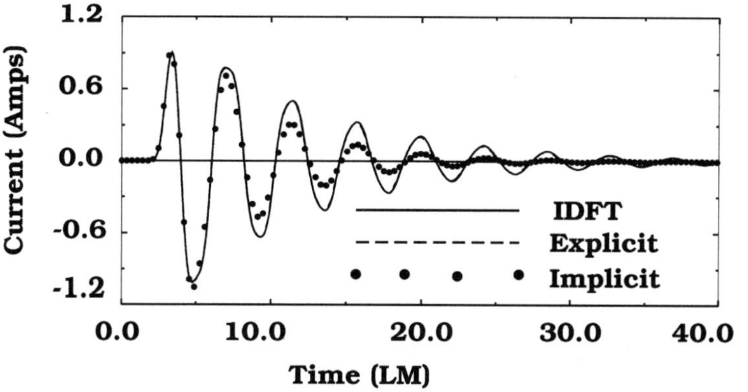

FIGURE 2.12
Transient current induced on a stepped-radius wire scatterer lying along the x-axis. Each section of the wire is 1 m long. The radius of the first section is 10^{-6} m and the second section has a radius of 10^{-3} m.

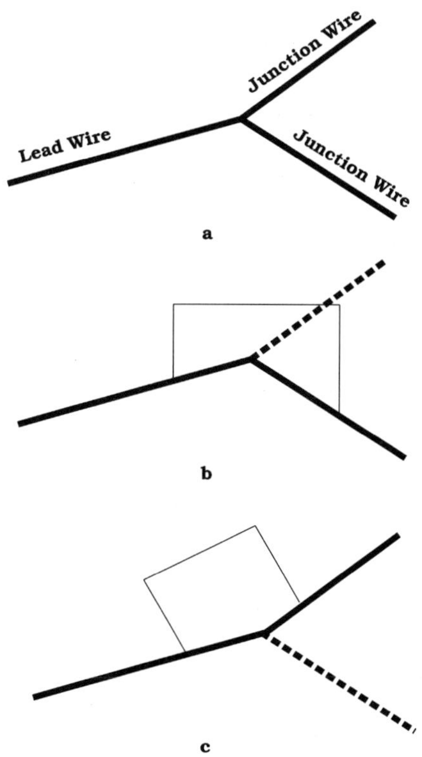

FIGURE 2.13
The wire junction problem.

two junction wires. Now, we define multiplicity of a junction as the number of basis functions associated with junction wires, which in this case is two.

In Figs. 2.13b and 2.13c, we show the placement of current basis functions associated with a junction. Note that the current direction is from the lead wire to junction wire for both figures. Thus, for the case presented in Fig. 2.13, we have two basis functions at the junction. For a general case of N wires joining at a junction, we would have $N - 1$ basis functions at that point. Lastly, we note that Kirchoff's current law is automatically satisfied in such a scheme.

2.6 CONCLUDING REMARKS

In this chapter, we have developed detailed numerical procedures for calculating the transient electromagnetic scattering from thin-wire structures by solving the electric field time domain integral equation. We have presented three methods to solve the integral equation: (i) the CG method, (ii) the explicit solution scheme, and (iii) the implicit solution scheme. All three solutions are reasonably accurate and provide efficient solution. However, the explicit solution invariably becomes unstable at late times, as discussed in Chapter 1, although this problem did not occur for the simple wire scatterers presented here. Furthermore, the implicit solution method is presented as an alternative to overcome the instability problem. We observe that the implicit solution has a faster damping rate, thereby eliminating the possibility of instabilities. We also note that for the simple problems presented here, the explicit solution appears to be more accurate than the implicit method when compared to the IDFT solution.

As indicated earlier, since many complex bodies can be approximated with wire grids, the transient scattering of such structures can be obtained by developing a unified algorithm including wire junctions by using the methods described in this chapter. Finally, we note that one such algorithm based on the explicit solution scheme is available for the user [4].

BIBLIOGRAPHY

[1] G. J. Burke and A. J. Poggio, "Numerical Electromagnetic Code (NEC)-Method of Moments," Technical Document 116, Naval Ocean Systems Center, San Diego, 1981.

[2] J. Rockway, J. Logan, D. Tam, and S. Li, *The MININEC SYSTEM: Microcomputer Analysis of Wire Antennas*, Artech House, Norwood, MA, 1988.

[3] A. R. Djordjevic, M. B. Bazdar, T. K. Sarkar, and R. F. Harrington, *AWAS for Windows: Analysis of Wire Antennas and Scatterers*, Artech House, Norwood, MA, 1995.

[4] J. A. Landt, E. K. Miller, and M. Van Blaricum, *WT-MBA/LLL1B: A Computer Program for the Time Domain Electromagnetic Response of Thin-Wire Structures*, Lawrence Livermore Laboratory Report UCRL-51585, 1974.

[5] R. F. Harrington, *Time Harmonic Electromagnetic Fields*, McGraw-Hill, New York, 1961.

[6] R. W. P. King, *The Theory of Linear Antennas*, Harvard Univ. Press, Cambridge, MA, 1956.

[7] G. F. Carrier and C. E. Pearson, *Partial Differential Equations*, Academic Press, New York, 1976.

[8] R. F. Harrington, *Field Computation by Moment Methods*, Macmillan, New York, 1968.

[9] S. M. Rao, T. K. Sarkar, and S. A. Dianat, "A Novel Technique to the Solution of Transient Electromagnetic Scattering from Thin Wires," *IEEE Trans. Antennas Propagat.*, vol. 34, pp. 630–634, 1986.

[10] C. L. Bennett and W. L. Weeks, "Electromagnetic Pulse Response of Cylindrical Scatterers," *1968 IEEE G-AP International Symposium*, Northeastern University, Boston, pp. 176–183, 1968.

[11] C. L. Bennett and W. L. Weeks, "Transient Scattering from Conducting Cylinders," *IEEE Trans. Antennas Propagat.*, vol. AP-18, pp. 627–633, 1970.

[12] K. M. Mitzner, "Numerical Solution for Transient Scattering from a Hard Surface of Arbitrary Shape-Retarded Potential Technique," *J. Acoust. Soc. Am.*, vol. 42, pp. 391–397, 1967.

[13] E. K. Miller and J. A. Landt, "Direct Time Domain Techniques for Transient Radiation and Scattering from Wires," *Proc. IEEE*, vol. 68, pp. 1396–1423, 1980.

[14] T. K. Sarkar, "The Application of the Conjugate Gradient Method to the Solution of Operator Equations Arising in Electromagnetic Scattering from Wire Antennas," *Radio Sci.*, vol. 19, pp. 1156–1172, 1984.

[15] S. M. Rao, T. K. Sarkar, and S. A. Dianat, "The Application of the Conjugate Gradient Method to the Solution of Transient Electromagnetic Scattering from Thin Wires," *Radio Sci.*, vol. 19, pp. 1319–1326, 1984.

CHAPTER 3

Infinite Conducting Cylinders: TDIE Solution

S. M. RAO
Department of Electrical Engineering, Auburn University

D. A. VECHINSKI
Nichols Research Corporation

T. K. SARKAR
Department of Electrical Engineering and Computer Science, Syracuse University

In the previous chapter, we discussed the time domain solution to wire problems. Although only simple wire geometries are presented in Chapter 2, it is possible to model fairly complex geometries using wire meshes. This type of modeling is known as wire-grid modeling and there exist user-oriented general-purpose computer codes, such as thin-wire time domain code [1], for modeling complex bodies. However, in another type of modeling, known as patch modeling, a given structure is covered with some predetermined, simple-shaped patches to obtain the scattering solution. In some situations, patch modeling works better than wire-grid modeling, specifically when calculating near-field quantities. In this chapter and Chapters 4 and 5, we present the patch modeling solution to solve different types of problems.

In this chapter, we discuss the numerical solution methods to solve the time domain integral equation for infinitely long, conducting cylinders of arbitrary cross section illuminated by a Gaussian plane wave pulse. We consider both the transverse magnetic (TM) and transverse electric (TE) solutions to this problem. It should be noted that the solution to the infinite cylinder problem, also popularly known as a two-dimensional problem, provides a precursor for the more difficult three-dimensional finite body solution. Here, we use simple square patches to model the surface of the cylinder. For this purpose, we first approximate the contour of the cylinder by straight line segments. Then, the length of the cylinder is divided in a similar manner to develop the square patches. Details of the modeling are discussed further in Section 3.2. The numerical solution procedure is obtained using the well-known method of moments (MoM) [2]. In the following sections, we derive the mathematical equations to calculate the current at a certain time instant as a function of currents of past instances and the incident field. The electric field integral equation (EFIE), obtained by enforcing the boundary condition on the tangential component of the electric

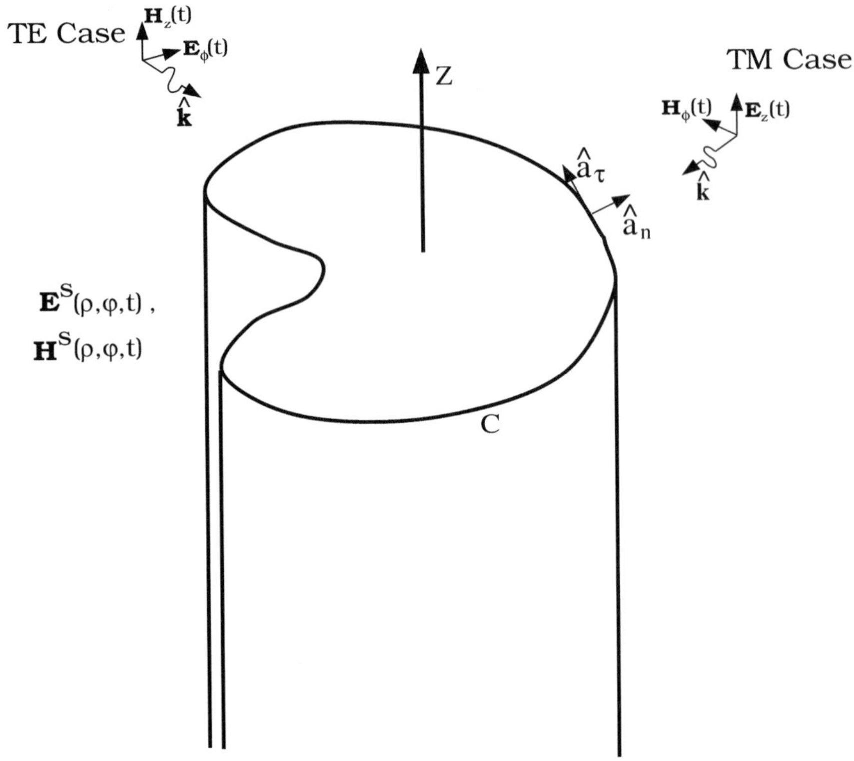

FIGURE 3.1
Arbitrarily shaped cylinder with an incident TM or TE pulse.

field, is used for both TM and TE incidence. It should be noted that the EFIE is applicable to both open and closed structures. We also present the magnetic H-field integral equation (HFIE), obtained by enforcing the boundary condition on the magnetic field, for TE incidence whose solution is useful in dealing with dielectric scatterers discussed in Chapter 5. However, the HFIE is valid only for closed cylinders.

3.1 INTEGRAL EQUATION FORMULATION

The scattering geometry under consideration is shown in Fig. 3.1. Let C denote the cross section of an open or closed perfectly electric conducting cylinder parallel to the z-axis. At each point on C let $\boldsymbol{a}_n$ represent an outward-directed unit vector normal to the contour. The circumferential vector, $\boldsymbol{a}_\tau$, is then obtained by $\boldsymbol{a}_\tau = \boldsymbol{a}_z \times \boldsymbol{a}_n$.

The incident field is a plane wave with either its electric field polarized in the z direction (TM incidence) or its magnetic field polarized in the z direction (TE incidence). The incident electric field, $\boldsymbol{E}^{\text{i}}(\boldsymbol{\rho}, t)$, defined in the absence of the scatterer, induces a surface current, $\boldsymbol{J}(\boldsymbol{\rho}, t)$, on the scatterer. The scattered field $\boldsymbol{E}^{\text{s}}(\boldsymbol{\rho}, t)$ is produced by $\boldsymbol{J}(\boldsymbol{\rho}, t)$. The boundary conditions require that the total tangential electric field, which is the vector sum of the incident and scattered electric fields, be zero on the conducting surface, or

$$[\boldsymbol{E}^{\text{s}}[\boldsymbol{J}] + \boldsymbol{E}^{\text{i}}]_{\text{tan}} = 0 \quad \text{on } C. \tag{3.1.1}$$

The scattered field, radiated by the current $\boldsymbol{J}$, may be written in terms of the magnetic vector and electric scalar potentials as

$$\boldsymbol{E}^{\mathrm{s}}[\boldsymbol{J}] = -\frac{\partial \boldsymbol{A}}{\partial t} - \nabla \Phi, \tag{3.1.2}$$

where

$$\boldsymbol{A}(\boldsymbol{\rho}, t) = \frac{\mu}{4\pi} \int_C \int_{z'=-\infty}^{\infty} \frac{\boldsymbol{J}(\boldsymbol{\rho}', t - R/c)}{R} \, dz' \, dC', \tag{3.1.3}$$

$$\Phi(\boldsymbol{\rho}, t) = \frac{1}{4\pi\epsilon} \int_C \int_{z'=-\infty}^{\infty} \frac{q_{\mathrm{s}}(\boldsymbol{\rho}', t - R/c)}{R} \, dz' \, dC', \tag{3.1.4}$$

and $R = \sqrt{|\boldsymbol{\rho} - \boldsymbol{\rho}'|^2 + z'^2}$, the distance from the field point, $\boldsymbol{\rho}$, to the source point $(\boldsymbol{\rho}', z')$. In Eqs. (3.1.3) and (3.1.4), μ and ϵ are the permeability and permittivity of the surrounding medium and c is the velocity of propagation of the electromagnetic wave. The electric surface charge density, q_{s}, may be related to the electric surface current density by the continuity equation

$$\nabla_{\mathrm{s}} \cdot \boldsymbol{J} = -\frac{\partial q_{\mathrm{s}}}{\partial t}. \tag{3.1.5}$$

Combining Eqs. (3.1.1) and (3.1.2) gives

$$\left[\frac{\partial \boldsymbol{A}}{\partial t} + \nabla \Phi\right]_{\tan} = [\boldsymbol{E}^{\mathrm{i}}]_{\tan} \tag{3.1.6}$$

and represents the EFIE formulation.

We may also develop an integral equation that uses the boundary condition on the magnetic field. From boundary conditions, we obtain

$$\boldsymbol{J} = \boldsymbol{a}_n \times [\boldsymbol{H}^{\mathrm{s}}[\boldsymbol{J}] + \boldsymbol{H}^{\mathrm{i}}], \tag{3.1.7}$$

where $\boldsymbol{H}^{\mathrm{i}}$ is the incident magnetic field and $\boldsymbol{H}^{\mathrm{s}}$ is the scattered magnetic field due to the induced currents $\boldsymbol{J}$. As before, the scattered field can be written in terms of the potential functions and is given by

$$\boldsymbol{H}^{\mathrm{s}}[\boldsymbol{J}] = \frac{1}{\mu} \nabla \times \boldsymbol{A}, \tag{3.1.8}$$

where $\boldsymbol{A}$ is given by Eq. (3.1.3). Equation (3.1.7), along with Eq. (3.1.8), represents the HFIE. Again, the HFIE is only applicable to closed geometries since a unique outward normal $\boldsymbol{a}_n$ can be defined for closed geometries only.

3.2 DISCRETIZATION SCHEME

Let the two dimensional cylinder be divided with square patches as shown in Fig. 3.2. To obtain this patch scheme, we first approximate the contour C into straight-line segments, each of length $\Delta\tau_m$. Then, we divide the whole mth column, i.e., from $(-\infty < z < \infty)$ associated with linear

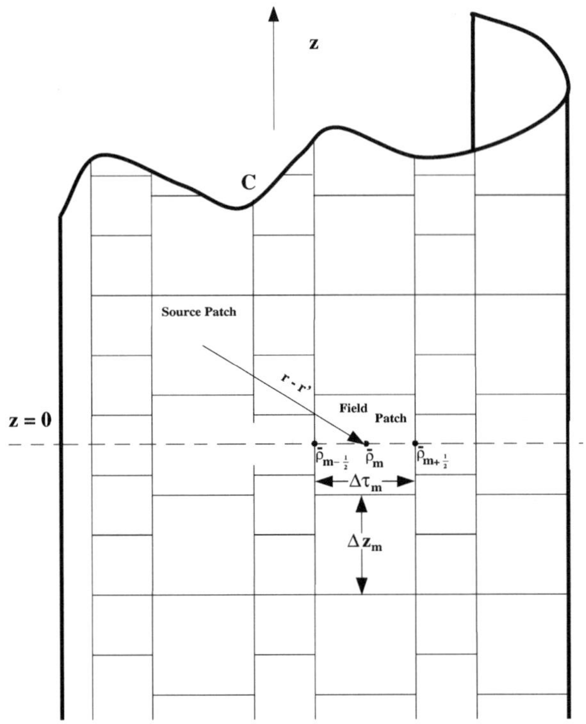

FIGURE 3.2
The grid scheme used for dividing the cylinder.

segment m, into zones with $\Delta z = \Delta\tau_m$ to obtain a square grid. It should be noted that the patch heights from one column to another are generally different with this scheme. Next, we note that, because the cylinder is infinite along the z-axis, all the field quantities are invariant with respect to the z variable. Hence, for simplicity, we will restrict ourselves to observing the current at $z = 0$.

Next, we note that the contour of the cylinder is divided in a slightly different manner for a TM incident field than for a TE incident field. This difference is necessary in order to model the behavior of the current correctly at any bends or, for open structures, at the edges. For TM incidence the current flow is in the z direction. This current is parallel to any bends or edges which may occur on the structure and hence is discontinuous along the contour. Therefore, in order to avoid any numerical stability problems, we restrict the expansion functions, as defined in the following section, to not extend over any bend. However, for TE incidence the current flow is in the transverse direction and is normal to any bends or edges. This current is continuous along C. Thus, the expansion functions around the contour may extend across any bend.

Thus, for TM incidence the m, lth patch is located between ρ_{m-1} and ρ_m along the contour and between $z_\ell - \Delta z_m/2$ and $z_\ell + \Delta z_m/2$ along the z-axis, where $z_\ell = \ell\Delta z_m$. The patch centers are located at $(\rho_{m-1/2}, z_\ell)$. Figure 3.3a shows the placement of the pulse expansion functions for a 90° bent strip.

For TE incidence the m, lth patch is located between $\rho_{m-1/2}$ and $\rho_{m+1/2}$ along the contour and between $z_\ell - \Delta z_m/2$ and $z_\ell + \Delta z_m/2$ along the z-axis, where $z_\ell = \ell\Delta z_m$. The patch centers are located at (ρ_m, z_ℓ). Figure 3.3b shows the placement of the pulse expansion functions for a 90° bent strip. Note that with this type of discretization, the corner of the strip shown in Fig. 3.3b is actually modeled with a flat rectangular patch extending from $\rho_{3/2}$ to $\rho_{5/2}$.

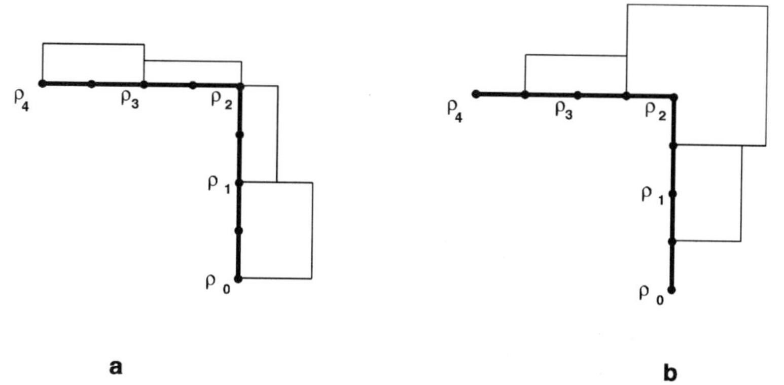

FIGURE 3.3
Placement of the expansion functions for (a) TM and (b) TE structures.

3.3 TM INCIDENCE: EFIE FORMULATION

For a TM incident field, the electric field only has a component in the $\boldsymbol{a}_z$ direction, and therefore the current is also only z directed. Due to the invariance with z, all derivatives with respect to z are zero. Thus, the z component of $\nabla\Phi$, i.e., $\nabla\Phi \cdot \boldsymbol{a}_z = \partial\Phi/\partial z = 0$ and Eq. (3.1.6) reduces to

$$\left[\frac{\partial \boldsymbol{A}}{\partial t}\right]_{\tan} = [\boldsymbol{E}^{\mathrm{i}}]_{\tan}, \tag{3.3.1}$$

where the subscript tan refers to the tangential component along the length of the cylinder.

By integrating both sides with respect to time, we obtain

$$\boldsymbol{A}_{\tan} = \int_0^t \boldsymbol{E}^{\mathrm{i}}_{\tan}\, dt, \tag{3.3.2}$$

which is the time domain electric field integral equation for the infinite conducting cylinder problem excited by a transient TM plane wave. In the following section, we discuss the numerical solution procedure to solve Eq. (3.3.2).

3.3.1 Explicit Solution Procedure

A set of basis functions for expansion purposes is defined as follows:

$$f_m(\rho) \equiv \begin{cases} 1 & \rho \in \rho_{m-1} \text{ to } \rho_m \\ 0 & \text{otherwise.} \end{cases} \tag{3.3.3}$$

These expansion functions are the standard pulse functions on a rectangular patch. Furthermore, the same expansion functions are also used as testing functions, and the inner product is defined as

$$\langle \boldsymbol{a}, \boldsymbol{b} \rangle = \int_C \boldsymbol{a} \cdot \boldsymbol{b}\, dC'. \tag{3.3.4}$$

The next step in applying the MoM is to test Eq. (3.3.2) to obtain

$$\langle f_m \boldsymbol{a}_z, \boldsymbol{A} \rangle = \left\langle f_m \boldsymbol{a}_z, \int_0^t \boldsymbol{E}^{\mathrm{i}}\, dt \right\rangle. \tag{3.3.5}$$

Let

$$\boldsymbol{J}(\boldsymbol{\rho}, t) = \boldsymbol{a}_z \sum_{k=1}^{N} I_k(t) f_k(\boldsymbol{\rho}), \tag{3.3.6}$$

where N is the number of linear segments along the contour C and $I_k(t)$ is the unknown coefficient at the kth zone. Let us now look at the evaluation of $\boldsymbol{A}(\boldsymbol{\rho}, t)$ at a time t_n and an observation point $\boldsymbol{\rho}$ in the mth patch. Combining Eqs. (3.1.3) and (3.3.6) gives

$$\begin{aligned} \boldsymbol{A}(\boldsymbol{\rho}, t_n) &= \frac{\mu}{4\pi} \int_C \int_{z'=-\infty}^{\infty} \frac{\boldsymbol{a}_z \sum_{k=1}^{N} I_k(t_n - R/c) f_k(\boldsymbol{\rho}')}{R}\, dz'\, dC' \\ &\approx \frac{\mu}{4\pi} \sum_{k=1}^{N} \sum_{\ell=-\infty}^{\infty} I_k\left(t_n - \frac{R_{mk\ell}}{c}\right) \kappa_{k,\ell}\, \boldsymbol{a}_z, \end{aligned} \tag{3.3.7}$$

where

$$\kappa_{k,\ell} = \int_{k,\ell} \int_{\text{patch}} \frac{ds'}{R}, \tag{3.3.8}$$

$$R = \sqrt{|\boldsymbol{\rho} - \boldsymbol{\rho}'|^2 + z'^2}, \tag{3.3.9}$$

$$R_{mk\ell} = \sqrt{|\boldsymbol{\rho}_{m-1/2} - \boldsymbol{\rho}_{k-1/2}|^2 + z_\ell^2}. \tag{3.3.10}$$

In Eq. (3.3.7), a time-consuming space–time integral is removed by replacing $I_k(t_n - R/c) \Rightarrow I_k(t_n - R_{mk\ell}/c)$. This assumes that the current in a patch does not vary appreciably with time.

For non-self terms, i.e., where $\boldsymbol{r}$ and $\boldsymbol{r}'$ do not belong to the same patch, the $1/R$ integral may be approximated by the patch area multiplied by the distance between the patch centers. For self terms, the integration is carried out analytically [3]. The values of $\kappa_{k,\ell}$ are then given by

$$\kappa_{k,\ell} = \begin{cases} \dfrac{\Delta\tau_k \Delta z_l}{\sqrt{|\boldsymbol{\rho}_{m-1/2} - \boldsymbol{\rho}_{k-1/2}|^2 + z_\ell^2}} & \text{for } k \neq m \text{ and } \ell \neq 0 \\ 4\Delta\tau_k \ln(1+\sqrt{2}) & \text{for } k = m \text{ and } \ell = 0. \end{cases} \tag{3.3.11}$$

It should be noted that the value for the self term shown previously is true only if $\Delta\tau_k = \Delta z_k$. A more general expression is provided by Wilton *et al.* [4] The infinite summation on ℓ in Eq. (3.3.7) can actually be truncated to $-\ell_{\max} \leq \ell \leq \ell_{\max}$. Due to the retarded nature of the currents, we only need to extend the summation on ℓ to a point where $t_n - R_{mk\ell}/c$ becomes negative. Whenever $t_n - R_{mk\ell}/c < 0$, it implies that the solution has not yet progressed out far enough in time for those currents to have an effect at the observation point (assuming a causal system). As time progresses, $\ell_{\max}$ will increase as more currents need to be included due to the infinite length of the cylinder and computational times begin to increase tremendously. However, due to the the symmetry about $z = 0$, the ℓ summations may be cut down to $0 \leq \ell \leq \ell_{\max}$.

If Δt is chosen appropriately, i.e., if $\Delta t \leq R_{\min}/c$, where $R_{\min}$ is the minimum distance between any two distinct patch centers, then there is only one current in Eq. (3.3.7) which is unknown, viz., when $k=m$ and $\ell=0$, which implies $R_{mk\ell}=0$. All the other currents are assumed known because they occur at some earlier time step. Thus, Eq. (3.3.7) may be rewritten as

$$\begin{aligned} \boldsymbol{A}(\boldsymbol{\rho}, t_n) &= \frac{\mu\kappa_{m,0}}{4\pi} I_m(t_n)\, \boldsymbol{a}_z + \frac{\mu}{4\pi} \sum_{\substack{k=1 \\ k\neq m \text{ and } \ell\neq 0}}^{N} \sum_{\ell=-\infty}^{\infty} I_k(t_n - R_{mk\ell}/c)\kappa_{k,\ell}\, \boldsymbol{a}_z \\ &= \frac{\mu\kappa_m}{4\pi} I_m(t_n)\, \boldsymbol{a}_z + \not{\boldsymbol{A}}(\boldsymbol{\rho}, t_n), \end{aligned} \tag{3.3.12}$$

where the $\ell=0$ subscripts have been omitted, and $\not{\boldsymbol{A}}(\boldsymbol{\rho}, t_n)$ represents $\boldsymbol{A}(\boldsymbol{\rho}, t_n)$ with the self term (i.e., $m=k$ and $\ell=0$) deleted.

Finally, combining Eqs. (3.3.5) and (3.3.12), taking all known quantities (those involving $n-1, n-2, \cdots$) to the right side, and using a 1-point integration for the testing procedure, we obtain

$$\frac{\mu\kappa_m \Delta\tau_m}{4\pi} I_m(t_n) = \Delta\tau_m \int_0^{t_n} E_z^{\mathrm{i}}\, dt - \not{\boldsymbol{A}}(\boldsymbol{\rho}, t_n) \cdot \Delta\tau_m \boldsymbol{a}_z. \tag{3.3.13}$$

Equation (3.3.13) can be written as a matrix equation:

$$[\alpha][\boldsymbol{I}(t_n)] = [\boldsymbol{F}(t_n)] + [\beta][\boldsymbol{I}(t_n - R_{mk\ell}/c)], \tag{3.3.14}$$

where

$$\alpha_{m,m} = \frac{\mu\kappa_m \Delta\tau_m}{4\pi}, \tag{3.3.15}$$

$$F_m(t_n) = \Delta\tau_m \int_0^{t_n} \boldsymbol{E}_z^{\mathrm{i}}\, dt, \tag{3.3.16}$$

$$\beta_{m,k} = \kappa_{k,\ell}\Delta\tau_m \quad \text{for } k\neq m \text{ and } \ell\neq 0. \tag{3.3.17}$$

Note that in Eq. (3.3.15), the $[\alpha]$ matrix contains vector potential terms with current coefficients at $t=t_n$ which are unknown. Also, note that the $[\alpha]$ matrix is a diagonal matrix.

By examining Eq. (3.3.14) it is evident that the left-hand side involves only terms at $t=t_n$, whereas the right-hand side contains the terms for $t \leq t_{n-1}$. This is due to $R_{mkl} \geq R_{\min}$, which makes the retarded time $t_n - R_{mk\ell}/c \leq t_{n-1}$. Once the currents at t_n are found for all $1 \leq m \leq N$, the time step is incremented, and the currents at t_{n+1} can then be found in the same manner since all the previous currents are known. This constitutes the explicit time marching procedure. The main advantage of the explicit solution is that no matrix storage/inversion is needed. However, an important limiting factor is the duration of the time step which must be less than or equal to $R_{\min}/c$. It is obvious that, with this restriction, one needs a large number of time steps to obtain a specific duration of the time domain signature. Furthermore, numerical results tend to grow unstable at late times, necessitating averaging procedures as discussed in Chapter 1, Section 1.6.1.

3.3.2 Implicit Solution Procedure

For the implicit solution, we remove the restriction on the time step. In this case, the time step is solely chosen based on the high-frequency content in the incident pulse and the structure dimensions. Thus, we rewrite Eq. (3.3.12) as

$$A(\rho, t_n) = A_1(\rho, t_n) + A_2(\rho, t_n), \tag{3.3.18}$$

where $A_1(\rho, t_n)$ represents the terms which satisfy the criterion given by $\Delta t > R_{mk\ell}/c$ where $R_{mk\ell}$ is given by Eq. (3.3.10). Obviously, $A_1(\rho, t_n)$ is unknown since the currents are known only until $t = t_{n-1}$.

We can write $A_1(\rho, t_n)$ as

$$A_1(\rho, t_n) = \frac{\mu}{4\pi} \sum_{k=1}^{M} \sum_{\ell=-P}^{P} I_k(t_R) \kappa_{k,\ell}\, a_z, \tag{3.3.19}$$

where $t_{n-1} < t_R \leq t_n, t_R = t_n - R_{mkl}/c$, and M and P represent those terms for which $R_{mkl} < c\Delta t$. Notice that $I(t_R)$, may be written as follows using linear interpolation in time and carrying out a few steps of algebra:

$$I(t_R) = \left[\frac{R_{mkl}}{c\Delta t}\right] I(t_{n-1}) + \left[1 - \frac{R_{mkl}}{c\Delta t}\right] I(t_n). \tag{3.3.20}$$

Using Eq. (3.3.20), Eq. (3.3.19) may be written as

$$\begin{aligned} A_1(\rho, t_n) = {} & \frac{\mu}{4\pi} \sum_{k=1}^{M} \sum_{\ell=-P}^{P} I_k(t_{n-1}) \left[\frac{R_{mkl}}{c\Delta t}\right] \kappa_{k,\ell}\, a_z \\ & + \frac{\mu}{4\pi} \sum_{k=1}^{M} \sum_{\ell=-P}^{P} I_k(t_n) \left[1 - \frac{R_{mkl}}{c\Delta t}\right] \kappa_{k,\ell}\, a_z. \end{aligned} \tag{3.3.21}$$

Thus, for the implicit scheme, we can rewrite Eq. (3.3.14) as

$$[\alpha][I(t_n)] = [F(t_n)] + [\beta][I(t_R)]. \tag{3.3.22}$$

Note that the elements of the $[\alpha]$ matrix in Eq. (3.3.22) are formed by the second term in Eq. (3.3.21). Also note that the $[\alpha]$ matrix is a sparse matrix and its sparsity depends on the choice of Δt. In addition, the matrix elements of $[\alpha]$ and $[\beta]$ are not functions of time and hence need to be computed only once at the first time step. Lastly, by solving Eq. (3.3.22) at each time step, the time domain current induced on the cylinder may be obtained iteratively.

3.3.3 Numerical Examples

In this section, numerical results for the induced currents on four geometrical shapes—a straight strip, a bent strip, a circular cylinder, and a square cylinder—illuminated by a TM Gaussian plane wave are presented. The results are compared with data obtained in the frequency domain and transformed into the time domain by using inverse Fourier transform techniques. The MoM was employed for calculating the frequency domain solutions unless stated otherwise. The results for the explicit solution are obtained by employing the averaging technique described in Chapter 1, Section 1.6.1.

For the figures that follow, the incident field is a Gaussian plane wave, given by Eq. (1.4.14), with $E_o = 120\,\pi a_z, a_k = -a_x, T = 4$ LM, and $ct_o = 6$ LM. The time domain results for the explicit

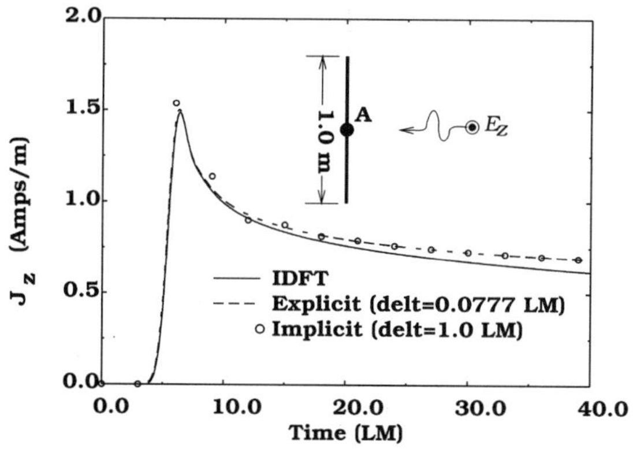

FIGURE 3.4
The induced current at the center (A) of a straight strip illuminated by a TM Gaussian plane wave. The strip is divided into nine segments.

solution were obtained with $c\Delta t = 0.7\ R_{\min}$. However, for the implicit solution, we have used relatively large values for Δt to illustrate the efficiency of this procedure. The frequency domain results were obtained by using 256 frequency samples between 0 and 250 MHz and performing an inverse discrete Fourier transform (IDFT).

As a first case, consider a straight strip of width $W = 1.0$ m located symmetrically with respect to the origin along the y-axis. The strip is divided uniformly into nine segments and the time steps for the explicit and implicit solutions are 0.0777 and 1 LM, respectively. Figure 3.4 shows the current induced at the center of the strip and this is compared with the transformed result from the frequency domain. The agreement is very good in the early time. The late-time discrepancy is attributed to an insufficient frequency sampling. Higher frequency sampling provides much better agreement even at late times [5].

The case of a 90° bent strip located along the x and y axes is considered next. Each strip is 1 m long and is divided into five uniform segments. Figure 3.5 shows the current at (−0.5, 0.0) for the three solutions. The time steps for the explicit and implicit solutions are 0.1 and 3 LM,

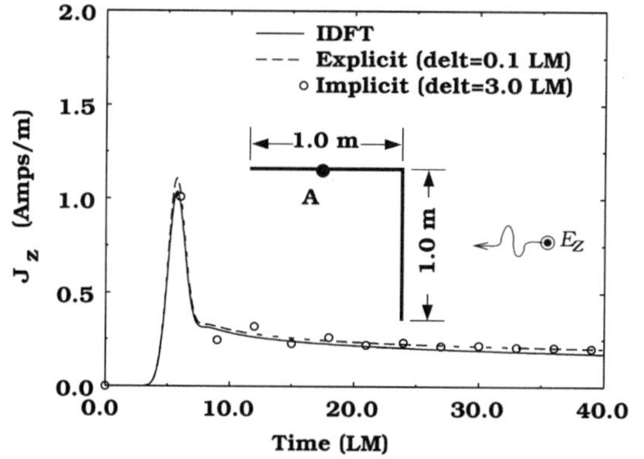

FIGURE 3.5
The induced current at (−0.5, 0.0) (A) of a 90° bent strip illuminated by a TM Gaussian plane wave. The strip is divided into a total of 10 segments.

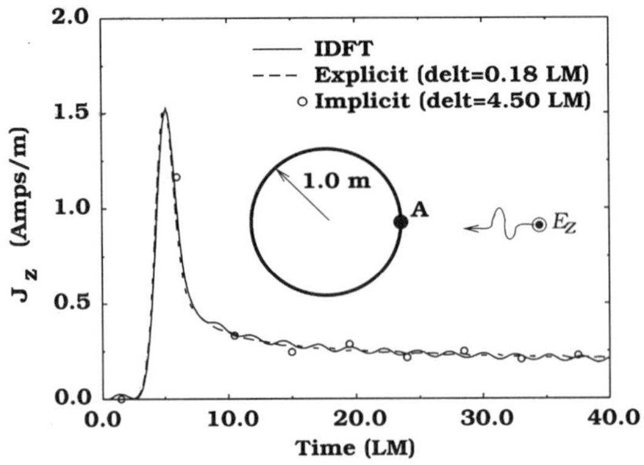

FIGURE 3.6
The induced current at $\phi = 0°$ (A) of a circular cylinder illuminated by a TM Gaussian plane wave. The cylinder is divided into 24 segments.

respectively. The current is not obtained at the corner because it experiences a singularity at the bend. The agreement is good, and it is noted that there is an insufficient frequency sampling for the frequency domain result.

Now, consider the case of a circular cylinder centered about the origin. The radius of the cylinder is 1 m, and for the numerical solution the circumference is approximated by 24 linear segments. Figure 3.6 shows the current induced at $\phi = 0°$. The time steps for the explicit and implicit solutions are 0.18 and 4.5 LM, respectively. Notice the large time step used for this calculation. Also for this case, the frequency domain solution is obtained by using the eigenfunction expansion method [6]. The agreement is good, both at early time and at late time, for the solutions considered.

As the final case for TM incidence, a square cylinder of side length 1 m is considered. The cylinder is centered about the origin and divided into 36 uniform segments. The time steps for the explicit and implicit solutions are 0.055 and 3 LM, respectively. The current at the center of the illuminated side is shown in Fig. 3.7. Again, the agreement is good for the solutions considered.

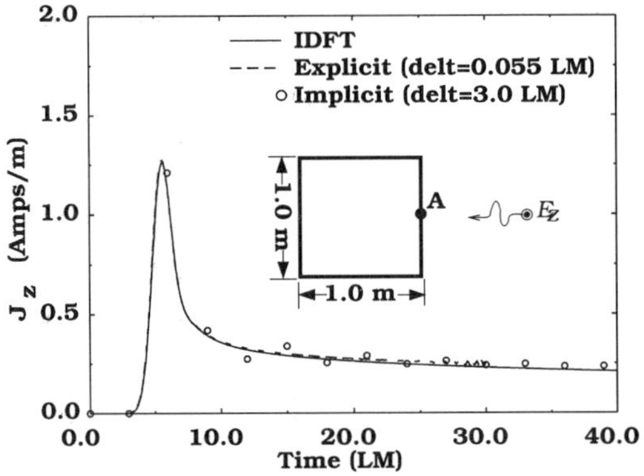

FIGURE 3.7
The induced current at point A of a square cylinder illuminated by a TM Gaussian plane wave. The square is divided into 36 segments.

3.4 TE INCIDENCE: EFIE FORMULATION

For TE incidence we use Eq. (3.1.6) and note that the induced current is in the transverse direction to the cylinder axis. For the TE case, unlike the TM case, the scalar potential term is not zero. However, as before, all the parameters are invariant with respect to the cylinder axis, i.e., z-axis, which enables us to compute the currents only at $z = 0$.

3.4.1 Explicit Solution Procedure

The basis functions used here are the same as those used for TM incidence except that they are shifted by a half zone, as shown in Fig. 3.3b, and are defined as

$$f_m(\rho) \equiv \begin{cases} 1 & \rho \in \rho_{m-1/2} \text{ to } \rho_{m+1/2} \\ 0 & \text{otherwise.} \end{cases} \tag{3.4.1}$$

As before, the same functions will be used for testing and the inner product is defined by Eq. (3.3.4). First, let us consider the testing procedure.

Applying the testing procedure to Eq. (3.1.6) gives

$$\left\langle f_m \boldsymbol{a}_\tau, \left[\frac{\partial \boldsymbol{A}}{\partial t} + \nabla\Phi\right]\right\rangle = \langle f_m \boldsymbol{a}_\tau, \boldsymbol{E}^{\mathrm{i}}\rangle. \tag{3.4.2}$$

Approximating the time derivative of the potential function by forward-finite difference,[1] we get

$$\langle f_m \boldsymbol{a}_\tau, \boldsymbol{L}[\boldsymbol{J}]\rangle = \langle f_m \boldsymbol{a}_\tau, \boldsymbol{E}^{\mathrm{i}}\rangle, \tag{3.4.3}$$

where

$$\boldsymbol{L}[\boldsymbol{J}] = \frac{\boldsymbol{A}(\rho, t_{n+1}) - \boldsymbol{A}(\rho, t_n)}{\Delta t} + \nabla\Phi(\rho, t_n). \tag{3.4.4}$$

The testing of the vector potential may be approximated by using a 1-point integration. This is done by evaluating $\boldsymbol{A}$ at ρ_m and multiplying by the width of the patch, $\Delta\tau_m$. Thus,

$$\langle f_m \boldsymbol{a}_\tau, \boldsymbol{A}(\rho)\rangle \approx \boldsymbol{A}(\rho_m) \cdot \Delta\tau_m \boldsymbol{a}_\tau. \tag{3.4.5}$$

The testing of the incident field is performed in a manner similar to that of the vector potential. We therefore make the approximation

$$\langle f_m \boldsymbol{a}_\tau, \boldsymbol{E}^{\mathrm{i}}(\rho, t_n)\rangle \approx \boldsymbol{E}^{\mathrm{i}}(\rho_m, t_n) \cdot \Delta\tau_m \boldsymbol{a}_\tau. \tag{3.4.6}$$

We now consider the evaluation of $\nabla\Phi(\rho, t_n)$. By using the fact that the line integral of the gradient of a potential function is the function evaluated at its endpoints, we can rewrite the testing

[1] It is also possible to obtain the solution by using backward difference, which is considered in Section 3.4.2. Furthermore, another solution is possible by taking an extra derivative with respect to time of Eq. (3.1.6) and using central difference, as done in Ref. [5].

of the gradient term as

$$\langle f_m \boldsymbol{a}_\tau, \nabla\Phi(\boldsymbol{\rho}, t_n)\rangle = \int_C \nabla\Phi(\boldsymbol{\rho}, t_n)\cdot f_m \boldsymbol{a}_\tau dC' \\ = \Phi(\boldsymbol{\rho}_{m+1/2}, t_n) - \Phi(\boldsymbol{\rho}_{m-1/2}, t_n). \tag{3.4.7}$$

Next, we consider the expansion procedure. Let the surface current be approximated as

$$\boldsymbol{J}(\boldsymbol{\rho}, t) = \boldsymbol{a}_\tau \sum_{k=1}^{N} I_k(t) f_k(\boldsymbol{\rho}), \tag{3.4.8}$$

where N is the number of linear segments along the contour C and $I_k(t)$ is the unknown coefficient at the kth zone. Substituting Eq. (3.4.8) into Eq. (3.1.3) results in

$$\boldsymbol{A}(\boldsymbol{\rho}, t_n) = \frac{\mu}{4\pi}\int_C\int_{z'=-\infty}^{\infty} \frac{\boldsymbol{a}_\tau \sum_{k=1}^{N} I_k(t_n - R/c) f_k(\boldsymbol{\rho}')}{R}\, dz'\, dC' \\ \approx \frac{\mu}{4\pi}\sum_{k=1}^{N}\sum_{\ell=-\infty}^{\infty} I_k(t_n - R_{mk\ell}/c)\kappa_{k,\ell}\, \boldsymbol{a}_\tau, \tag{3.4.9}$$

where

$$\kappa_{k,\ell} = \int_{k,\ell}\int_{\text{patch}} \frac{ds'}{R}, \tag{3.4.10}$$

$$R = \sqrt{|\boldsymbol{\rho} - \boldsymbol{\rho}'|^2 + z'^2}, \tag{3.4.11}$$

$$R_{mk\ell} = \sqrt{|\boldsymbol{\rho}_m - \boldsymbol{\rho}_k|^2 + z_\ell^2}. \tag{3.4.12}$$

As discussed in Section 3.3, the current is assumed to be constant with time within a patch so that the integral may be simplified. The calculation of the $1/R$ integral is performed in a similar manner as that in the TM case and is given by

$$\kappa_{k,\ell} = \begin{cases} \frac{\Delta\tau_k \Delta z_l}{\sqrt{|\boldsymbol{\rho}_m - \boldsymbol{\rho}_k|^2 + z_\ell^2}} & \text{for } k \neq m \text{ and } \ell \neq 0 \\ 4\Delta\tau_k \ln(1+\sqrt{2}) & \text{for } k = m \text{ and } \ell = 0. \end{cases} \tag{3.4.13}$$

Extracting the self term, Eq. (3.4.9) may be rewritten as

$$\boldsymbol{A} = \frac{\mu\kappa_m}{4\pi} I_m(t_n)\, \boldsymbol{a}_\tau + \not{\boldsymbol{A}}(\boldsymbol{\rho}, t_n), \tag{3.4.14}$$

where the $\not{\boldsymbol{A}}$ represents the rest of the terms in the summation.

Now, consider the scalar potential term given by Eq. (3.1.4). Substituting the continuity equation, given by Eq. (3.1.5), into Eq. (3.1.4), we get

$$\Phi(\boldsymbol{\rho}_m, t_n) = \frac{-1}{4\pi\epsilon}\int_C\int_{z'=-\infty}^{\infty} \frac{\int_{t'=0}^{t_n - R/c} \nabla_s \cdot \boldsymbol{J}(\boldsymbol{\rho}, t')\, dt'}{R}\, dz'\, dC', \tag{3.4.15}$$

where $R = \sqrt{|\rho_m - \rho'|^2 + z'^2}$. Using Eq. (3.4.8) for the expansion of $\boldsymbol{J}$ in Eq. (3.4.15) results in

$$\Phi(\rho_m, t_n) = \frac{-1}{4\pi\epsilon} \sum_{k=1}^{N} \sum_{\ell=-\infty}^{\infty} \int_{t'=0}^{t_n - R_{mk\ell}/c} I_k(t')\, dt' \int_{k,\ell}\int_{\text{patch}} \frac{\partial f_k/\partial\tau}{R}\, ds'. \tag{3.4.16}$$

Approximating the spatial derivative, $\partial f_k/\partial\tau$ in Eq. (3.4.16), by finite difference, results in

$$\Phi(\rho_m, t_n) = \sum_{k=1}^{N} \sum_{\ell=-\infty}^{\infty} \Phi^+(\rho_m, t_n) - \Phi^-(\rho_m, t_n), \tag{3.4.17}$$

where

$$\Phi^+ = \frac{-1}{4\pi\epsilon\Delta\tau_k^+} \int_{t'=0}^{t_R^+} I_k(t')\, dt' \int_{\rho_k}^{\rho_{k+1}} \int_{z_1}^{z_2} \frac{d\ell'\, dz'}{\sqrt{|\rho_m - \rho'|^2 + z'^2}}, \tag{3.4.18}$$

$$\Phi^- = \frac{-1}{4\pi\epsilon\Delta\tau_k^-} \int_{t'=0}^{t_R^-} I_k(t')\, dt' \int_{\rho_{k-1}}^{\rho_k} \int_{z_1}^{z_2} \frac{d\ell'\, dz'}{\sqrt{|\rho_m - \rho'|^2 + z'^2}}, \tag{3.4.19}$$

with

$$t_R^\pm = t_n - \frac{R_{mk\ell}^\pm}{c}, \tag{3.4.20}$$

$$R_{mk\ell}^\pm = \sqrt{|\rho_m - \rho_{k\pm 1/2}|^2 + z_l^2}, \tag{3.4.21}$$

$$\Delta\tau_k^- = |\rho_k - \rho_{k-1}|, \tag{3.4.22}$$

$$\Delta\tau_k^+ = |\rho_{k+1} - \rho_k|, \tag{3.4.23}$$

$$z_1 = z_\ell - \frac{\Delta z_l}{2}, \tag{3.4.24}$$

$$z_2 = z_\ell + \frac{\Delta z_l}{2}. \tag{3.4.25}$$

This completes the numerical processing of potential terms.

For an explicit solution, we assume that the time step is less than or equal to $R_{\min}/c$. Thus, if we replace $n \Rightarrow n-1$, utilize Eqs. (3.4.5), (3.4.6), and (3.4.7), and take all known quantities $(n-1, n-2, \cdots)$ to the right side, we can rewrite Eq. (3.4.2) as

$$\begin{aligned} \frac{\mu\kappa_m\Delta\tau_m}{4\pi\,\Delta t} I_m(t_n) = \boldsymbol{E}^{\mathrm{i}}(\rho_m, t_{n-1}) \cdot \boldsymbol{a}_\tau \Delta\tau_m - \frac{\cancel{\boldsymbol{A}}(\rho_m, t_n) - \boldsymbol{A}(\rho_m, t_{n-1})}{\Delta t} \cdot \boldsymbol{a}_\tau \Delta\tau_m \\ - \Phi(\rho_{m+1/2}, t_{n-1}) + \Phi(\rho_{m-1/2}, t_{n-1}). \end{aligned} \tag{3.4.26}$$

Now consider Eq. (3.4.26). The left-hand side of Eq. (3.4.26) involves only terms at $t = t_n$, whereas the right-hand side contains the terms retarded in time. Therefore, the currents may be obtained by the marching on in time procedure. The numerical procedure starts at $n = 2$, assuming the current to be zero at $n = 0$ and $n = 1$, and then marches in time for $n = 3, 4, 5, \cdots$. Furthermore, as in the TM case, we can write Eq. (3.4.26) in matrix form as

$$[\alpha][\boldsymbol{I}(t_n)] = [\boldsymbol{F}(t_n)] + [\beta][\boldsymbol{I}(t_R)], \tag{3.4.27}$$

and, again, $[\alpha]$ is a diagonal matrix.

3.4.2 Implicit Solution Procedure

First, for the implicit scheme, we approximate the time derivative of the potential function by backward-finite difference. This is an important step in the numerical solution procedure since the backward-finite difference allows the scalar potential terms to be included in the $[\alpha]$ matrix. Now, we rewrite Eq. (3.4.2) as

$$\langle f_m \boldsymbol{a}_\tau, \boldsymbol{L}[\boldsymbol{J}]\rangle = \langle f_m \boldsymbol{a}_\tau, \boldsymbol{E}^{\mathrm{i}}\rangle, \tag{3.4.28}$$

where

$$\boldsymbol{L}[\boldsymbol{J}] = \frac{\boldsymbol{A}(\boldsymbol{\rho}, t_n) - \boldsymbol{A}(\boldsymbol{\rho}, t_{n-1})}{\Delta t} + \nabla\Phi(\boldsymbol{\rho}, t_n). \tag{3.4.29}$$

The testing and expansion procedures for the vector potential term are identical to those of the explicit scheme. However, notice that for the implicit scheme, the time step Δt is chosen independently and not related to the spatial discretization. Thus, we can write $\boldsymbol{A}(\boldsymbol{\rho}_m, t_n)$ as

$$\boldsymbol{A}(\boldsymbol{\rho}_m, t_n) = \boldsymbol{A}_1(\boldsymbol{\rho}_m, t_n) + \boldsymbol{A}_2(\boldsymbol{\rho}_m, t_n), \tag{3.4.30}$$

where

$$\boldsymbol{A}_{1,2}(\boldsymbol{\rho}_m, t_n) \approx \frac{\mu}{4\pi} \sum_{k=1}^{N} \sum_{\ell=-\infty}^{\infty} I_k(t_R)\kappa_{k,\ell}\, \boldsymbol{a}_\tau, \tag{3.4.31}$$

and $t_R = t_n - R_{mk\ell}/c$. For $\boldsymbol{A}_1(\boldsymbol{\rho}_m, t_n)$, $t_{n-1} < t_R \le t_n$. Thus, the current values at $t = t_R$ are unknown. However, for $\boldsymbol{A}_2(\boldsymbol{\rho}_m, t_n)$, the retarded time $t_R = t_n - R_{mk\ell}/c$ is less than t_{n-1}, and therefore the current values at these instants are known or can be calculated in an easy manner.

We now consider the evaluation of the scalar potential. Again, we use the same expansion and testing procedure as in the explicit scheme. Furthermore, as in the case of vector potential, the scalar potential term may also be split into two parts:

$$\Phi(\boldsymbol{\rho}_m, t_n) = \Phi_1(\boldsymbol{\rho}_m, t_n) + \Phi_2(\boldsymbol{\rho}_m, t_n), \tag{3.4.32}$$

where $\Phi_1(\boldsymbol{\rho}_m, t_n)$ and $\Phi_2(\boldsymbol{\rho}_m, t_n)$ represent the terms involving unknown and known current values, respectively, depending on the value of the retarded time. Thus, we can write Eq. (3.4.28) in a matrix form as

$$[\alpha][\boldsymbol{I}(t_n)] = [\boldsymbol{F}(t_n)] + [\beta][\boldsymbol{I}(t_R)]. \tag{3.4.33}$$

Referring to Eq. (3.4.33), we make the following observations.

- The elements of the $[\alpha]$ matrix are formed by the potential terms when the retarded time t_R satisfies the condition $t_{n-1} < t_R \le t_n$.
- In the current formulation both the vector potential and the scalar potentials contribute to the $[\alpha]$ matrix.
- The $[\alpha]$ matrix is a sparse matrix, but never diagonal, and its sparsity depends on the choice of Δt.

- The matrix elements, α_{mn}, are not functions of time and hence need to be computed only once at the first time step.
- The elements of the $[\boldsymbol{F}]$ matrix are obtained from incident field calculations as given in Eq. (3.4.6).
- The product term $[\beta][\boldsymbol{I}(t_R)]$ can be easily computed since $t_R \leq t_{n-1}$ and hence the currents at these instants are known.

Lastly, by solving Eq. (3.4.33) at each time step, the time domain current induced on the scatterer may be obtained iteratively.

3.4.3 Numerical Examples

In this section, we present the numerical results for four representative two-dimensional scatterers: a straight strip, a 90° bent strip, a circular cylinder, and a square cylinder. The scatterers are illuminated by a Gaussian plane wave, given by

$$\boldsymbol{H}^{\mathrm{i}}(\rho, t) = \boldsymbol{a}_z H_{\mathrm{o}} \frac{4.0}{\sqrt{\pi} T} e^{-\gamma^2}, \tag{3.4.34}$$

where

$$\gamma = \frac{4.0}{T} \left[ct - ct_{\mathrm{o}} - \boldsymbol{\rho} \cdot \boldsymbol{k}\right], \tag{3.4.35}$$

with $H_{\mathrm{o}} = 1.0$, $\boldsymbol{k} = -\boldsymbol{a}_{\rho}$, $T = 4$ LM, and $ct_{\mathrm{o}} = 6$ LM. For comparison, we present the results obtained by taking the IDFT of the frequency domain solution calculated at 1 MHz interval with 256 samples. Furthermore, all the results reported in this work for the explicit solution method are obtained using the averaging procedure described in Chapter 1, Section 1.6.1 to prevent the late-time instabilities.

First, consider a straight strip of width $W = 0.5$ m located symmetrically with respect to the origin along the y-axis. The strip is divided uniformly into 10 segments, and the time steps for the explicit and implicit solutions are 0.035 and 0.5 LM, respectively. Figure 3.8 shows the

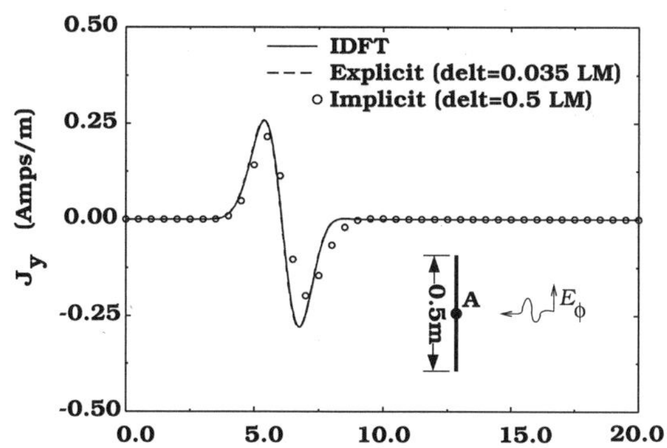

FIGURE 3.8
The induced current at the center (A) of a straight strip illuminated by a TE Gaussian plane wave. The strip is divided into 10 segments.

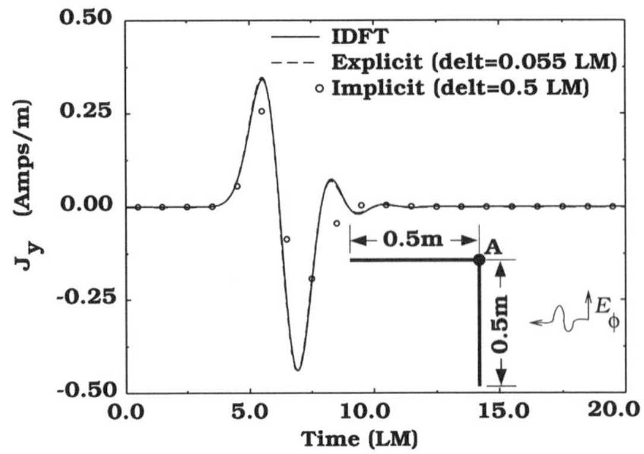

FIGURE 3.9
The induced current at the center (A) of a bent strip illuminated by a TE Gaussian plane wave. The strip is divided into 10 segments.

current induced at the center of the strip and this is compared with the transformed result from the frequency domain. The agreement is good both in the early time and in the late time.

The case of a 90° bent strip located along the x and y axes is considered next. Each strip is 0.5 m long and is divided into 5 uniform segments. Figure 3.9 shows the current at the bend for the three solutions. The time steps for the explicit and implicit solutions are 0.055 and 0.5 LM, respectively. The current is obtained at the corner because, for the TE case, the current around the bend is continuous. Again, all three solutions agree very well.

Next, we consider the circular cylinder with a 0.1 m radius. The contour of the cylinder is divided into 12 linear segments. Figure 3.10 shows the currents at $\phi = 0°$ for both explicit and implicit schemes and these are compared with the frequency domain Fourier transformed solution. The time steps used in this case are 0.034 LM for the explicit solution and 0.5 LM for the implicit solution. The eigenfunction method is used to obtain the frequency domain results. It is evident that the solutions in all three cases agree very well. Also, notice the relatively large time step used for the implicit solution which enables us to compute the time domain signature in much fewer time steps.

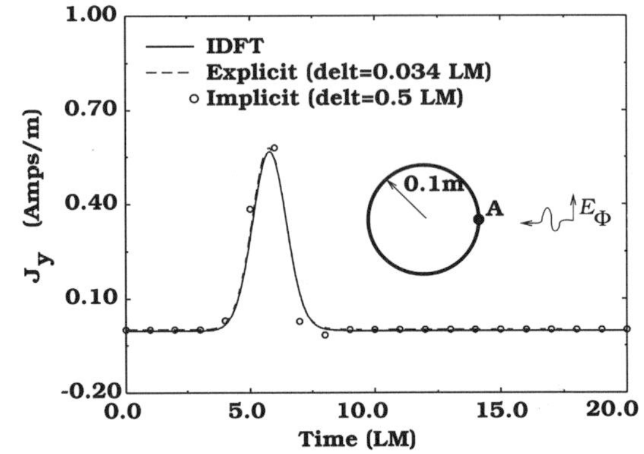

FIGURE 3.10
The induced current at $\phi = 0°$ (A) of a circular cylinder illuminated by a TE Gaussian plane wave. The circle is divided into 12 segments.

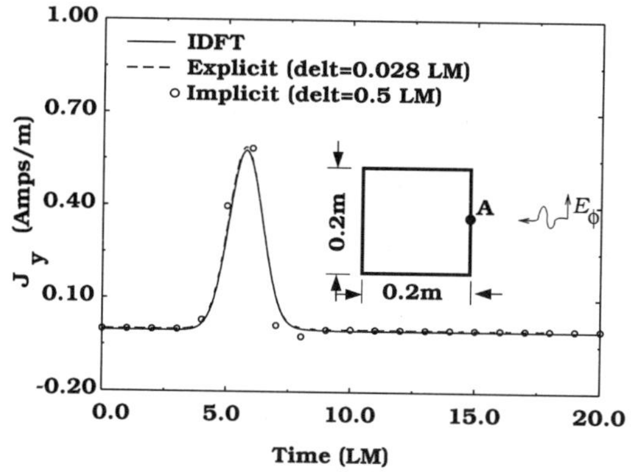

FIGURE 3.11
The induced current at point A of a square cylinder illuminated by a TE Gaussian plane wave. The square is divided into 16 segments.

Finally, we consider a square cylinder with a 0.2-m side length. Each side of the cylinder is divided into four zones thus giving 16 unknowns in the solution scheme. The time steps for explicit and implicit solution for this case are 0.028 and 0.5 LM, respectively. The current at the center of the illuminated side is shown in Fig. 3.11 and is compared with that of the IDFT solution. Again, the agreement is excellent in all solutions.

3.5 TE INCIDENCE: HFIE FORMULATION

In this section, we consider the solution of the H-field integral equation given by Eq. (3.1.7). This solution is useful when dealing with dielectric scatterers. Again, for the MoM solution, we use the same set of basis functions as defined in the EFIE case. Also note that this equation is valid only for closed contour scatterers. First, we consider the explicit scheme.

3.5.1 Explicit Solution Procedure

Combining Eq. (3.1.7) and Eq. (3.1.8) and applying the testing procedure results in the following integral equation:

$$\langle f_m \boldsymbol{a}_\tau, \boldsymbol{J} \rangle = \left\langle f_m \boldsymbol{a}_\tau, \boldsymbol{a}_n \times \left[\boldsymbol{H}^{\mathrm{i}} + \frac{1}{\mu} \nabla \times \boldsymbol{A} \right] \right\rangle. \tag{3.5.1}$$

Using Eq. (3.1.3) and extracting the Cauchy principal value from the curl term, we may rewrite the curl of the vector potential as

$$\frac{1}{\mu} \nabla \times \boldsymbol{A}(\boldsymbol{\rho}, t_n) = \frac{J_m(t_n)}{2} \boldsymbol{a}_z + \frac{1}{\mu} \nabla \times \boldsymbol{A\!\!\!/}(\boldsymbol{\rho}, t_n). \tag{3.5.2}$$

By taking the curl operator inside the integral, which is allowed because $R \neq 0$, and using the vector identity $\nabla \times (w\boldsymbol{A}) = w\nabla \times \boldsymbol{A} - \boldsymbol{A} \times \nabla w$, gives

$$\frac{1}{\mu} \nabla \times \boldsymbol{A\!\!\!/}(\boldsymbol{\rho}, t_n) = \frac{1}{4\pi} \int\!\!\int_S \frac{\nabla \times \boldsymbol{J}(\boldsymbol{\rho}', t_n - R/c)}{R} - \boldsymbol{J}(\boldsymbol{\rho}', t_n - R/c) \times \nabla \frac{1}{R}\, ds'. \tag{3.5.3}$$

Note that

$$\nabla \frac{1}{R} = -\frac{\boldsymbol{\rho} - \boldsymbol{r}'}{R^3} = -\frac{\boldsymbol{a}_R}{R^2}, \tag{3.5.4}$$

where $\boldsymbol{a}_R$ is a unit vector in the direction of $\boldsymbol{\rho} - \boldsymbol{r}'$. Letting $t_r = t_n - R/c$ results in

$$\begin{aligned} \nabla \times \boldsymbol{J}(\boldsymbol{\rho}', t_n - R/c) &= \nabla \times J(\boldsymbol{\rho}', t_r)\boldsymbol{a}'_\tau \\ &= [J(\boldsymbol{\rho}', t_r)\nabla \times \boldsymbol{a}'_\tau - \boldsymbol{a}'_\tau \times \nabla J(\boldsymbol{\rho}', t_r)]. \end{aligned} \tag{3.5.5}$$

Note that the ∇ operator is operating on unprimed coordinates, which implies $\nabla \times \boldsymbol{a}'_\tau = 0$. Using the vector identity $\nabla f(c(t_r)) = (\partial f/\partial c(t_r)) \cdot \nabla t_r$, we get

$$\nabla \times \boldsymbol{J}\left(\boldsymbol{\rho}', t_n - \frac{R}{c}\right) = -\boldsymbol{a}'_\tau \times \frac{\partial J}{\partial t_r} \cdot \nabla t_r$$

$$\nabla t_r = \nabla\left(t_n - \frac{R}{c}\right) = \nabla t_n - \nabla \frac{R}{c} = -\frac{1}{c}\nabla R = -\frac{1}{c}\boldsymbol{a}_R.$$

Thus,

$$\nabla \times \boldsymbol{J}\left(\boldsymbol{\rho}', t_n - \frac{R}{c}\right) = \frac{1}{c}\left(\frac{\partial J}{\partial t_r}\boldsymbol{a}'_\tau \times \boldsymbol{a}_R\right). \tag{3.5.6}$$

Combining Eqs. (3.5.3)–(3.5.6) results in

$$\frac{1}{\mu}\nabla \times \boldsymbol{A}(\boldsymbol{\rho}, t_n) = \frac{1}{4\pi}\iint_S \frac{1}{Rc}\frac{\partial J(\boldsymbol{\rho}', t_r)}{\partial t_r}\boldsymbol{a}'_\tau \times \boldsymbol{a}_R + \frac{J(\boldsymbol{\rho}', t_n - R/c)}{R^2}\boldsymbol{a}'_\tau \times \boldsymbol{a}_R\, ds'. \tag{3.5.7}$$

By substituting Eq. (3.4.8) into Eq. (3.5.7), we obtain

$$\frac{1}{\mu}\nabla \times \boldsymbol{A}(\boldsymbol{\rho}, t_n) = \sum_{\substack{k=1 \\ k\neq m \text{ and } \ell\neq 0}}^{N} \sum_{\ell=-\infty}^{\infty} \frac{\partial J_k(t_r)}{\partial t_r} I_p + \sum_{\substack{k=1 \\ k\neq m \text{ and } \ell\neq 0}}^{N} \sum_{\ell=-\infty}^{\infty} J_k\left(t_n - \frac{R_{mk\ell}}{c}\right) I_q, \tag{3.5.8}$$

where

$$I_p = \frac{1}{4\pi c}\int_{k,\ell}\int_{\text{patch}} \frac{\boldsymbol{a}'_\tau \times \boldsymbol{a}_R}{R}\, ds' \tag{3.5.9}$$

and

$$I_q = \frac{1}{4\pi}\int_{k,\ell}\int_{\text{patch}} \frac{\boldsymbol{a}'_\tau \times \boldsymbol{a}_R}{R^2}\, ds'. \tag{3.5.10}$$

As before, the non-self term integrals may be approximated by the patch area times the integrand evaluated at the patch center. The $\partial J_k/\partial t_r$ term is approximated by a first-order backward difference given by

$$\frac{\partial J_k(t_r)}{\partial t_r} \approx \frac{J_k(t_n - R_{mk\ell}/c) - J_k(t_{n-1} - R_{mk\ell}/c)}{\Delta t}.$$

By combining Eqs. (3.5.1) and (3.5.2), collecting terms involving n on the left and those prior to n on the right, and using a 1-point integration for the testing procedure, the following time marching equation is obtained:

$$\frac{\Delta\tau_m}{2} I_m(t_n) = \left\langle f_m \boldsymbol{a}_\tau, \boldsymbol{a}_n \times \left[\boldsymbol{H}^{\mathrm{i}} + \frac{1}{\mu}\nabla \times \mathcal{A}(\boldsymbol{\rho}, t_n)\right]\right\rangle. \tag{3.5.11}$$

The testing on the right-hand side may be performed in a manner similar to that of Eq. (3.4.5). The right side may also be expanded by using Eq. (3.5.8).

Finally, the explicit solution involves choosing Δt less than $R_{\min}/c$ as in the EFIE case. Similar to the EFIE case, this solution does not involve a matrix storage or inversion. However, the HFIE explicit solution also becomes unstable at late times but can be controlled using the averaging procedure described in Chapter 1, Section 1.6.1.

3.5.2 Implicit Solution Procedure

For the implicit solution, we first define a much larger value for the time step. Then, following the numerical procedure described for the explicit solution, we develop a matrix equation given by

$$[\alpha][\boldsymbol{I}(t_n)] = [\boldsymbol{F}(t_n)] + [\beta][\boldsymbol{I}(t_R)]. \tag{3.5.12}$$

As in the EFIE case, the $[\alpha]$ matrix is a sparse matrix. The elements of the $[\alpha]$ matrix are given by

$$\alpha_{mk} = \begin{cases} 0.5 + \left[1 - \frac{R_{mkl}}{c\Delta t}\right]\left[\frac{I_p}{\Delta t} + I_q\right] & \text{for } c\Delta t \geq R_{mkl} \\ 0 & \text{otherwise.} \end{cases} \tag{3.5.13}$$

In Eq. (3.5.13), R_{mkl} is given by Eq. (3.3.10) and I_p and I_q are defined in Eq. (3.5.9) and Eq. (3.5.10), respectively. These elements are obtained by collecting the terms which satisfy the condition $R_{mkl}/c \leq \Delta t$ in Eq. (3.5.8). The computation of the $[\beta]$ matrix is trivial and this is obtained from Eq. (3.5.8). Furthermore, the remarks made for the implicit version of the EFIE solution regarding the time step and stability are also applicable for the HFIE implicit solution.

3.5.3 Numerical Examples

In this section, we consider the numerical results obtained by the H-field solution. Note that only closed cylinders can be solved using this method. Thus, we consider only two examples: a circular cylinder and a square cylinder. The results are shown in Figs. 3.12 and 3.13, respectively.

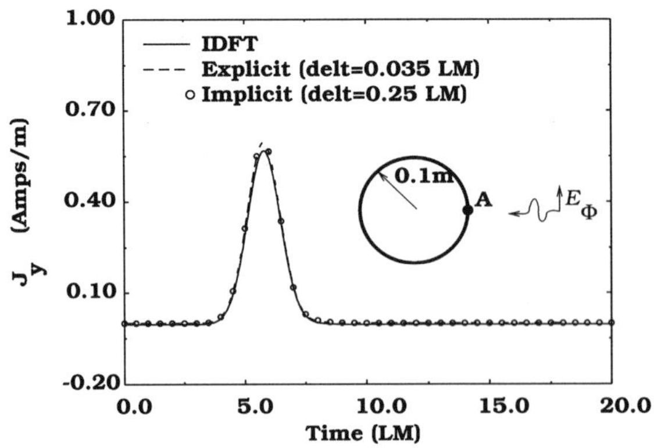

FIGURE 3.12
The induced current (HFIE solution) at $\phi = 0°$ (A) of a circular cylinder illuminated by a TE Gaussian plane wave. The circle is divided into 12 segments.

The time steps used for each solution are shown in the insets of the figures. Suffice it to say that the agreement is excellent between each method.

3.6 CONCLUDING REMARKS

In this chapter, we developed the numerical algorithms for infinitely long conducting cylinders by solving time domain integral equations. Admittedly, applications for this type of problems are few and include periodic arrays of infinite elements, long structures in which one can ignore the reflections from the ends, and structures with cross-sectional dimensions much smaller than axial length. However, the numerical procedures developed in this chapter form the basis for more practical, and more involved, three-dimensional structures of finite size and arbitrary shape. All the known structures are of finite size, and the modeling of these structures in an accurate and

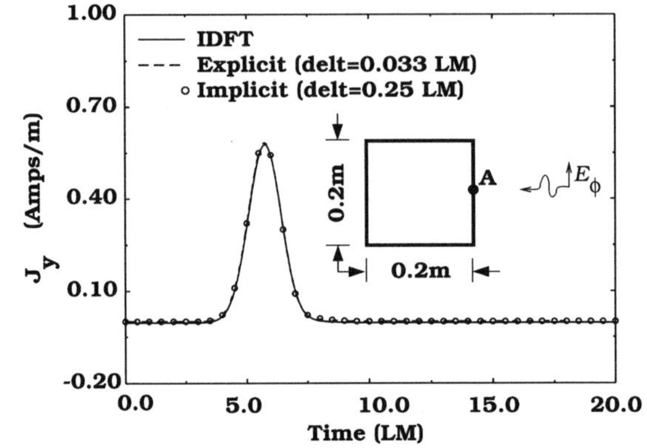

FIGURE 3.13
The induced current (HFIE solution) at point A of a square cylinder illuminated by a TE Gaussian plane wave. The square is divided into 16 segments.

efficient manner is always considered an important field of research. We shall consider these algorithms in Chapter 4.

BIBLIOGRAPHY

[1] J. A. Landt, E. K. Miller, and M. Van Blaricum, *WT-MBA/LLL1B: A Computer Program for the Time Domain Electromagnetic Response of Thin-Wire Structures*, Lawrence Livermore Laboratory Report UCRL-51585, 1974.

[2] R. F. Harrington, *Field Computation by Moment Methods*, Macmillan, New York, 1968.

[3] N. J. Damaskos, R. T. Brown, J. R. Jameson, and P. L. R. Uslenghi, "Transient Scattering by Resistive Cylinders," *IEEE Trans. Antennas Propagat.*, vol. 33, pp. 21–25, 1985.

[4] D. R. Wilton, S. M. Rao, A. W. Glisson, D. H. Schaubert, O. M. Al-Bundak, and C. M. Butler, "Potential Integrals for Uniform and Linear Source Distributions on Polygonal and Polyhedral Domains," *IEEE Trans. Antennas Propagat.*, vol. 32, pp. 276–281, 1984.

[5] D. A. Vechinski, *Direct Time Domain Analysis of Arbitrarily Shaped Conducting or Dielectric Structures Using Patch Modeling Techniques*, PhD dissertation, Auburn University, Auburn, AL, 1992.

[6] R. F. Harrington, *Time Harmonic Electromagnetic Fields*, McGraw-Hill, New York, 1961.

CHAPTER 4

Finite Conducting Bodies: TDIE Solution

S. M. RAO
Department of Electrical Engineering, Auburn University

D. A. VECHINSKI
Nichols Research Corporation

In this chapter, we discuss the numerical procedures to calculate the transient scattering from perfectly conducting, arbitrarily shaped, three-dimensional, finite-sized bodies. There are many practical applications to treat such bodies. First, many systems are employing short-pulse radar systems. These systems are being used for target identification, high range resolution, and wide-band digital communications. In all these situations, the structures involved are finite-sized bodies. Second, numerous military and space applications often involve the calculation of radar signatures of complex bodies over a wide frequency range which can be easily accomplished by the techniques discussed in this chapter. We remark that, although in this work we present numerical examples for standard canonical shapes only, the techniques are general and can be used for a wide variety of geometrical shapes. We first develop the basic integral equations to obtain the transient scattered field.

4.1 INTEGRAL EQUATION FORMULATION

Let S denote the surface of a closed or open perfectly conducting body illuminated by a transient electromagnetic pulse as shown in Fig. 4.1. This pulse induces a surface current, $\boldsymbol{J}(\boldsymbol{r}, t)$, on S which then reradiates a scattered field. If S is open, $\boldsymbol{J}(\boldsymbol{r}, t)$ is regarded as the sum of the currents on opposite sides of the surface. The boundary conditions require that the total tangential electric field on the conducting surface be zero or

$$[\boldsymbol{E}^{s}(\boldsymbol{J}) + \boldsymbol{E}^{i}]_{\tan} = 0 \quad \text{on } S, \tag{4.1.1}$$

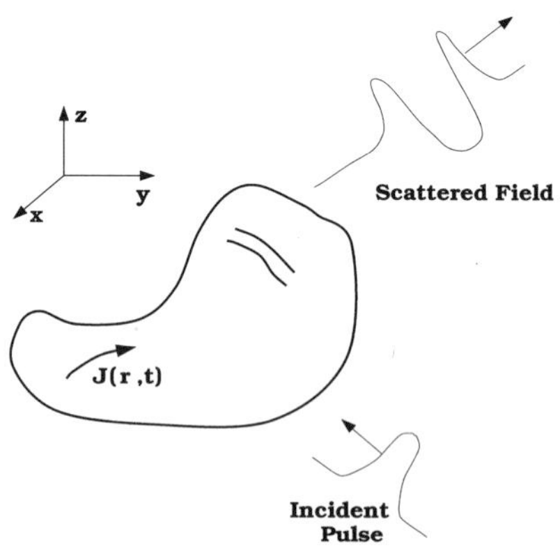

FIGURE 4.1
Transient pulse incident upon an arbitrarily shaped body.

where $\boldsymbol{E}^{\mathrm{s}}(\boldsymbol{J})$ and $\boldsymbol{E}^{\mathrm{i}}$ represent the scattered electric field due to the induced current $\boldsymbol{J}$ and the incident electric field, respectively. The scattered field radiated by the current $\boldsymbol{J}$ may be written in terms of the magnetic vector and electric scalar potentials as

$$\boldsymbol{E}^{\mathrm{s}}(\boldsymbol{J}) = -\frac{\partial \boldsymbol{A}}{\partial t} - \nabla\Phi, \tag{4.1.2}$$

where

$$\boldsymbol{A}(\boldsymbol{r}, t) = \mu \int_S \frac{\boldsymbol{J}(\boldsymbol{r}', t - R/c)}{4\pi R}\, dS', \tag{4.1.3}$$

$$\Phi(\boldsymbol{r}, t) = \frac{1}{\epsilon} \int_S \frac{q_{\mathrm{s}}(\boldsymbol{r}', t - R/c)}{4\pi R}\, dS', \tag{4.1.4}$$

and $R = |\boldsymbol{r} - \boldsymbol{r}'|$. In Eqs. (4.1.1)–(4.1.4), μ and ϵ are the permeability and permittivity of the surrounding medium, c is the velocity of propagation of the electromagnetic wave, and $\boldsymbol{r}$ and $\boldsymbol{r}'$ are the arbitrarily located observation point and source point on the scatterer, respectively. Also, note that the incident field, $\boldsymbol{E}^{\mathrm{i}}$, is defined in the absence of the scatterer, as in the two-dimensional case.

Upon using the continuity equation, given by

$$\nabla \cdot \boldsymbol{J} = -\frac{\partial q_{\mathrm{s}}}{\partial t} \Rightarrow q_{\mathrm{s}} = -\int_{\tau=0}^{t} \nabla \cdot \boldsymbol{J}\, d\tau, \tag{4.1.5}$$

we may rewrite the scalar potential term as

$$\Phi(\boldsymbol{r}, t) = \frac{-1}{\epsilon} \int_S \int_{\tau=0}^{t-R/c} \frac{\nabla \cdot \boldsymbol{J}(\boldsymbol{r}', \tau)}{4\pi R}\, dS'\, d\tau. \tag{4.1.6}$$

Using Eqs. (4.1.1) and (4.1.2), we may write the time domain electric field integral equation as

$$\left[\frac{\partial \boldsymbol{A}(\boldsymbol{r}, t)}{\partial t} + \nabla\Phi(\boldsymbol{r}, t)\right]_{\tan} = \boldsymbol{E}^{\mathrm{i}}_{\tan}(\boldsymbol{r}, t). \tag{4.1.7}$$

An alternate form of the TDIE may also be obtained by differentiating Eq. (4.1.7) with respect to time, given by

$$\left[\frac{\partial^2 \boldsymbol{A}(\boldsymbol{r},t)}{\partial t^2} + \nabla\Psi(\boldsymbol{r},t)\right]_{\tan} = \left[\frac{\partial \boldsymbol{E}^{\mathrm{i}}(\boldsymbol{r},t)}{\partial t}\right]_{\tan}, \tag{4.1.8}$$

where

$$\Psi(\boldsymbol{r},t) = \frac{-1}{\epsilon}\int_S \frac{\nabla \cdot \boldsymbol{J}(\boldsymbol{r}',t-R/c)}{4\pi R}\,dS'. \tag{4.1.9}$$

Note that both versions, i.e., Eqs. (4.1.7) and (4.1.8), are popular and have been used in the literature [1–3].

Also, Equation (4.1.7) is similar to the transverse electric (TE) cylinder integral equation derived in Chapter 3, Section 3.4. Hence, the solution follows along similar lines but is also quite different because of the finite size of the scattering object. Lastly, we present the analysis of Eq. (4.1.7), which seems to have certain advantages in numerical calculations as discussed in the following sections.

4.2 NUMERICAL SOLUTION SCHEME

The first step in the numerical scheme is to adequately describe the geometry involved to the digital computer. This task is most easily accomplished by covering the body surface with planar triangles to generate a "patch model" of the actual body. We choose the planar triangular patches to model the body because they have the ability to conform to any geometrical surface or boundary. In fact, simple and complex bodies can be easily modeled by planar triangular patches and can be described to the computer using automated schemes. Figure 4.2 shows several bodies modeled by triangular patches. Furthermore, for numerical purposes, it is very easy to increase the patch density in areas where more resolution is required.

The next task in the numerical solution procedure is to develop an algorithm to solve the integral equation (Eq. 4.1.7). We accomplish this task by selecting the well-known method of moments (MoM) [4] and extending the numerical procedures developed in Chapters 2 and 3 to

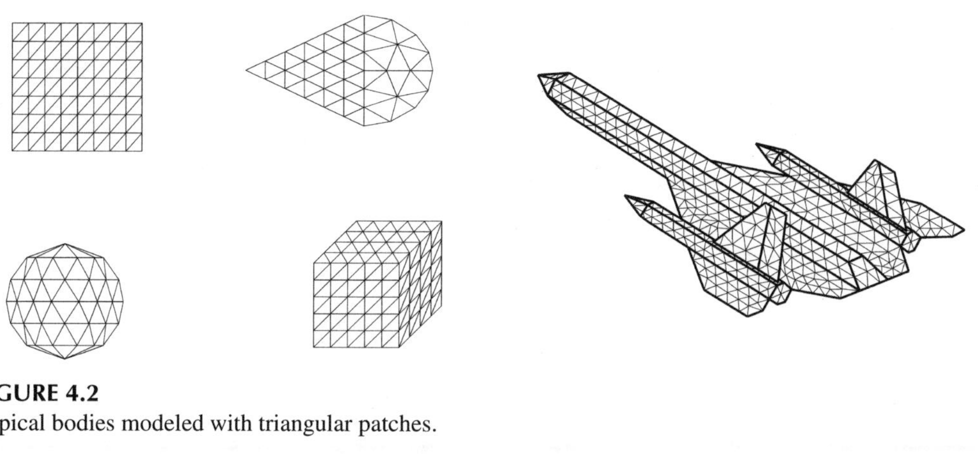

FIGURE 4.2
Typical bodies modeled with triangular patches.

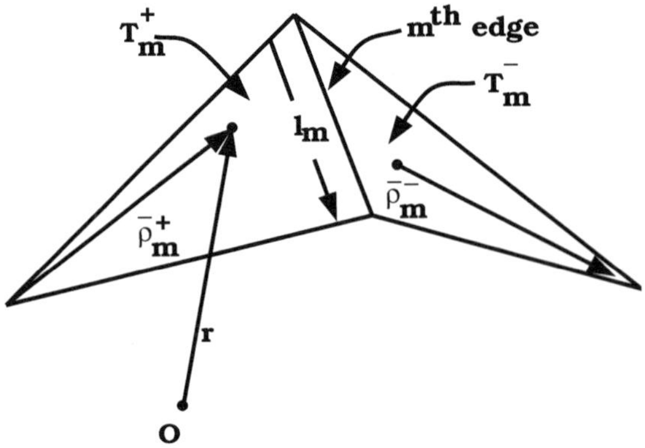

FIGURE 4.3
Triangle pair and geometrical parameters associated with the mth interior edge.

obtain an accurate and efficient numerical method. For the MoM solution, the first step is to define a set of basis functions. Assuming that the body is modeled with triangular patches, we define the basis function for any edge m common to the two triangles $T_m^{\pm}$ (referring to Fig. 4.3) given by

$$\boldsymbol{f}_m(\boldsymbol{r}) = \begin{cases} \frac{l_m}{2A_m^+}\boldsymbol{\rho}_m^+ & \text{for } \boldsymbol{r} \in T_m^+ \\ \frac{l_m}{2A_m^-}\boldsymbol{\rho}_m^- & \text{for } \boldsymbol{r} \in T_m^- \\ 0 & \text{otherwise.} \end{cases} \tag{4.2.1}$$

In Eq. (4.2.1), l_m denotes the length of the mth edge and $A_m^{\pm}$ is the area of triangle $T_m^{\pm}$. Furthermore, an arbitrary point in T_m^+ may be located by the position vector $\boldsymbol{r}$, relative to the origin, O, or by $\boldsymbol{\rho}_m^+$, referenced at the free vertex of T_m^+. For an arbitrary point in T_m^-, the position vector $\boldsymbol{\rho}_m^-$ is similarly defined except that it is directed toward the free vertex of T_m^-. The "+" or "−" convention is determined by choosing a reference direction for positive current flow for the mth edge. This current is assumed to flow from T_m^+ to T_m^-. Also, we follow the convention in which superscripts refer to the faces and subscripts refer to the edges. For example, T_m^+ is the positive triangle associated with edge m.

Some advantages of expressing the current in terms of the $\boldsymbol{f}_m$'s have been noted [5]. For example, along the outer boundary of the triangle pair, the current has no component normal to the boundary so no line charges exist along the boundary. The current normal to the mth edge, i.e., the edge common to T_m^+ and T_m^-, is constant and continuous across the edge. These two results imply that all the edges of T_m^+ and T_m^- are free of line charges. This is an important consideration in developing an efficient and robust numerical procedure since the presence of such line charges involves extra unknowns and cumbersome evaluation of singular integrals. Lastly, the surface divergence of the basis functions is given by

$$\nabla_s \cdot \boldsymbol{f}_m(\boldsymbol{r}) = \begin{cases} \frac{l_m}{A_m^+} & \text{for } \boldsymbol{r} \in T_m^+ \\ -\frac{l_m}{A_m^-} & \text{for } \boldsymbol{r} \in T_m^- \\ 0 & \text{otherwise.} \end{cases} \tag{4.2.2}$$

The surface divergence is proportional to the charge density. Therefore, the charge density is constant on each triangle, and the total charge associated with each triangle pair is zero.

A suitable testing procedure is also needed for the MoM solution. In this work, we choose the expansion functions $\boldsymbol{f}_m$, presented in Eq. (4.2.1), as testing functions. The inner product is chosen as

$$\langle \boldsymbol{a}, \boldsymbol{b} \rangle = \int_S \boldsymbol{a} \cdot \boldsymbol{b} \, ds, \tag{4.2.3}$$

where $\boldsymbol{a}$ and $\boldsymbol{b}$ are real-valued vector or scalar functions. For scalar functions, the "dot" may be simply interpreted as the multiplication sign.

Once we define the suitable basis and testing functions for the moment method solution, we divide the time axis into equal intervals of Δt and refer $t_j = j\Delta t$. Note that it is not necessary to divide the time axis into equal intervals.

Now, we have all the basic ingredients for the numerical solution using the MoM. In the following subsections, we develop two algorithms—the explicit method and the implicit method—and discuss their relative advantages and disadvantages.

4.2.1 Explicit Numerical Method

We begin by approximating the time derivative in Eq. (4.1.7) with a forward-finite difference scheme to obtain

$$\left[\frac{\boldsymbol{A}(\boldsymbol{r}, t_{j+1}) - \boldsymbol{A}(\boldsymbol{r}, t_j)}{\Delta t} + \nabla\Phi(\boldsymbol{r}, t_j)\right]_{\tan} = [\boldsymbol{E}^{\mathrm{i}}(\boldsymbol{r}, t_j)]_{\tan}, \tag{4.2.4}$$

which may be rewritten as

$$[\boldsymbol{A}(\boldsymbol{r}, t_{j+1})]_{\tan} = [(\Delta t)\boldsymbol{E}^{\mathrm{i}}(\boldsymbol{r}, t_j) - (\Delta t)\nabla\Phi(\boldsymbol{r}, t_j) + \boldsymbol{A}(\boldsymbol{r}, t_j)]_{\tan}. \tag{4.2.5}$$

By applying the testing procedure to Eq. (4.2.5) and recognizing that the current testing scheme extracts the tangential component, we have

$$\langle \boldsymbol{f}_m, \boldsymbol{A}(\boldsymbol{r}, t_{j+1})\rangle = (\Delta t)\langle \boldsymbol{f}_m, \boldsymbol{E}^{\mathrm{i}}(\boldsymbol{r}, t_j)\rangle - (\Delta t)\langle \boldsymbol{f}_m, \nabla\Phi(\boldsymbol{r}, t_j)\rangle + \langle \boldsymbol{f}_m, \boldsymbol{A}(\boldsymbol{r}, t_j)\rangle. \tag{4.2.6}$$

Next, we consider the evaluation of each inner-product term.

First, we consider the testing procedure with the vector potential term. Using Eq. (4.2.1), we can write $\langle \boldsymbol{f}_m, \boldsymbol{A}\rangle$ at any time instant as

$$\langle \boldsymbol{f}_m, \boldsymbol{A}\rangle = \int_{T_m^+} \frac{l_m}{2A_m^+} \boldsymbol{\rho}_m^+ \cdot \boldsymbol{A} \, dS + \int_{T_m^-} \frac{l_m}{2A_m^-} \boldsymbol{\rho}_m^- \cdot \boldsymbol{A} \, dS. \tag{4.2.7}$$

The integrals will be approximated by evaluating $\boldsymbol{A}$ at the center of the mth edge. Therefore,

$$\langle \boldsymbol{f}_m, \boldsymbol{A}\rangle \approx \boldsymbol{A}(\boldsymbol{r}_m) \cdot \left[\frac{l_m}{2A_m^+}\int_{T_m^+} \boldsymbol{\rho}_m^+ \, dS + \frac{l_m}{2A_m^-}\int_{T_m^-} \boldsymbol{\rho}_m^- \, dS\right]. \tag{4.2.8}$$

The integrations in Eq. (4.2.8) are trivial and the result is given by

$$\langle \boldsymbol{f}_m, \boldsymbol{A}\rangle \approx \boldsymbol{A}(\boldsymbol{r}_m) \cdot \frac{l_m}{2}\left(\boldsymbol{\rho}_m^{c+} + \boldsymbol{\rho}_m^{c-}\right), \tag{4.2.9}$$

where $\boldsymbol{\rho}_m^{c+}$ is the vector from the free vertex to the centroid of T_m^+ and $\boldsymbol{\rho}_m^{c-}$ is the vector from the centroid to the free vertex of T_m^-.

Next, we consider the testing of the scalar potential term. Using the vector identity $\nabla \cdot (\Phi \boldsymbol{A}) = \boldsymbol{A} \cdot \nabla\Phi + \Phi\nabla \cdot \boldsymbol{A}$ and using the properties of the basis function $\boldsymbol{f}_m$, we have

$$\langle \boldsymbol{f}_m, \nabla\Phi\rangle = -\int_S \Phi\nabla \cdot \boldsymbol{f}_m \, dS. \tag{4.2.10}$$

Using Eq. (4.2.2) and approximating the integral by evaluating it at the centroids of the triangles, Eq. (4.2.10) becomes

$$\begin{aligned} -\int_S \Phi\nabla \cdot \boldsymbol{f}_m \, dS &= -\left[\frac{l_m}{A_m^+}\int_{T_m^+} \Phi \, dS - \frac{l_m}{A_m^-}\int_{T_m^-} \Phi \, dS\right], \\ &\approx -l_m\left[\Phi\left(\boldsymbol{r}_m^{c+}\right) - \Phi\left(\boldsymbol{r}_m^{c-}\right)\right] \end{aligned} \tag{4.2.11}$$

where $\boldsymbol{r}_m^{c\pm}$ are the centroids of triangles $T_m^{\pm}$.

Finally, the testing procedure is applied to the incident field term. As in the vector potential case, we approximate the integral by using the field evaluated at the center of the mth edge. This assumes that the incident field does not vary in time and space over the triangles $T_m^{\pm}$. This gives,

$$\langle \boldsymbol{f}_m, \boldsymbol{E}^{\mathrm{i}}\rangle \approx \frac{l_m}{2}\left(\boldsymbol{\rho}_m^{c+} + \boldsymbol{\rho}_m^{c-}\right) \cdot \boldsymbol{E}^{\mathrm{i}}(\boldsymbol{r}_m, t_j). \tag{4.2.12}$$

Thus, using Eqs. (4.2.9), (4.2.11), and (4.2.12), by replacing $j \Rightarrow j-1$ and rearranging the terms, we can write Eq. (4.2.6) as

$$\begin{aligned} \frac{l_m}{2}\left(\boldsymbol{\rho}_m^{c+} + \boldsymbol{\rho}_m^{c-}\right) \cdot \boldsymbol{A}(\boldsymbol{r}_m, t_j) = {} & \frac{l_m}{2}\left(\boldsymbol{\rho}_m^{c+} + \boldsymbol{\rho}_m^{c-}\right) \cdot (\Delta t)\boldsymbol{E}^{\mathrm{i}}(\boldsymbol{r}_m, t_{j-1}) \\ & + (\Delta t) l_m\left[\Phi\left(\boldsymbol{r}_m^{c+}, t_{j-1}\right) - \Phi\left(\boldsymbol{r}_m^{c-}, t_{j-1}\right)\right] \\ & + \frac{l_m}{2}\left(\boldsymbol{\rho}_m^{c+} + \boldsymbol{\rho}_m^{c-}\right) \cdot \boldsymbol{A}(\boldsymbol{r}_m, t_{j-1}). \end{aligned} \tag{4.2.13}$$

We now consider the expansion procedure. Here, we approximate the current on S by

$$\boldsymbol{J}(\boldsymbol{r}, t) = \sum_{k=1}^{N} I_k(t) \, \boldsymbol{f}_k(\boldsymbol{r}), \tag{4.2.14}$$

where $I_k(t)$ is the unknown current coefficient at edge k and N is the number of nonboundary edges. We define a boundary edge as an edge which is associated with only one triangular patch. Each patch may have up to three nonzero basis functions. However, at a particular edge, k, the basis function associated with that edge, $\boldsymbol{f}_k$, is the only one to have a component of current normal to the edge. The currents associated with the other edges of T_k^+ and T_k^- are parallel to the kth edge, and therefore their contribution to the normal component at edge k is zero. Thus, the current coefficient, $I_k(t)$, may be interpreted as the current density flowing normal to the kth edge at time t since the normal component of $\boldsymbol{f}_k$ at the kth edge is 1. The boundary edges may be omitted from Eq. (4.2.14) because the current normal to a surface boundary is required to be zero.

We now discuss the determination of the vector potential at some observation point $\boldsymbol{r}$ associated with the mth edge at time $t = t_j$. Substituting Eq. (4.2.14) into Eq. (4.1.3) gives

$$\boldsymbol{A}(\boldsymbol{r}, t_j) = \mu \int_S \frac{\sum_{k=1}^{N} I_k(t_j - R/c)\, \boldsymbol{f}_k(\boldsymbol{r}')}{4\pi R}\, dS' \approx \sum_{k=1}^{N} I_k(t_j - R_{mk}/c)\boldsymbol{\kappa}_{mk} \tag{4.2.15}$$

with

$$\boldsymbol{\kappa}_{mk} = \int_S \frac{\mu \boldsymbol{f}_k}{4\pi R_m}\, dS', \tag{4.2.16}$$

$$R_{mk} = |\boldsymbol{r}_m - \boldsymbol{r}_k|, \tag{4.2.17}$$

$$R_m = |\boldsymbol{r}_m - \boldsymbol{r}'|, \tag{4.2.18}$$

where $\boldsymbol{r}_m$ and $\boldsymbol{r}_k$ are the position vectors to the center of the mth and kth edges, respectively. In obtaining Eq. (4.2.15) we removed a cumbersome integral on space–time by replacing $I_k(t_j - R/c) \Rightarrow I_k(t_j - R_{mk}/c)$. This makes the assumption that the current does not vary appreciably with time within the pair of triangles.

Next, we notice that for the self term, i.e., when $k = m$, we have $R_{mk} = 0$. Separating this term from Eq. (4.2.15), we have

$$\boldsymbol{A}(\boldsymbol{r}, t_j) = \boldsymbol{\kappa}_{mm} I_m(t_j) + \not{\boldsymbol{A}}(\boldsymbol{r}, t_j), \tag{4.2.19}$$

where $\not{\boldsymbol{A}}(\boldsymbol{r}, t_j)$ represents $\boldsymbol{A}(\boldsymbol{r}, t_j)$ with the self term deleted.

Now, we consider the evaluation of the scalar potential at some observation point $\boldsymbol{r}$ associated with the mth edge and time $t = t_j$. Following steps similar to the evaluation of the vector potential, we combine Eqs. (4.1.4) and (4.2.14) to get

$$\begin{aligned}
\Phi(\boldsymbol{r}, t_j) &= -\int_S \int_{\tau=0}^{t_j - R/c} \frac{\sum_{k=1}^{N} I_k(\tau)\nabla' \cdot \boldsymbol{f}_k}{4\pi\epsilon R}\, dS'\, d\tau' \\
&\approx -\sum_{k=1}^{N} \int_{\tau=0}^{t_R^+} I_k(\tau)\, d\tau \int_{T_k^+} \frac{\nabla' \cdot \boldsymbol{f}_k}{4\pi\epsilon R_m^+}\, dS' - \sum_{k=1}^{N} \int_{\tau=0}^{t_R^-} I_k(\tau)\, d\tau \int_{T_k^-} \frac{\nabla' \cdot \boldsymbol{f}_k}{4\pi\epsilon R_m^-}\, dS' \\
&\approx \sum_{k=1}^{N} \left[\int_{\tau=0}^{t_R^+} I_k(\tau)\, d\tau\, \phi_{mk}^+ + \int_{\tau=0}^{t_R^-} I_k(\tau)\, d\tau\, \phi_{mk}^- \right],
\end{aligned} \tag{4.2.20}$$

where

$$\phi_{mk}^{\pm} = \frac{-l_k}{A_k^{\pm}} \int_{T_k^{\pm}} \frac{dS'}{4\pi\epsilon R_m^{\pm}}, \tag{4.2.21}$$

$$R_{mk}^{\pm} = \left|\boldsymbol{r}_m - \boldsymbol{r}_k^{c\pm}\right|, \tag{4.2.22}$$

$$R_m^{\pm} = |\boldsymbol{r}_m - \boldsymbol{r}'|, \tag{4.2.23}$$

$$t_R^{\pm} = t_j - \frac{R_{mk}^{\pm}}{c}. \tag{4.2.24}$$

The Φ's may then be evaluated at the centroids of the triangles by replacing $\boldsymbol{r}$ with $\boldsymbol{r}_m^{c+}$ or $\boldsymbol{r}_m^{c-}$.

Here, we have evaluated all the terms in the Eq. (4.2.13). Upon careful examination of Eq. (4.2.13), there is only one term, i.e., the vector potential term $\boldsymbol{A}(r_m, t_j)$, which needs to be calculated at $t = t_j$, where as all other terms are evaluated at $t = t_{j-1}$, which we assume to be known. Furthermore, choosing $\Delta t < R_{mk}/c$ for all values of m and k ($m \neq k$), we can write an explicit solution for the TDIE, using Eq. (4.2.19), as

$$\left[\kappa_{mm} \cdot \frac{l_m}{2}\left(\rho_m^{c+} + \rho_m^{c-}\right)\right] I_m(t_j) = \frac{l_m}{2}\left(\rho_m^{c+} + \rho_m^{c-}\right) \cdot (\Delta t)\boldsymbol{E}^{\mathrm{i}}(\boldsymbol{r}_m, t_{j-1}) - \frac{l_m}{2}\left(\rho_m^{c+} + \rho_m^{c-}\right) \cdot [\not{\boldsymbol{A}}(\boldsymbol{r}_m, t_j) - \boldsymbol{A}(\boldsymbol{r}_m, t_{j-1})] + l_m(\Delta t)\left[\Phi\left(\boldsymbol{r}_m^{c+}, t_{j-1}\right) - \Phi\left(\boldsymbol{r}_m^{c-}, t_{j-1}\right)\right]. \tag{4.2.25}$$

Equation (4.2.25) is a recursion formula relating present time (unknown) currents in terms of retarded (known) currents. The currents for $m = 1, 2, \cdots, N$ may now be found at $t = t_j$. Then, we move forward a time step and reevaluate Eq. (4.2.25).

Lastly, we can write Eq. (4.2.25) in matrix form as

$$[\alpha][\boldsymbol{I}(t_j)] = [\boldsymbol{F}(t_j)] + [\beta][\boldsymbol{I}(t_R)], \tag{4.2.26}$$

where t_R represents the retarded time less than or equal to t_{j-1}. Also, note that the $[\alpha]$ matrix is a diagonal matrix.

4.2.2 Implicit Numerical Method

Similar to the case of implicit solution for the TE cylinder, we begin by approximating the time derivative in Eq. (4.1.7) by a backward-finite difference scheme to obtain

$$\left[\frac{\boldsymbol{A}(\boldsymbol{r}, t_j) - \boldsymbol{A}(\boldsymbol{r}, t_{j-1})}{\Delta t} + \nabla\Phi(\boldsymbol{r}, t_j)\right]_{\mathrm{tan}} = [\boldsymbol{E}^{\mathrm{i}}(\boldsymbol{r}, t_j)]_{\mathrm{tan}}, \tag{4.2.27}$$

which may be rewritten as

$$[\boldsymbol{A}(\boldsymbol{r}, t_j) + (\Delta t)\nabla\Phi(\boldsymbol{r}, t_j)]_{\mathrm{tan}} = [(\Delta t)\boldsymbol{E}^{\mathrm{i}}(\boldsymbol{r}, t_j) + \boldsymbol{A}(\boldsymbol{r}, t_{j-1})]_{\mathrm{tan}}. \tag{4.2.28}$$

Applying the testing procedure to Eq. (4.2.28), we get

$$\begin{aligned}\langle \boldsymbol{f}_m, \boldsymbol{A}(\boldsymbol{r}, t_j)\rangle &+ \langle \boldsymbol{f}_m, (\Delta t)\nabla\Phi(\boldsymbol{r}, t_j)\rangle \\ &= \langle \boldsymbol{f}_m, (\Delta t)\boldsymbol{E}^{\mathrm{i}}(\boldsymbol{r}, t_j)\rangle + \langle \boldsymbol{f}_m, \boldsymbol{A}(\boldsymbol{r}, t_{j-1})\rangle,\end{aligned} \tag{4.2.29}$$

which may be rewritten as

$$\begin{aligned}\langle \boldsymbol{f}_m, \boldsymbol{A}(\boldsymbol{r}, t_j)\rangle &- \langle \nabla \cdot \boldsymbol{f}_m, (\Delta t)\Phi(\boldsymbol{r}, t_j)\rangle \\ &= \langle \boldsymbol{f}_m, (\Delta t)\boldsymbol{E}^{\mathrm{i}}(\boldsymbol{r}, t_j)\rangle + \langle \boldsymbol{f}_m, \boldsymbol{A}(\boldsymbol{r}, t_{j-1})\rangle.\end{aligned} \tag{4.2.30}$$

In the following, we consider the evaluation of each inner-product term similar to the case of the explicit solution. First, we consider the testing procedure with the vector potential term.

Using Eq. (4.2.1), we can write $\langle \boldsymbol{f}_m, \boldsymbol{A} \rangle$, at any time instant, as

$$\langle \boldsymbol{f}_m, \boldsymbol{A} \rangle = \int_{T_m^+} \frac{l_m}{2A_m^+} \boldsymbol{\rho}_m^+ \cdot \boldsymbol{A}\, dS + \int_{T_m^-} \frac{l_m}{2A_m^-} \boldsymbol{\rho}_m^- \cdot \boldsymbol{A}\, dS. \tag{4.2.31}$$

The integrals in the implicit solution, in contrast to those in the explicit method, are approximated by evaluating $\boldsymbol{A}$ at the centroids of the triangles $T_m^\pm$. This procedure is more efficient than evaluating at edge centers since the same integration can be used for all three edges corresponding to the triangle. Therefore,

$$\langle \boldsymbol{f}_m, \boldsymbol{A} \rangle \approx \boldsymbol{A}(\boldsymbol{r}_m^{c+}) \cdot \left[\frac{l_m}{2A_m^+} \int_{T_m^+} \boldsymbol{\rho}_m^+\, dS \right] + \boldsymbol{A}(\boldsymbol{r}_m^{c-}) \cdot \left[\frac{l_m}{2A_m^-} \int_{T_m^-} \boldsymbol{\rho}_m^-\, dS \right]. \tag{4.2.32}$$

The integrations in Eq. (4.2.32) may be carried out analytically and the result is given by

$$\langle \boldsymbol{f}_m, \boldsymbol{A} \rangle \approx \frac{l_m}{2} \left[\boldsymbol{A}(\boldsymbol{r}_m^{c+}) \cdot \boldsymbol{\rho}_m^{c+} + \boldsymbol{A}(\boldsymbol{r}_m^{c-}) \cdot \boldsymbol{\rho}_m^{c-} \right]. \tag{4.2.33}$$

Notice that the approximation in Eq. (4.2.33) is the same as that in the case of the frequency domain solution of finite-sized conducting scatterers [5].

Next, consider the testing of the scalar potential term which is, in fact, the same as that in the case of the explicit method. Thus, we have

$$\langle \boldsymbol{f}_m, \nabla \Phi \rangle \approx -l_m \left[\Phi(\boldsymbol{r}_m^{c+}) - \Phi(\boldsymbol{r}_m^{c-}) \right]. \tag{4.2.34}$$

Finally, the testing procedure is applied to the incident field term. As in the vector potential case, we approximate the integral by using the field evaluated at the centroids of the triangles $T_m^\pm$. This gives

$$\langle \boldsymbol{f}_m, \boldsymbol{E}^{\mathrm{i}} \rangle \approx \frac{l_m}{2} \left[\boldsymbol{\rho}_m^{c+} \cdot \boldsymbol{E}^{\mathrm{i}}(\boldsymbol{r}_m^{c+}) + \boldsymbol{\rho}_m^{c-} \cdot \boldsymbol{E}^{\mathrm{i}}(\boldsymbol{r}_m^{c-}) \right]. \tag{4.2.35}$$

Thus, using Eqs. (4.2.33)–(4.2.35), replacing $j \Rightarrow j-1$, and rearranging the terms, we can write Eq. (4.2.29) as

$$\begin{aligned} &\frac{l_m}{2} \left[\boldsymbol{\rho}_m^{c+} \cdot \boldsymbol{A}(\boldsymbol{r}_m^{c+}, t_j) + \boldsymbol{\rho}_m^{c-} \cdot \boldsymbol{A}(\boldsymbol{r}_m^{c-}, t_j) \right] - (\Delta t) l_m \left[\Phi(\boldsymbol{r}_m^{c+}, t_j) - \Phi(\boldsymbol{r}_m^{c-}, t_j) \right] \\ &\quad = \frac{\Delta t\, l_m}{2} \left[\boldsymbol{\rho}_m^{c+} \cdot \boldsymbol{E}^{\mathrm{i}}(\boldsymbol{r}_m^{c+}, t_j) + \boldsymbol{\rho}_m^{c-} \cdot \boldsymbol{E}^{\mathrm{i}}(\boldsymbol{r}_m^{c-}, t_j) \right] \\ &\quad + \frac{l_m}{2} \left[\boldsymbol{\rho}_m^{c+} \cdot \boldsymbol{A}(\boldsymbol{r}_m^{c+}, t_{j-1}) + \boldsymbol{\rho}_m^{c-} \cdot \boldsymbol{A}(\boldsymbol{r}_m^{c-}, t_{j-1}) \right]. \end{aligned} \tag{4.2.36}$$

We now consider the expansion procedure which is identical to that of the explicit procedure. Thus, substituting Eqs. (4.2.15) and (4.2.20) into Eq. (4.2.36) yields, after a few steps of algebra, the following set of equations, given by

$$\sum_{k=1}^{N} Z_{mk}^{a}(t_j) + \Delta t \sum_{k=1}^{N} Z_{mk}^{b}(t_j) = F_m(t_j) + \sum_{k=1}^{N} Z_{mk}^{a}(t_{j-1}); \qquad m = 1, 2, \ldots, N, \tag{4.2.37}$$

where

$$
\begin{aligned}
Z^a_{mk}(t_j) = {} & \left[I_k(t_R^{++})\right]\left[\frac{\mu l_m l_k}{2A_k^+}\boldsymbol{\rho}_m^{c+}\cdot\int_{T_k^+}\frac{\boldsymbol{\rho}_k^+}{4\pi\left|\boldsymbol{r}_m^{c+}-\boldsymbol{r}'\right|}\,ds'\right] \\
& +\left[I_k(t_R^{+-})\right]\left[\frac{\mu l_m l_k}{2A_k^-}\boldsymbol{\rho}_m^{c+}\cdot\int_{T_k^-}\frac{\boldsymbol{\rho}_k^-}{4\pi\left|\boldsymbol{r}_m^{c+}-\boldsymbol{r}'\right|}\,ds'\right] \\
& +\left[I_k(t_R^{-+})\right]\left[\frac{\mu l_m l_k}{2A_k^+}\boldsymbol{\rho}_m^{c-}\cdot\int_{T_k^+}\frac{\boldsymbol{\rho}_k^+}{4\pi\left|\boldsymbol{r}_m^{c-}-\boldsymbol{r}'\right|}\,ds'\right] \\
& +\left[I_k(t_R^{--})\right]\left[\frac{\mu l_m l_k}{2A_k^-}\boldsymbol{\rho}_m^{c-}\cdot\int_{T_k^-}\frac{\boldsymbol{\rho}_k^-}{4\pi\left|\boldsymbol{r}_m^{c-}-\boldsymbol{r}'\right|}\,ds'\right],
\end{aligned}
\tag{4.2.38}
$$

$$
\begin{aligned}
Z^b_{mk}(t_j) = {} & \left[\int_0^{t_R^{++}} I_k(\tau)\,d\tau\right]\left[\frac{l_m l_k}{\epsilon A_k^+}\int_{T_k^+}\frac{ds'}{4\pi\left|\boldsymbol{r}_m^{c+}-\boldsymbol{r}'\right|}\right] \\
& -\left[\int_0^{t_R^{+-}} I_k(\tau)\,d\tau\right]\left[\frac{l_m l_k}{\epsilon A_k^-}\int_{T_k^-}\frac{ds'}{4\pi\left|\boldsymbol{r}_m^{c+}-\boldsymbol{r}'\right|}\right] \\
& -\left[\int_0^{t_R^{-+}} I_k(\tau)\,d\tau\right]\left[\frac{l_m l_k}{\epsilon A_k^+}\int_{T_k^+}\frac{ds'}{4\pi\left|\boldsymbol{r}_m^{c+}-\boldsymbol{r}'\right|}\right] \\
& +\left[\int_0^{t_R^{--}} I_k(\tau)\,d\tau\right]\left[\frac{l_m l_k}{\epsilon A_k^-}\int_{T_k^-}\frac{ds'}{4\pi\left|\boldsymbol{r}_m^{c+}-\boldsymbol{r}'\right|}\right].
\end{aligned}
\tag{4.2.39}
$$

In Eqs. (4.2.38) and (4.2.39), $A_k^{\pm}$ represents the area of the triangle $T_k^{\pm}$, $t_R^{\pm\pm} = t_j - R_{mk}^{\pm\pm}/c$, and $R_{mk}^{\pm\pm} = |\boldsymbol{r}_m^{c\pm} - \boldsymbol{r}_k^{c\pm}|$. Also, $F_m(t_j)$ is given by Eq. (4.2.35).

Next, we consider Eq. (4.2.37). As in the case of wire and cylinder problems using the implicit scheme, here we also select the time step, Δt, independent of triangulation of the body. We also note that the current coefficients are known up to $t = t_{j-1}$ when we are calculating the currents at $t = t_j$. Thus, it is clear that many terms in Eqs. (4.2.38) and (4.2.39) are known and can be moved to the right-hand side of Eq. (4.2.37). These are the terms for which $R_{mk}^{\pm\pm}/c \geq \Delta t$. Moving these terms to the right-hand side of Eq. (4.2.37) and retaining the unknown terms on the left-hand side, we may rewrite Eq. (4.2.37) as

$$\sum_{k=1}^{N} \alpha_{mk} I_k(t_j) = F_{m,j} + \sum_{k=1}^{N} \beta_{mk} I_k\left(t_{j-1} - \frac{R_{mk}^{\pm\pm}}{c}\right). \tag{4.2.40}$$

Note that the elements of the $[\alpha]$ matrix in Eq. (4.2.38) are formed by the potential terms when $R_{mk}^{\pm\pm}/c < \Delta t$. Also, in the present formulation, both the vector potential and the scalar potentials (Z^a_{mk} and Z^b_{mk}) contribute to the $[\alpha]$ matrix. However, the $[\alpha]$ matrix is a sparse matrix and its sparsity depends on the choice of Δt. Also, the matrix elements of $[\alpha]$ are not functions of time and hence need to be computed only once at the first time step. Lastly, by solving Eq. (4.2.40) at each time step, the time domain current induced on the scatterer may be obtained iteratively.

4.2.3 Efficiency Considerations

In this section, we discuss some important numerical features applicable to time-marching solutions in general and for the implicit scheme in particular. By incorporating these special features in the algorithm development, one can achieve a significant reduction in computer resource requirements and at the same time make the algorithm fast and efficient.

First, for the implicit scheme, all the field quantities are evaluated at the centroid of a given triangular patch, although the unknown current coefficients are associated with edges. This implies that a general matrix element Z_{mk} in Eq. (4.2.37) is associated with the pair of edges m and k; however, in actuality, this element is related to a source triangle attached to edge k with an observation point at the centroid of a triangle attached to edge m. The integrals to evaluate Z_{mk} (both vector and scalar potential integrals) are the same integrals for any other pair of edges connected to the same pair of triangles. Thus, rather than individually compute each element of Z_{mk}, it is more efficient to compute all vector and scalar potentials associated with each observation and source triangle combination and then place these quantities, appropriately weighted, into the elements of Z corresponding to the various edges associated with these triangular patch pairs. Doing computations in this fashion results in up to a ninefold increase in efficiency in computing the matrix elements over the direct edge-by-edge approach. Furthermore, this approach is similar to the one adopted for the frequency domain solution using triangular patches [5].

When the object under consideration is several pulse widths (measured in light-meters) long, it is more efficient to first compute the retarded time and the current at this time instant before computing the matrix element Z_{mk}. In fact, it is even possible to eliminate the computation of Z_{mk} if the current is below a specific threshold dictated by the peak of the current pulse.

Finally, the algorithm can be made even more efficient by making intelligent guesses to account for the shadow region when electrically large bodies are considered. One can develop a simple algorithm, using geometrical optics, to check for the physical shadow region and eliminate computations in this region. By adopting all these measures, it is possible to make the TDIE implicit solution comparable to or even surpass other numerical techniques in terms of efficiency and electrically large problem-handling capability.

4.2.4 Numerical Examples

In this section, we discuss currents induced on various geometries illuminated by a Gaussian plane wave, given by

$$\boldsymbol{E}^{\mathrm{i}}(\boldsymbol{r}, t) = \boldsymbol{E}_{\mathrm{o}} \frac{4.0}{\sqrt{\pi} T} e^{-\gamma^2}, \tag{4.2.41}$$

where

$$\gamma = \frac{4.0}{T} \left[ct - ct_{\mathrm{o}} - \boldsymbol{r} \cdot \boldsymbol{a}_k\right], \tag{4.2.42}$$

with $\boldsymbol{E}_{\mathrm{o}} = 120\,\pi \boldsymbol{a}_x$, $\boldsymbol{a}_k = -\boldsymbol{a}_z$, $T = 4$ LM, and $ct_{\mathrm{o}} = 6$ LM. We obtain the results using both the explicit and implicit schemes. For comparison, we also present the results obtained by the inverse discrete Fourier transform (IDFT) of the frequency domain solution obtained using the frequency domain solution [5]. For all the data presented, the range of frequencies considered was $0 < f \leq 0.5$ GHz, with 128 sample points. The averaging process discussed in Chapter 1, Section 1.6.1 is used while calculating the currents in the explicit solution scheme. Note that no such averaging is needed for the implicit method because the method generates a stable solution even at late times. Furthermore, the implicit schemes are more efficient since one can use a larger time step in the solution procedure, thus requiring fewer steps to obtain a given duration of the time domain signature.

We first consider a flat, 0.5×0.5-m square plate, located in the xy plane and centered about the origin. The plate is divided into six and five equal divisions along x and y directions, respectively,

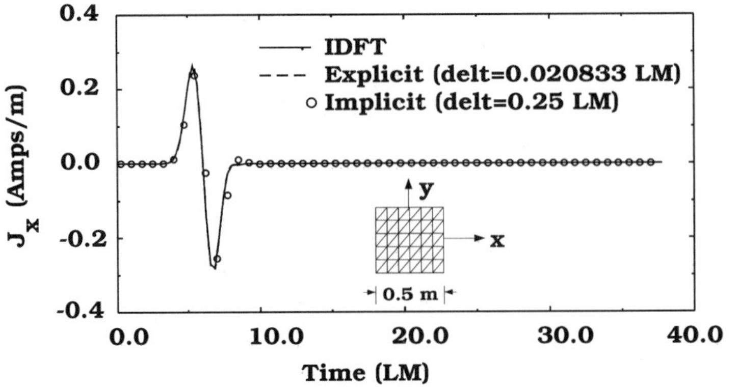

FIGURE 4.4
Transient current at the center of a square plate (side = 0.5 m) located in the xy plane. Number of unknowns = 79.

resulting in 30 rectangular patches. This division allows us to obtain the current at the center of the plate directly. Joining the diagonals results in 60 triangular patches with 79 unknowns. Figure 4.4 shows the x-directed center current as a function of time. Again, note that the explicit solution is obtained by employing the averaging technique to control the late-time instabilities. For the implicit solution, the time step is almost 12 times the time step of the explicit scheme. It is evident from Fig. 4.4 that the implicit solution remains stable even at very late times. All three solutions, IDFT, explicit, and implicit, agree reasonably well as is evident from the figure.

As a second example, consider a pie-shaped plate. The geometry consists of a equilateral triangular plate, 0.5 m on a side, joined to a semicircular disk with a 0.5-m diameter. The plate lies in the xy plane with the "center" of the disk located at the origin. The triangular portion is divided into 36 equilateral triangles. Also, dividing the disk portion into 20 triangular patches resulted in a total of 56 patches with 74 unknowns. Figure 4.5 shows the x component of the induced current density for both the explicit and implicit solutions. The current density shown is located at (0.0, 0.04167 m). The time steps used are 0.02 and 0.25 LM for the explicit and implicit solutions, respectively. The results are in good agreement with each other and also with the IDFT solution.

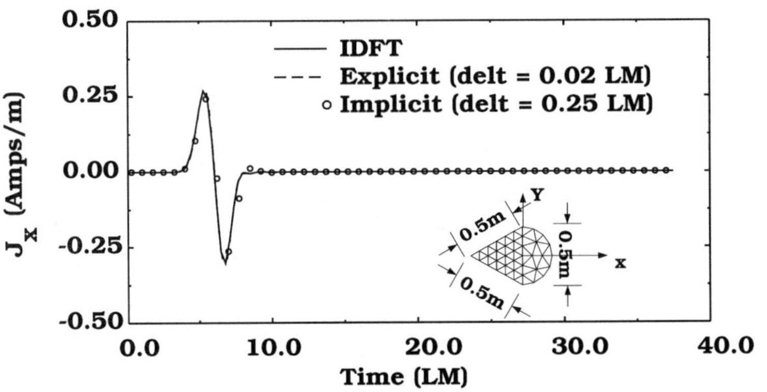

FIGURE 4.5
Transient current at (0, 0.04167 m) of a pie-shaped plate located in the xy plane. Number of unknowns = 74.

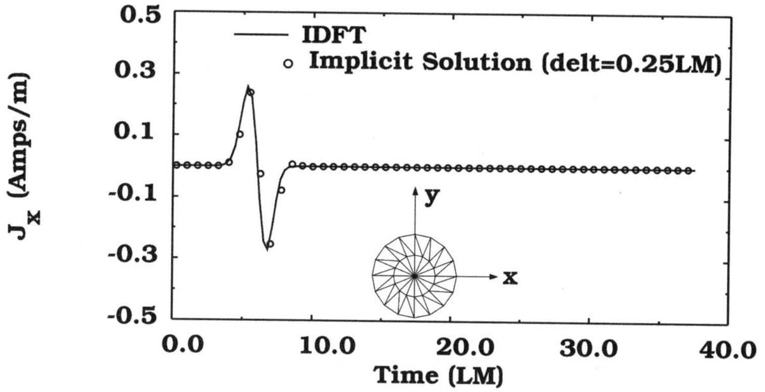

FIGURE 4.6
Transient current on a circular disk, radius = 0.25 m, located in the xy plane centered about the origin. The current is shown at (0.0, 0.1875 m). Number of unknowns = 64.

Next, consider a circular disk, 0.5 m in diameter, located in the xy plane and centered about the origin. The disk is divided into 16 segments along the circumference and 2 segments along the radius resulting in 48 triangular patches with 64 unknowns. Figure 4.6 shows the x-directed current induced at (0.0, 0.1875 m) as a function of time obtained using the implicit solution and compared with the IDFT method. The explicit solution is not shown since it requires a very large number of iterations (>2500). It is evident from the figure that these two solutions compare very well. Also, note the absence of any late-time instability, which has been the major problem with these methods until now.

Next, consider a circular cylinder, 0.5 m in length and 0.2 m in diameter, open at both ends. For this body, the circumference of the cylinder is divided into eight linear segments and the length into four equal divisions resulting in 32 rectangular patches. Joining the diagonals results in 64 triangular patches with 88 unknowns. Figure 4.7 shows the axial component of the induced current density at the middle of the cylinder ($z = 0$) and 22.5° from the x-axis obtained using the implicit solution. Again, the explicit solution requires excessively large time steps and hence

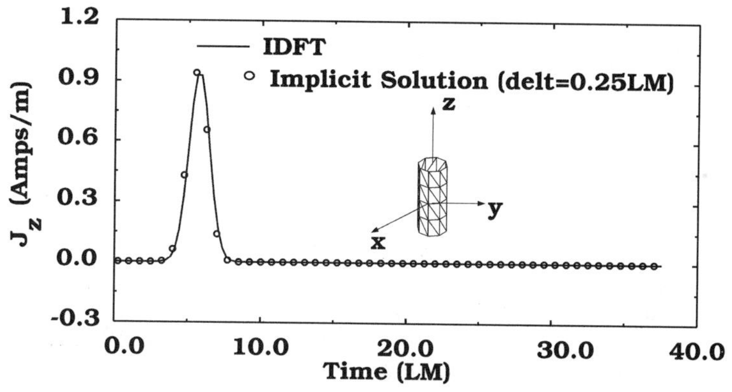

FIGURE 4.7
Transient current at the center of an open cylinder ($l = 0.5$ m, radius = 0.1 m) located along the z-axis. Number of unknowns = 88.

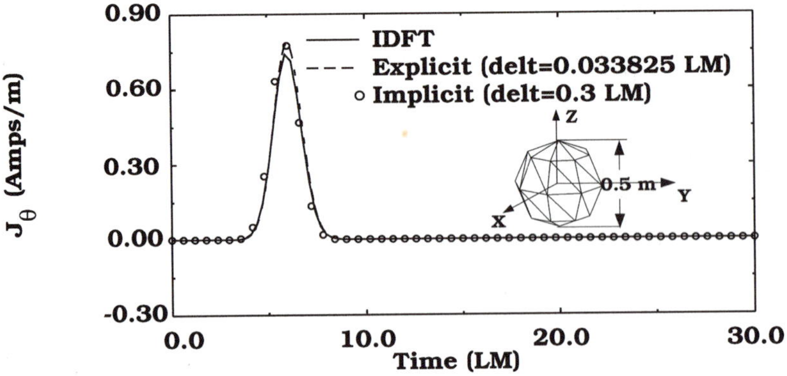

FIGURE 4.8
Transient current at the equator of a sphere, radius = 0.25 m, centered at the origin. Number of unknowns = 72.

is not included. For comparison, see the IDFT solution. The results are in good agreement as evident from the figure.

Next, we consider a sphere of radius 0.25 m centered around the origin. The sphere is divided into four and eight equal divisions along θ and ϕ directions, respectively, resulting in 72 unknowns. Figure 4.8 shows the current induced (J_θ) as a function of time for all three solution schemes. The current is located on the equator at 22.5° from the x-axis. The time step in the explicit solution method, constrained by the grid scheme, is very small, given by 0.033825 LM. As a result, a very large number of iterations are needed (>450) to calculate a modest time domain signature of 15 LM. However, for the same grid scheme, a time step of 0.3 LM can be used for the implicit scheme requiring only 50 iterations. Thus, the advantages of the implicit scheme can clearly be seen from this example.

Finally, consider a 0.2 m cube centered at the origin. The cube is divided into 64 patches with 96 unknowns. Figure 4.9 shows the current density as in the earlier examples. Only the implicit solution scheme and the IDFT solution are presented in this example. The current is located at the center of the top plate. A good agreement in both results is evident in this example. Furthermore, note the absence of any trace of instabilities as the currents in the late times are computed.

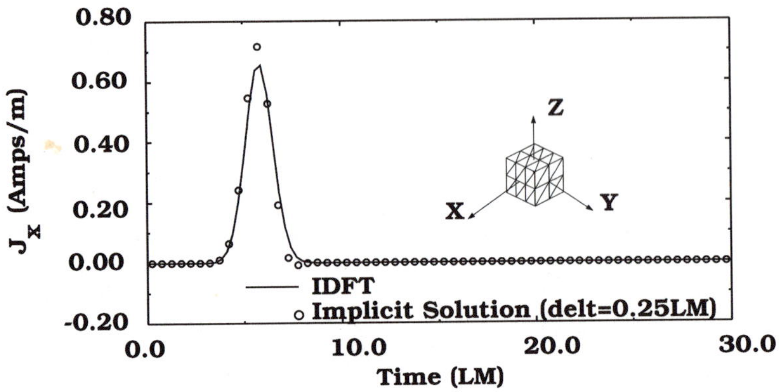

FIGURE 4.9
Transient current at the center of the top side of a cube, side length = 0.2 m, centered at the origin. Number of unknowns = 96.

4.3 FAR-SCATTERED FIELDS

Once the transient currents on the scatterer have been determined, we can calculate the electric and magnetic fields anywhere outside the scatterer. Note that the fields inside the scatterer are zero. In this section, we derive the equations to compute the far-scattered fields. The far-scattered fields represent the radar signature and are easier to derive. In the next section, we will consider the near-field calculations.

First, the scattered magnetic field, $\boldsymbol{H}^s(\boldsymbol{r}, t)$, at a point $\boldsymbol{r}$ is related to the induced currents by

$$\boldsymbol{H}^s(\boldsymbol{r}, t) = \frac{1}{\mu}\nabla \times \boldsymbol{A} = \frac{1}{\mu}\nabla \times \mu \int_S \frac{\boldsymbol{J}(\boldsymbol{r}', t - R/c)}{4\pi R}\, dS', \tag{4.3.1}$$

where $R = |\boldsymbol{r} - \boldsymbol{r}'|$. Taking the curl operator inside the integral and using the vector identity, $\nabla \times (w\boldsymbol{A}) = w\nabla \times \boldsymbol{A} - \boldsymbol{A} \times \nabla w$, results in

$$\boldsymbol{H}^s(\boldsymbol{r}, t) = \int_S \frac{\nabla \times \boldsymbol{J}(\boldsymbol{r}', t - R/c)}{4\pi R} - \frac{\boldsymbol{a}_r}{4\pi R^2} \times \boldsymbol{J}(\boldsymbol{r}', t - R/c)\, dS'. \tag{4.3.2}$$

Since we are restricting ourselves to the far field, $R^2 \gg R$; therefore, we can neglect the second term in Eq. (4.3.2). Using simple algebra,

$$\nabla \times \boldsymbol{J}\left(\boldsymbol{r}', t - \frac{R}{c}\right) = \frac{1}{c}\frac{\partial \boldsymbol{J}}{\partial t_R} \times \boldsymbol{a}_R, \tag{4.3.3}$$

where $t_R = t - R/c$, and $\boldsymbol{a}_R$ is a unit vector in the direction $\boldsymbol{r} - \boldsymbol{r}'$. Substituting Eqs. (4.2.14) and (4.3.3) into Eq. (4.3.2) gives

$$\boldsymbol{H}^s(\boldsymbol{r}, t_n) \approx \sum_{k=1}^{N} \frac{1}{c}\frac{\partial I_k(t_R)}{\partial t_R} \int_{T_k^+ + T_k^-} \frac{\boldsymbol{f}_k \times \boldsymbol{a}_R}{4\pi R}\, dS'. \tag{4.3.4}$$

For far-field calculations, we can make the following approximations: $R \approx r$ for magnitude terms in which $r = |\boldsymbol{r}|$, $R \approx r - \boldsymbol{r}' \cdot \boldsymbol{a}_r$ for time retardation terms, and $\boldsymbol{a}_R \approx \boldsymbol{a}_r$. The time derivative of the current is approximated with a finite difference as

$$\frac{\partial I_k(t_R)}{\partial t_R} \approx \frac{I_k\left(t_{n+1/2} - \frac{r - \boldsymbol{r}' \cdot \boldsymbol{a}_r}{c}\right) - I_k\left(t_{n-1/2} - \frac{r - \boldsymbol{r}' \cdot \boldsymbol{a}_r}{c}\right)}{\Delta t}, \tag{4.3.5}$$

and the integral may be carried out analytically to give

$$\int_{T_k^+ + T_k^-} \frac{\boldsymbol{f}_k \times \boldsymbol{a}_r}{4\pi r}\, dS' = \frac{l_k}{8\pi r}(\boldsymbol{\rho}_k^{c+} + \boldsymbol{\rho}_k^{c-}) \times \boldsymbol{a}_r. \tag{4.3.6}$$

Finally, combining Eqs. (4.3.4)–(4.3.6), the normalized far magnetic field is given by

$$r\boldsymbol{H}^s(\boldsymbol{r}, t_n) = \frac{1}{4\pi}\sum_{k=1}^{N}\left[\frac{I_k\left(t_{n+1/2} - \frac{r - \boldsymbol{r}' \cdot \boldsymbol{a}_r}{c}\right) - I_k\left(t_{n-1/2} - \frac{r - \boldsymbol{r}' \cdot \boldsymbol{a}_r}{c}\right)}{c\Delta t}\right]$$
$$\bullet \frac{l_k}{2}(\boldsymbol{\rho}_k^{c+} + \boldsymbol{\rho}_k^{c-}) \times \boldsymbol{a}_r. \tag{4.3.7}$$

The far-scattered electric field may then be obtained with

$$\boldsymbol{E}^s(\boldsymbol{r}, t_n) = \eta \boldsymbol{H}^s(\boldsymbol{r}, t_n) \times \boldsymbol{a}_r, \tag{4.3.8}$$

where η is the wave impedance in the medium surrounding the scatterer.

4.3.1 Numerical Examples

In this section, we consider a few examples of far-field computation using the TDIE algorithm. For the examples shown, the currents are obtained by the explicit scheme, although the same results may be obtained with the implicit scheme. We have already established that both the implicit and explicit schemes give virtually same results. Further, all results are obtained for the incident field given in Eq. (4.2.41) with $\boldsymbol{E_o} = 1.0\boldsymbol{a}_x$.

As a first example, consider a flat, 2 × 2-m square plate located in the xy plane and centered about the origin. Eight divisions are made in the x direction and seven in the y direction, resulting in 56 rectangular patches. Joining diagonals of each rectangle results in 112 triangular patches with 153 unknowns. The normalized, x-directed, far-scattered, electric field is shown in Fig. 4.10

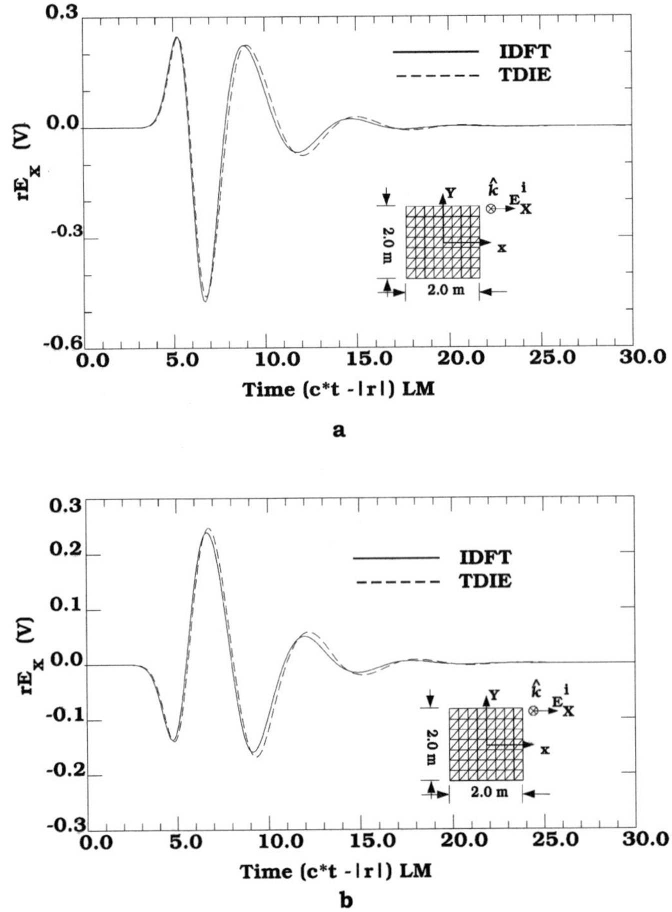

FIGURE 4.10
Normalized (a) backscattered ($\theta = 0°, \phi = 0°$) and (b) side-scattered ($\theta = 90°, \phi = 90°$) far-field response of a conducting 2 × 2-m plate illuminated by a Gaussian plane wave. Number of unknowns = 153.

for the backscattered direction ($\theta = 0°$, $\phi = 0°$) and for the "side-scattered" direction ($\theta = 90°$, $\phi = 90°$). The time scale on the far-field plots is shifted so that if a portion of the incident field at a time t_1 was scattered from the origin, the response would be seen at the observation point at t_1. The time domain curves agree well with the results obtained from the frequency domain.

Next, a pie-shaped plate is considered. This geometry consists of an equilateral triangular plate, 2 m on a side, joined to a semicircular disk with a 2-m diameter. The plate lies in the xy plane with the "center" of the disk portion located at the origin. The triangular portion was divided into 36 uniform equilateral patches and the semicircular disk into 20 triangular patches. The discretization resulted in a total of 56 patches with 74 unknowns. Figure 4.11 compares the far-scattered fields at $\theta = 0°$, $\phi = 0°$ and $\theta = 90°$, $\phi = 90°$ with the frequency domain results. The results are in good agreement.

Solid bodies are considered next and we begin with a sphere of radius 1 m centered about the origin. There were six divisions made in the θ direction and each "ring" had (starting from top) 13, 27, 29, 29, 27, and 13 patches, respectively, for a total of 138 patches and 207 unknowns. This

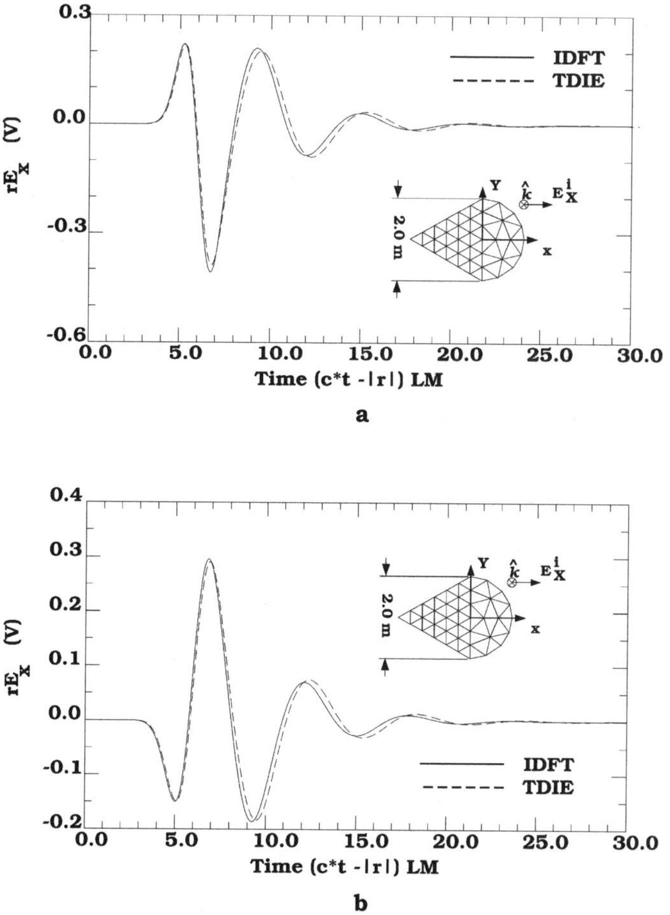

FIGURE 4.11
Normalized (a) backscattered ($\theta = 0°$, $\phi = 0°$) and (b) side-scattered ($\theta = 90°$, $\phi = 90°$) far-field response of a conducting pie-shaped plate illuminated by a Gaussian plane wave. Number of unknowns = 74.

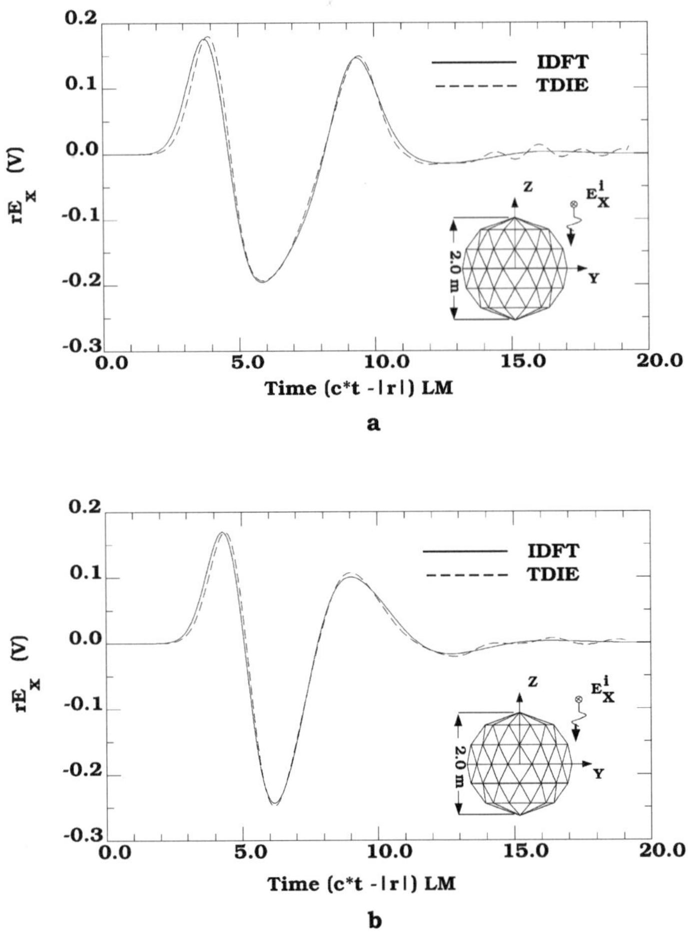

FIGURE 4.12
Normalized (a) backscattered ($\theta = 0°, \phi = 0°$) and (b) side-scattered ($\theta = 90°, \phi = 90°$) far-field response of a conducting sphere 1 m in radius illuminated by a Gaussian plane wave. Number of unknowns = 207.

scheme was chosen so that the triangles would be closer to being equilateral. If the ϕ direction were also divided uniformly, the patches on the top and bottom ring would be more skewed. The normalized, x-directed, far-scattered electric fields are shown in Fig. 4.12 for $\theta = 0°, \phi = 0°$ and $\theta = 90°, \phi = 90°$. In both cases there is a good agreement.

The transient far-field response from a cube, 1 m on a side, centered about the origin is shown next. The x, y, and z directions were divided into 4, 5, and 5 uniform segments, respectively. By connecting the diagonals of the resulting rectangular patches, a total of 260 triangular patches with 390 unknowns were obtained. The backscattered and side-scattered far electric fields are shown in Fig. 4.13. The agreement with the frequency domain data is very good.

Finally, an example of a multiple body problem is provided. Two square plates 2×2 m lie parallel to the xy plane located at $z = 1$ m and $z = -1$ m, respectively. Each plate is divided in the manner described for the first plate example. A total of 306 unknowns result with this grid scheme. Figure 4.14 shows the far-scattered fields. As can be seen, the responses are more oscillatory due to the multiple interaction between the plates. The results are in agreement with

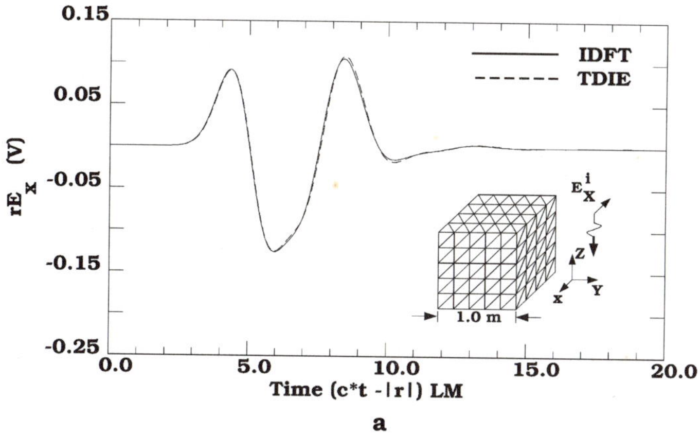

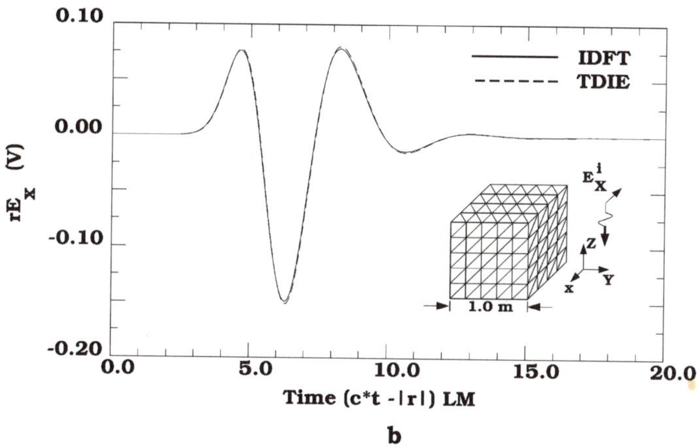

FIGURE 4.13
Normalized (a) backscattered ($\theta = 0°$, $\phi = 0°$) and (b) side-scattered ($\theta = 90°$, $\phi = 90°$) far-field response of a conducting cube 1 m on a side illuminated by a Gaussian plane wave. Number of unknowns = 390.

the frequency domain data. In the next section, we consider the field calculations in the near zone of the scatterer.

4.4 NEAR-SCATTERED FIELDS

The near-scattered fields may be determined from the relations

$$\boldsymbol{E}^{s}(\boldsymbol{r}, t) = -\frac{\partial \boldsymbol{A}}{\partial t} - \nabla\Phi, \tag{4.4.1}$$

$$\boldsymbol{H}^{s}(\boldsymbol{r}, t) = \frac{1}{\mu}\nabla \times \boldsymbol{A}. \tag{4.4.2}$$

The curl and gradient operator are approximated by finite differences in order to determine the near electric and magnetic fields at a point $\boldsymbol{r}$ (x, y, z). Taking an extra time derivative of Eq. (4.4.1)

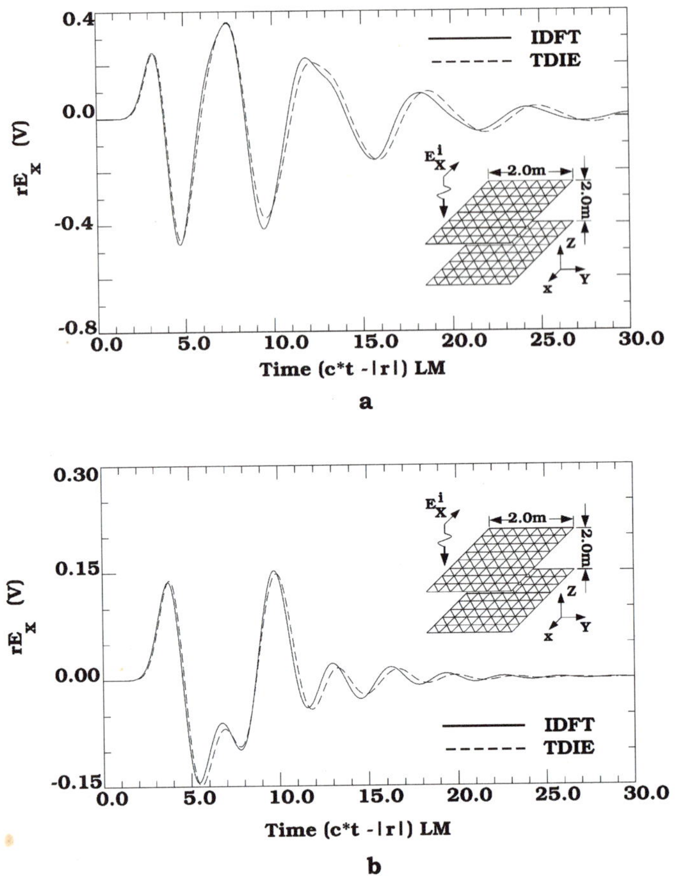

FIGURE 4.14
Normalized (a) backscattered ($\theta = 0°, \phi = 0°$) and (b) side-scattered ($\theta = 90°, \phi = 90°$) far-field response of a parallel plate structure illuminated by a Gaussian plane wave. Number of unknowns = 306.

and expanding the gradient term gives

$$\frac{\partial \boldsymbol{E}^s(t_n)}{\partial t} = -\left(\frac{\partial^2 A_x(t_n)}{\partial t^2} + \frac{\partial \Psi(t_n)}{\partial x}\right)\boldsymbol{a}_x - \left(\frac{\partial^2 A_y(t_n)}{\partial t^2} + \frac{\partial \Psi(t_n)}{\partial y}\right)\boldsymbol{a}_y$$
$$- \left(\frac{\partial^2 A_z(t_n)}{\partial t^2} + \frac{\partial \Psi(t_n)}{\partial z}\right)\boldsymbol{a}_z. \tag{4.4.3}$$

Approximating the derivatives with finite differences gives

$$\frac{\partial \boldsymbol{E}^s(t_n)}{\partial t} \approx \left\{\frac{1}{2}\left[\widetilde{\boldsymbol{A}}_n^{+\Delta x} + \widetilde{\boldsymbol{A}}_n^{-\Delta x}\right]\cdot \boldsymbol{a}_x + \frac{1}{2\Delta x}\left[\Psi_n^{+\Delta x} - \Psi_n^{-\Delta x}\right]\right\}\boldsymbol{a}_x$$
$$+ \left\{\frac{1}{2}\left[\widetilde{\boldsymbol{A}}_n^{+\Delta y} + \widetilde{\boldsymbol{A}}_n^{-\Delta y}\right]\cdot \boldsymbol{a}_y + \frac{1}{2\Delta y}\left[\Psi_n^{+\Delta y} - \Psi_n^{-\Delta y}\right]\right\}\boldsymbol{a}_y$$
$$+ \left\{\frac{1}{2}\left[\widetilde{\boldsymbol{A}}_n^{+\Delta z} + \widetilde{\boldsymbol{A}}_n^{-\Delta z}\right]\cdot \boldsymbol{a}_z + \frac{1}{2\Delta z}\left[\Psi_n^{+\Delta z} - \Psi_n^{-\Delta z}\right]\right\}\boldsymbol{a}_z, \tag{4.4.4}$$

where

$$\tilde{\boldsymbol{A}}_n^{\pm\Delta x} = \frac{\boldsymbol{A}_{n+1}(x \pm \Delta x, y, z) - 2\,\boldsymbol{A}_n(x \pm \Delta x, y, z) + \boldsymbol{A}_{n-1}(x \pm \Delta x, y, z)}{\Delta t^2},$$
$$\tilde{\boldsymbol{A}}_n^{\pm\Delta y} = \frac{\boldsymbol{A}_{n+1}(x, y \pm \Delta y, z) - 2\,\boldsymbol{A}_n(x, y \pm \Delta y, z) + \boldsymbol{A}_{n-1}(x, y \pm \Delta y, z)}{\Delta t^2},$$
$$\tilde{\boldsymbol{A}}_n^{\pm\Delta z} = \frac{\boldsymbol{A}_{n+1}(x, y, z \pm \Delta z) - 2\,\boldsymbol{A}_n(x, y, z \pm \Delta z) + \boldsymbol{A}_{n-1}(x, y, z \pm \Delta z)}{\Delta t^2},$$
$$\Psi^{\pm\Delta x} = \Psi_n(x \pm \Delta x, y, z),$$
$$\Psi^{\pm\Delta y} = \Psi_n(x, y \pm \Delta y, z),$$
$$\Psi^{\pm\Delta z} = \Psi_n(x, y, z \pm \Delta z).$$

The scattered electric field may then be found by numerically integrating Eq. (4.4.4) to obtain

$$\boldsymbol{E}^{\mathrm{s}}(t_n) \approx \Delta t \sum_{k=1}^{n} \frac{\partial \boldsymbol{E}(t_n)}{\partial t}. \tag{4.4.5}$$

Expanding the curl in Eq. (4.4.2) results in

$$\boldsymbol{H}^{\mathrm{s}}(\boldsymbol{r}, t) = \frac{1}{\mu}\left\{\left(\frac{\partial \boldsymbol{A}_{z,n}}{\partial y} - \frac{\partial \boldsymbol{A}_{y,n}}{\partial z}\right)\boldsymbol{a}_x + \left(\frac{\partial \boldsymbol{A}_{x,n}}{\partial z} - \frac{\partial \boldsymbol{A}_{z,n}}{\partial x}\right)\boldsymbol{a}_y + \left(\frac{\partial \boldsymbol{A}_{y,n}}{\partial x} - \frac{\partial \boldsymbol{A}_{x,n}}{\partial y}\right)\boldsymbol{a}_z\right\}, \tag{4.4.6}$$

which may be written as

$$\boldsymbol{H}^{\mathrm{s}} = H_x^{\mathrm{s}}\boldsymbol{a}_x + H_y^{\mathrm{s}}\boldsymbol{a}_y + H_z^{\mathrm{s}}\boldsymbol{a}_z, \tag{4.4.7}$$

where

$$H_x^{\mathrm{s}}(\boldsymbol{r}, t) \approx \frac{1}{\mu}\left\{\left[\frac{\boldsymbol{A}_{z,n}(x, y+\Delta y, z) - \boldsymbol{A}_{z,n}(x, y-\Delta y, z)}{2\Delta y} - \frac{\boldsymbol{A}_{y,n}(x, y, z+\Delta z) - \boldsymbol{A}_{y,n}(x, y, z-\Delta z)}{2\Delta z}\right]\right\}, \tag{4.4.8}$$

$$H_y^{\mathrm{s}}(\boldsymbol{r}, t) \approx \frac{1}{\mu}\left\{\left[\frac{\boldsymbol{A}_{x,n}(x, y, z+\Delta z) - \boldsymbol{A}_{x,n}(x, y, z-\Delta z)}{2\Delta z} - \frac{\boldsymbol{A}_{z,n}(x+\Delta x, y, z) - \boldsymbol{A}_{z,n}(x-\Delta x, y, z)}{2\Delta x}\right]\right\}, \tag{4.4.9}$$

and

$$H_z^{\mathrm{s}}(\boldsymbol{r}, t) \approx \frac{1}{\mu}\left\{\left[\frac{\boldsymbol{A}_{y,n}(x+\Delta x, y, z) - \boldsymbol{A}_{y,n}(x-\Delta x, y, z)}{2\Delta x} - \frac{\boldsymbol{A}_{x,n}(x, y+\Delta y, z) - \boldsymbol{A}_{x,n}(x, y-\Delta y, z)}{2\Delta y}\right]\right\}. \tag{4.4.10}$$

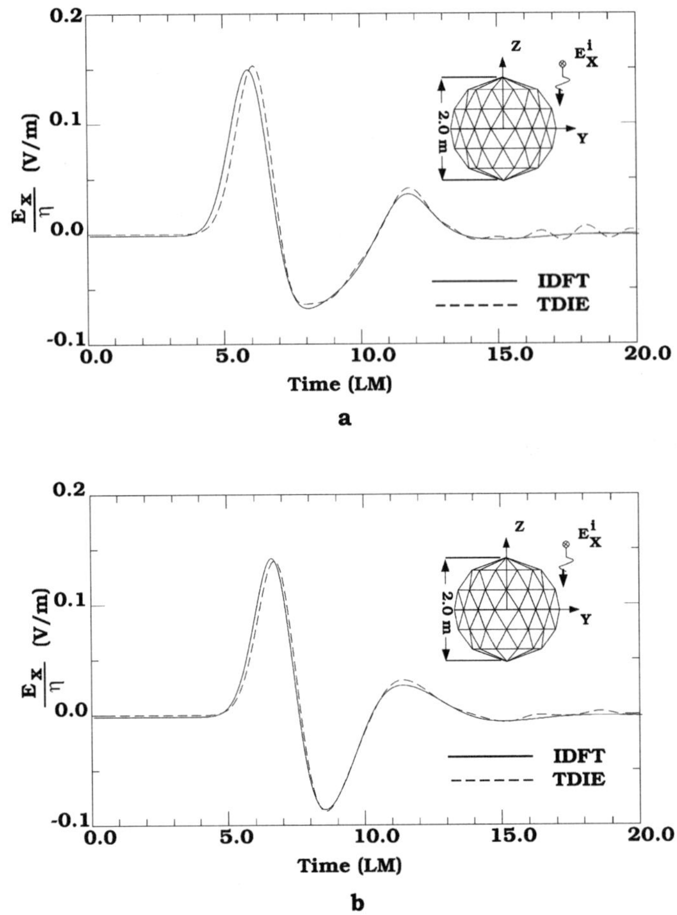

FIGURE 4.15
Near scattered electric field at (a) (0,0,2) and (b) (0,2,0) due to a conducting sphere 1 m in radius illuminated by a Gaussian plane wave. Number of unknowns = 207.

The field quantities may now be found by evaluating $\boldsymbol{A}$ and Ψ in Eqs. (4.4.4) and (4.4.8)–(4.4.10) at the coordinates shown.

As an example, consider the case of a conducting sphere with a 1.0-m radius illuminated by the transient Gaussian plane wave given by Eq. (4.2.41). Figure 4.15 compares the x component of the electric field at the points (0,0,2) and (0,2,0) with those obtained from the analytical solution. Figure 4.16 show the x- and z-directed near fields for the point (2,0,0). In both cases there is good agreement with the analytical solution.

4.5 EXTRAPOLATION OF TIME DOMAIN RESPONSE

In this section, we present an alternate way of computing the time domain signature at late times. This technique, known as the matrix pencil approach, efficiently extrapolates the time domain response from three-dimensional conducting objects that arise in the numerical solution of

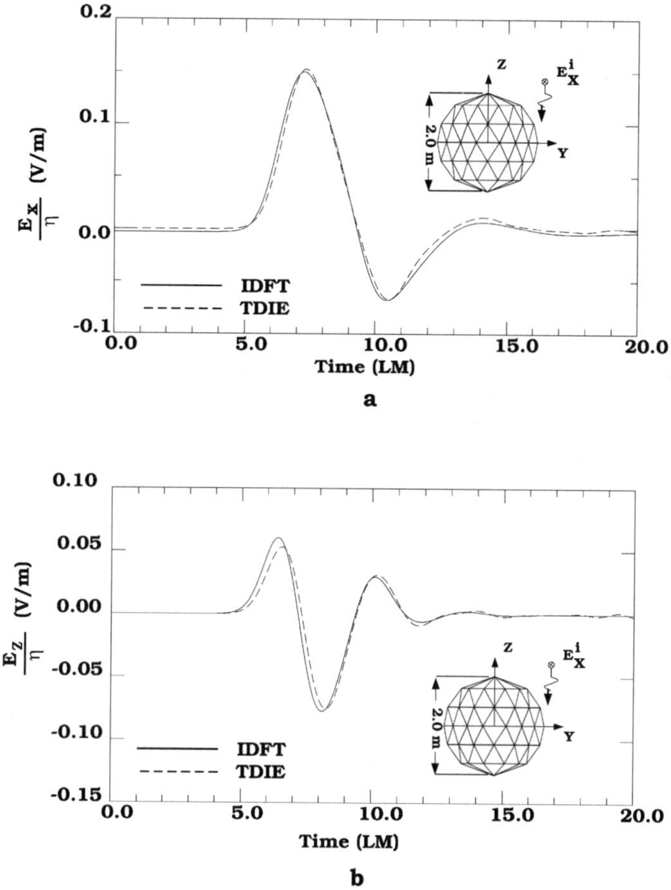

FIGURE 4.16
Near scattered (a) x component and (b) z component of the electric field at (2,0,0) due to a conducting sphere 1 m in radius illuminated by a Gaussian plane wave. Number of unknowns = 207.

electromagnetic field problems. By modeling the time functions as a sum of complex exponentials, we can accurately extrapolate the time domain response very efficiently in terms of computer resources.

In the matrix pencil approach, we model the free response, i.e., the time domain response after the excitation has died down, as a sum of complex exponentials. The input to the matrix pencil algorithm is the output from the IE solution using either the explicit or the implicit method (see Section 4.2) for a short period of time after the excitation has died down. Modeling the free response as a sum of complex exponentials results in a stable time domain response for all times. Since the integral equation solution code needs to be run only for a short duration of time after the excitation dies down, this approach can lead to significant savings in program execution time. In the following, we describe this approrach in detail and validate it with several numerical example problems. Assume that the currents on the scatterer are available for a certain duration as a response to a known excitation. These currents have been calculated as a function of time over a limited region. First, we discuss the matrix pencil as a mathematical tool to model a time domain sequence as a sum of complex exponentials.

4.5.1 Matrix Pencil Method

Consider a function $y(t)$ that represents the current at a particular position on a three-dimensional conducting scatterer as a function of time. This current is the transient response to some known excitation. We model the function, after the excitation dies-down, as a sum of complex exponentials:

$$y(t) = \sum_{i=1}^{M} a_i e^{s_i t}. \tag{4.5.1}$$

Such a model is valid because the scatterer can be treated as a linear, time-invariant (LTI) system. It is well-known that for a LTI system, the eigenfunctions of the transfer operator are of the form $e^{s_i t}$, where s_i are the poles of the system. Also, these eigenfunctions are complete in the output space, i.e., any output response can be modeled as a weighted sum of these eigenfunctions.

As a result of the IE solution algorithm, N samples of this function are available at intervals of T_s. Therefore, Eq. (4.5.1) can be written as

$$\begin{aligned} y_k &= y(kT_s + T_0) \\ &= \sum_{i=1}^{M} R_i z_i^k \qquad k = 0, \cdots, N-1, \end{aligned} \tag{4.5.2}$$

where T_0 has been introduced to make sure that the response is the free response of the system, after the excitation has died down. In addition,

$$\begin{aligned} z_i &= e^{s_i T_s} \\ &= e^{(\alpha_i + j\Omega_i)T_s}, \end{aligned} \tag{4.5.3}$$

and α_i is the negative of the damping factor of the ith pole, Ω_i is the angular frequency of the ith pole, R_i is the complex amplitude of the ith pole, N is the number of data samples, and M is the number of poles of the signal.

The problem reduces to finding the best estimates for M, R_i, and z_i, $i = 1, \cdots, M$. This problem can be solved in various ways. Prony's method [7] and the matrix pencil [8] are among the most popular methods. The matrix pencil method is computationally more efficient and more robust to noise. Details of the proof are available in Pereira-Filho and Sarkar [9].

For the matrix pencil method, define two matrices Y_1 and Y_2:

$$[Y_1] = \begin{bmatrix} y_1 & y_2 & \cdots & y_L \\ y_2 & y_3 & \cdots & y_{L+1} \\ \vdots & \vdots & \ddots & \vdots \\ y_{N-L} & y_{N-L+1} & \cdots & y_{N-1} \end{bmatrix}_{(N-L)\times L} \tag{4.5.4}$$

$$[Y_2] = \begin{bmatrix} y_0 & y_1 & \cdots & y_{L-1} \\ y_1 & y_2 & \cdots & y_L \\ \vdots & \vdots & \ddots & \vdots \\ y_{N-L-1} & y_{N-L} & \cdots & y_{N-2} \end{bmatrix}_{(N-L)\times L}. \tag{4.5.5}$$

These matrices can be written as

$$[Y_1] = [Z_1]\,[R]\,[Z_0]\,[Z_2] \tag{4.5.6}$$

$$[Y_2] = [Z_1]\,[R]\,[Z_2]\,, \tag{4.5.7}$$

where

$$[Z_1] = \begin{bmatrix} 1 & 1 & \cdots & 1 \\ z_1 & z_2 & \cdots & z_M \\ \vdots & \vdots & \ddots & \vdots \\ z_1^{(N-L-1)} & z_2^{(N-L-1)} & \cdots & z_M^{(N-L-1)} \end{bmatrix}_{(N-L)\times M} \tag{4.5.8}$$

$$[Z_2] = \begin{bmatrix} 1 & z_1 & \cdots & z_1^{L-1} \\ 1 & z_2 & \cdots & z_2^{L-1} \\ \vdots & \vdots & \ddots & \vdots \\ 1 & z_M & \cdots & z_M^{L-1} \end{bmatrix}_{M\times L} \tag{4.5.9}$$

$$[Z_0] = \begin{bmatrix} z_1 & 0 & \cdots & 0 \\ 0 & z_2 & \cdots & 0 \\ \vdots & \vdots & \ddots & \vdots \\ 0 & 0 & \cdots & z_M \end{bmatrix}_{M\times M} \tag{4.5.10}$$

$$[R] = \begin{bmatrix} R_1 & 0 & \cdots & 0 \\ 0 & R_2 & \cdots & 0 \\ \vdots & \vdots & \ddots & \vdots \\ 0 & 0 & \cdots & R_M \end{bmatrix}_{M\times M} . \tag{4.5.11}$$

Now, consider the matrix pencil

$$[Y_1] - \lambda\,[Y_2] = [Z_1]\,[R]\,\{[Z_0] - \lambda\,[I]\}\,[Z_2]\,. \tag{4.5.12}$$

Provided $M \leq L \leq N - M$, the matrix $[Y_1] - \lambda\,[Y_2]$ has rank M [10]. However, if $\lambda = z_i$, $i = 1, \cdots, M$, the rank is reduced to $M - 1$. This implies that z_i's are the generalized eigenvalues of the matrix pair $\{[Y_1], [Y_2]\}$. Therefore,

$$[Y_1]\,[r_i] = z_i\,[Y_2]\,[r_i]\,, \tag{4.5.13}$$

where r_i is the generalized eigenvector corresponding to z_i or, in the equivalent form,

$$\left\{[Y_2]^{\dagger}\,[Y_1] - z_i\,[I]\right\}[r_i] = [0]\,, \tag{4.5.14}$$

where $[Y_2]^{\dagger}$ is the Moore–Penrose pseudo-inverse of $[Y_2]$ [11]. From Eq. (4.5.14), we can obtain z_i's from the eigenvalues of $[Y_2]^{\dagger}\,[Y_1]$. Hence, for the matrix pencil method, the poles are obtained directly as a one-step process.

Once M and z_i's are known, the amplitudes of the modes R_i's are easily solved from the following least squares problem:

$$\begin{bmatrix} y_0 \\ y_1 \\ \vdots \\ y_{N-1} \end{bmatrix} = \begin{bmatrix} 1 & 1 & \cdots & 1 \\ z_1 & z_2 & \cdots & z_M \\ \vdots & \vdots & \ddots & \vdots \\ z_1^{N-1} & z_2^{N-1} & \cdots & z_M^{N-1} \end{bmatrix} \begin{bmatrix} R_1 \\ R_2 \\ \vdots \\ R_M \end{bmatrix}. \tag{4.5.15}$$

4.5.2 Total Least Squares Matrix Pencil

The procedure detailed previously is efficient and yields good results in the absence of numerical errors and random noise in the available data. However, in the applications of matrix pencil to real-life problems, the given data are perturbed from their true value due to numerical errors or noise. In this case, the perturbations corrupt the eigenvalues. This results in errors in all aspects of the solution method—the choice of the number of poles (M), the solution for the poles (z_i), and the amplitudes (R_i).

In the case of noisy data, an alternative and more stable method exists—the total least squares matrix pencil method. To explain this method, we begin by defining the matrix

$$[Y] = \begin{bmatrix} y_0 & y_1 & \cdots & y_L \\ y_1 & y_2 & \cdots & y_{L+1} \\ \vdots & \vdots & \ddots & \vdots \\ y_{N-L-1} & y_{N-L} & \cdots & y_{N-1} \end{bmatrix}_{(N-L)\times(L+1)} \tag{4.5.16}$$

Define the singular value decomposition (SVD) of Y as

$$[Y] = [U][\Sigma][V]^T, \tag{4.5.17}$$

where U is the $(N-L)\times(N-L)$ unitary matrix whose columns are the eigenvectors of YY^T, V is the $(L+1)\times(L+1)$ unitary matrix whose columns are the eigenvectors of Y^TY, and Σ is the $(N-L)\times(L+1)$ diagonal matrix with the singular values of Y (square root of the eigenvalues of Y^TY) in its main diagonal in descending order.

If the given data y_n were noise free, $[Y]$ would have exactly M nonzero singular values. However, due to the noise, the zero singular values are perturbed. This results in several small nonzero singular values. This error due to the noise can be suppressed by eliminating these spurious singular values from $[\Sigma]$ and the corresponding left and right singular vectors. Define $[\Sigma']$ as the $M\times M$ diagonal matrix with the M largest singular values of $[Y]$ on its main diagonal. Further define $[U']$ and $[V']$ as submatrices of U and V corresponding to these singular values. Since the singular values of $[Y]$ appear in descending order in $[\Sigma]$, we can write in MATLAB notation:

$$[U'] = [U(:, 1:M)] \tag{4.5.18}$$

$$[V'] = [V(:, 1:M)] \tag{4.5.19}$$

$$[\Sigma'] = [\Sigma(1:M, 1:M)] \tag{4.5.20}$$

$$[Y'] = [U'][\Sigma'][V']^T. \tag{4.5.21}$$

Comparing the definition of the matrices $[Y]$, $[Y_1]$, and $[Y_2]$ from Eqs. (4.5.16), (4.5.4), and (4.5.5), we obtain

$$[Y] = [c_1, Y_1] = [Y_2, c_{L+1}], \tag{4.5.22}$$

where c_i represents the ith column of $[Y]$. Therefore, using $[Y']$ instead of $[Y]$ in Eq. (4.5.22) results in filtering the noise in both $[Y_1]$ and $[Y_2]$. From Eqs. (4.5.21) and (4.5.22) we can write

$$[Y_1] = [U'][\Sigma'][V_2']^T \tag{4.5.23}$$

$$[Y_2] = [U'][\Sigma'][V_1']^T, \tag{4.5.24}$$

where $[V_1']$ and $[V_2']$ are equal to $[V']$ without the last and the first row, respectively. Using Eqs. (4.5.23) and (4.5.24) the poles of the signal (eigenvalues of $[Y_2]^\dagger [Y_1]$) are given by the nonzero eigenvalues of

$$\{[V_1']^T\}^\dagger [V_2']^T,$$

which are the same as the eigenvalues of

$$[V_2']^T \{[V_1']^T\}^\dagger.$$

The number of modes M is chosen by the number of dominant singular values in the range

$$\sigma_{\max} > \sigma > 10^{-p}\sigma_{\max}$$

where p is the number of significant decimal digits in the data.

The ratio $\sigma : \sigma_{\max}$ as a function of the index (singular value number) can be used to determine the proper value of M, for the assumed precision. Practically, if we overestimate M, we find spurious modes of small magnitude. These do not severely affect the solution. On the other hand, underestimating M would lead to large errors. Hence, it is always preferable to overestimate M. Using this better choice of M, we can evaluate the poles z_i and the amplitudes R_i using the previously detailed approach. In the following section, we present some simple numerical results obtained using this method.

4.5.3 Numerical Examples

For the examples considered in this section, the excitation is given by

$$\boldsymbol{E}^{\mathrm{i}}(\boldsymbol{r}, t) = \boldsymbol{E}_{\mathrm{o}} e^{-0.5\gamma^2}, \tag{4.5.25}$$

where

$$\gamma = \frac{4.0}{T}\,[ct - ct_{\mathrm{o}} - \boldsymbol{r} \cdot \boldsymbol{k}]. \tag{4.5.26}$$

Furthermore, the amplitude of the incident pulse is $120\,\pi$ V/m for all the examples considered in this section.

The first example is a square plate of zero thickness and side 1 m centered at the origin. The plate is located in the xy plane. Eight divisions are made in the x direction and nine in the y direction. By joining the diagonals of each resulting rectangle, 144 triangular patches with 199 unknowns are obtained. This division scheme allows us to evaluate the current at the center of the plate. The excitation arrives along the negative z-axis and polarized along the the x-axis. Here, $T = 2.4$ LM, $ct_o = 3$ LM, and the time step $c\Delta t = 0.0277$ LM.

In this example, the IE solution evaluates the current for the first 233 time steps. Time samples from numbers 188 to 233 are used as the input to the matrix pencil program, i.e., $N = 46$. At the 188th sample, the value of the excitation has fallen to less than 1000th its maximum value. L in the matrix pencil method is chosen to be 24. After filtering the singular values, the estimate for the number of modes is $M = 2$. The numerical values of the amplitudes and the exponent of the modes are

$$R_1 = -0.167024 + j0.293158 \quad z_1 = -2.24472 + j6.43712$$
$$R_2 = -0.167024 - j0.293158 \quad z_2 = -2.24472 - j6.43712.$$

The amplitudes and the exponents of the two modes are complex conjugates of each other, guaranteeing that the resulting extrapolation is real. Also, the real part of the exponents are negative, hence guaranteeing a stable extrapolation. Using the values for (R_1, z_1) and (R_2, z_2) the current is evaluated from 233 to 1500 time samples. Figure 4.17 shows the results for this extrapolation. Although not shown here, the results from direct IE solution and the extrapolation using the matrix pencil algorithm are identical. Furthermore, this extrapolation can be done beyond 1500 time steps with little computation.

The next example is a disk of zero thickness, located in the xy plane and centered at the origin. It has a radius of 0.3 m. The triangulation uses 128 triangles resulting in 208 edges, 32 of which are boundary edges yielding 176 unknowns. The triangulation of the disk is shown in Fig. 4.18. The excitation arrives along the negative z direction and polarized along the x-axis. For this case, $T = 1.2$ LM, $ct_o = 3$ LM, and the time step $c\Delta t = 0.0144$ LM.

The IE solution evaluates the current for the first 333 time steps. Then, 66 time samples from numbers 268 to 333 are used as input into the matrix pencil program. The program uses these data to extrapolate the current from the 334th time step to 1500 time steps. L is chosen to be 34. The

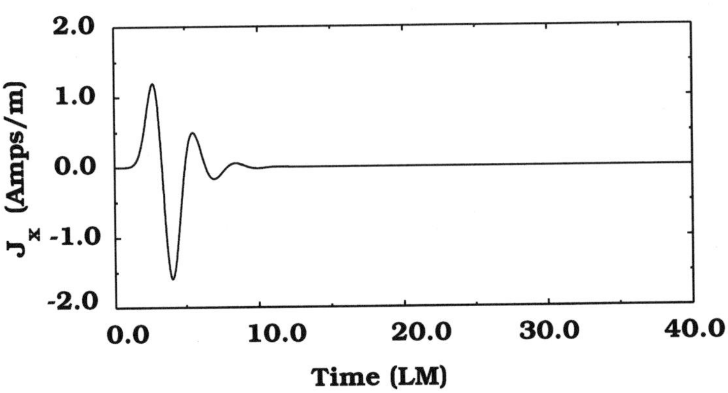

FIGURE 4.17
Transient current at the center of a square plate obtained by the matrix pencil method.

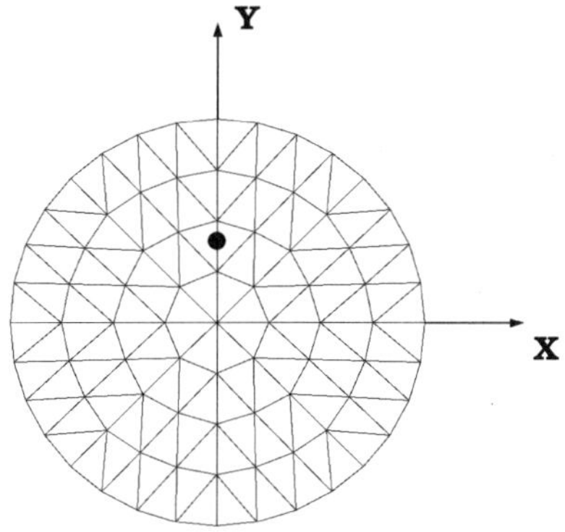

FIGURE 4.18
Triangular patch model of a disk.

required number of modes is $M = 4$. The values of the amplitudes and exponents of the modes are

$$\begin{aligned}
R_1 &= -0.448306 - j0.459212 \quad & z_1 &= -4.62863 + j12.6044 \\
R_2 &= -0.448306 + j0.459212 \quad & z_2 &= -4.62863 - j12.6044 \\
R_3 &= -0.01355 - j0.007086 \quad & z_3 &= -20.2747 - j16.7159 \\
R_4 &= -0.01355 + j0.007086 \quad & z_4 &= -20.2747 + j16.7159.
\end{aligned}$$

The amplitudes of modes 1 and 2 are conjugates of each other, as are their exponents. This also holds true for modes 3 and 4. Hence, the extrapolation is real. Also, all modes have exponents with a negative real part, thus guaranteeing a stable extrapolation. Modes 3 and 4 have relatively low amplitude and a very high damping factor. The extrapolated result is shown in Fig. 4.19.

The next example is a sphere of radius 0.5 m. The sphere is centered at the origin. The "top" half of the sphere ($\theta = 0$ to $\theta = \pi/2$) has six divisions in the θ direction. The first ring

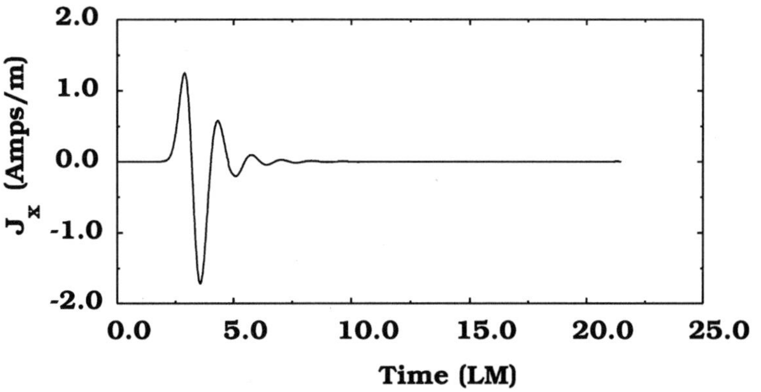

FIGURE 4.19
Transient current at the point indicated in Fig. 4.18 of a circular disk obtained by the matrix pencil method.

extends from $\theta = 0$ to $\theta = \pi/16$. The other five rings are equispaced in θ from $\theta = \pi/16$ to $\theta = \pi/2$. The rings, starting from the top, have 6, 16, 20, 24, 28, and 32 triangular patches, respectively. The sphere is symmetric with respect to the xy plane. This scheme is chosen so that all triangles are as close to equilateral as possible. If the ϕ direction were also divided uniformly, the triangles would be skewed. Also, this scheme allows us to evaluate the current $\boldsymbol{J}_\theta$ at the point $(-0.5, 0.0, 0.0)$. The excitation arrives along the x direction, and the polarization is along the z-axis. In this example, $T = 3.6$ LM, $ct_o = 6.6$ LM, and the time step $c\Delta t = 0.06$ LM.

The IE program evaluated the current for thc first 183 time steps. Then, 61 time samples from numbers 123 to 183 are used as the input to the matrix pencil program. L is chosen to be 32. The estimate for the number of modes is $M = 4$. The numerical values of the amplitudes and the exponent of the modes are

$$
\begin{aligned}
R_1 &= 0.154758 + j0.004330 & z_1 &= -1.90179 + j5.45202\\
R_2 &= 0.154758 - j0.004330 & z_2 &= -1.90179 - j5.45202\\
R_3 &= 0.136469 - j0.4115999 & z_3 &= -1.29296 + j3.03859\\
R_4 &= 0.136469 + j0.4115999 & z_4 &= -1.29296 - j3.03859.
\end{aligned}
$$

Using these values of (R_i, z_i), $i = 1, \cdots, 4$, the current is evaluated from time steps 184 to 500. The extrapolated result is shown in Fig. 4.20.

The fourth example is a cube with a 1-m side centered at the origin. The faces of the cube are lined along the three coordinate axes. The faces at $x = 0.5$ m and $x = -0.5$ m have five divisions in the y and z direction. All other faces have four divisions in one direction and five in the other. This allows us to find the current at the center of the top face. The excitation arrives along the $-z$-axis and polarized along the x-axis. In this example, $T = 2.8$ LM, $ct_o = 6$ LM, and the time step $c\Delta t = 0.047$ LM.

The IE program evaluated the current for the first 193 time steps. Then, 64 time samples from numbers 130 to 193 are provided as the input to the matrix pencil program. L is chosen to be 32. The estimate for the number of modes is $M = 4$. The numerical values of the amplitudes and

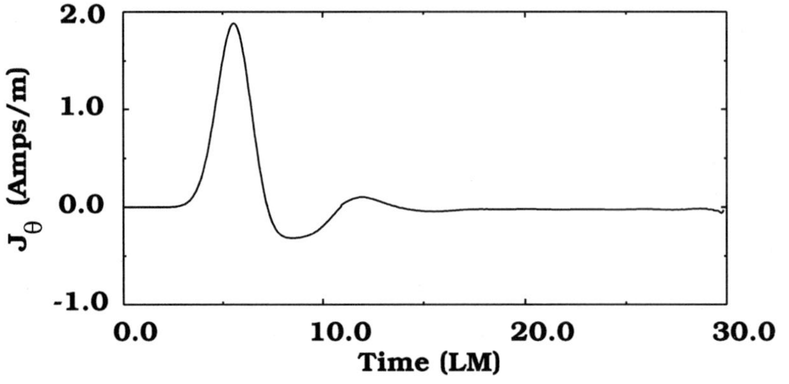

FIGURE 4.20
Transient current at (0.5, 0, 0) on a conducting sphere obtained by the matrix pencil method.

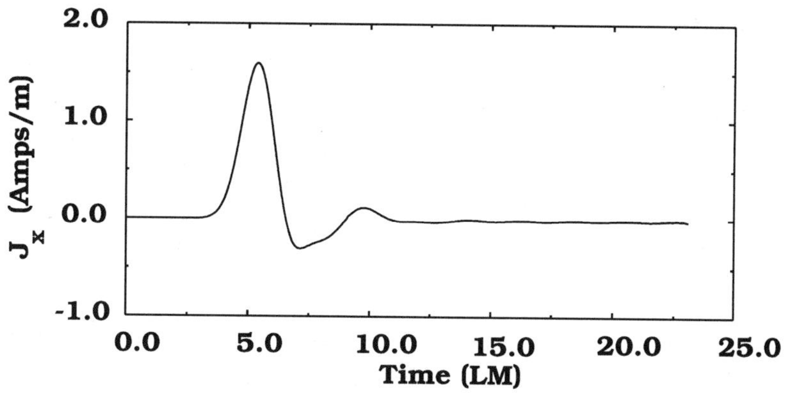

FIGURE 4.21
Transient current at the center of the top plate of a conducting cube obtained by the matrix pencil method.

exponents of the modes are

$$
\begin{aligned}
R_1 &= -0.04925 + j0.130665 & z_1 &= -2.21533 + j8.95793\\
R_2 &= -0.04925 - j0.130665 & z_2 &= -2.21533 - j8.95793\\
R_3 &= -0.311722 + j0.251171 & z_3 &= -1.58959 + j4.42677\\
R_4 &= -0.311722 - j0.251171 & z_4 &= -1.58959 - j4.42677.
\end{aligned}
$$

Using these values for (R_i, z_i), $i = 1, \cdots 4$, the current at the center of the top face is evaluated from time steps 194 to 500. The extrapolated result is shown in Fig. 4.21.

The final example is a combination of a cone and a hemisphere. The hemisphere is attached to the base of the cone forming one compound three-dimensional object. The base of the cone and hemisphere is centered at the origin and it has a radius of 1 m. The height of the cone is 2 m. The central axis of the combination lies on the z-axis. The triangular patch approximation for the cone has six divisions in the z direction. The planes defining the rings are at $z = 2.0$, $z = 1.75$, $z = 1.4$, $z = 1.05$, $z = 0.7$, $z = 0.35$, and $z = 0$, respectively. Each ring, starting from the top, has 7, 16, 20, 24, 28 and 32 triangles, respectively. The hemisphere has three divisions in the θ direction. The rings extend from $\theta = \pi$ to $\theta = 2\pi/3$, $\theta = 5\pi/6$ to $\theta = 2\pi/3$, and $\theta = 2\pi/3$ to $\theta = \pi/2$. Each ring, starting from the bottom, has 13, 28, and 32 triangular patches, respectively. Such a triangulation scheme allows for the current at the point $(-0.1, 0.0, 0.0)$ to be evaluated. The excitation arrives along the x direction and the electric field is polarized along the z-axis. In this example, $T = 7.2$ LM, $ct_o = 7.5$ LM, and the time step $c\Delta t = 0.027$ LM.

The IE program evaluated the current for the first 482 time steps. Then, 160 time samples from numbers 323 to 482 are used as input to the matrix pencil program. L is chosen to be 60. The estimate for the number of modes is $M = 4$. The numerical values of the mode amplitudes and exponents are

$$
\begin{aligned}
R_1 &= -0.205264 + j0.364285 & z_1 &= -0.907949 + j2.55135\\
R_2 &= -0.205264 - j0.364285 & z_2 &= -0.907949 - j2.55135\\
R_3 &= -0.002816 + j0.002763 & z_3 &= -5.64474 + j9.75556\\
R_4 &= -0.002816 - j0.002763 & z_4 &= -5.64474 - j9.75556.
\end{aligned}
$$

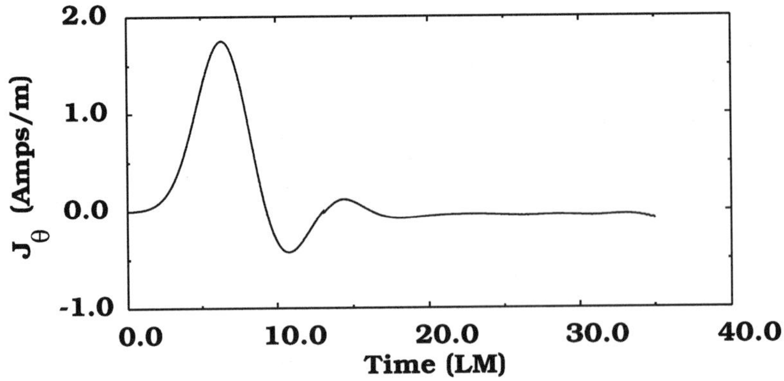

FIGURE 4.22
Transient current at (−0.1, 0.0, 0.0) of a conducting cone–sphere obtained by the matrix pencil method.

Using these values for the amplitudes and exponents, the current values are extrapolated from time steps 483 to 1300. The extrapolated result is shown in Fig. 4.22.

4.6 CONCLUDING REMARKS

In this chapter, we developed the numerical algorithms to calculate the transient electromagnetic scattering from finite-sized conducting bodies of arbitrary shape by solving time domain integral equations. It is easy to visualize that many practical structures fall into this category and applications for this type of problems are numerous. However, the present-day applications involve not only conductors but also nonconducting material bodies, and the electromagnetic interaction phenomenon with these structures is an important research area. We shall consider the scattering from material bodies in Chapter 5.

BIBLIOGRAPHY

[1] S. M. Rao and D. R. Wilton, "Transient Scattering by Conducting Surfaces of Arbitrary Shape," *IEEE Trans. Antennas Propagat.*, vol. 39, pp. 56–61, 1991.

[2] D. A. Vechinski and S. M. Rao, "A Stable Procedure to Calculate the Transient Scattering by Conducting Surfaces of Arbitrary Shape," *IEEE Trans. Antennas Propagat.*, vol. 40, pp. 661–665, 1992.

[3] S. M. Rao and T. K. Sarkar, "An Alternate Version of the Time-Domain Electric Field Integral Equation for Arbitrarily Shaped Conductors," *IEEE Trans. Antennas Propagat.*, vol. 41, pp. 831–834, 1993.

[4] R. F. Harrington, *Field Computation by Moment Methods*, Macmillan, New York, 1968.

[5] S. M. Rao, D. R. Wilton, and A. W. Glisson, "Electromagnetic Scattering by Surfaces of Arbitrary Shape," *IEEE Trans. Antennas Propagat.*, vol. 30, pp. 409–418, 1982.

[6] D. R. Wilton, S. M. Rao, A. W. Glisson, D. H. Schaubert, O. M. Al-Bundak, and C. M. Butler, "Potential Integrals for Uniform and Linear Source Distributions on Polygonal and Polyhedral Domains," *IEEE Trans. Antennas Propagat.*, vol. 32, pp. 276–281, 1984.

[7] R. Prony, "Essai Experimental et Analytique sur les Lois de la Dilatabilite de Fluides Elastiques et sur Celles del la Force Expansive de la Vapeur de l'Alkool a Differentes Temparatures," *Paris J. l'Ecole Polytech.*, vol. 1, pp. 24–76, 1795.

[8] Y. Hua and T. K. Sarkar, "Matrix Pencil Method for Estimating Parameters of Exponentially, Damped/Undamped Sinusoids in Noise," *IEEE Trans. Acoust. Speech Signal Processing*, vol. 38, pp. 814–824, 1990.

[9] O. M. Pereira-Filho and T. K. Sarkar, "Using the Matrix Pencil Method to Estimate the Parameters of a Sum of Complex Exponentials," *IEEE Antennas Propagat. Mag.*, vol. 37, pp. 48–55, 1995.

[10] F. Hu, "The Band-Pass Matrix Pencil Method for Parameter Estimation of Exponentially Damped/ Undamped Sinusoidal Signals in Noise," PhD thesis, Syracuse University, Syracuse, NY, 1990.

[11] G. H. Golub and C. F. Van Loan, *Matrix Computations*, Johns Hopkins Univ. Press, Baltimore, MD, 1989.

CHAPTER 5

Dielectric Bodies: TDIE Solution

S. M. RAO
Department of Electrical Engineering, Auburn University

D. A. VECHINSKI
Nichols Research Corporation

In this chapter, we develop the numerical solution methods to solve the time domain integral equations involving dielectric bodies. We consider the solution schemes for infinite cylinders of arbitrary cross section (two-dimensional problems) and also finite-sized arbitrary-shaped bodies (three-dimensional problems). Here, we consider the explicit solution schemes only since implicit solution schemes are currently not available for dielectric bodies. However, it should be a relatively easy task to extend the explicit solution schemes to develop implicit algorithms.

There exist many situations in which the scattering from dielectric material bodies is important. Most of the present-day antenna structures involve some kind of dielectric material for the purpose of either feed support or ground plane coverage. Furthermore, many conducting scatterers are either selectively or entirely coated with radar-absorbing materials to reduce the electromagnetic visibility. Lastly, almost all the biological applications involve dielectric bodies.

The formulation of the integral equations is presented in Section 5.1 using the equivalence principle [1]. We present the solution to two- and three-dimensional problems along with some representative examples in Sections 5.2 and 5.3, respectively.

5.1 INTEGRAL EQUATION FORMULATION

Consider an arbitrarily shaped, dielectric body described by a surface S as shown in Fig. 5.1. The scatterer has material parameters of μ_d and ϵ_d, and exterior to the body is a homogeneous medium with parameters μ_e and ϵ_e. Exterior to the body, the total fields are designated by $\boldsymbol{E}_e$ and $\boldsymbol{H}_e$, whereas interior to the object the fields are given by $\boldsymbol{E}_d$ and $\boldsymbol{H}_d$. The structure is illuminated by an arbitrary electromagnetic, plane wave pulse. Also, the tangential components of the electric and magnetic fields must be continuous at the dielectric interface S as dictated by the boundary conditions.

We assume that the dielectric body is a closed body so that a unique outward normal vector can be defined unambiguously. Employing the equivalence principle, the body may be replaced

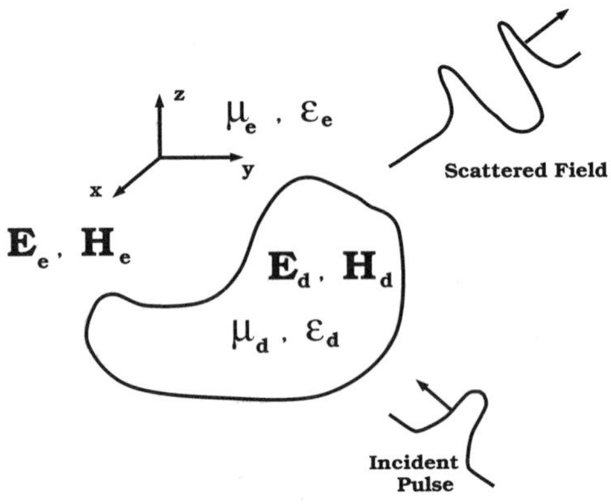

FIGURE 5.1
Arbitrarily shaped dielectric body illuminated by an electromagnetic pulse.

with two sets of electric ($\boldsymbol{J}_e$ and $\boldsymbol{J}_d$) and magnetic ($\boldsymbol{M}_e$ and $\boldsymbol{M}_d$) currents. Each set radiates in an infinite homogeneous medium having constitutive parameters associated with medium "e" or "d." It can be easily proved that the continuity of the tangential fields requires that

$$\boldsymbol{J}_e = -\boldsymbol{J}_d \equiv \boldsymbol{J} \tag{5.1.1}$$

$$\boldsymbol{M}_e = -\boldsymbol{M}_d \equiv \boldsymbol{M}. \tag{5.1.2}$$

Using the potential theory [1], the scattered fields radiated by the equivalent electric and magnetic currents may be written in terms of potential functions as

$$\boldsymbol{E}_\nu^{\mathrm{s}}[\boldsymbol{J}, \boldsymbol{M}] = \mp \frac{\partial \boldsymbol{A}_\nu}{\partial t} \mp \nabla \Phi_\nu^{\mathrm{e}} \mp \frac{1}{\epsilon_\nu} \nabla \times \boldsymbol{F}_\nu \tag{5.1.3}$$

$$\boldsymbol{H}_\nu^{\mathrm{s}}[\boldsymbol{J}, \boldsymbol{M}] = \mp \frac{\partial \boldsymbol{F}_\nu}{\partial t} \mp \nabla \Phi_\nu^{\mathrm{m}} \mp \frac{(-1)}{\mu_\nu} \nabla \times \boldsymbol{A}_\nu, \tag{5.1.4}$$

where $\boldsymbol{A}_\nu$ and $\boldsymbol{F}_\nu$ are the magnetic and electric vector potentials, respectively, and Φ_ν^{e} and Φ_ν^{m} are the electric and magnetic scalar potentials, respectively, given by

$$\boldsymbol{A}_\nu(\boldsymbol{r}, t) = \frac{\mu_\nu}{4\pi} \int_S \frac{\boldsymbol{J}(\boldsymbol{r}', t - R/c_\nu)}{R} \, dS' \tag{5.1.5}$$

$$\boldsymbol{F}_\nu(\boldsymbol{r}, t) = \frac{\epsilon_\nu}{4\pi} \int_S \frac{\boldsymbol{M}(\boldsymbol{r}', t - R/c_\nu)}{R} \, dS' \tag{5.1.6}$$

$$\Phi_\nu^{\mathrm{e}}(\boldsymbol{r}, t) = \frac{1}{4\pi \epsilon_\nu} \int_S \frac{q_{\mathrm{s}}^{\mathrm{e}}(\boldsymbol{r}', t - R/c_\nu)}{R} \, dS' \tag{5.1.7}$$

$$\Phi_\nu^{\mathrm{m}}(\boldsymbol{r}, t) = \frac{1}{4\pi \mu_\nu} \int_S \frac{q_{\mathrm{s}}^{\mathrm{m}}(\boldsymbol{r}', t - R/c_\nu)}{R} \, dS', \tag{5.1.8}$$

for $\nu = e$ or $\nu = d$; the "$-$" sign is for $\nu = e$, and the "$+$" is for $\nu = d$. In Eqs. (5.1.5)–(5.1.8), $R = |\boldsymbol{r} - \boldsymbol{r}'|$, the distance from the field point, $\boldsymbol{r}$, to the source point, $\boldsymbol{r}'$. The electric and magnetic surface charge density, q_s^{e} and q_s^{m}, respectively, are related to the electric and magnetic surface current density by the continuity equations

$$\nabla \cdot \boldsymbol{J} = -\frac{\partial q_s^e}{\partial t} \tag{5.1.9}$$

and

$$\nabla \cdot \boldsymbol{M} = -\frac{\partial q_s^m}{\partial t}, \tag{5.1.10}$$

respectively. Note that the time retardation, R/c_ν, depends on which medium the field is in.

We may eliminate q_s^e and q_s^m from Eqs. (5.1.7) and (5.1.8), respectively, by defining

$$\Psi_\nu^e(\boldsymbol{r}, t) \equiv \frac{\partial \Phi_\nu^e(\boldsymbol{r}, t)}{\partial t} = \frac{1}{4\pi\epsilon_\nu} \int_S \frac{\partial q_s^e(\boldsymbol{r}', t - R/c_\nu)}{\partial t} \frac{1}{R} dS' \tag{5.1.11}$$

$$= \frac{-1}{4\pi\epsilon_\nu} \int_S \frac{\nabla' \cdot \boldsymbol{J}(\boldsymbol{r}', t - R/c_\nu)}{R} dS' \tag{5.1.12}$$

$$\Psi_\nu^m(\boldsymbol{r}, t) \equiv \frac{\partial \Phi_\nu^m(\boldsymbol{r}, t)}{\partial t} = \frac{1}{4\pi\mu_\nu} \int_S \frac{\partial q_s^m(\boldsymbol{r}', t - R/c_\nu)}{\partial t} \frac{1}{R} dS' \tag{5.1.13}$$

$$= \frac{-1}{4\pi\mu_\nu} \int_S \frac{\nabla' \cdot \boldsymbol{M}(\boldsymbol{r}', t - R/c_\nu)}{R} dS' . \tag{5.1.14}$$

By enforcing the continuity of the tangential electric and magnetic fields at the dielectric interface and taking an extra derivative with respect to time, we derive the following integral equations:

$$\frac{\partial}{\partial t}\left[\boldsymbol{E}_e^s - \boldsymbol{E}_d^s\right]_{\tan} = -\frac{\partial}{\partial t}[\boldsymbol{E}^i]_{\tan} \qquad \boldsymbol{r} \in S \tag{5.1.15}$$

$$\frac{\partial}{\partial t}\left[\boldsymbol{H}_e^s - \boldsymbol{H}_d^s\right]_{\tan} = -\frac{\partial}{\partial t}[\boldsymbol{H}^i]_{\tan} \qquad \boldsymbol{r} \in S, \tag{5.1.16}$$

which can be rewritten as

$$\left[\frac{\partial^2}{\partial t^2}(\boldsymbol{A}_e + \boldsymbol{A}_d) + \nabla\left(\Psi_e^e + \Psi_d^e\right) + \nabla \times \left(\frac{1}{\epsilon_e}\frac{\partial \boldsymbol{F}_e}{\partial t} + \frac{1}{\epsilon_d}\frac{\partial \boldsymbol{F}_d}{\partial t}\right)\right]_{\tan} = \left[\frac{\partial \boldsymbol{E}^i}{\partial t}\right]_{\tan} \tag{5.1.17}$$

$$\left[\frac{\partial^2}{\partial t^2}(\boldsymbol{F}_e + \boldsymbol{F}_d) + \nabla\left(\Psi_e^m + \Psi_d^m\right) - \nabla \times \left(\frac{1}{\mu_e}\frac{\partial \boldsymbol{A}_e}{\partial t} + \frac{1}{\mu_d}\frac{\partial \boldsymbol{A}_d}{\partial t}\right)\right]_{\tan} = \left[\frac{\partial \boldsymbol{H}^i}{\partial t}\right]_{\tan} . \tag{5.1.18}$$

The formulation described so far is popularly known as the PMCHW (Poggio, Miller, Chang, Harrington, and & Wu) formulation [2].

Several other formulations can be derived via the equivalence principle [3]. Furthermore, it is a well-known fact that for the problems in the frequency domain, certain types of formulations have distinct advantages in the numerical solution compared to other types of formulations. Although numerical solutions to several different formulations have been developed in the time domain [4], there seems to be no real advantage in using one formulation over the other. Thus, we limit our treatment to the PMCHW formulation.

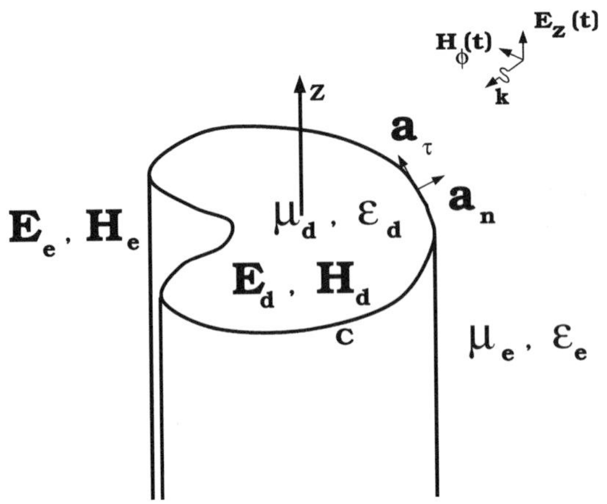

FIGURE 5.2
Two-dimensional dielectric cylinder illuminated by an transverse magnetic electromagnetic pulse.

5.2 TWO-DIMENSIONAL CYLINDERS

For the two-dimensional cylinder shown in Fig. 5.2, we consider only transverse magnetic (TM) incidence. The transverse electric (TE) case may be obtained through the use of duality [1].

For the TM case, we note that $\boldsymbol{J}=J_z\boldsymbol{a}_z$ and $\boldsymbol{M}=M_\tau\boldsymbol{a}_\tau$. We also note that for TM incidence, the electric current density only has a z component, and since all derivatives with respect to z are zero this implies $\nabla\Phi_\nu^{\mathrm{e}}=0$. Thus, Eqs. (5.1.17) and (5.1.18) reduce to

$$\left[\frac{\partial^2}{\partial t^2}(\boldsymbol{A}_e+\boldsymbol{A}_d)+\nabla\times\left(\frac{1}{\epsilon_e}\frac{\partial \boldsymbol{F}_e}{\partial t}+\frac{1}{\epsilon_d}\frac{\partial \boldsymbol{F}_d}{\partial t}\right)\right]_{\tan}=\left[\frac{\partial \boldsymbol{E}^{\mathrm{i}}}{\partial t}\right]_{\tan} \tag{5.2.1}$$

and

$$\left[\frac{\partial^2}{\partial t^2}(\boldsymbol{F}_e+\boldsymbol{F}_d)+\nabla\left(\Psi_e^{\mathrm{m}}+\Psi_d^{\mathrm{m}}\right)-\nabla\times\left(\frac{1}{\mu_e}\frac{\partial \boldsymbol{A}_e}{\partial t}+\frac{1}{\mu_d}\frac{\partial \boldsymbol{A}_d}{\partial t}\right)\right]_{\tan}=\left[\frac{\partial \boldsymbol{H}^{\mathrm{i}}}{\partial t}\right]_{\tan}, \tag{5.2.2}$$

where

$$\boldsymbol{A}_\nu(\boldsymbol{\rho},t)=\frac{\mu_\nu}{4\pi}\int_C\int_{z'=-\infty}^{\infty}\frac{\boldsymbol{J}(\boldsymbol{\rho}',t-R/c_\nu)}{R}\,dz'\,dC', \tag{5.2.3}$$

$$\boldsymbol{F}_\nu(\boldsymbol{\rho}',t)=\frac{\epsilon_\nu}{4\pi}\int_C\int_{z'=-\infty}^{\infty}\frac{\boldsymbol{M}(\boldsymbol{\rho}',t-R/c_\nu)}{R}\,dz'\,dC', \tag{5.2.4}$$

$$\Phi_\nu^{\mathrm{m}}(\boldsymbol{\rho},t)=\frac{1}{4\pi\mu_\nu}\int_C\int_{z'=-\infty}^{\infty}\frac{q_{\mathrm{s}}^{\mathrm{m}}(\boldsymbol{\rho}',t-R/c_\nu)}{R}\,dz'\,dC', \tag{5.2.5}$$

and $R=\sqrt{|\boldsymbol{\rho}-\boldsymbol{\rho}'|^2+z'^2}$, the distance from the field point to the source point, and $\nu=e$ or d. Finally, note that the actual cylinder is closed for penetrable objects.

Here, the evaluation of the electric current, $\boldsymbol{J}=J_z\boldsymbol{a}_z$, for the dielectric cylinder problem is similar to the evaluation of the current induced on an infinite conducting cylinder illuminated by a TM incident pulse (see Chapter 3, Section 3.3). In both cases, the current is along the z-axis and the governing equations are similar. Hence, the mathematical treatment in this case is very similar to that discussed in Chapter 3, Section 3.3. Furthermore, the magnetic current $\boldsymbol{M} = M_\tau \boldsymbol{a}_\tau$ is similar to the TE scattering from conducting cylinders. In this case, however, the governing equations are similar to those of the H-field integral equation (HFIE) solution presented in Chapter 3, Section 3.5.

5.2.1 Numerical Solution Procedure

The grid scheme that will be used is the same as that used for the TE conducting case in Chapter 3 (see Fig. 3.2 and 3.3b).

For the numerical solution of Eqs. (5.2.1) and (5.2.2), we start by defining the the basis functions given by

$$f_m(\rho) \equiv \begin{cases} 1 & \rho \in \rho_{m-1/2} \text{ to } \rho_{m+1/2} \\ 0 & \text{otherwise,} \end{cases} \tag{5.2.6}$$

and approximating the electric and magnetic currents as

$$\boldsymbol{J}(\boldsymbol{\rho}, t) = \boldsymbol{a}_z \sum_{k=1}^{N} I_k(t) f_k(\boldsymbol{\rho}) \tag{5.2.7}$$

$$\boldsymbol{M}(\boldsymbol{\rho}, t) = \boldsymbol{a}_\tau \sum_{k=1}^{N} M_k(t) f_k(\boldsymbol{\rho}). \tag{5.2.8}$$

Using Eq. (5.2.6) as a testing function, we apply the testing procedure to Eqs. (5.2.1) and (5.2.2) to obtain

$$\left\langle f_m \boldsymbol{a}_z, \frac{\partial^2}{\partial t^2}(\boldsymbol{A}_e + \boldsymbol{A}_d) + \nabla \times \left(\frac{1}{\epsilon_e} \frac{\partial \boldsymbol{F}_e}{\partial t} + \frac{1}{\epsilon_d} \frac{\partial \boldsymbol{F}_d}{\partial t} \right) \right\rangle = \left\langle f_m \boldsymbol{a}_z, \frac{\partial \boldsymbol{E}^{\mathrm{i}}}{\partial t} \right\rangle \tag{5.2.9}$$

and

$$\left\langle f_m \boldsymbol{a}_\tau, \frac{\partial^2}{\partial t^2}(\boldsymbol{F}_e + \boldsymbol{F}_d) + \nabla \left(\Psi_e^{\mathrm{m}} + \Psi_d^{\mathrm{m}} \right) - \nabla \times \left(\frac{1}{\mu_e} \frac{\partial \boldsymbol{A}_e}{\partial t} + \frac{1}{\mu_d} \frac{\partial \boldsymbol{A}_d}{\partial t} \right) \right\rangle$$
$$= \left\langle f_m \boldsymbol{a}_\tau, \frac{\partial \boldsymbol{H}^{\mathrm{i}}}{\partial t} \right\rangle. \tag{5.2.10}$$

Defining

$$\boldsymbol{L}_\nu^E[\boldsymbol{J}, \boldsymbol{M}] = \frac{\boldsymbol{A}_\nu(\boldsymbol{\rho}, t_{n+1}) - 2\boldsymbol{A}_\nu(\boldsymbol{\rho}, t_n) + \boldsymbol{A}_\nu(\boldsymbol{\rho}, t_{n-1})}{\Delta t^2} + \frac{1}{\epsilon_\nu} \nabla \times \left[\frac{\boldsymbol{F}_\nu(\boldsymbol{\rho}, t_{n+1}) - \boldsymbol{F}_\nu(\boldsymbol{\rho}, t_n)}{\Delta t} \right] \tag{5.2.11}$$

and

$$\boldsymbol{L}_\nu^H[\boldsymbol{J}, \boldsymbol{M}] = \frac{\boldsymbol{F}_\nu(\boldsymbol{\rho}, t_{n+1}) - 2\boldsymbol{F}_\nu(\boldsymbol{\rho}, t_n) + \boldsymbol{F}_\nu(\boldsymbol{\rho}, t_{n-1})}{\Delta t^2} + \nabla \Psi_\nu^{\mathrm{m}}(\boldsymbol{\rho}, t_n) - \frac{1}{\mu_\nu} \nabla \times \left[\frac{\boldsymbol{A}_\nu(\boldsymbol{\rho}, t_{n+1}) - \boldsymbol{A}_\nu(\boldsymbol{\rho}, t_n)}{\Delta t} \right], \tag{5.2.12}$$

Eqs. (5.2.9) and (5.2.10) may be written as

$$\langle f_m \boldsymbol{a}_z, \boldsymbol{L}_e^E[\boldsymbol{J}, \boldsymbol{M}] + \boldsymbol{L}_d^E[\boldsymbol{J}, \boldsymbol{M}] \rangle = \left\langle f_m \boldsymbol{a}_z, \frac{\partial \boldsymbol{E}^{\mathrm{i}}(t_n)}{\partial t} \right\rangle \tag{5.2.13}$$

$$\langle f_m \boldsymbol{a}_\tau, \boldsymbol{L}_e^H[\boldsymbol{J}, \boldsymbol{M}] + \boldsymbol{L}_d^H[\boldsymbol{J}, \boldsymbol{M}] \rangle = \left\langle f_m \boldsymbol{a}_\tau, \frac{\partial \boldsymbol{H}^{\mathrm{i}}(t_n)}{\partial t} \right\rangle. \tag{5.2.14}$$

The incident field is assumed to be known so that its derivative may be evaluated either analytically or numerically. Now, we consider the evaluation of $\boldsymbol{L}_\nu^E$ which implies the evaluation of $\boldsymbol{A}_\nu(\boldsymbol{\rho}, t_n)$ and $\nabla \times \boldsymbol{F}_\nu(\boldsymbol{\rho}, t_n)$.

The evaluation of $\boldsymbol{A}_\nu(\boldsymbol{\rho}, t_n)$ is the same as that in Chapter 3, Section 3.3 except that we make the following substitutions: $\boldsymbol{A} \Rightarrow \boldsymbol{A}_\nu$, $\mu \Rightarrow \mu_\nu$, and $c \Rightarrow c_\nu$. Making these substitutions in Eq. (3.3.12), we obtain

$$\begin{aligned} \boldsymbol{A}_\nu(\boldsymbol{\rho}, t_n) &= \frac{\mu_\nu \kappa_m}{4\pi} I_m(t_n)\, \boldsymbol{a}_z + \frac{\mu_\nu}{4\pi} \sum_{\substack{k=1 \\ k\neq m \text{ and } \ell\neq 0}}^{N} \sum_{\ell=-\infty}^{\infty} I_k\left(t_n - \frac{R_{mk\ell}}{c_\nu}\right) \kappa_{k,\ell}\, \boldsymbol{a}_z \\ &= \frac{\mu_\nu \kappa_m}{4\pi} I_m(t_n)\, \boldsymbol{a}_z + \not{\boldsymbol{A}}_\nu(\boldsymbol{\rho}, t_n), \end{aligned} \tag{5.2.15}$$

where $\kappa_{k,\ell}$, R, and $R_{mk\ell}$ are given by Eqs. (3.3.8)–(3.3.10). The $1/R$ integral may be approximated according to Eq. (3.3.11). Note that in order for the first term of Eq. (5.2.15) to be the only unknown, Δt must be chosen according to $\Delta t \leq R_{\min}/(\max\{c_e, c_d\})$.

The determination of $\frac{1}{\epsilon_\nu}\nabla \times \boldsymbol{F}_\nu$ follows the discussion of $\frac{1}{\mu}\nabla \times \boldsymbol{A}$ in Chapter 3 Section 3.5. Thus, using Eq. (3.5.8) and making the substitutions $\boldsymbol{A} \Rightarrow \boldsymbol{F}_\nu$, $\boldsymbol{J} \Rightarrow \boldsymbol{M}$, and $c \Rightarrow c_\nu$, we write $\frac{1}{\epsilon_\nu}\nabla \times \not{\boldsymbol{F}}_\nu$ as

$$\begin{aligned} \frac{1}{\epsilon_\nu}\nabla \times \not{\boldsymbol{F}}_\nu(\boldsymbol{\rho}, t_n) &= \frac{1}{4\pi c_\nu} \sum_{\substack{k=1 \\ k\neq m \text{ and } \ell\neq 0}}^{N} \sum_{\ell=-\infty}^{\infty} \frac{\partial M_k(t_r)}{\partial t_r} \int_{k,\ell} \int_{\text{patch}} \frac{\boldsymbol{a}'_\tau \times \boldsymbol{a}_R}{R}\, ds' \\ &+ \frac{1}{4\pi} \sum_{\substack{k=1 \\ k\neq m \text{ and } \ell\neq 0}}^{N} \sum_{\ell=-\infty}^{\infty} M_k\left(t_n - \frac{R_{mk\ell}}{c_\nu}\right) \int_{k,\ell} \int_{\text{patch}} \frac{\boldsymbol{a}'_\tau \times \boldsymbol{a}_R}{R^2}\, ds'. \end{aligned} \tag{5.2.16}$$

As a first-order approximation, a 1-point integration may be used for the non-self term integrals. The $\partial M_k/\partial t_r$ term is approximated by a first-order backward difference given by

$$\frac{\partial M_k(t_r)}{\partial t_r} \approx \frac{M_k(t_n - R_{mk\ell}/c_\nu) - M_k(t_{n-1} - R_{mk\ell}/c_\nu)}{\Delta t}. \tag{5.2.17}$$

Finally, the curl of the electric vector potential may be written as

$$\frac{1}{\epsilon_\nu}\nabla \times \boldsymbol{F}_\nu(\boldsymbol{\rho}, t_n) = \pm \frac{M_m(t_n)}{2} \boldsymbol{a}_z + \frac{1}{\epsilon_\nu}\nabla \times \not{\boldsymbol{F}}_\nu(\boldsymbol{\rho}, t_n). \tag{5.2.18}$$

Note that the $\pm M_m/2$ is obtained by extracting the Cauchy principal value from the curl term. The "+" sign is for $\nu = e$, and the "–" sign is for $\nu = d$.

Next, we consider the evaluation of L_ν^H which implies the evaluation of $\boldsymbol{F}_\nu(\boldsymbol{\rho}, t_n)$, $\nabla\Psi_\nu^{\mathrm{m}}(\boldsymbol{\rho}, t_n)$, and $\nabla \times \boldsymbol{A}_\nu(\boldsymbol{\rho}, t_n)$. The evaluation of $\boldsymbol{F}_\nu$ and $\nabla \times \boldsymbol{A}_\nu$ is very similar to the determination of $\boldsymbol{A}_\nu$ and $\nabla \times \boldsymbol{F}_\nu$, respectively. Therefore,

$$\begin{aligned}\boldsymbol{F}_\nu(\boldsymbol{\rho}, t_n) &= \frac{\epsilon_\nu}{4\pi}\sum_{k=1}^{N}\sum_{\ell=-\infty}^{\infty} M_k\left(t_n - \frac{R_{mk\ell}}{c_\nu}\right)\kappa_{k,\ell}\,\boldsymbol{a}_\tau \\ &= \frac{\epsilon_\nu \kappa_m}{4\pi} M_m(t_n)\,\boldsymbol{a}_\tau + \not{\boldsymbol{F}}_\nu(\boldsymbol{\rho}, t_n),\end{aligned} \tag{5.2.19}$$

and

$$\frac{1}{\mu_\nu}\nabla \times \boldsymbol{A}_\nu = \mp\frac{I_m(t_n)}{2}\boldsymbol{a}_\tau + \frac{1}{\mu_\nu}\nabla \times \not{\boldsymbol{A}}_\nu(\boldsymbol{\rho}, t_n), \tag{5.2.20}$$

where

$$\begin{aligned}\frac{1}{\mu_\nu}\nabla \times \not{\boldsymbol{A}}_\nu(\boldsymbol{\rho}, t_n) = {}& \frac{1}{4\pi c_\nu}\sum_{\substack{k=1\\ k\neq m \text{ and }}}^{N}\sum_{\substack{\ell=-\infty\\ \ell\neq 0}}^{\infty} \frac{\partial I_k(t_r)}{\partial t_r}\int_{k,\ell}\int_{\text{patch}} \frac{\boldsymbol{a}_z \times \boldsymbol{a}_R}{R}\,ds' \\ &+ \frac{1}{4\pi}\sum_{\substack{k=1\\ k\neq m \text{ and }}}^{N}\sum_{\substack{\ell=-\infty\\ \ell\neq 0}}^{\infty} I_k\left(t_n - \frac{R_{mk\ell}}{c_\nu}\right)\int_{k,\ell}\int_{\text{patch}} \frac{\boldsymbol{a}_z \times \boldsymbol{a}_R}{R^2}\,ds'.\end{aligned}$$

The scalar potential term, $\nabla\Psi_\nu^{\mathrm{m}}(\boldsymbol{\rho}, t_n)$, can be found by using the same mathematical steps for $\nabla\Phi(\boldsymbol{\rho}, t_n)$ in Chapter 3, Section 3.4. Note that $\Psi = \partial\Phi/\partial t$. By changing the order of the scalar product, we can rewrite the gradient term as

$$\begin{aligned}\left\langle f_m\boldsymbol{a}_z, \boldsymbol{a}_n \times \nabla\Psi_\nu^{\mathrm{m}}(\boldsymbol{\rho}, t_n)\right\rangle &= \int_S \nabla\Psi_\nu^{\mathrm{m}}(\boldsymbol{\rho}, t_n)\cdot f_m\boldsymbol{a}_z \times \boldsymbol{a}_n\,ds' \\ &= \Psi_\nu^{\mathrm{m}}(\boldsymbol{\rho}_{m+1/2}, t_n) - \Psi_\nu^{\mathrm{m}}(\boldsymbol{\rho}_{m-1/2}, t_n).\end{aligned} \tag{5.2.21}$$

If we replace $\boldsymbol{J} \Rightarrow \boldsymbol{M}$, $\epsilon \Rightarrow \mu_\nu$, and $c \Rightarrow c_\nu$ in Eqs. (3.4.17)–(3.4.19), the magnetic scalar potential is given by

$$\Psi_\nu^{\mathrm{m}}(\boldsymbol{\rho}_{m+1/2}, t_n) = \sum_{k=1}^{N}\sum_{\ell=-\infty}^{\infty} \Psi_\nu^{\mathrm{m}+}(\boldsymbol{\rho}_{m+1/2}, t_n) - \Psi_\nu^{\mathrm{m}-}(\boldsymbol{\rho}_{m+1/2}, t_n), \tag{5.2.22}$$

where

$$\Psi_\nu^{\mathrm{m}+}(\boldsymbol{\rho}_{m+1/2}, t_n) = \frac{-M_k(t_R^+)}{4\pi\mu_\nu\Delta\tau_k^+}\int_{\rho_k}^{\rho_{k+1}}\int_{z_1}^{z_2} \frac{d\ell'\,dz'}{\sqrt{|\boldsymbol{\rho}_{m+1/2} - \boldsymbol{\rho}'|^2 + z'^2}}, \tag{5.2.23}$$

$$\Psi_\nu^{\mathrm{m}-}(\boldsymbol{\rho}_{m+1/2}, t_n) = \frac{-M_k(t_R^-)}{4\pi\mu_\nu\Delta\tau_k^-}\int_{\rho_{k-1}}^{\rho_k}\int_{z_1}^{z_2} \frac{d\ell'\,dz'}{\sqrt{|\boldsymbol{\rho}_{m+1/2} - \boldsymbol{\rho}'|^2 + z'^2}}, \tag{5.2.24}$$

and $t_R^\pm$, $R_{mk\ell}^\pm$, $\Delta\tau_k^\pm$, z_1, and z_2 are defined by Eqs. (3.4.20)–(3.4.25), respectively.

After evaluating the operators $\boldsymbol{L}_{\nu}^{E}(\boldsymbol{J},\boldsymbol{M})$ and $\boldsymbol{L}_{\nu}^{H}(\boldsymbol{J},\boldsymbol{M})$ for $\nu = e$ or d, we turn our attention to Eqs. (5.2.9) and (5.2.10). Replacing $n \Rightarrow n-1$, taking all known quantities to the right, and using a 1-point integration for the testing procedure gives

$$\frac{\kappa_m \Delta\tau_m}{4\pi\,\Delta t^2}(\mu_e+\mu_d)I_m(t_n) = \langle f_m\boldsymbol{a}_z, \boldsymbol{Y}^E\rangle \tag{5.2.25}$$

$$\frac{\kappa_m \Delta\tau_m}{4\pi\,\Delta t^2}(\epsilon_e+\epsilon_d)M_m(t_n) = \langle f_m\boldsymbol{a}_\tau, \boldsymbol{Y}^H\rangle, \tag{5.2.26}$$

where

$$\boldsymbol{Y}^E = \frac{\partial \boldsymbol{E}^{\rm inc}(t_{n-1})}{\partial t} - \not{\boldsymbol{L}}_e^E[\boldsymbol{J},\boldsymbol{M}] - \not{\boldsymbol{L}}_d^E[\boldsymbol{J},\boldsymbol{M}], \tag{5.2.27}$$

$$\boldsymbol{Y}^H = \frac{\partial \boldsymbol{H}^{\rm inc}(t_{n-1})}{\partial t} - \not{\boldsymbol{L}}_e^H[\boldsymbol{J},\boldsymbol{M}] - \not{\boldsymbol{L}}_d^H[\boldsymbol{J},\boldsymbol{M}]. \tag{5.2.28}$$

Notice that the equations have been decoupled in the "present-time" sense. That is, Eq. (5.2.25) only has $I_m(t_n)$ as the unknown, whereas Eq. (5.2.26) only has $M_m(t_n)$ as the unknown. They are, however, still coupled in terms of past history currents. This decoupling comes about by self term cancelation which is an advantage of the PMCHW formulation. The currents may now be obtained by the marching on in time procedure.

5.2.2 Numerical Examples

In this section, we consider two examples: a circular dielectric cylinder and a square cylinder with an $\epsilon_r = 2.0$ immersed in air. The time domain results were obtained with an incident field of the form of Eq. (1.4.14) with $\boldsymbol{E}_\circ = \boldsymbol{a}_z$, $\boldsymbol{a}_k = -\boldsymbol{a}_x$, $T = 2$ LM, and $ct_o = 3$ LM. Furthermore, to overcome the instabilities prevalent in the explicit solution scheme, all the time domain results used the averaging scheme described in Chapter 1, Section 1.6.1. The time step used in all calculations is equal to $R_{\min}/2c_{\rm e}$. These results are compared with data obtained in the frequency domain and inverse Fourier transformed. For the inverse discrete Fourier transform (IDFT) solution, 128 frequencies between 0 and 0.5 GHz were used. The frequency domain results were obtained by using the method of moments (MoM) in conjunction with the PMCHW formulation.

The currents were found to sometimes grow quite quickly (without averaging), even before the peak of the incident pulse had occurred. This was believed to be attributed to numerical problems with the relative difference in the magnitude of the electric and magnetic currents, with the magnetic currents generally being approximately a factor of η larger. Therefore, if we scale $\boldsymbol{M}$ by the constant $\eta_{\rm e}$, (i.e., let $M = \eta_{\rm e} M'$) and solve for I and M', the currents will be on the same order. When scaling was used, better results were obtained.

Two different sizes of circular cylinders are considered initially. The first cylinder has a radius of 0.25 m and is centered about the origin. The circumference of the cylinder was approximated by 28 linear segments. The minimum distance between patch centers, $R_{\min}$, is 0.05528 m. Therefore, a time step of 0.02764 LM was chosen. Although an eigenfunction series solution exists for the frequency domain solution of a dielectric circular cylinder, the results shown were obtained by the MoM. The current at $\phi = 0°$ is used for comparison. Figure 5.3 shows the equivalent electric and magnetic currents as a function of time, and it is evident that good comparison is obtained in both cases. The second major peak may be attributed to the reflection of the

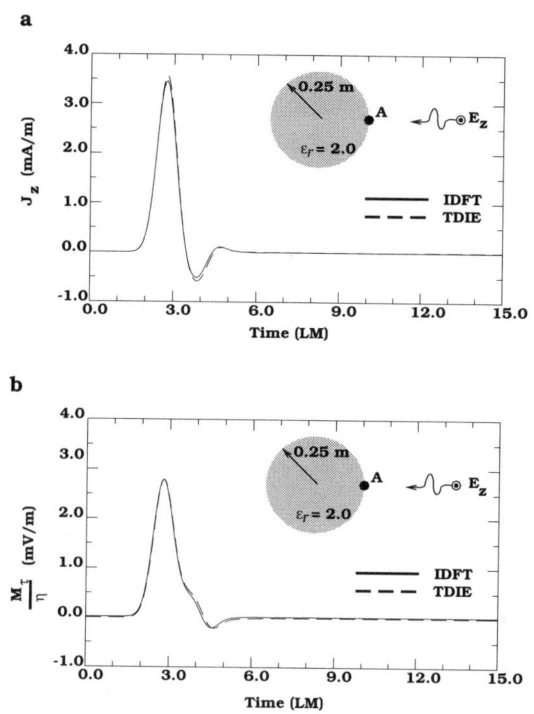

FIGURE 5.3
Equivalent (a) electric and (b) magnetic currents at point A on a dielectric circular cylinder, 0.25 m in radius, illuminated by a TM Gaussian plane wave. $N = 28$.

transmitted pulse from the back side of the cylinder. This peak is expected to occur at a time $ct = 2.75 + 2(.5)\sqrt{2} = 4.16$ LM; 2.75 is derived from the time when the peak of the incident field first strikes the cylinder, 2(.5) represents the total round trip distance the wave has to travel, and the factor of $\sqrt{2}$ arises from the fact that the field is traveling in a medium with $\epsilon_r = 2.0$.

Next, a much larger cylinder (radius, 1 m) is approximated with 44 linear segments and the results are shown in Fig. 5.4. The next peak in the current occurs 5.22 LM from the main peak. This is close to the time, $2(2)\sqrt{2} = 5.65$ LM, that it takes for a wave to propagate through the cylinder, reflect, and travel back.

Finally, a square cylinder, centered about the origin, with side length of 1 m is considered. A total of 40 segments, 10 on each side, are used for the time domain method, with $c\Delta t = 0.03536$ LM. Due to a slight difference in the way the two methods are computed, which results in a half-zone shift in the discretized contours, 80 zones are used in the frequency solution. The current at the center of the illuminated side is used for comparison. Figure 5.5 shows the equivalent electric and magnetic currents obtained using this formulation. The agreement is very good during the time interval of the main peak. There appears to be a slight DC shift in the electric current at late times. This shift may be attributed to the number of unknowns used, the difference of the discretization scheme, and that the corners are "rounded" in this method. The "down and back" time should be approximately $2\sqrt{2} = 2.82$ LM. The time (distance) from the first major peak to the second is about 2.56 LM.

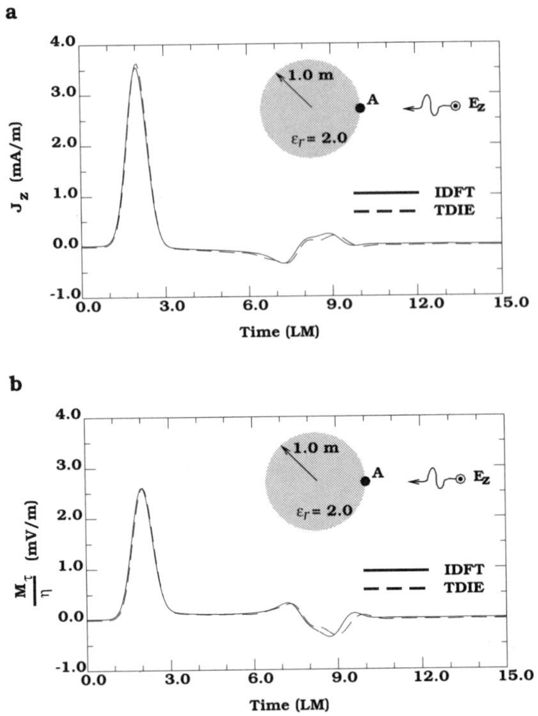

FIGURE 5.4
Equivalent (a) electric and (b) magnetic currents at point A on a dielectric circular cylinder, 1 m in radius, illuminated by a TM Gaussian plane wave. $N = 44$.

5.3 THREE-DIMENSIONAL BODIES

For the three-dimensional body problem we follow numerical procedures similar to those discussed in Chapter 4. For this purpose, we approximate the arbitrary body, assumed to be closed, by a set of planar triangular patches. As noted earlier, triangular patch modeling is capable of approximating any arbitrary body accurately and efficiently. Furthermore, for the application of the numerical procedures, we use the basis functions defined in Eq. (4.2.1). In fact, we use these functions to approximate both $\boldsymbol{J}$ and $\boldsymbol{M}$ which results in a very efficient algorithm. Also, we use the basis functions as the testing functions. In the following, we consider the numerical analysis in detail.

5.3.1 Numerical Solution Procedure

As the first step in the application of the MoM, we test Eqs. (5.1.17) and (5.1.18) to obtain

$$\left\langle \boldsymbol{f}_m, \frac{\partial^2}{\partial t^2}(\boldsymbol{A}_e + \boldsymbol{A}_d) + \nabla\left(\Psi_e^{\mathrm{e}} + \Psi_d^{\mathrm{e}}\right) + \nabla \times \left(\frac{1}{\epsilon_e}\frac{\partial \boldsymbol{F}_e}{\partial t} + \frac{1}{\epsilon_d}\frac{\partial \boldsymbol{F}_d}{\partial t}\right)\right\rangle = \left\langle \boldsymbol{f}_m, \frac{\partial \boldsymbol{E}^{\mathrm{i}}}{\partial t}\right\rangle \tag{5.3.1}$$

$$\left\langle \boldsymbol{f}_m, \frac{\partial^2}{\partial t^2}(\boldsymbol{F}_e + \boldsymbol{F}_d) + \nabla\left(\Psi_e^{\mathrm{m}} + \Psi_d^{\mathrm{m}}\right) - \nabla \times \left(\frac{1}{\mu_e}\frac{\partial \boldsymbol{A}_e}{\partial t} + \frac{1}{\mu_d}\frac{\partial \boldsymbol{A}_d}{\partial t}\right)\right\rangle = \left\langle \boldsymbol{f}_m, \frac{\partial \boldsymbol{H}^{\mathrm{i}}}{\partial t}\right\rangle, \tag{5.3.2}$$

for $m = 1, 2, \cdots, N$, where N is the total number of edges.

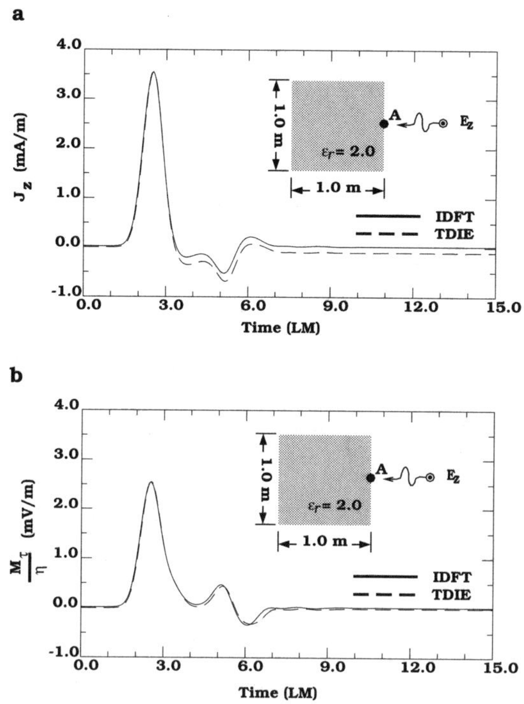

FIGURE 5.5
Equivalent (a) electric and (b) magnetic currents at point A on a dielectric square cylinder, illuminated by a TM Gaussian plane wave. $N = 40$ ($N = 80$ for the IDFT).

Approximating the time derivatives of the potential functions by finite differences, we get

$$\langle \boldsymbol{f}_m, \boldsymbol{L}_e^E[\boldsymbol{J}, \boldsymbol{M}] + \boldsymbol{L}_d^E[\boldsymbol{J}, \boldsymbol{M}]\rangle = \left\langle \boldsymbol{f}_m, \frac{\partial \boldsymbol{E}^{\mathrm{i}}(t_n)}{\partial t} \right\rangle \tag{5.3.3}$$

$$\langle \boldsymbol{f}_m, \boldsymbol{L}_e^H[\boldsymbol{J}, \boldsymbol{M}] + \boldsymbol{L}_d^H[\boldsymbol{J}, \boldsymbol{M}]\rangle = \left\langle \boldsymbol{f}_m, \frac{\partial \boldsymbol{H}^{\mathrm{i}}(t_n)}{\partial t} \right\rangle, \tag{5.3.4}$$

where

$$\boldsymbol{L}_\nu^E[\boldsymbol{J}, \boldsymbol{M}] = \frac{\boldsymbol{A}_\nu(\boldsymbol{r}, t_{n+1}) - 2\boldsymbol{A}_\nu(\boldsymbol{r}, t_n) + \boldsymbol{A}_\nu(\boldsymbol{r}, t_{n-1})}{\Delta t^2} + \nabla \Psi_\nu^{\mathrm{e}}(\boldsymbol{r}, t_n) + \frac{1}{\epsilon_\nu} \nabla \times \left[\frac{\boldsymbol{F}_\nu(\boldsymbol{r}, t_{n+1}) - \boldsymbol{F}_\nu(\boldsymbol{r}, t_n)}{\Delta t} \right] \tag{5.3.5}$$

$$\boldsymbol{L}_\nu^H[\boldsymbol{J}, \boldsymbol{M}] = \frac{\boldsymbol{F}_\nu(\boldsymbol{r}, t_{n+1}) - 2\boldsymbol{F}_\nu(\boldsymbol{r}, t_n) + \boldsymbol{F}_\nu(\boldsymbol{r}, t_{n-1})}{\Delta t^2} + \nabla \Psi_\nu^{\mathrm{m}}(\boldsymbol{r}, t_n) - \frac{1}{\mu_\nu} \nabla \times \left[\frac{\boldsymbol{A}_\nu(\boldsymbol{r}, t_{n+1}) - \boldsymbol{A}_\nu(\boldsymbol{r}, t_n)}{\Delta t} \right] \tag{5.3.6}$$

for $\nu = e$ or d. First, we consider the testing of the vector potential terms. Using the approximations of Eqs. (4.2.7)–(4.2.9), we can write $\langle \boldsymbol{f}_m, \boldsymbol{A}_\nu \rangle$ and $\langle \boldsymbol{f}_m, \boldsymbol{F}_\nu \rangle$ as

$$\langle \boldsymbol{f}_m, \boldsymbol{A}_\nu \rangle \approx \boldsymbol{A}_\nu(\boldsymbol{r}_m) \cdot \frac{l_m}{2}(\boldsymbol{\rho}_m^{c+} + \boldsymbol{\rho}_m^{c-}) \tag{5.3.7}$$

$$\langle \boldsymbol{f}_m, \boldsymbol{F}_\nu \rangle \approx \boldsymbol{F}_\nu(\boldsymbol{r}_m) \cdot \frac{l_m}{2}(\boldsymbol{\rho}_m^{c+} + \boldsymbol{\rho}_m^{c-}). \tag{5.3.8}$$

Next, the testing of the scalar potential terms is considered. Applying steps similar to those for the scalar potential in Eqs. (4.2.10) and (4.2.11), results in

$$\langle \boldsymbol{f}_m, \nabla \Psi_\nu^{\mathrm{e}} \rangle \approx -l_m\left[\Psi_\nu^{\mathrm{e}}(\boldsymbol{r}_m^{c+}) - \Psi_\nu^{\mathrm{e}}(\boldsymbol{r}_m^{c-})\right] \tag{5.3.9}$$

$$\langle \boldsymbol{f}_m, \nabla \Psi_\nu^{\mathrm{m}} \rangle \approx -l_m\left[\Psi_\nu^{\mathrm{m}}(\boldsymbol{r}_m^{c+}) - \Psi_\nu^{\mathrm{m}}(\boldsymbol{r}_m^{c-})\right]. \tag{5.3.10}$$

Next, the incident fields are tested in the same manner as the vector potential. Thus,

$$\left\langle \boldsymbol{f}_m, \frac{\partial \boldsymbol{E}^{\mathrm{i}}}{\partial t} \right\rangle \approx \frac{l_m}{2}(\boldsymbol{\rho}_m^{c+} + \boldsymbol{\rho}_m^{c-}) \cdot \frac{\partial \boldsymbol{E}^{\mathrm{i}}(\boldsymbol{r}_m, t_n)}{\partial t} \tag{5.3.11}$$

$$\left\langle \boldsymbol{f}_m, \frac{\partial \boldsymbol{H}^{\mathrm{i}}}{\partial t} \right\rangle \approx \frac{l_m}{2}(\boldsymbol{\rho}_m^{c+} + \boldsymbol{\rho}_m^{c-}) \cdot \frac{\partial \boldsymbol{H}^{\mathrm{i}}(\boldsymbol{r}_m, t_n)}{\partial t}. \tag{5.3.12}$$

This completes the testing procedure. Now, we consider the expansion procedure. For the expansion procedure, as noted earlier, we approximate the equivalent electric and magnetic currents on S by

$$\boldsymbol{J}(\boldsymbol{r}, t) = \sum_{k=1}^{N} I_k(t)\boldsymbol{f}_k(\boldsymbol{r}) \tag{5.3.13}$$

$$\boldsymbol{M}(\boldsymbol{r}, t) = \sum_{k=1}^{N} M_k(t)\boldsymbol{f}_k(\boldsymbol{r}). \tag{5.3.14}$$

For the dielectric body case, the number of boundary edges is equal to zero since a dielectric body is assumed to be closed. Combining Eqs. (5.1.5) and (5.3.13) and Eqs. (5.1.6) and (5.3.14) gives

$$\boldsymbol{A}_\nu(\boldsymbol{r}, t_n) = \frac{\mu_\nu}{4\pi}\sum_{k=1}^{N} I_k\left(t_n - \frac{R_{mk}}{c_\nu}\right)\boldsymbol{\kappa}_{mk} \tag{5.3.15}$$

$$\boldsymbol{F}_\nu(\boldsymbol{r}, t_n) = \frac{\epsilon_\nu}{4\pi}\sum_{k=1}^{N} M_k\left(t_n - \frac{R_{mk}}{c_\nu}\right)\boldsymbol{\kappa}_{mk}, \tag{5.3.16}$$

for a particular field point $\boldsymbol{r}$ associated with the mth edge at time t_n and where $\boldsymbol{\kappa}_{mk}$, R_{mk}, and R_m are given by Eqs. (4.2.16)–(4.2.18). If Δt is chosen suitably ($\Delta t \le R_{\min}/\max\{c_e, c_d\}$), then a single nonretarded current may be separated in Eqs. (5.3.15) and (5.3.16) from currents which occur at earlier time instances. The nonretarded currents occur when $k = m$, and Eqs. (5.3.15) and (5.3.16) may be rewritten as

$$\boldsymbol{A}_\nu(\boldsymbol{r}, t_n) = \frac{\mu_\nu \boldsymbol{\kappa}_{mm}}{4\pi} I_m(t_n) + \not{\boldsymbol{A}}_\nu(\boldsymbol{r}, t_n) \tag{5.3.17}$$

$$\boldsymbol{F}_\nu(\boldsymbol{r}, t_n) = \frac{\epsilon_\nu \boldsymbol{\kappa}_{mm}}{4\pi} M_m(t_n) + \not{\boldsymbol{F}}_\nu(\boldsymbol{r}, t_n), \tag{5.3.18}$$

with $\not{\boldsymbol{A}}_\nu$ and $\not{\boldsymbol{F}}_\nu$ given by Eqs. (5.3.15) and (5.3.16) with the $k = m$ terms omitted.

Next, the evaluation of the scalar potential terms is considered. By combining Eqs. (5.1.12) and (5.3.13) and Eqs. (5.1.14) and (5.3.14), we get

$$\Psi_\nu^{\mathrm{e}}(\boldsymbol{r}, t_n) = \frac{-1}{4\pi\epsilon_\nu}\sum_{k=1}^{N}\left[I_k\left(t_n - \frac{R_{mk}^{+}}{c_\nu}\right)\psi_{mk}^{+} + I_k\left(t_n - \frac{R_{mk}^{-}}{c_\nu}\right)\psi_{mk}^{-}\right] \tag{5.3.19}$$

$$\Psi_\nu^{\mathrm{m}}(\boldsymbol{r}, t_n) = \frac{-1}{4\pi\mu_\nu}\sum_{k=1}^{N}\left[M_k\left(t_n - \frac{R_{mk}^{+}}{c_\nu}\right)\psi_{mk}^{+} + M_k\left(t_n - \frac{R_{mk}^{-}}{c_\nu}\right)\psi_{mk}^{-}\right], \tag{5.3.20}$$

where $\psi_{mk}^{\pm}$, $R_{mk}^{\pm}$, and $R_m^{\pm}$ are given by Eqs. (4.2.21)–(4.2.24). The Ψ's may be evaluated at the centroids by replacing $\boldsymbol{r}$ with $\boldsymbol{r}_m^{c+}$ or $\boldsymbol{r}_m^{c-}$.

We next consider the expansion of the terms involving the curl of the vector potentials. As long as $R \neq 0$, we can take the curl operator inside the integral of Eq. (5.1.6) and, by using the vector identity

$$\nabla \times (w\boldsymbol{A}) = w\nabla \times \boldsymbol{A} - \boldsymbol{A} \times \nabla w,$$

we then obtain

$$\frac{1}{\epsilon_\nu}\nabla \times \boldsymbol{F}_\nu(\boldsymbol{r}, t_n) = \frac{1}{4\pi}\int_S \left[\frac{\nabla \times \boldsymbol{M}(\boldsymbol{r}', t_n - R/c_\nu)}{R} - \boldsymbol{M}(\boldsymbol{r}', t_n - R/c_\nu) \times \nabla\frac{1}{R}\right] dS'. \tag{5.3.21}$$

Following similar steps as those in Chapter 3, Section 3.5 and letting $t_r = t_n - R/c$, we may write

$$\begin{aligned}\frac{1}{\epsilon_\nu}\nabla \times \boldsymbol{F}_\nu(\boldsymbol{r}, t_n) = {} & \frac{1}{4\pi c_\nu}\sum_{\substack{k=1\\k\neq m}}^{N}\frac{\partial M_k(t_r)}{\partial t_r}\int_{T_k^+ + T_k^-}\frac{\boldsymbol{f}_k \times \boldsymbol{a}_R}{R}\,dS' \\ & + \frac{1}{4\pi}\sum_{\substack{k=1\\k\neq m}}^{N} M_k\left(t_n - \frac{R_{mk}}{c_\nu}\right)\int_{T_k^+ + T_k^-}\frac{\boldsymbol{f}_k \times \boldsymbol{a}_R}{R^2}\,dS',\end{aligned} \tag{5.3.22}$$

where $R_{mk} = |\boldsymbol{r}_m - \boldsymbol{r}_k|$. The $\partial M_k/\partial t_r$ term is approximated by the first-order backward difference of Eq. (5.2.17). Similarly,

$$\begin{aligned}\frac{1}{\mu_\nu}\nabla \times \boldsymbol{A}_\nu(\boldsymbol{r}, t_n) = {} & \frac{1}{4\pi c_\nu}\sum_{\substack{k=1\\k\neq m}}^{N}\frac{\partial I_k(t_r)}{\partial t_r}\int_{T_k^+ + T_k^-}\frac{\boldsymbol{f}_k \times \boldsymbol{a}_R}{R}\,dS' \\ & + \frac{1}{4\pi}\sum_{\substack{k=1\\k\neq m}}^{N} I_k\left(t_n - \frac{R_{mk}}{c_\nu}\right)\int_{T_k^+ + T_k^-}\frac{\boldsymbol{f}_k \times \boldsymbol{a}_R}{R^2}\,dS',\end{aligned} \tag{5.3.23}$$

with a similar approximation for $\partial I_k/\partial t_r$.

Finally, by replacing $n \Rightarrow n-1$ and taking all the quantities involving $n-1, n-2, \cdots$ to the right side, we can rewrite Eqs. (5.3.3) and (5.3.4) as

$$\alpha_{\mathrm{e}} I_m(t_n) = \left\langle \boldsymbol{f}_m, \frac{\partial \boldsymbol{E}^{\mathrm{i}}}{\partial t} - \boldsymbol{L}_e^E[\boldsymbol{J}, \boldsymbol{M}] - \boldsymbol{L}_d^E[\boldsymbol{J}, \boldsymbol{M}]\right\rangle \tag{5.3.24}$$

$$\alpha_{\mathrm{m}} M_m(t_n) = \left\langle \boldsymbol{f}_m, \frac{\partial \boldsymbol{H}^{\mathrm{i}}}{\partial t} - \boldsymbol{L}_e^H[\boldsymbol{J}, \boldsymbol{M}] - \boldsymbol{L}_d^H[\boldsymbol{J}, \boldsymbol{M}]\right\rangle \tag{5.3.25}$$

where

$$\mathcal{L}_{\nu}^{E}[\boldsymbol{J},\boldsymbol{M}] = \frac{\boldsymbol{\mathcal{A}}_{\nu}(\boldsymbol{r}_m, t_n) - 2\boldsymbol{A}_{\nu}(\boldsymbol{r}_m, t_{n-1}) + \boldsymbol{A}_{\nu}(\boldsymbol{r}_m, t_{n-2})}{\Delta t^2} + \nabla\Psi_{\nu}^{\mathrm{e}}(\boldsymbol{r}_m, t_{n-1})$$
$$+ \frac{1}{\epsilon_{\nu}}\nabla \times \left[\frac{\boldsymbol{\mathcal{F}}_{\nu}(\boldsymbol{r}_m, t_n) - \boldsymbol{F}_{\nu}(\boldsymbol{r}_m, t_{n-1})}{\Delta t}\right], \tag{5.3.26}$$

$$\mathcal{L}_{\nu}^{H}[\boldsymbol{J},\boldsymbol{M}] = \frac{\boldsymbol{\mathcal{F}}_{\nu}(\boldsymbol{r}_m, t_n) - 2\boldsymbol{F}_{\nu}(\boldsymbol{r}_m, t_{n-1}) + \boldsymbol{F}_{\nu}(\boldsymbol{r}_m, t_{n-2})}{\Delta t^2} + \nabla\Psi_{\nu}^{\mathrm{m}}(\boldsymbol{r}_m, t_{n-1})$$
$$- \frac{1}{\mu_{\nu}}\nabla \times \left[\frac{\boldsymbol{\mathcal{A}}_{\nu}(\boldsymbol{r}_m, t_n) - \boldsymbol{A}_{\nu}(\boldsymbol{r}_m, t_{n-1})}{\Delta t}\right], \tag{5.3.27}$$

$$\alpha_{\mathrm{e}} = \frac{(\mu_e + \mu_d)}{4\pi\,\Delta t^2}\kappa_{mm} \cdot \frac{\ell_m}{2}\left(\rho_m^{\mathrm{c}+} + \rho_m^{\mathrm{c}-}\right), \tag{5.3.28}$$

$$\alpha_{\mathrm{m}} = \frac{(\epsilon_e + \epsilon_d)}{4\pi\,\Delta t^2}\kappa_{mm} \cdot \frac{\ell_m}{2}\left(\rho_m^{\mathrm{c}+} + \rho_m^{\mathrm{c}-}\right). \tag{5.3.29}$$

In Eqs. (5.3.24) and (5.3.25) the electric and magnetic currents have been decoupled in the present-time sense. Equation (5.3.24) only has $I_m(t_n)$ as the unknown, whereas Eq. (5.3.25) only has $M_m(t_n)$ as the unknown. These equations are still, however, coupled in terms of previously occurring currents. This decoupling comes about by self term cancelation of the curl terms. This is an advantage of the PMCHW formulation because now we do not need to calculate the self terms of the curl operators.

As mentioned earlier, the time step should be $\Delta t \leq R_{\min}/\max\{c_e, c_d\}$ in order to obtain an explicit solution. An advantage of the time domain method over the frequency domain method is that the maximum allowable time step does not decrease as the permittivity or permeability of the body increases as long as the velocity of propagation is greater in the exterior material. In the frequency domain case, as ϵ_d or μ_d increase, the integrals, which are of the form e^{-jk_dR}/R or $(1+\mathrm{e}^{-jk_dR})/R$, need to be more accurately determined since the phase variation increases.

5.3.2 Far-Scattered Fields

Once the equivalent currents on the body have been determined, the far-scattered fields may be calculated. These fields may be thought of as the superposition of the fields due to the electric currents with the fields due to the magnetic currents. The far-scattered fields due to the electric currents are given in Chapter 4, Section 4.3. Here, we may rewrite Eqs. (4.3.7) and (4.3.8) with c replaced by c_{e} and η replaced with η_{e}, and we designate them as follows:

$$\boldsymbol{H}_J^{\mathrm{s}} = H_{J,\theta}^{\mathrm{s}}\boldsymbol{a}_\theta + H_{J,\phi}^{\mathrm{s}}\boldsymbol{a}_\phi \tag{5.3.30}$$

$$\boldsymbol{E}_J^{\mathrm{s}} = \eta_{\mathrm{e}}\left(H_{J,\phi}^{\mathrm{s}}\boldsymbol{a}_\theta - H_{J,\theta}^{\mathrm{s}}\boldsymbol{a}_\phi\right), \tag{5.3.31}$$

where the subscript J refers to the electric currents. The scattered electric field due to the magnetic currents is given by

$$\boldsymbol{E}_M^{\mathrm{s}}(\boldsymbol{r}, t) = \frac{-1}{\epsilon_{\mathrm{e}}}\nabla \times \boldsymbol{F}_{\mathrm{e}} = \frac{-1}{\epsilon_{\mathrm{e}}}\nabla \times \frac{\epsilon_{\mathrm{e}}}{4\pi}\int_S \frac{\boldsymbol{M}(\boldsymbol{r}', t - R/c_{\mathrm{e}})}{R}\,dS'. \tag{5.3.32}$$

Following steps similar to those in Chapter 4, Section 4.3, we can show that

$$rE_M^s(r, t_n) = \frac{-1}{4\pi c_e \Delta t} \sum_{k=1}^{N} \left[M_k \left(t_{n+1/2} - \frac{r - r' \cdot a_r}{c_e} \right) - M_k \left(t_{n-1/2} - \frac{r - r' \cdot a_r}{c_e} \right) \right] \bullet \frac{l_k}{2} (\rho_k^{c+} + \rho_k^{c-}) \times a_r \quad (5.3.33)$$

and

$$H_M^s(r, t_n) = \frac{1}{\eta_e} a_r \times E_M^s(r, t_n). \quad (5.3.34)$$

We may represent Eqs. (5.3.33) and (5.3.34) as

$$E_M^s = E_{M,\theta}^s a_\theta + E_{M,\phi}^s a_\phi \quad (5.3.35)$$

$$H_M^s = \frac{1}{\eta_e} \left(-E_{M,\phi}^s a_\theta + E_{M,\theta}^s a_\phi \right), \quad (5.3.36)$$

where the subscript M refers to the magnetic currents. Finally, the total fields scattered from the dielectric body may be obtained by adding the incident fields to the scattered fields computed by Eqs. (5.3.30), (5.3.31), (5.3.35), and (5.3.36).

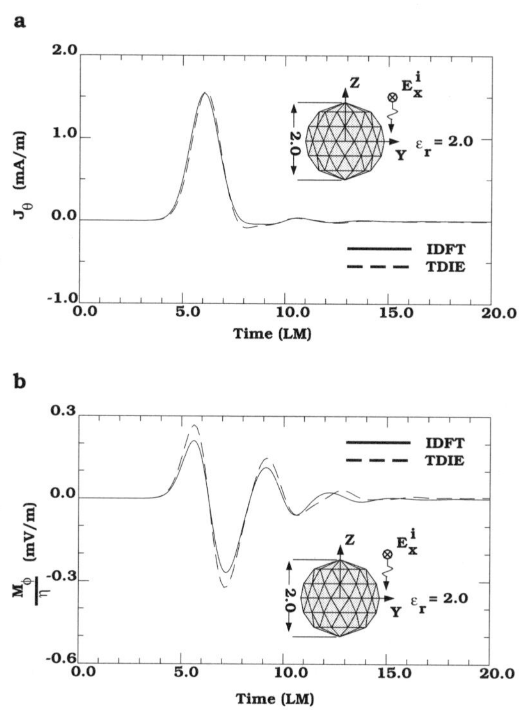

FIGURE 5.6
Equivalent currents (a) J_θ and (b) M_ϕ on a dielectric ($\epsilon_r = 2.0$) sphere of radius 1 m at $\theta = 90°$, $\phi = 0°$ illuminated by a Gaussian plane wave. Number of edges = 207.

5.4 NUMERICAL EXAMPLES

In this section, we present the numerical results for a homogeneous, dielectric sphere, cube, and thick circular disk. The bodies are illuminated with the Gaussian plane wave of Eq. (1.4.14) with $\boldsymbol{E}_o = -\boldsymbol{a}_x, \boldsymbol{a}_k = -\boldsymbol{a}_z$, $T = 4$ LM, and $ct_o = 6$ LM. As before, the averaging process was used. These results are compared with data which were transformed from the frequency domain. The frequency domain solutions were calculated in the 0 to 250 MHz range.

We first consider a dielectric sphere 1 m in radius with an $\epsilon_r = 2.0$. The sphere is triangulated in the same manner as described for Fig. 4.12 and has 207 edges. The time step was set to 0.14776 LM. Figure 5.6 depicts the equivalent electric and magnetic currents J_θ and M_ϕ, respectively, at $\theta = 90°$, $\phi = 0°$. The results are in good agreement with the frequency domain result which was obtained by using 128 frequency samples of the Mie series solution. The equivalent J_ϕ and M_θ currents at $\theta = 90°$, $\phi = 90°$ are shown in Fig. 5.7. Both J_θ and M_θ agree very well, whereas the time domain peaks are slightly larger than the frequency domain peaks for J_ϕ and M_ϕ. Next, the far-scattered, x-directed components of the electric fields are compared.

Figure 5.8a shows the backscattered field at $\theta = 90°$, $\phi = 0°$, and Fig. 5.8b shows the side-scattered field at $\theta = 90°$, $\phi = 90°$. As is evident from the figure, both IDFT and TDIE results compare quite well.

Next, a dielectric cube of side length 1 m is considered. The cube is discretized in the same manner as the conducting cube of Fig. 4.13 and has a relative permittivity of 2.0. The time step

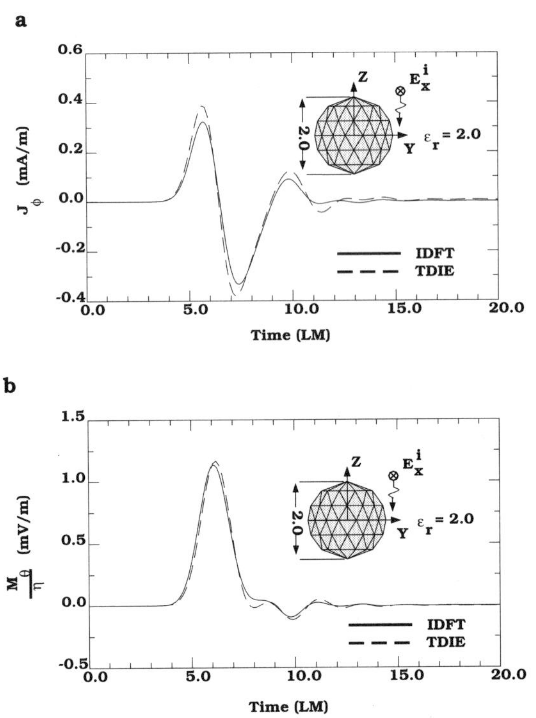

FIGURE 5.7
Equivalent currents (a) J_ϕ and (b) M_θ on a dielectric ($\epsilon_r = 2.0$) sphere of radius 1 m at $\theta = 90°$, $\phi = 90°$ illuminated by a Gaussian plane wave. Number of edges = 207.

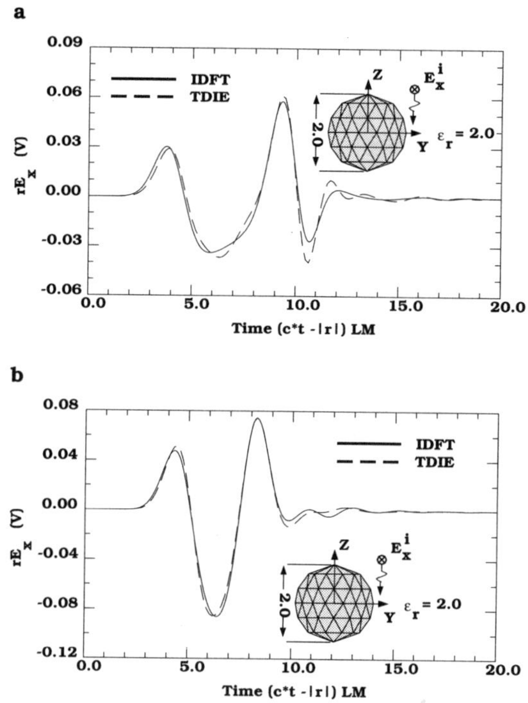

FIGURE 5.8
Normalized (a) backscattered ($\theta = 90°, \phi = 0°$) and (b) side-scattered ($\theta = 90°, \phi = 90°$) far-field response of a dielectric ($\epsilon_r = 2.0$) sphere 1 m in radius illuminated by a Gaussian plane wave. Number of edges = 207.

is $c\Delta t = 0.08957$ LM. The frequency domain result was obtained from a lossy dielectric cube which had a conductivity of 1 mS and 128 sample points. This conductivity was included to develop a smooth frequency response. Figure 5.9 compares the backscattered and side-scattered electric fields with the frequency domain results. They agree fairly well and show the same trends.

As a last example, consider a thick circular disk. The disk is 1 m in diameter, 0.2 m thick, centered about the origin, and has $\epsilon_r = 2.0$. The top and bottom portions of the disk were divided into three rings in the radial direction. The first, second, and third ring were constructed with 4, 12, and 24 triangular patches, respectively, for a total of 40 patches on each face. The height of the disk was modeled with one edge and there were 32 patches around the circumference of the disk. The time step was equal to 0.0664 LM. The results from the frequency domain were obtained with 128 frequency points and with a conductivity of 1.0 mS. Figure 5.10 shows the backscattered and side-scattered electric fields, which again compare well with frequency domain results.

5.5 CONCLUDING REMARKS

In this chapter, we developed the numerical algorithms to calculate the transient electromagnetic scattering from material bodies of arbitrary shape by solving TDIEs. We considered the numerical

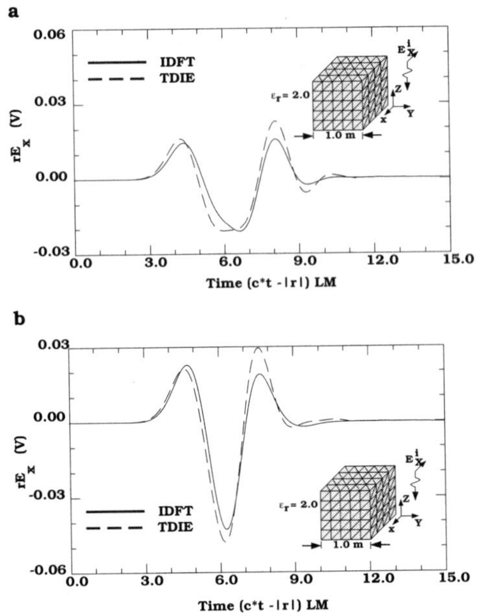

FIGURE 5.9
Normalized (a) backscattered ($\theta = 90°, \phi = 0°$) and (b) side-scattered ($\theta = 90°, \phi = 90°$) far-field response of a dielectric ($\epsilon_r = 2.0$) cube 1 m on a side illuminated by a Gaussian plane wave. Number of edges = 390.

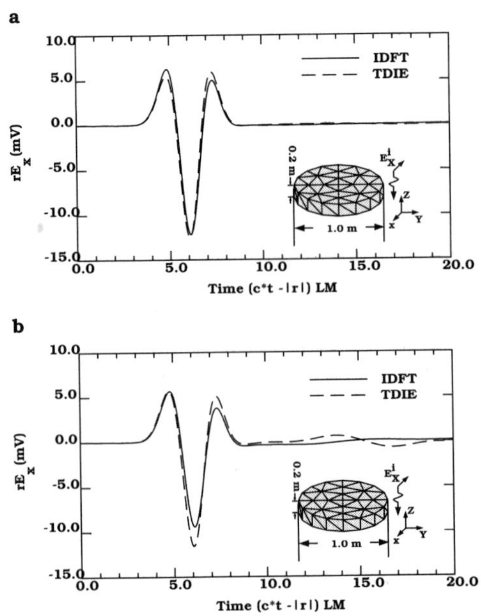

FIGURE 5.10
Normalized (a) backscattered ($\theta = 90°, \phi = 0°$) and (b) side-scattered ($\theta = 90°, \phi = 90°$) far-field response of a thick dielectric ($\epsilon_r = 2.0$) disk 1 m in diameter and 0.2 m thick illuminated by a Gaussian plane wave. Number of edges = 168.

methods for infinite cylinders and also finite-sized arbitrary bodies. It is easy to visualize that many applications exist for this type of problem, particularly when combined with conducting bodies treated in Chapters 3 and 4. In fact, the present-day applications involve not only conductors but also other material bodies. However, we did not explicitly present the treatment of composite structures, i.e., a combination of conducting and dielectric bodies, although the current methods can easily be extended to deal with such a case.

BIBLIOGRAPHY

[1] R. F. Harrington, *Time-Harmonic Electromagnetic Fields*, McGraw-Hill, New York, 1961.
[2] R. Mittra, *Computer Techniques for Electromagnetics*, Pergamon, Oxford, 1973.
[3] A. A. Kishk and L. Shafai, "Different Formulations for Numerical Solution of Single and Multi-Bodies of Revolution with Mixed Boundary Conditions," *IEEE Trans. Antennas Propagat.*, vol. 34, pp. 666–673, 1986.
[4] D. A. Vechinski and S. M. Rao, "Transient Scattering from Dielectric Cylinders—E-Field, H-Field, and Combined Field Solutions," *Radio Sci.*, vol. 27, pp. 611–622, 1992.
[5] D. A. Vechinski, S. M. Rao, and T. K. Sarkar, "Transient Scattering from Three-Dimensional Arbitrarily Shaped Dielectric Bodies," *J. Optical Soc. Am.*, vol. 11, pp. 1458–1470, 1994.

CHAPTER 6

Finite-Difference Time Domain Method

K. R. UMASHANKAR
Department of Electrical Engineering and Computer Science
University of Illinois at Chicago

A major contemporary thrust in computational electromagnetics, the finite-difference time domain (FDTD) technique, is presented for the direct numerical solution of Maxwell's equations in the time domain. No study related to the electromagnetic scattering, propagation, coupling, and interaction phenomena is complete without an insight into the real-time knowledge of the electromagnetic fields. In this chapter, we study in detail the formulation and application of FDTD to various electromagnetic field problems.

6.1 INTRODUCTION TO FDTD

The FDTD technique is a computationally efficient means of directly solving Maxwell's time-dependent curl equations or their equivalent integral equations using the finite-difference technique. In this extensively computer-based numerical method, the continuous distribution of electromagnetic fields in a finite volume of space is sampled at distinct points in a space and time lattice. The electromagnetic wave propagation, scattering, and penetration phenomena are modeled in a self-consistent manner by marching in time step and repeatedly implementing the finite-difference numerical analog of Maxwell's equations at each spatial lattice point. This approach basically results in a simulation of the actual coupled electromagnetic field full-wave solution by the sampled data numerical analogs propagating in a data space stored in a computer. Space and time sampling increments are selected to avoid aliasing of the continuous field distribution and to guarantee stability of the time marching algorithm. Time marching is completed when the desired late time or steady-state field behavior is observed. An important application of the latter is the sinusoidal steady state which eventually results if a continuous sinusoidal incident excitation is selected.

Overall computer storage and running time requirements for the FDTD technique are linearly proportional to N, the number of field unknowns in the finite volume of space being modeled. The current supercomputers and variety of super workstations provide tools with sufficient central memory and computational speed to contain the FDTD models of the three-dimensional structures spanning from 10 to 100 wavelengths. There are extensive validations of the FDTD models at all levels for important canonical problems against the method of moments (MoM) results, and selected experimental validation cases indicate that the FDTD and MoM are providing comparable degrees of modeling detail and accuracy as required in many engineering design and applications.

In the case of the three-dimensional time-dependent boundary value problem, various electromagnetic vector field quantities vary with respect to the three spatial coordinate variables and the time parameter variable. We consider the case of a linear, inhomogeneous, and isotropic lossy medium. The medium has two types of conductivity which represent the electric and magnetic lossy situation in the medium. The driving source term in Maxwell's equations, in fact, should include all existing current density distributions, both the primary and the secondary source terms. In the case of lossy medium, according to the generalized Ohm's law, the conduction-type electric and magnetic currents flow everywhere in the medium given by the product of respective medium conductivity and field intensity distribution. For the case of a source-free region, Maxwell's time-dependent equations in integral form are given by

$$\oint_C \boldsymbol{E}(\boldsymbol{r},t)\cdot d\boldsymbol{l} = -\frac{\partial}{\partial t}\int\!\!\int_S \boldsymbol{B}(\boldsymbol{r},t)\cdot d\boldsymbol{s} - \int\!\!\int_S \boldsymbol{M}_{\mathrm{c}}(\boldsymbol{r},t)\cdot d\boldsymbol{s} \tag{6.1.1}$$

$$\oint_C \boldsymbol{H}(\boldsymbol{r},t)\cdot d\boldsymbol{l} = \frac{\partial}{\partial t}\int\!\!\int_S \boldsymbol{D}(\boldsymbol{r},t)\cdot d\boldsymbol{s} + \int\!\!\int_S \boldsymbol{J}_{\mathrm{c}}(\boldsymbol{r},t)\cdot d\boldsymbol{s}, \tag{6.1.2}$$

where S is an arbitrary open surface bounded by a close contour C along its edge. In the previous integral expressions, the various time-varying electromagnetic vector field quantities are as follows:

- $\boldsymbol{E}(\boldsymbol{r},t)$: electric field distribution, in volts per meter
- $\boldsymbol{D}(\boldsymbol{r},t)$: electric flux density distribution, in coulomb per square meter
- $\boldsymbol{H}(\boldsymbol{r},t)$: magnetic field distribution, in ampere per meter
- $\boldsymbol{B}(\boldsymbol{r},t)$: magnetic flux density distribution, in weber per square meter

For the case of linear, inhomogeneous, and isotropic lossy medium, the following constitutive relationships are utilized for the electric and magnetic field densities and for the secondary source terms due to the conduction-type current densities

$$\boldsymbol{D}(\boldsymbol{r},t) = \epsilon(\boldsymbol{r})\boldsymbol{E}(\boldsymbol{r},t) \tag{6.1.3}$$

$$\boldsymbol{B}(\boldsymbol{r},t) = \mu(\boldsymbol{r})\boldsymbol{H}(\boldsymbol{r},t) \tag{6.1.4}$$

$$\boldsymbol{J}_{\mathrm{c}}(\boldsymbol{r},t) = \sigma^{(\mathrm{e})}(\boldsymbol{r})\boldsymbol{E}(\boldsymbol{r},t) \tag{6.1.5}$$

$$\boldsymbol{M}_{\mathrm{c}}(\boldsymbol{r},t) = \sigma^{(\mathrm{m})}(\boldsymbol{r})\boldsymbol{H}(\boldsymbol{r},t), \tag{6.1.6}$$

where $\epsilon(\boldsymbol{r})$ is the permittivity of medium (in farads per meter), $\mu(\boldsymbol{r})$ is the permeability of medium (in henrys per meter), $\sigma^{(\mathrm{e})}(\boldsymbol{r})$ is the electric conductivity of medium (in siemens per meter), and $\sigma^{(\mathrm{m})}(\boldsymbol{r})$ is the magnetic conductivity of medium (in ohms per meter).

Equations (6.1.1) and (6.1.2) are further analyzed in the rectangular coordinate system with the following components of the electric and magnetic fields:

$$\boldsymbol{E}(\boldsymbol{r},t) = E_x(\boldsymbol{r},t)\boldsymbol{a}_x + E_y(\boldsymbol{r},t)\boldsymbol{a}_y + E_z(\boldsymbol{r},t)\boldsymbol{a}_z \tag{6.1.7}$$

$$\boldsymbol{H}(\boldsymbol{r},t) = H_x(\boldsymbol{r},t)\boldsymbol{a}_x + H_y(\boldsymbol{r},t)\boldsymbol{a}_y + H_z(\boldsymbol{r},t)\boldsymbol{a}_z. \tag{6.1.8}$$

6.2 PULSE PROPAGATION IN A LOSSY, INHOMOGENEOUS, LAYERED MEDIUM

The electromagnetic-coupled field equations (Eqs. 6.1.1 and 6.1.2) are now reduced to a one-dimensional form by substituting the rectangular field components of Eqs. (6.1.7) and (6.1.8). At any field point in the three-dimensional medium, it is assumed that the electric field and the magnetic field components do not vary with respect to x and z coordinate variables with

$$\frac{\partial}{\partial x} \to 0 \tag{6.2.1}$$

and

$$\frac{\partial}{\partial z} \to 0. \tag{6.2.2}$$

For the case of one-dimensional layered medium, the electric and magnetic field components vary only with respect to a single y coordinate variable and time parameter t. For this case, the two nonzero field components are given by the electric field in z-coordinate direction and the magnetic field in x-coordinate direction. Thus, we have the following two coupled integral form of equations:

$$\oint_C E_z(y,t)\,dl = -\frac{\partial}{\partial t}\int\int_S \mu(y)H_x(y,t)\,ds - \int\int_S \sigma^{(\mathrm{m})}(y)H_x(y,t)\,ds \tag{6.2.3}$$

$$\oint_C H_x(y,t)\,dl = \frac{\partial}{\partial t}\int\int_S \epsilon(y)E_z(y,t)\,ds + \int\int_S \sigma^{(\mathrm{e})}(y)E_z(y,t)\,ds. \tag{6.2.4}$$

Faraday's and Ampere's integral expressions (Eqs. 6.2.3 and 6.2.4) are quite elegant to use in the regions consisting of boundary layers separating different media with spatially varying layered media parameters. For the one-dimensional case, a systematic numerical solution technique based on the time marching and finite-difference approximation is considered in the following. Since the previous two coupled integral forms of equations are valid for every value of y and t, it is assumed that the two components of electric and magnetic fields are continuous or at least piecewise continuous with respect to y and t variables. Referring to Fig. 6.1, along the y-coordinate variable the inhomogeneous region is divided into parallel thin boundary layers. Let Δy and Δt represent the discretized spatial increment (in meters) and discretized time increment (in seconds), respectively.

An attempt is made to enforce the validity of Eqs. (6.2.3) and (6.2.4) and analyze them for the solution of two unknown field components, E_z and H_x, at m discrete spatial points and at n discrete time steps. For convenience in defining the first-order derivatives in the finite-difference calculation the electric field and the magnetic field components are not calculated at the same spatially discretized locations but are staggered alternately as shown in Fig. 6.2.

The medium is first discretized uniformly into M number of thin layers each having constant spatial width Δy. The boundary layers are designated with appropriate boundary surface or interface numbers. The unknown electric field component $E_z(m)$, for $m = 1, 2, 3, \cdots$, and the unknown magnetic field component $H_x(m - 1/2)$, for $m = 1, 2, 3, \cdots$, are specifically located at the boundary interfaces, which automatically ensures continuity of the tangential components of fields. The final layer M is such that it covers the complete inhomogeneous region of interest to be analyzed. It is also assumed that the thickness of each layer is very small, and for all practical purpose each layer can be assumed to be a piecewise linear, homogeneous, and isotropic medium. Furthermore, the electric field distribution $E_z(m)$ located at the mid-point within a layer, in between the two adjacent magnetic field components $H_x(m - 1/2)$ and

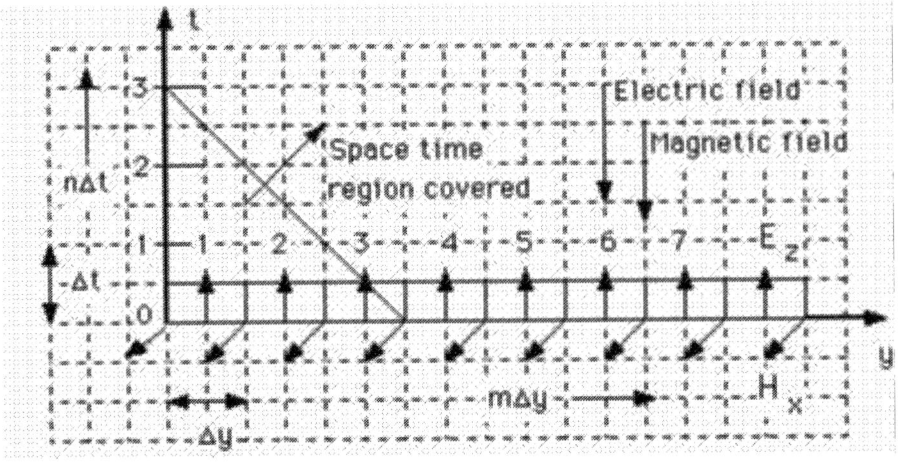

FIGURE 6.1
Space–time resolution for one-dimensional plane wave fields.

$H_x(m + 1/2)$, is assumed to be piecewise constant. Thus, under the spatial discretization, corresponding to the component of the electric field defined at the midpoint of each layer, the constant permittivity and electric conductivity of each layer are selected to be the permittivity and electric conductivity at the midpoint of the layer where the electric field component is located.

Similarly, the magnetic field distribution $H_x(m + 1/2)$ located at the midpoint within a layer, in between the two adjacent electric field components $E_z(m)$ and $E_z(m + 1)$, is assumed to be piecewise constant. Again, under the spatial discretization, corresponding to the component of the magnetic field defined at the midpoint of each layer, the constant permeability and magnetic conductivity of each layer are selected to be the permeability and magnetic conductivity at the midpoint of the layer where the magnetic field component is located. Since the electric field and the magnetic field components are staggered, Figs. 6.1 and 6.2 depict a piecewise linear

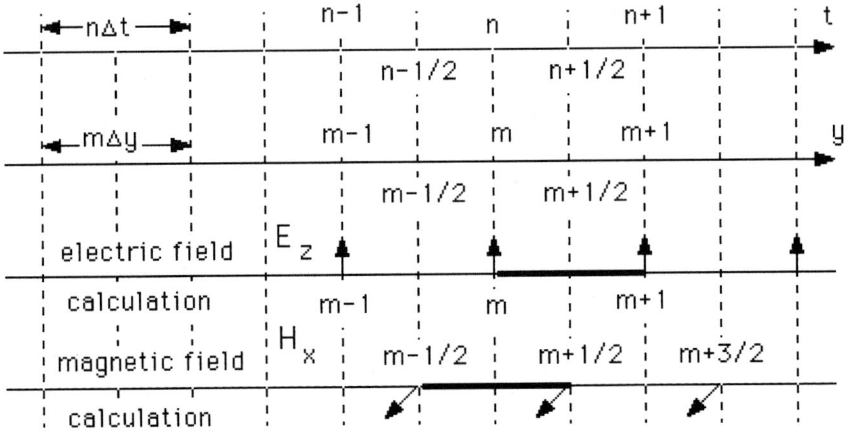

FIGURE 6.2
Finite-difference calculation of electric and magnetic fields.

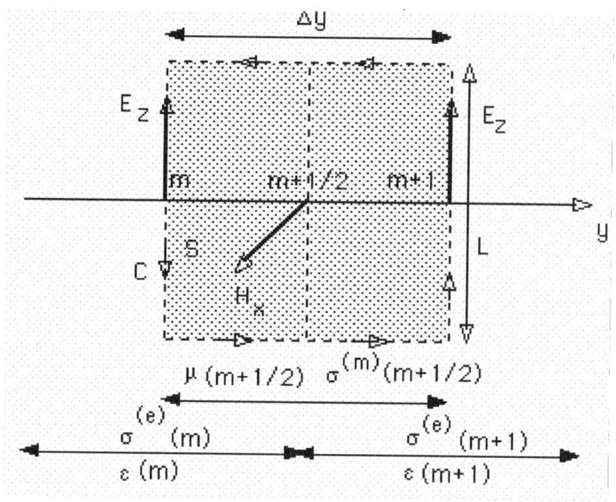

FIGURE 6.3
Discretization of one-dimensional Faraday's integral expression.

distribution of the field components for defining the first-order central differences required for solving the coupled integral form or the differential form of equations.

The integral forms of one-dimensional Maxwell's equations (Eqs. 6.2.3 and 6.2.4) are now approximated to analyze the unknown electric and magnetic field components based on the spatial discretization by selecting a suitable contour C bounding a corresponding suitable open surface S. Figure 6.3 shows the Faraday's law relationship between the discretized piecewise constant electric field components along a closed rectangular elemental contour C bounding the corresponding open rectangular elemental surface S with the discretized piecewise constant magnetic field distribution.

The following expressions define the discretized one-dimensional electric field component and the magnetic field components:

$$E_z(y,t) = E_z(m\Delta y, n\Delta t) = E_z^n(m) \tag{6.2.5}$$

$$H_x(y,t) = H_x(m\Delta y, n\Delta t) = H_x^n(m) \tag{6.2.6}$$

for space discretization $m = 1, 2, 3, \cdots, M$ and for time discretization $n = 1, 2, 3, \cdots, N$.

The previous discretization, referring to Fig. 6.2, basically yields a centered finite-difference approximation with the second-order accuracy with respect to either y-coordinate variable or time variable t. The spatial and temporal partial derivatives of the field components are given by

$$\frac{\partial E_z^n(m)}{\partial t} = \frac{E_z^{n+1/2}(m) - E_z^{n-1/2}(m)}{\Delta t} + \bigcirc(\Delta t^2) \tag{6.2.7}$$

$$\frac{\partial H_x^n(m+1/2)}{\partial t} = \frac{H_x^{n+1/2}(m+1/2) - H_x^{n-1/2}(m+1/2)}{\Delta t} + \bigcirc(\Delta t^2) \tag{6.2.8}$$

$$\frac{\partial E_z^n(m+1/2)}{\partial y} = \frac{E_z^n(m+1) - E_z^n(m)}{\Delta y} + \bigcirc(\Delta y^2) \tag{6.2.9}$$

$$\frac{\partial H_x^n(m)}{\partial y} = \frac{H_x^n(m+1/2) - H_x^n(m-1/2)}{\Delta y} + \bigcirc(\Delta y^2). \tag{6.2.10}$$

Following the right-hand rule and referring to Fig. 6.3, the integral expression (Eq. 6.2.3) simplifies to

$$E_z^n(m+1)L - E_z^n(m)L = -(L\Delta y)\left[\frac{\partial}{\partial t}\mu(m+1/2)H_x^n(m+1/2) + \sigma^{(\mathrm{m})}(m+1/2)H_x^n(m+1/2)\right]. \quad (6.2.11)$$

For the case of linear and nondispersive media, assuming the discretized medium parameters to be independent of time variation, Eq. (6.2.11) can be rearranged at the nth time step as

$$\frac{\partial H_x^n(m+1/2)}{\partial t} = -\left[\frac{1}{\mu_0\mu_{\mathrm{r}}(m+1/2)}\right]\left[\frac{E_z^n(m+1)-E_z^n(m)}{\Delta y}\right] - \frac{\sigma^{(\mathrm{m})}(m+1/2)}{\mu_0\mu_{\mathrm{r}}(m+1/2)}H_x^n(m+1/2). \quad (6.2.12)$$

The spatially discretized expression (Eq. 6.2.12) is further simplified. Using the finite-difference approximation (Eq. 6.2.8) corresponding to $m+1/2$ and n, Eq. (6.2.12) can be written as

$$\begin{aligned}\frac{H_x^{n+1/2}(m+1/2) - H_x^{n-1/2}(m+1/2)}{\Delta t} &= -\left[\frac{1}{\mu_0\mu_{\mathrm{r}}(m+1/2)}\right]\left[\frac{E_z^n(m+1)-E_z^n(m)}{\Delta y}\right] \\ &\quad - \frac{\sigma^{(\mathrm{m})}(m+1/2)}{\mu_0\mu_{\mathrm{r}}(m+1/2)}\left[\frac{H_x^{n+1/2}(m+1/2)+H_x^{n-1/2}(m+1/2)}{2}\right],\end{aligned} \quad (6.2.13)$$

which can be rewritten as

$$\begin{aligned}H_x^{n+1/2}(m+1/2)\left[1+\frac{\sigma^{(\mathrm{m})}(m+1/2)\Delta t}{2\mu_0\mu_{\mathrm{r}}(m+1/2)}\right] &= H_x^{n-1/2}(m+1/2)\left[1-\frac{\sigma^{(\mathrm{m})}(m+1/2)\Delta t}{2\mu_0\mu_{\mathrm{r}}(m+1/2)}\right] \\ &\quad + \left[\frac{\Delta t}{\mu_0\mu_{\mathrm{r}}(m+1/2)\Delta y}\right]\left[E_z^n(m)-E_z^n(m+1)\right].\end{aligned} \quad (6.2.14)$$

Equation (6.2.14) forms a convenient time-stepping numerical algorithm for calculating the magnetic field component in terms of the adjacent electric field components. Similarly, utilizing the one-dimensional integral expression (Eq. 6.2.4), another time-stepping numerical algorithm for calculating the electric field component in terms of the adjacent magnetic field components can be obtained. Figure 6.4 shows Ampere's law relationship between the discretized piecewise constant magnetic field components along a closed rectangular elemental contour C bounding the corresponding open rectangular elemental surface S with the discretized piecewise constant electric field distribution. Following the right-hand rule, the integral expression (Eq. 6.2.4) simplifies to

$$-H_x^n(m+1/2)L + H_x^n(m-1/2)L = \frac{\partial}{\partial t}\epsilon(m)E_z^n(m)(L\Delta y) + \sigma^{(\mathrm{e})}(m)E_z^n(m)(L\Delta y). \quad (6.2.15)$$

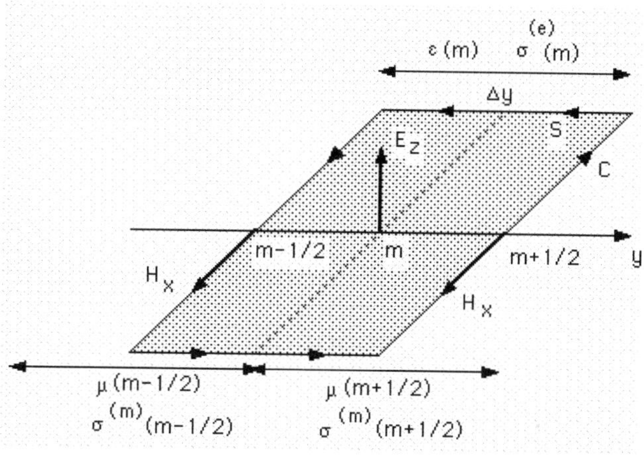

FIGURE 6.4
Discretization of one-dimensional Ampere's integral expression.

For the case of linear and nondispersive media, assuming the discretized medium parameters to be independent of time variation, Eq. (6.2.15) can be rearranged at the $(n+1/2)$th time step as

$$\frac{\partial}{\partial t}E_z^{n+1/2}(m) = -\frac{\sigma^{(e)}(m)}{\epsilon_0\epsilon_r(m)}E_z^{n+1/2}(m) - \frac{1}{\epsilon_0\epsilon_r(m)\Delta y}\left[H_x^{n+1/2}(m+1/2)\right] + \frac{1}{\epsilon_0\epsilon_r(m)\Delta y}\left[H_x^{n+1/2}(m-1/2)\right]. \tag{6.2.16}$$

The spatially discretized expression (Eq. 6.2.16) is further simplified. Using the finite-difference approximation (Eq. 6.2.7) corresponding to m and $n+1/2$, Eq. (6.2.16) can be written as

$$\begin{aligned}\frac{E_z^{n+1}(m) - E_z^n(m)}{\Delta t} &= -\left[\frac{1}{\epsilon_0\epsilon_r(m)\Delta y}\right]\left[H_x^{n+1/2}(m+1/2)\right] \\ &\quad + \left[\frac{1}{\epsilon_0\epsilon_r(m)\Delta y}\right]\left[H_x^{n+1/2}(m-1/2)\right] \\ &\quad - \frac{\sigma^{(e)}(m)}{\epsilon_0\epsilon_r(m)}\left[\frac{E_z^{n+1}(m) + E_z^n(m)}{2}\right]\end{aligned} \tag{6.2.17}$$

$$\begin{aligned}E_z^{n+1}(m)\left[1 + \frac{\sigma^{(e)}(m)\Delta t}{2\epsilon_0\epsilon_r(m)}\right] &= +E_z^n(m)\left[1 - \frac{\sigma^{(e)}(m)\Delta t}{2\epsilon_0\epsilon_r(m)}\right] \\ &\quad + \left[\frac{\Delta t}{\epsilon_0\epsilon_r(m)\Delta y}\right]\left[H_x^{n+1/2}(m-1/2)\right] \\ &\quad - \left[\frac{\Delta t}{\epsilon_0\epsilon_r(m)\Delta y}\right]\left[H_x^{n+1/2}(m+1/2)\right].\end{aligned} \tag{6.2.18}$$

The two time-stepping coupled difference equations (Eqs. 6.2.14 and 6.2.18) are rewritten in terms of the discretized spatial media coefficients as follows:

$$E_z^{n+1}(m) = C_a(m)E_z^n(m) + C_b(m)\left[+H_x^{n+1/2}(m-1/2) - H_x^{n+1/2}(m+1/2)\right]. \tag{6.2.19}$$

where

$$C_a(m) = \frac{C_{na}}{C_{da}}, \tag{6.2.20}$$

$$C_b(m) = \frac{C_{nb}}{C_{da}}, \tag{6.2.21}$$

$$C_{na} = 1 - \frac{\sigma^{(e)}(m)\Delta t}{2\epsilon_0\epsilon_r(m)}, \tag{6.2.22}$$

$$C_{da} = 1 + \frac{\sigma^{(e)}(m)\Delta t}{2\epsilon_0\epsilon_r(m)}, \tag{6.2.23}$$

$$C_{nb} = \frac{\Delta t}{\epsilon_0\epsilon_r(m)\Delta y}, \tag{6.2.24}$$

and

$$H_x^{n+1/2}(m+1/2) = D_a(m+1/2)H_x^{n-1/2}(m+1/2) + D_b(m+1/2)\left[-E_z^n(m+1) + E_z^n(m)\right], \tag{6.2.25}$$

where

$$D_a(m+1/2) = \frac{D_{na}}{D_{da}}, \tag{6.2.26}$$

$$D_b(m+1/2) = \frac{D_{nb}}{D_{da}}, \tag{6.2.27}$$

$$D_{na} = 1 - \frac{\sigma^{(m)}(m+1/2)\Delta t}{2\mu_0\mu_r(m+1/2)}, \tag{6.2.28}$$

$$D_{da} = 1 + \frac{\sigma^{(m)}(m+1/2)\Delta t}{2\mu_0\mu_r(m+1/2)}, \tag{6.2.29}$$

$$D_{nb} = \frac{\Delta t}{\mu_0\mu_r(m+1/2)\Delta y}. \tag{6.2.30}$$

Equations (6.2.19) and (6.2.25) are the two coupled one-dimensional basic time-stepping algorithms for the calculation of electric and magnetic field components. With the field solutions obtained in the space–time difference equations, the field components for a given time step can be calculated iteratively. The updated new value of a field component at any layer depends only on its value in the previous time step and the previous values of components of the other field at adjacent spatial points. Hence, at any given time step, the computation of the electric or magnetic field component will proceed one point at a time.

The electromagnetic field solution discussed previously is generally referred as the one-dimensional finite-difference time domain technique, which is often used when coupled partial differential equations or their equivalent integral form of equations are to be solved based on the numerical time-stepping scheme. Use of this technique to study two-dimensional and three-dimensional Maxwell's equations is discussed later. The limitation at this stage seems to be the discretization of the complete space and time. For the case of a one-dimensional problem, y-coordinate variable varies from $-\infty$ to ∞. Hence, for studying wave propagation through a finite-width inhomogeneous lossy material slab, proper field terminations are required on either side where the space discretization is discontinued. In fact, proper field boundary conditions are enforced based on zero field reflection at the terminating layers so that the electric field and the magnetic field plane waves propagating along the y-coordinate direction do not return but rather continue to propagate into the unbounded medium.

The choice of space increment Δy and time increment Δt is dictated by accuracy and algorithm stability, respectively. To ensure the numerical stability of the time-stepping equations for the computed electric and magnetic fields, Δt is chosen to satisfy the following inequality for the one-dimensional layer model:

$$\Delta t \leq \frac{\Delta y}{c_{\max}}, \tag{6.2.31}$$

where $c_{\max}$ is the maximum wave velocity within the model.

In the discretized one-dimensional layered model discussed previously, each layer with the staggered electric and magnetic field components is numerically simulated in terms of its relative permittivity, relative permeability, electric conductivity, and magnetic conductivity. The maximum wave velocity within the one-dimensional model occurs corresponding to the free space medium permittivity and permeability. It should be noted that the actual wave velocity of the electric and magnetic field components in a layer is implicit within the time-stepping algorithm.

6.2.1 Propagation of Half-Sine Pulse

The FDTD numerical algorithm discussed previously is quite useful for the study of one-dimensional, time-dependent electric and magnetic fields in a layered, inhomogeneous, lossy medium having different conductivity, permittivity, and permeability characteristics. In this approach, there is no restriction on the selection of time-dependent incident field excitation. The plane wave incident field excitation can be stated in terms of an incident electric field or a magnetic field. Even an incident electric current excitation can be introduced by invoking the relationship between the electric current and magnetic field boundary condition at a planar interface.

To illustrate the basic concept of numerical simulation of a time-dependent plane wave incident electric field, the propagation of a half-wave sinusoidal time pulse in a layered medium is considered. Figure 6.5 shows a large region of one-dimensional space that is linear and isotropic. Along the y coordinate, the one-dimensional region is divided into equally spaced spatial cells of width Δy. For convenience, the spatial layers are numbered $m = 1, 2, 3, \cdots, M$, with the total number of layers $M = 100$. The z component of the electric field and x component of the magnetic field must be calculated with respect to the coordinate variables (y, t) in the layered medium for a specified field excitation. Referring to Fig. 6.5, the one-dimensional space is spatially truncated at the electric field locations corresponding to the spatial layers $m = 1$ with $E_z(1)$ and $m = M$ with $E_z(M)$. Now, the layered region is divided into two separate zones consisting of a total-field zone and a scattered-field zone that are separated by a planar connecting interface $m = s$. In the total-field zone, the layered scattering object under study is embedded by appropriately specifying the material permittivity, permeability, and conductivity parameters for each layer. The connecting interface between the total-field and scattered-field zones is ideal for numerically simulating the time-dependent incident electric field excitation. It is assumed that the incident electric field is turned on at an arbitrary reference time $t = 0$. The incident electric field, E_z^{i}, located at the interface $s = 5$, develops a half-wave sinusoidal time pulse

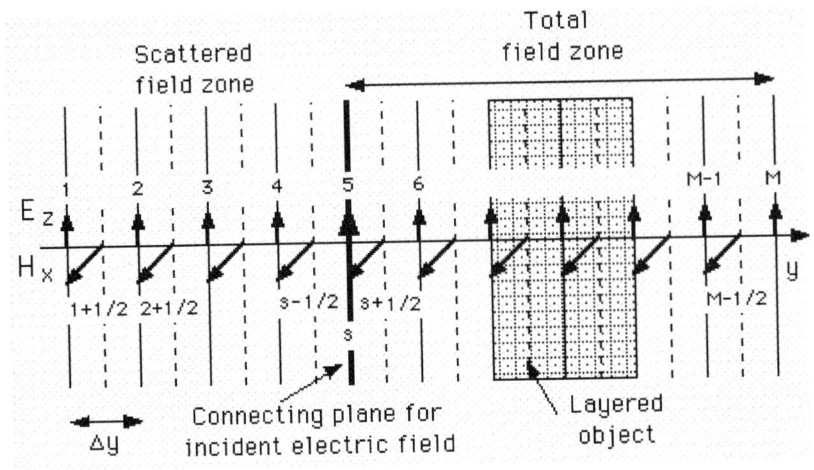

FIGURE 6.5
Interface connecting total and scattered-field zones.

given by

$$E_z^{\mathrm{i}}(t) = E_0 \sin(\omega t)\delta(m - s) \tag{6.2.32}$$

$$\omega t = 2\pi f n \Delta t \tag{6.2.33}$$

for $n = 1, 2, 3, \cdots, N_{\mathrm{p}}$, where N_{p} represents the number of time steps needed to simulate the half-wave sinusoidal time pulse and $f = 300$ MHz represents the frequency. Also, E_0 is the peak amplitude of half-wave sinusoidal excitation, given by

$$E_0 = \begin{cases} 100 \text{ V/m} & \text{for } N_{\mathrm{p}} \leq 50 \\ 0 & \text{for } N_{\mathrm{p}} > 50, \end{cases} \tag{6.2.34}$$

and $\delta(m - s)$ is the delta distribution.

The time-dependent incident electric field propagates in the total-field zone from left to right only for $m \geq s$. There is no incident electric field in the scattered-field zone. For the incident half-wave sinusoidal time pulse, $c_0 = 3 \times 10^8$ m/s is the maximum velocity of incident wave and $w = 0.5$ m is the width of half-wave sinusoidal time pulse.

Hence, by selecting a spatial cell resolution of $\Delta y = 0.01$, a total of 50 cells are required to completely span the incident electric field excitation in the medium. Furthermore, the time step resolution is selected based on Eq. (6.2.31), and it is given by

$$\Delta t = \frac{\Delta y}{c_0}. \tag{6.2.35}$$

The time-dependent incident electromagnetic field pulse impinges on the one-dimensional layered scattering object embedded in the total-field zone. Part of the incident electromagnetic field is reflected back and the remaining field penetrates into the layered scattering object. Referring to Fig. 6.5, at the truncation boundaries corresponding to the two spatial layers $m = 1$ with $E_z(1)$ and $m = M$ with $E_z(M)$, the electromagnetic propagation and radiation boundary condition is enforced by simulating numerically plane wave electric field reflection coefficients. The electric and magnetic fields reaching the truncation boundaries will continue to propagate with no reflection back into the numerical layered model.

Connecting Condition at Interface of Total-Field and Scattered-Field Zones With the incident electric field turned on at the interface $m = s$, the two finite-difference, time-stepping coupled expressions (Eqs. 6.2.19 and 6.2.25) are iteratively solved at each spatial cell for every time step increment. At the planar interface separating the total-field zone and the scattered-field zone, there arises a problem of inconsistency in utilizing directly the time-stepping expressions. On the right side of the connecting interface, the field component to be used in the finite-difference expressions is the total-field component, E_z^{t}, consisting of the incident field and the scattered field. Similarly, on the left side of the connecting interface, the field component to be used in the finite-difference expressions is the scattered field component, E_z^{s}, only. Thus, it is inconsistent to perform a finite-difference calculation between a total-field value and a scattered-field value at $m = s$.

The inconsistency problem on the right side and the left side of the connecting interface $m = s$ is addressed systematically in the following. In fact, the two finite-difference, time-stepping coupled expressions (Eqs. 6.2.19 and 6.2.25) are valid solely for the calculation of total fields in the total-field zone and, similarly, are valid solely for the calculation of scattered fields in the scattered-field zone for every time step increment. Noting that there exists no incident magnetic field in the scattered-field zone at the layer $s - 1/2$, the calculation of the scattered electric field at the connecting interface $m = s$ is accomplished by

$$\begin{aligned} E_z^{\mathrm{s}\,n+1}(s) = {} & C_a(s)\left[E_z^{\mathrm{t}\,n}(s) - E_z^{\mathrm{i}\,n}(s)\right] + C_b(s)\left[-H_x^{\mathrm{t}\,n+1/2}(s+1/2)\right. \\ & \left. + H_x^{\mathrm{i}\,n+1/2}(s+1/2) + H_x^{\mathrm{s}\,n+1/2}(s-1/2)\right] \end{aligned} \tag{6.2.36}$$

and the successive time stepping of the incident electric field only at the connecting interface $m = s$ yields the total electric field

$$E_z^{\mathrm{t}\,n+1}(s) = E_z^{\mathrm{s}\,n+1}(s) + E_z^{\mathrm{i}\,n+1}(s). \tag{6.2.37}$$

Similarly, the calculation of the scattered magnetic field at the layer $s - 1/2$ adjacent to the connecting interface is accomplished by

$$\begin{aligned} H_x^{\mathrm{s}\,n+1/2}(s-1/2) = {} & D_a(s-1/2)H_x^{\mathrm{s}\,n-1/2}(s-1/2) \\ & + D_b(s-1/2)\left[-E_z^{\mathrm{t}\,n}(s) + E_z^{\mathrm{i}\,n}(s) + E_z^{\mathrm{s}\,n}(s-1)\right]. \end{aligned} \tag{6.2.38}$$

To study the case of electromagnetic field propagation by half-wave sinusoidal time pulse in a free-space medium, it is assumed that each spatial cell is lossless, homogeneous, and isotropic, having the following medium parameters:

$$\epsilon_{\mathrm{r}}(m) = 1 \tag{6.2.39}$$

$$\mu_{\mathrm{r}}(m+1/2) = 1 \tag{6.2.40}$$

$$\sigma^{(\mathrm{e})}(m) = 0 \tag{6.2.41}$$

$$\sigma^{(\mathrm{m})}(m+1/2) = 0 \tag{6.2.42}$$

for $m = 1, 2, 3, \cdots, M$.

The spatial electric field distribution of the propagating half-wave sinusoidal time pulse is shown in Fig. 6.6. For $t < 0$, the incident electric field excitation is not turned on. According to

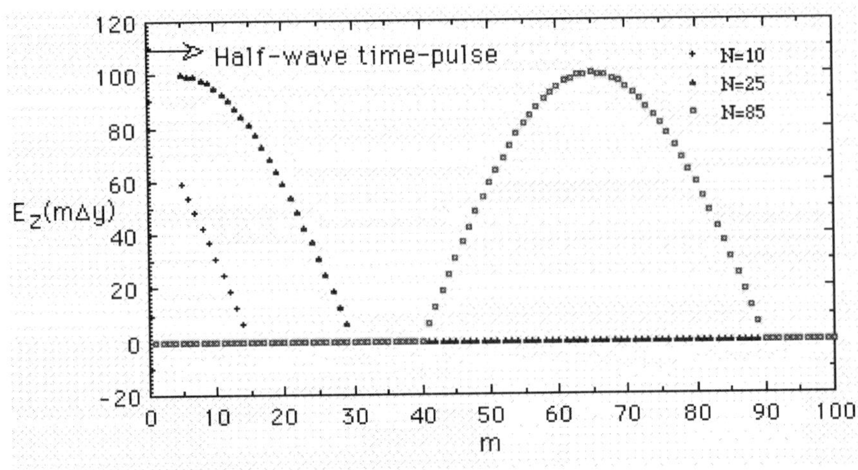

FIGURE 6.6
Propagation of half-wave sinusoidal time pulse in lossless free space medium.

the time causality, the electric and magnetic field distributions are zero everywhere in the medium. The incident electric field excitation is now turned on at $t = 0$ at the connecting interface $s = 5$ and is turned off as soon as the half-wave pulse generation is completed. This occurs at the time step for the pulse $N_p = 50$. Figure 6.6 shows three cases depicting the spatial distribution of the z component of the electric field in the free-space medium at the end of specific time steps $N = 10$, 25, and 85. For the time step $N = 10$, only a small portion of the wave front has appeared in the medium. For time step $N = 25$, only the first half of the half-wave time pulse has appeared in the medium. For time step $N = 85$, the complete half-wave sinusoidal time pulse has appeared in the medium and, in fact, has already traveled a certain distance in the positive y direction away from the source location. Since the medium is homogeneous and isotropic, there is no variation of the medium parameters at any layered cells and thus the half-wave time pulse travels in the forward direction with no change in amplitude nor in shape at a constant phase velocity given by the velocity of the electromagnetic wave in the free-space medium.

In fact, this is not the case in the electric lossy and the magnetic lossy, homogeneous, and isotropic media. To illustrate the effects of medium conductivity, the propagation of the half-wave sinusoidal time pulse discussed earlier is now considered in a lossy half-space medium. This case study is similar to that discussed previously except certain electric and magnetic conductivities are introduced into the half-space lossy layered medium. For this special propagation case study, the two conductivity parameters are selected by forcing the two media coefficients given by Eqs. (6.2.22) and (6.2.28) equal to zero. Hence,

$$\frac{\sigma^{(e)}(m)\Delta t}{2\epsilon_0\epsilon_r(m)} = 1 \tag{6.2.43}$$

$$\frac{\sigma^{(m)}(m+1/2)\Delta t}{2\mu_0\mu_r(m+1/2)} = 1. \tag{6.2.44}$$

On equating the previous two relationships,

$$\sigma^{(m)}(m+1/2) = (120\pi)^2 \frac{\mu_r(m+1/2)}{\epsilon_r(m)} \sigma^{(e)}(m). \tag{6.2.45}$$

The modeling of the layered media in terms of medium parameters is similar to that for the previously discussed numerical case. The electric conductivity parameter is specified at the spatial locations where the electric field components are located and, similarly, the magnetic conductivity parameter is specified at the spatial locations where the staggered magnetic field components are located. It is assumed that each spatial cell by itself is lossy, homogeneous, and isotropic, having the following medium parameters:

$$\epsilon_r(m) = 1 \qquad (6.2.46)$$

$$\mu_r(m + 1/2) = 1 \qquad (6.2.47)$$

for $m = 1, 2, 3, \cdots, M$,

$$\sigma^{(e)}(m) = 0 \qquad (6.2.48)$$

$$\sigma^{(m)}(m + 1/2) = 0 \qquad (6.2.49)$$

for $m = 1, 2, 3, \cdots, 30$, and

$$\sigma^{(e)}(m) = 0.005 \qquad (6.2.50)$$

$$\sigma^{(m)}(m + 1/2) = 710.61 \qquad (6.2.51)$$

for $m = 31, 32, 33, \cdots, M$.

Using the previous media parameters, the z component of electric field distribution at specific spatial points is plotted as a function of the number of time steps. The excitation is turned on at $t = 0$ at the location $s = 5$. The two finite-difference coupled expressions (Eqs. 6.2.19 and 6.2.25) are solved iteratively at each spatial cell for every time step. Figure 6.7 shows the distribution of the electric field as the half-wave sinusoidal time pulse passes the spatial locations $y = 0.1, 0.2, 0.3, 0.4, 0.5$, and 0.6 corresponding to the location of free-space layers $m = 10, 20,$

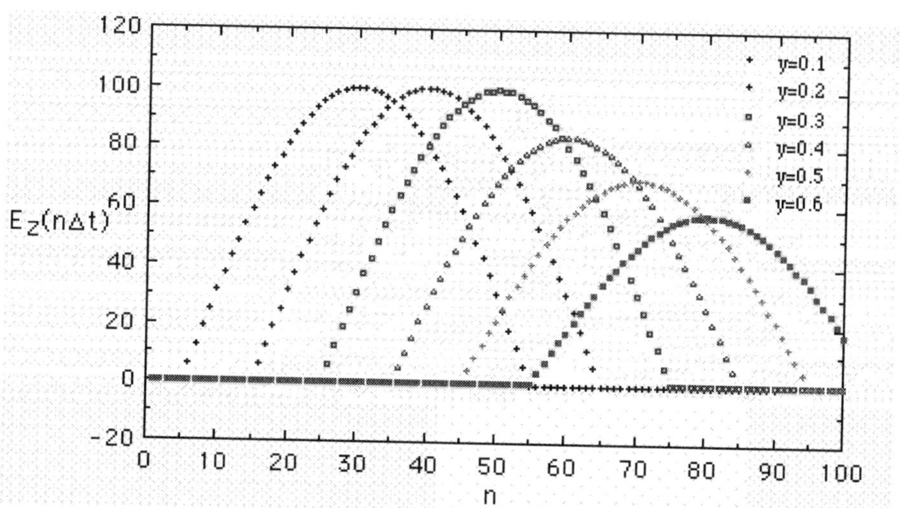

FIGURE 6.7
Propagation of half-wave sinusoidal time pulse in electric and magnetic half-space medium.

and 30 and half-space lossy conductivity layers $m = 40$, 50, and 60. Due to the conductivity of medium, the wave amplitude gradually attenuates as the wave propagates along the y-coordinate direction.

6.3 REMOTE SENSING OF INHOMOGENEOUS, LOSSY, LAYERED MEDIA

The one-dimensional time-dependent electromagnetic wave propagation model discussed earlier is ideal for inverse scattering studies as well. The electromagnetic medium characteristics, namely, conductivity, permittivity, and permeability (σ, ϵ, and μ), profiles of an inhomogeneous, one-dimensional, lossy layered propagation medium, can be obtained in the time domain by utilizing the layer-stripping procedure. Despite its simplicity, this time domain inverse scattering technique is often considered a useful model for many practical applications involving the electromagnetic material synthesis and fabrication of layered structures, remote sensing of biological media and natural geophysical media, and target identification and object reconstruction problems.

An efficient profile-inversion technique suitable for simultaneous extraction of the medium parameters is discussed that uses an optimization scheme and a real-time wave propagation and scattering model based on the FDTD method. In Fig. 6.8, the one-dimensional, lossy, inhomogeneous medium is excited by a linearly polarized pulse-transient plane wave normally incident at its interface in the outer medium. The resulting time history of the total field at the receiver site R is recorded and discretized into a suitable number of M time steps depending on the required profile resolution. Next, using the one-dimensional FDTD formulation discussed earlier, the optimization scheme of Fig. 6.9 is utilized to construct the (σ, ϵ, μ) profile of the layered medium such that at the point R, the resulting total field fits the recorded field (input data) in a least square error sense at M discrete instants in time. This essentially means that the unknown inhomogeneous medium is modeled by M layers of media, each of which is homogeneous within itself. It should be noted that because the time discretization Δt is uniform, the electrical width of each layer d_e is also uniform. However, the physical width of the nonpermeable layers is

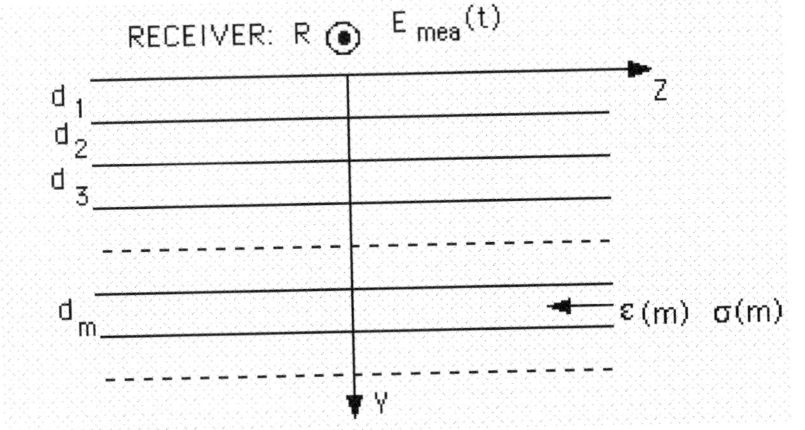

FIGURE 6.8
One-dimensional lossy layered medium for remote sensing.

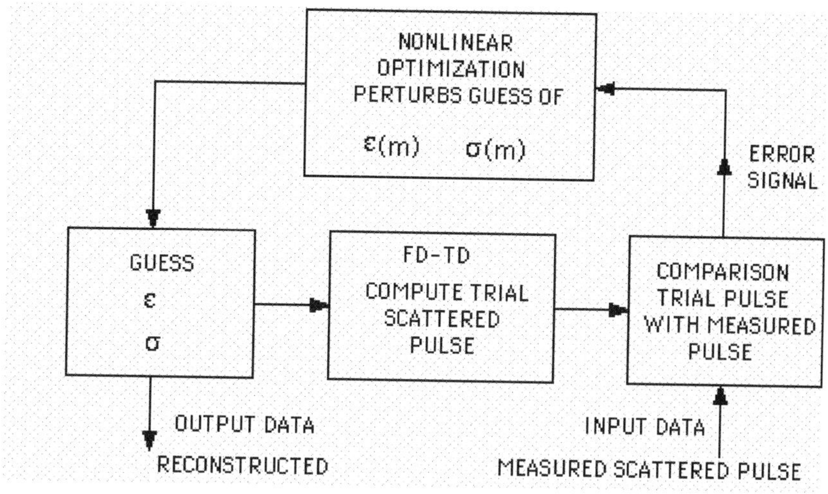

FIGURE 6.9
Optimization scheme for inverse scattering.

dependent on the corresponding relative permittivity ϵ_r, and hence it is nonuniform given by

$$d(m) = \frac{d_e}{\sqrt{\epsilon_r(m)}}. \tag{6.3.1}$$

Since the FDTD formulation deals with electrical lengths as opposed to physical lengths, the causality of the marching in uniform time steps is assured, even with the different wave velocity of propagation at the different layers of the unknown medium. This is a requirement for the proposed inverse scattering formulation.

To initiate the optimization scheme for the layer stripping (Fig. 6.9), the causality of the time marching is exploited. If the receiver is located at the media interface, as shown in Fig. 6.8, then at $t = 2\Delta t$ only the first layer d_1 will contribute to the total field at the point R, assuming the time reference $t = 0$ is considered to be the instant when the leading edge of the incident field reaches the receiver location. Thus, the only unknowns in the field calculation with FDTD up to the time instant $t = 2\Delta t$ are the parameters $[\epsilon(1), \sigma(1)]$ of the first layer.

For determining the unknowns, an error function $W(y, t)$ at the location R is now defined as

$$W|_{t=2n\Delta t} = [E_{\text{mea}}(2n\Delta t) - E_{\text{cal}}(2n\Delta t, \sigma(m), \epsilon(m))]^2, \tag{6.3.2}$$

where E_{mea} and E_{cal} are the total electric fields measured and calculated by FDTD using assumed values of $\sigma(m)$ and $\epsilon(m)$, respectively, at R.

The numerical nonlinear optimization algorithm minimizes the error function of Eq. (6.3.2) at the location R for $m = 1$ by iteratively adjusting the values of the profile characteristics $[\epsilon(1), \sigma(1)]$ of the layer d_1. The current nonlinear optimization scheme requires the error function at the location R as defined in Eq. (6.3.2) and also the corresponding field gradients of the error function at the location R calculated with respect to the unknown characteristic parameters $[\epsilon(m), \sigma(m)]$ of the mth layer being probed, as defined later. Hence, at location R, the gradient of the error function with respect to the unknown conductivity and permittivity parameters is given by

$$\left.\frac{\partial W(j=R,m)}{\partial \sigma(m)}\right|_{t=2n\Delta t} = -2[E_{\text{mea}} - E_{\text{cal}}]G_z(j=R,m) \tag{6.3.3}$$

$$P_x^n(j,m) = \frac{\partial H_x^n(j)}{\partial \sigma(m)} \tag{6.3.4}$$

$$G_z^n(j,m) = \frac{\partial E_z^n(j)}{\partial \sigma(m)} \tag{6.3.5}$$

and

$$\left.\frac{\partial W(j=R,m)}{\partial \epsilon(m)}\right|_{t=2n\Delta t} = -2[E_{\text{mea}} - E_{\text{cal}}]T_z(j=R,m) \tag{6.3.6}$$

$$S_x^n(j,m) = \frac{\partial H_x^n(j)}{\partial \epsilon(m)} \tag{6.3.7}$$

$$T_z^n(j,m) = \frac{\partial E_z^n(j)}{\partial \epsilon(m)}. \tag{6.3.8}$$

Similarly, at the time instant $t = 4\Delta t$, the first two layers $m = 1, 2$ with resolution depths d_1 and d_2 take part in the electromagnetic interaction. Now, the characteristics of the first layer d_1 are already known, and thus by iterative optimization the characteristic parameters of the second layer d_2 are recovered. The process is continued for the remaining layers. The following are some observations regarding the numerical implementation of the previous layer-stripping scheme:

- As a first step, the time interval Δt can be made large and a coarse profile estimate of the electromagnetic characteristics of the medium is obtained. In this estimate, the region where the profile appears to be changing rapidly can be recalculated by using finer subintervals of the time step in the corresponding region of the measured time history of the total electric field at point R. This variable resolution is effective and accurate.
- When the mth layer's profile is being recovered, all the nonzero field information up to the time instant $2(m-1)\Delta t$ is known from the previous optimization step, stripping of the $(m-1)$th layer, and these values are not affected by the changes in $\epsilon(m)$ and $\sigma(m)$ due to causality. Thus, in the mth optimization step, the field values obtained in the earlier optimization steps are of direct use.
- Since the FDTD formulation gives field values at all the grid points of the layered geometry, the field values inside the layer medium can be successively monitored to determine the depth at which the profile recovery scheme should be terminated, i.e., the depth into the medium at which the field due to the transmitted pulse has decayed sufficiently to produce no significant scattering contributions.

6.3.1 Profile Inversion Results

Based on the profile inversion technique presented previously involving the FDTD with the nonlinear optimization scheme, selected numerical results are presented to demonstrate applicability

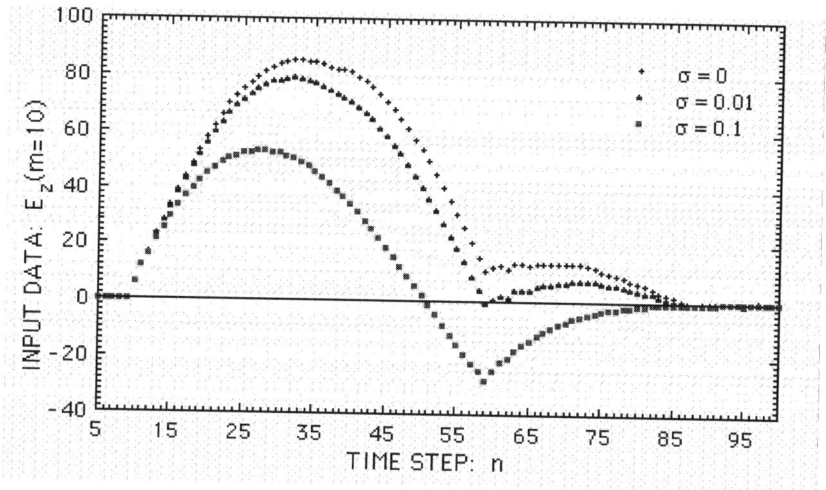

FIGURE 6.10
Electric field input data for staircase permittivity profile.

of the layer-stripping procedure for the one-dimensional time domain inverse scattering. Several case studies involving pure conductivity profile, a pure permittivity profile, and also combination of both the conductivity and permittivity profile have been tested and recovered including generic profile distribution, namely, the rectangular, triangular, and sinusoidal profiles.

For the profile inversion scheme, the input data consist of the recorded time history of the total electric field at the receiver site R located in front of the medium being probed. The recorded data at the receiver site R for the case of constant conductivity and staircase permittivity profile are shown in Fig. 6.10. In fact, the recorded data utilized are completely noise free and were obtained by the direct time domain scattering problem with a very high resolution of a sinusoidal half-pulse excitation. Generally, practical input data when measured will include random noise distribution depending on the transient measurements. No attempt has been made here to include any noise case studies. All questions concerning the input data with a random noise distribution, the effect of different types of time-pulse excitation, and the limits and accuracy of the data recovered play a major role in the development of efficient and accurate algorithms for the reconstruction of target shapes.

Figure 6.11 shows the result of inversion recovery of the staircase permittivity profile having uniform conductivity. The exciting incident time pulse used is a half-cycle sinusoid at 300 MHz. The results are recovered with a spatial resolution of 0.01 m in the region outside the medium probed; also, the same physical resolution is utilized within the medium being probed. The source is located at the 5th spatial layer and is turned on at the time instant $t = 0$. With a spatial resolution of 0.01 m, the field observation or measurement point R is situated at the 10th spatial layer, just one layer in front of the medium being probed. Figure 6.11 shows the results of stripping of the first 10 layers corresponding to a 0.1-m-thick layered slab, recovered based on the numerical nonlinear optimization algorithm described earlier. Due to low scattering returns, numerical difficulties are encountered when probing deeper layers. In order to overcome this limitation, it is necessary to use low-frequency half-sinusoid exciting time pulses or alternative pulse shapes such as the Gaussian and its derivatives.

In fact, the current numerical algorithm is also suited for other characteristic parameters such as the permeability distribution. The algorithm developed here is quite useful for the layered material synthesis because it specifies the required reflection coefficient at point R. The inversion

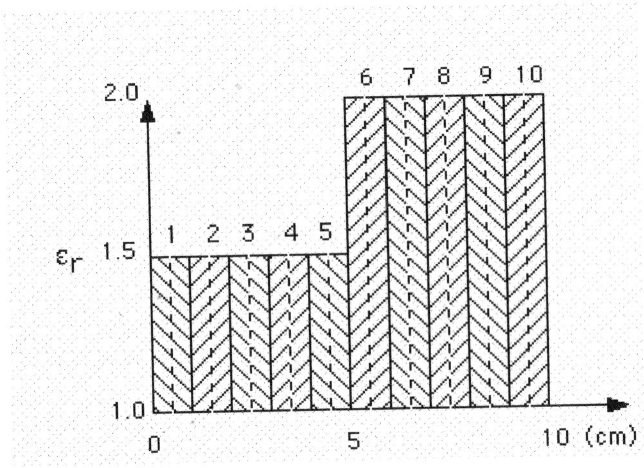

FIGURE 6.11
Recovery of staircase permittivity profile with uniform conductivity.

approach is quite general and can be easily extended to two- and three-dimensional cases by appropriately replacing the forward-scattering algorithm with the two- and three-dimensional FDTD scheme.

6.4 KEY ELEMENTS OF FDTD MODELING THEORY

The direct time domain solution of Maxwell's equations using the FDTD technique for the analysis of electromagnetic wave interactions with arbitrary material structures appears to be suitable for simulating structures that are too large or too complex for simulation by any other principal numerical techniques available today. The FDTD technique permits, in principle, the modeling of electromagnetic wave interactions with a high level of detail but has a dimensionally small computer resource requirement relative to popular numerical modeling approaches of comparable scope, such as the time domain MoM. In fact, it employs no electromagnetic potentials. Instead, it applies simple second-order accurate central-difference approximations for the space and time derivatives of the electric and magnetic fields directly to the respective differential operators of the curl equations. Space and time discretizations are selected to bound errors in the sampling process and to ensure numerical stability of the algorithm. The electric and magnetic field components are interleaved in space to permit a natural satisfaction of the tangential field continuity conditions at media interfaces. The resulting system of equations for the fields is fully explicit so that there is no need to solve a set of linear matrix equations, and the required computer storage and running time is proportional to the electrical size of the volume modeled.

The general concept of time domain wave tracking using the FDTD technique is illustrated in Fig. 6.12. A region of space within the lattice truncation planes is selected for numerical field sampling in space and time. At the time instant $t = 0$, it is assumed that all field quantities within the numerical sampling region are identically zero. An incident transient plane wave is assumed to enter the sampling region at this time instant. Propagation of the incident wave is modeled by the commencement of time stepping, which is simply the implementation of the finite-difference analog of Maxwell's equations. Time stepping continues as the analog of the incident wave strikes the arbitrary-shaped material structure embedded in the numerical sampling region. All outgoing scattered wave analogs ideally propagate through the lattice truncation planes with negligible

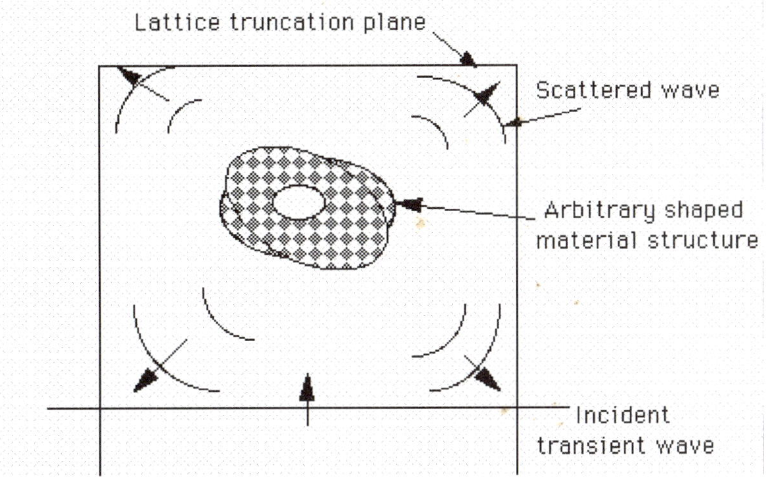

FIGURE 6.12
Time domain wave-tracking concept of FDTD technique.

reflection to exit the sampling region. Various electromagnetic phenomena, such as the induction of surface electric and magnetic currents, scattering and multiple-scattering effects, singular and edge effects, penetration of fields through apertures, and cavity field excitation, are systematically modeled by the difference equations analog. Self-consistency of these electromagnetic modeled phenomena is generally ensured if both their spatial and temporal variations are well resolved by the space and time sampling process.

Time stepping is continued until the desired late-time response or the steady-state behavior is achieved. An important application of the latter is the extraction of frequency-domain sinusoidal steady-state data, wherein the incident plane wave is assumed to have a sinusoidal time dependence and the time stepping is continued until all transient behavior is completed and all fields in the sampling region exhibit sinusoidal time harmonic repetition. Extensive numerical experimentation with FDTD modeling has shown that the number of complete cycles of the incident wave required to be time stepped to achieve the sinusoidal steady state is approximately equal to the quality factor Q of the material structure being modeled.

The locations of interleaved electric and magnetic field components about a structured unit cell of the three-dimensional FDTD space lattice in the rectangular coordinate system are illustrated in Fig. 6.13. Each magnetic field component is surrounded by four circulating electric field components. By examining the adjacent structured unit cells, it is evident that each electric field component is surrounded by four circulating magnetic field components. This arrangement permits not only a centered-difference analog to the space derivatives of Maxwell's equations but also a natural structured lattice geometry for implementing the integral form of Faraday's law and Ampere's law at the space cell level. Similar to the one-dimensional study discussed earlier, this integral interpretation permits appropriate spatial and directional resolution of the medium permittivity, permeability, and conductivity parameters. In addition, it permits a simple but effective modeling of the physics of structures requiring subcell spatial resolution as in the case of thin slots, thin wires, and arbitrary curved surfaces.

Figure 6.14 illustrates how an arbitrarily shaped three-dimensional complex material scatterer is embedded in a FDTD space lattice composed of unit cells in the rectangular coordinate system. Simply, desired values of electrical permittivity and conductivity are assigned to each electric field component of the lattice. Correspondingly, desired values of magnetic permeability and

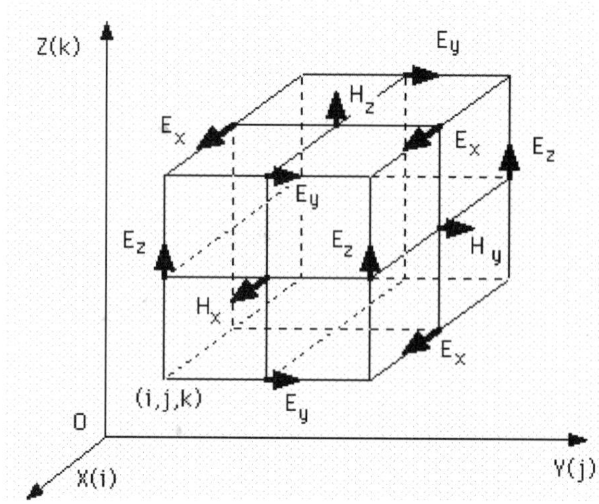

FIGURE 6.13
FDTD structures unit cell with locations of interleaved field components.

equivalent conductivity are assigned to each magnetic field component of the lattice. As will be shown, the media parameters are interpreted by the FDTD algorithm as local coefficients for the time-stepping scheme. Specification of media properties on this component-by-component basis results in a stepped-edge or staircase approximation of the curved surface. Continuity of the tangential fields at the interface of dissimilar media is ensured with this procedure. However, note that the stepped-edge approximation of the curved surface is not always an adequate structure model. Later, a detailed conformal surface model is introduced based on the subcell FDTD model to simulate actual arbitrary surface configuration. The conformal FDTD model requires application of a modified contour path algorithm only at special points adjacent to the curved surface and other specific features such as thin slots, thin plates, and single or multiple

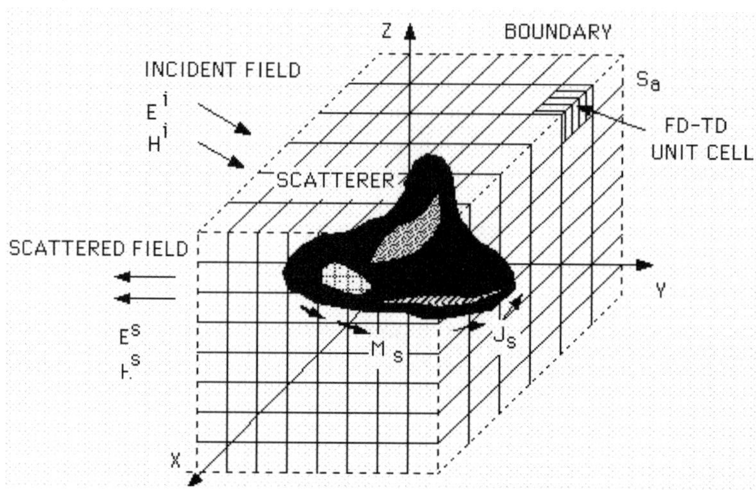

FIGURE 6.14
Arbitrarily shaped three-dimensional material scatterer embedded in a FDTD structured space lattice.

bundle of wires requiring subcell spatial resolution. Before presenting the detailed and rigorous numerical algorithm based on the FDTD structured space–time lattice for analyzing the arbitrarily shaped three-dimensional complex material scatterer (Fig. 6.14), simple and yet practical time domain propagation case studies are discussed to illustrate numerical modeling requirements for the two-dimensional close- and open-type electromagnetic scattering and interaction problems.

6.5 FDTD FORMULATION FOR TWO-DIMENSIONAL CLOSED-REGION PROBLEMS

A guided wave structure is either an open or a closed type and is generally three-dimensional, having a uniform cross section along its longitudinal direction representing the transmission of electromagnetic signal power. The close type of waveguide consists of an arbitrary geometrical cross section having a perfectly conducting boundary with no electromagnetic coupling to its exterior zone. To determine the propagation characteristics, a complete three-dimensional FDTD analysis must be carried out within the close waveguide geometry. In order to obtain insight into the FDTD full-wave analysis, the mode cutoff frequencies are numerically evaluated to determine types of modes which can be successfully excited and propagated along the waveguide structure. Using the numerical tool developed for the mode cutoff frequencies, other waveguide parameters, such as the mode characteristic impedance, the mode electric and magnetic field distribution, and the mode dispersion characteristic, can be numerically evaluated without any loss of generality. The results of waveguide mode cutoff frequencies, depending on the type of canonical geometry, are compared with respect to the rigorous analytical solution.

In the following, relevant time domain equations for the transverse magnetic (TM) and transverse electric (TE) polarizations are briefly discussed to expose their applicability to the full-wave FDTD analysis of the complex close waveguide geometries. The two Maxwell's curl equations in scalar form, using the rectangular coordinate system (x, y, z), are given by

$$\frac{\partial E_z}{\partial y} - \frac{\partial E_y}{\partial z} = -\mu \frac{\partial H_x}{\partial t} - \sigma^{(m)} H_x \tag{6.5.1}$$

$$\frac{\partial E_x}{\partial z} - \frac{\partial E_z}{\partial x} = -\mu \frac{\partial H_y}{\partial t} - \sigma^{(m)} H_y \tag{6.5.2}$$

$$\frac{\partial E_y}{\partial x} - \frac{\partial E_x}{\partial y} = -\mu \frac{\partial H_z}{\partial t} - \sigma^{(m)} H_z \tag{6.5.3}$$

$$\frac{\partial H_z}{\partial y} - \frac{\partial H_y}{\partial z} = \epsilon \frac{\partial E_x}{\partial t} + \sigma^{(e)} E_x \tag{6.5.4}$$

$$\frac{\partial H_x}{\partial z} - \frac{\partial H_z}{\partial x} = \epsilon \frac{\partial E_y}{\partial t} + \sigma^{(e)} E_y \tag{6.5.5}$$

$$\frac{\partial H_y}{\partial x} - \frac{\partial H_x}{\partial y} = \epsilon \frac{\partial E_z}{\partial t} + \sigma^{(e)} E_z. \tag{6.5.6}$$

Equations (6.5.1)–(6.5.6) can be specialized for TM and TE polarizations. Figure 6.15 shows a typical three-dimensional close-type waveguide structure having a uniform cross-sectional geometry along its axial z-coordinate direction representing the propagation of the electric and magnetic fields. Furthermore, for TM to z and TE to z polarizations, $H_z(x, y, z, t) = 0$ and $E_z(x, y, z, t) = 0$, respectively.

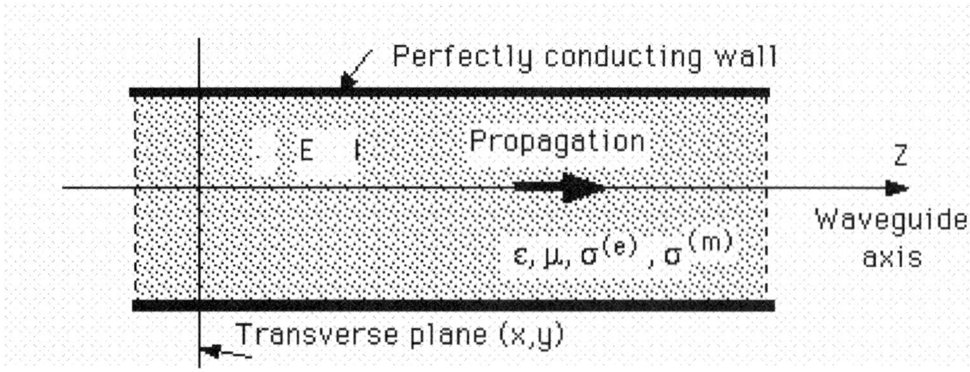

FIGURE 6.15
A typical closed-region waveguide geometry.

6.5.1 FDTD Formulation for TM and TE Cases

Further simplification of the scalar field equations (Eqs. 6.5.1–6.5.6) is obtained for the special case of the waveguide cutoff propagation condition. In fact, the same two-dimensional equations are obtained for the exterior electromagnetic wave scattering and interaction studies. Since the cross-sectional geometry is uniform along the z-axis, the various field components are considered only at the $z = 0$ transverse plane. For TM to z polarization, nonzero field components are given by the axial component of the electric field $E_z(x, y, t)$ and the two transverse components of the magnetic field, $H_x(x, y, t)$ and $H_y(x, y, t)$. For the dual TE to z polarization, nonzero field components are given by the axial component of the magnetic field $H_z(x, y, t)$, and the two transverse components of the electric field, $E_x(x, y, t)$ and $E_y(x, y, t)$.

For TM to z polarization, the following coupled scalar equations are obtained:

$$\frac{\partial E_z}{\partial y} = -\mu \frac{\partial H_x}{\partial t} - \sigma^{(\mathrm{m})} H_x \tag{6.5.7}$$

$$-\frac{\partial E_z}{\partial x} = -\mu \frac{\partial H_y}{\partial t} - \sigma^{(\mathrm{m})} H_y \tag{6.5.8}$$

$$\frac{\partial H_y}{\partial x} - \frac{\partial H_x}{\partial y} = \epsilon \frac{\partial E_z}{\partial t} + \sigma^{(\mathrm{e})} E_z. \tag{6.5.9}$$

The FDTD two-dimensional algorithm based on time marching and finite-difference approximation, similar to the one-dimensional algorithm, is considered here. Since the coupled scalar equations are valid for all values of x, y, and t, it is assumed that the three components of the electric and magnetic fields are continuous or at least piecewise continuous with respect to the x, y, and t variables. Let Δx and Δy represent the lattice spatial increment along the x and y coordinates, respectively, and Δt represent the discretized time increment measured in seconds. Thus,

$$(i, j) = (i\Delta x, j\Delta y) \tag{6.5.10}$$

$$F^n(i, j) = F(i\Delta x, j\Delta y, n\Delta t). \tag{6.5.11}$$

For convenience, in the finite-difference calculation, the electric and magnetic field components are not calculated at the same spatial locations but are staggered alternately as shown in Fig. 6.16.

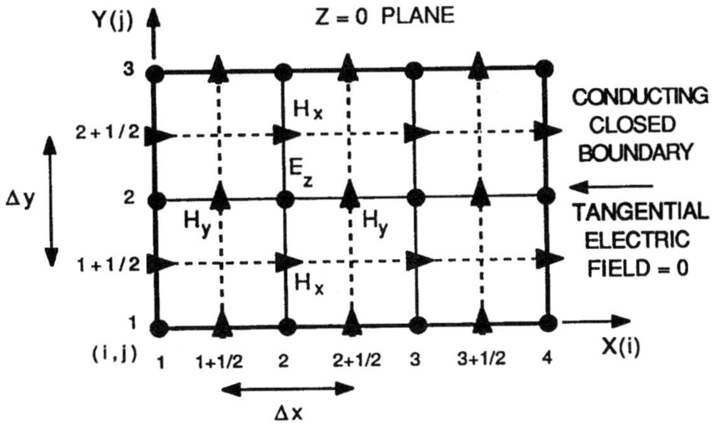

FIGURE 6.16
Position of interior field components for two-dimensional FDTD-TM to z polarization.

The rectangular interior region is first discretized uniformly into small rectangular cells having constant spatial widths and time steps. It is assumed that the thickness of each cell is electrically very small, and for all practical purposes the region is linear, homogeneous, and isotropic.

For the TM to z case, the electric field distribution within each cell is assumed to be constant. Thus, as shown in Fig. 6.16, the midpoint of each cell is selected under spatial discretization, and the z component of the axial electric field is defined at the midpoint of each cell. The constant permittivity and electrical conductivity for a given cell are selected to be the permittivity and electrical conductivity at the midpoint of the cell in which the electric field component is located. Since the electric and magnetic field components are staggered, the calculation of transverse components of the magnetic field is performed at the spatial locations in between any two adjacent electric field components. The spatial discretization for the transverse magnetic field overlaps half a cell on either side. The constant permeability and magnetic conductivity for a given cell are selected to be the permeability and magnetic conductivity in which the magnetic field is located. This natural arrangement of the field components is ideal for defining first-order central differences required for solving Eqs. (6.5.7)–(6.5.9).

The following finite-difference approximations for the derivatives of electric and magnetic fields are accurate to their second power with piecewise linear basis function discretization:

$$\frac{\partial F^n(i,j)}{\partial t} = \frac{F^{n+1/2}(i,j) - F^{n-1/2}(i,j)}{\Delta t} + O(\Delta t^2) \tag{6.5.12}$$

$$\frac{\partial F^n(i,j)}{\partial x} = \frac{F^n(i+1/2,j) - F^n(i-1/2,j)}{\Delta x} + O(\Delta x^2) \tag{6.5.13}$$

$$\frac{\partial F^n(i,j)}{\partial y} = \frac{F^n(i,j+1/2) - F^n(i,j-1/2)}{\Delta y} + O(\Delta y^2). \tag{6.5.14}$$

For TM to z polarization, using the previous difference approximations, the scalar partial differential equations (Eqs. 6.5.7–6.5.9) yield the following time-stepping numerical algorithm for the interior region or the exterior of the close-type geometry:

$$H_x^{n+1/2}(i, j+1/2) = C_a(i, j+1/2)H_x^{n-1/2}(i, j+1/2) + C_b(i, j+1/2)\left[\frac{E_z^n(i, j) - E_z^n(i, j+1)}{\Delta y}\right] \tag{6.5.15}$$

$$H_y^{n+1/2}(i+1/2, j) = C_a(i+1/2, j)H_y^{n-1/2}(i+1/2, j) + C_b(i+1/2, j)\left[\frac{E_z^n(i+1, j) - E_z^n(i, j)}{\Delta x}\right] \tag{6.5.16}$$

$$E_z^{n+1}(i, j) = C_c(i, j)E_z^n(i, j) + C_d(i, j)\left[\frac{H_y^{n+1/2}(i+1/2, j) - H_y^{n+1/2}(i-1/2, j)}{\Delta x}\right] + C_d(i, j)\left[\frac{H_x^{n+1/2}(i, j-1/2) - H_x^{n+1/2}(i, j+1/2)}{\Delta y}\right], \tag{6.5.17}$$

where the lattice media coefficients are given by

$$R_b(i, j) = \frac{\Delta t}{\mu_0 \mu_r(i, j)} \tag{6.5.18}$$

$$C_a(i, j) = \frac{\left[1 - \frac{\sigma^{(m)}(i,j)\Delta t}{2\mu_0\mu_r(i,j)}\right]}{\left[1 + \frac{\sigma^{(m)}(i,j)\Delta t}{2\mu_0\mu_r(i,j)}\right]} \tag{6.5.19}$$

$$C_b(i, j) = \frac{R_b(i, j)}{\left[1 + \frac{\sigma^{(m)}(i,j)\Delta t}{2\mu_0\mu_r(i,j)}\right]} \tag{6.5.20}$$

and

$$R_d(i, j) = \frac{\Delta t}{\epsilon_0 \epsilon_r(i, j)} \tag{6.5.21}$$

$$C_c(i, j) = \frac{\left[1 - \frac{\sigma^{(e)}(i,j)\Delta t}{2\epsilon_0\epsilon_r(i,j)}\right]}{\left[1 + \frac{\sigma^{(e)}(i,j)\Delta t}{2\epsilon_0\epsilon_r(i,j)}\right]} \tag{6.5.22}$$

$$C_d(i, j) = \frac{R_d(i, j)}{\left[1 + \frac{\sigma^{(e)}(i,j)\Delta t}{2\epsilon_0\epsilon_r(i,j)}\right]}. \tag{6.5.23}$$

For TE to z polarization, the following coupled scalar equations are obtained:

$$\frac{\partial E_y}{\partial x} - \frac{\partial E_x}{\partial y} = -\mu\frac{\partial H_z}{\partial t} - \sigma^{(m)}H_z \tag{6.5.24}$$

$$\frac{\partial H_z}{\partial y} = \epsilon\frac{\partial E_x}{\partial t} + \sigma^{(e)}E_x \tag{6.5.25}$$

$$-\frac{\partial H_z}{\partial x} = \epsilon\frac{\partial \boldsymbol{E}_y}{\partial t} + \sigma^{(e)}E_y. \tag{6.5.26}$$

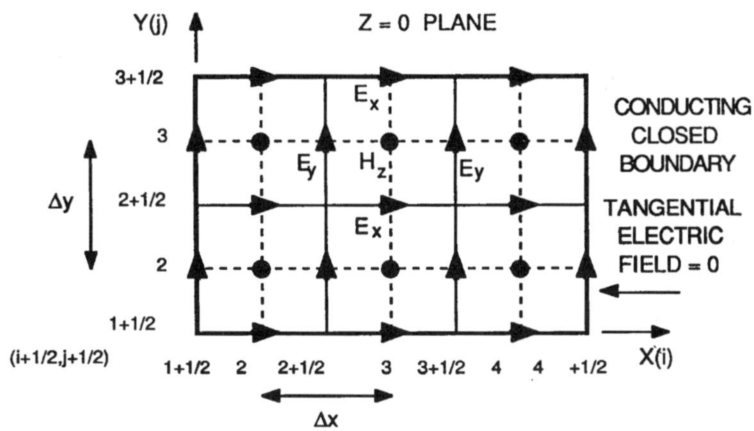

FIGURE 6.17
Position of interior field components for two-dimensional FDTD-TE to z polarization.

For TE to z polarization, the magnetic field distribution within each cell is assumed to be constant. Thus, as shown in Fig. 6.17, the midpoint of each cell is selected under spatial discretization, and the z component of the axial magnetic field is defined at the midpoint of each cell. The constant permeability and magnetic conductivity for a given cell are selected to be the permeability and magnetic conductivity at the midpoint of the cell in which the magnetic field component is located. Since the electric and magnetic field components are staggered, the calculation of transverse components of the electric field is performed at the spatial locations in between any two adjacent magnetic field components. The spatial discretization for the transverse electric field overlaps half a cell on either side. The constant permittivity and electric conductivity for a given cell are selected to be the permittivity and electric conductivity in which the electric field is located.

For TE to z polarization, Eqs. (6.5.24)–(6.5.26) yield the following time-stepping numerical algorithm for the interior region or the exterior of the close-type geometry:

$$E_x^{n+1/2}(i, j+1/2) = C_c(i, j+1/2)E_x^{n-1/2}(i, j+1/2) + C_d(i, j+1/2)\left[\frac{H_z^n(i, j+1) - H_z^n(i, j)}{\Delta y}\right] \tag{6.5.27}$$

$$E_y^{n+1/2}(i+1/2, j) = C_c(i+1/2, j)E_y^{n-1/2}(i+1/2, j) + C_d(i+1/2, j)\left[\frac{H_z^n(i, j) - H_z^n(i+1, j)}{\Delta x}\right] \tag{6.5.28}$$

$$H_z^{n+1}(i, j) = C_a(i, j)H_z^n(i, j) + C_b(i, j)\left[\frac{E_y^{n+1/2}(i-1/2, j) - E_y^{n+1/2}(i+1/2, j)}{\Delta x}\right] + C_b(i, j)\left[\frac{E_x^{n+1/2}(i, j+1/2) - E_x^{n+1/2}(i, j-1/2)}{\Delta y}\right]. \tag{6.5.29}$$

Equations (6.5.15)–(6.5.17) for the TM to z polarization and the dual expressions (Eqs. 6.5.27–6.5.29) for the TE to z polarization are the special time-stepping algorithms for the calculation of two-dimensional electric and magnetic field components. These equations can be utilized in an interior problem to study the waveguide propagation cutoff condition and, similarly, in an exterior problem to study the two-dimensional electromagnetic scattering and coupling phenomenon.

Using the field solutions obtained in the space–time difference equations, the field components for a given time step can be calculated iteratively. The new value of a field component at any cell point depends only on its value in the previous time step and the previous values of the components of the other field at the adjacent spatial points. Hence, at any given time step, the computation of a field component will proceed one point at a time. The choice of space increments Δx and Δy and time increment Δt is dictated by accuracy and algorithm stability, respectively. To ensure the stability of the computed fields, Δt is chosen to satisfy the following inequality for the two-dimensional lattice model:

$$\Delta t \leq \frac{1}{c_{\max}} \left[\frac{1}{\Delta x^2} + \frac{1}{\Delta y^2} \right]^{-1/2}, \tag{6.5.30}$$

where $c_{\max}$ is the maximum velocity within the model.

The numerical algorithms discussed previously can be utilized to study electric and magnetic fields within the interior region of the close-type waveguide. In Fig. 6.12, the perfectly conducting close-type boundary isolates the exterior region from the interior region and thus, no radiation boundary condition at the lattice truncation planes is required. The close-type waveguide region can be composed of any material with space-dependent properties. In the interior region, the waveguide excitation can be introduced in terms of either an electric field or a magnetic field component. Even though the structured numerical algorithm is ideally suited for modeling rectangular geometries, arbitrarily shaped curved boundaries can be modeled by appropriately conforming the lattice cells near the boundary. This modeling aspect of conforming the boundary cells to fit the shape of an arbitrary surface is discussed later.

In the following sections, the numerical results of the TM and TE mode cutoff frequencies are presented with analytical validation for the canonical close-type geometries, such as a hollow rectangular waveguide, a partially filled rectangular dielectric waveguide, and a shielded microstrip line.

6.5.2 Hollow Rectangular Waveguide

Figure 6.18 shows the cross section of a hollow (air-filled) rectangular close-type waveguide geometry. The side lengths along the x and y coordinates are as follows: $a = 0.01$ m and $b = 0.02$ m. This rectangular cross section is modeled by discretizing the inside region into structured square cells, $\Delta x = \Delta y = \Delta$, with $i = 51$ mesh points along the x-axis and $j = 101$ mesh points along the y-axis. For this spatial discretization, the spatial increment is $\Delta = 0.0002$ m and the corresponding time-stepping increment (in seconds) is $\Delta t = \Delta/\sqrt{2}c_{\max}$, where $c_{\max}$ is the maximum wave velocity in the waveguide region given by 3×10^8 m/s, with the waveguide region having free-space medium parameters. For TM to z polarization, as shown in Fig. 6.16, the z component of the axial electric field is defined at the midpoint of each square cell. Since the electric and magnetic field components are staggered, the calculation of transverse components of the magnetic field is performed at the spatial locations in between any two adjacent axial electric field components. The boundary condition at the rectangular, perfectly conducting walls is simulated by appropriately truncating cells at the axial components of the electric field, E_z. Thus, the axial electric field components at the rectangular conducting boundary are simulated with a very large electric conductivity; for example, the conductivity of copper material, $\sigma^{(e)} = 10^7$ S/m, is assigned

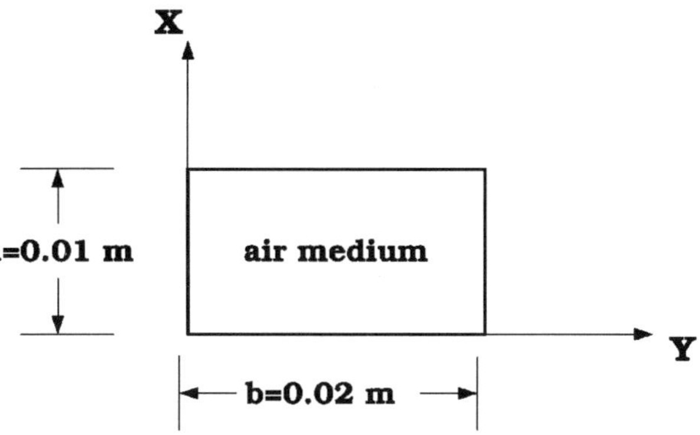

FIGURE 6.18
Rectangular waveguide.

to the corresponding boundary cells. This is also equivalent to forcing the axial components of the electric field, E_z, parallel to the conducting boundary equal to zero, to simulate a two-dimensional close-type cavity-like region.

Initially, there is no excitation field inside the waveguide region, and all the spatial electric and magnetic field components are assumed to be zero. The rectangular waveguide is excited by injecting a Gaussian tone burst. This can be accomplished with an initial arbitrary excitation field assigned to the axial components of the electric field at the time instant, $t = 0$, at the center region covering few specific cells of the waveguide. Even though any arbitrary initial field can be assigned to the cells inside the waveguide region, from the viewpoint of determining waveguide mode cutoff frequencies, the initial assigned field should be such that it is capable of exciting all possible spatial frequency harmonics inside the rectangular waveguide.

After performing many numerical experiments to excite various modal fields inside the rectangular region, the following spatially distributed Gaussian pulse is selected and assigned to the axial electric field cells to trigger every possible TM mode in the cavity-type close rectangular geometry:

$$G_f(x, y, t = 0) = Ae^{-\frac{[(x-x_c)^2+(y-y_c)^2]}{2\tau^2}} \tag{6.5.31}$$

for $x \geq x_c$ and $y \geq y_c$.

This initial axial electric field is in fact suitable for exciting the rectangular TM waveguide cutoff modes. Due to the spatial symmetry of Gaussian distribution, the excitation can be applied with its peak located at any cell inside the waveguide except at the center cell. To provide an efficient initial excitation with its peak located at the center of the waveguide, only a quarter of the distribution of Gaussian pulse is generally utilized. Using the previous initial axial electric field excitation, the time-stepping algorithm for the TM cutoff conditions (Eqs. 6.5.15–6.5.17) is implemented. As a function of time step, the components of the transverse magnetic and axial electric fields inside the rectangular waveguide are recorded. For determining the mode cutoff frequencies, only the temporal variation of a single component of a field at any general cell within the waveguide needs to be recorded. Figure 6.19 shows a plot of the temporal variation of the axial component of the electric field at the spatial cell ($i = 5$, $j = 5$) for $N = 1000$ time steps by applying one-fourth of a Gaussian pulse as the excitation, with the amplitude $A = 100$ and the decay factor $\tau = 0.0018$.

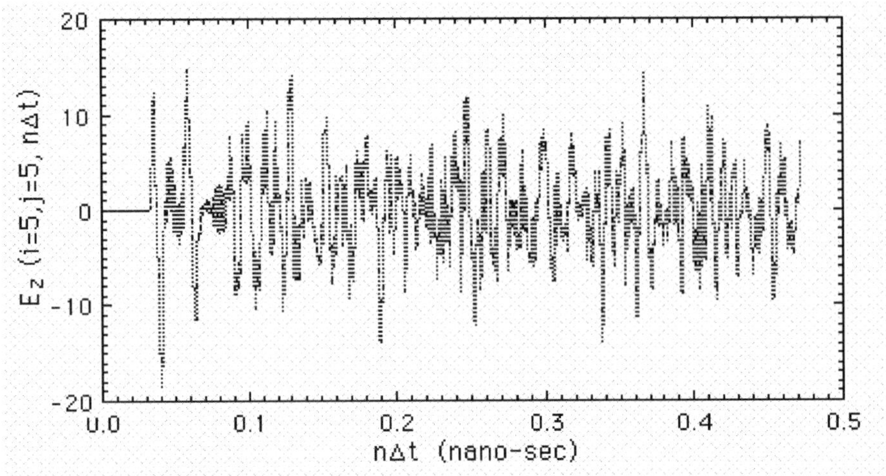

FIGURE 6.19
Axial z component of electric field as a function of time step.

In order to obtain the TM mode cutoff frequencies, it is necessary to obtain the discrete Fourier transform (DFT) of the recorded time domain data given by

$$F(k\Delta f) = \sum_{n=0}^{N-1} F(n\Delta t)e^{-j2\pi nk/N} \tag{6.5.32}$$

for $k = 0, 1, 2, \cdots, K - 1$. In Eq. (6.5.32), K and N represent the total number of sampling points in frequency and time, respectively, and

$$\Delta f = \frac{1}{N\Delta t}. \tag{6.5.33}$$

In order to speed up the numerical process of the DFT calculation, the fast-Fourier transform (FFT) algorithm is utilized, which always requires the total time domain sampling points, $N = 2^p$, where p is an integer. Now, the time domain data shown in Fig. 6.19 with an extended total number of time steps, $N = 16,384$, are Fourier transformed. A spatial resolution of $\Delta = 0.0002$ m gives a frequency resolution of $\Delta f = 0.129475$ GHz. Figure 6.20 shows a plot of the distribution of the cutoff frequency spectrum for the hollow rectangular waveguide. It is interesting to observe that the cutoff frequency spectrum of the axial component of the electric field shows isolated spike-like peaks which are in fact the mode cutoff frequencies of the hollow rectangular waveguide. These mode cutoff frequencies for TM polarization are tabulated in Table 6.1.

In order to validate the results of TM mode cutoff frequencies, an eigenvalue equation of the Helmholtz-type partial differential equation can be solved directly in the frequency domain. By expanding the axial electric field distribution in terms of a set of rectangular orthogonal functions and enforcing the boundary condition that the total axial electric field is equal to zero on the perfectly conducting boundary, an analytical expression for the distribution of TM_{mn} modes can be written as

$$E_z(x, y) = A_{mn} \sin\left(\frac{m\pi x}{a}\right) \sin\left(\frac{n\pi y}{b}\right) \tag{6.5.34}$$

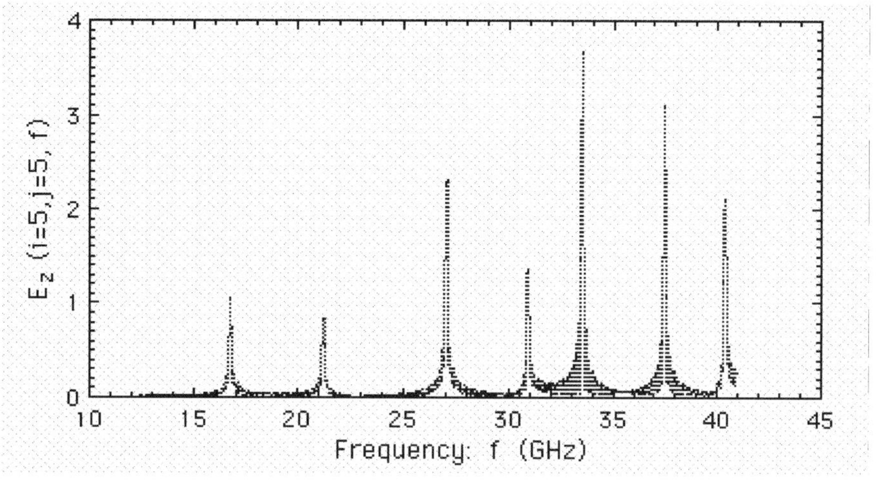

FIGURE 6.20
Frequency spectrum of the axial z component of the electric field.

for $m = 1, 2, \cdots, \infty$ and $n = 1, 2, \cdots, \infty$. In Eq. (6.5.34), a and b are the dimensions of the hollow rectangular waveguide, and A_{mn} is the TM modal amplitude. The TM mode cutoff frequencies are given by

$$f_{c,mn} = \frac{c_0}{2\pi}\left[\left(\frac{m\pi}{a}\right)^2 + \left(\frac{n\pi}{b}\right)^2\right]^{1/2}, \tag{6.5.35}$$

where c_0 is the wave speed in free space. The various cutoff frequencies for TM to z polarization based on Eq. (6.5.35) are also tabulated in Table 6.1. There is excellent agreement between the data obtained by analytical expression and the FDTD numerical results.

Similarly, the previous case is repeated for TE to z polarization. As shown in Fig. 6.17, the z component of the axial magnetic field is defined at the midpoint of each square cell. The boundary condition at the rectangular perfectly conducting walls is simulated by appropriately truncating cells at the transverse components of the electric field, E_x and E_y. The transverse electric field components are located parallel to the rectangular boundary and a very large conductivity, for example, the conductivity of copper material $\sigma^{(e)} = 10^7$ S/m, is assigned to the corresponding boundary

Table 6.1 Mode Cutoff Frequencies of Hollow Rectangular Waveguide, TM Polarization

m	n	Analytical (GHz)	Numerical (GHz)	Mode
1	1	16.75891	16.78012	TM11
2	1	21.19853	21.21923	TM21
3	1	27.02993	27.04158	TM31
1	2	30.90190	30.92315	TM12
2	2	33.51782	33.51086	TM22
4	1	33.51782	33.51086	TM41
3	2	37.47406	37.52181	TM32

cells. This is equivalent to forcing the transverse components of the electric field parallel to the conducting boundary equal to zero.

Initially, there is no field excitation inside the waveguide and all the spatial magnetic and electric field components are assumed to be zero. Now, the rectangular waveguide is excited by simulating one-fourth of a Gaussian pulse (discussed earlier), given by Eq. (6.5.31). The same initial excitation field is assigned to the axial component of the magnetic field at the time instant, $t = 0$, with its peak at the center cell of the rectangular waveguide. Using this initial axial magnetic field excitation, the time-stepping algorithm for the TE mode cutoff conditions, given by Eqs. (6.5.27)–(6.5.29), are implemented. As a function of time, the components of transverse electric and axial magnetic fields inside the rectangular waveguide are recorded. For determining the mode cutoff frequencies, only the temporal variation of the axial component of the magnetic field at the cell ($i = 5$, $j = 5$) needs to be recorded. To obtain the TE mode cutoff frequencies, the FFT algorithm is utilized to convert the time domain data, with the total number of time steps $N = 16,384$. The mode cutoff frequencies for TE to z polarization are tabulated in Table 6.2.

To validate the results of TE mode cutoff frequencies, an eigenvalue equation of the Helmholtz-type partial differential equation can be solved directly in the frequency domain. By expanding the axial magnetic field distribution in terms of a set of rectangular orthogonal functions and enforcing the boundary condition that the normal derivative of the axial component of magnetic field is equal to zero on the perfectly conducting boundary, an analytical expression for the distribution of TE_{mn} modes can be written as

$$H_z(x, y) = B_{mn} \cos\left(\frac{m\pi x}{a}\right) \cos\left(\frac{n\pi y}{b}\right) \tag{6.5.36}$$

for $m = 0, 2, \cdots, \infty$ and $n = 0, 1, 2, \cdots, \infty$. Note that the mode numbers m and n can be zero, but both cannot be zero at the same time. In Eq. (6.5.36), a and b are the dimensions of the hollow rectangular waveguide, and B_{mn} is the TE modal amplitude. The TE mode cutoff frequencies are

Table 6.2 Mode Cutoff Frequencies of Hollow Rectangular Waveguide, TE Polarization

m	n	Analytical (GHz)	Numerical (GHz)	Mode
1	0	7.49481	7.50436	TE10
0	1	14.98963	15.00873	TE01
2	0	14.98963	15.00873	TE20
1	1	16.75891	16.73074	TE11
2	1	21.19853	21.21923	TE21
3	0	22.48444	22.51309	TE30
3	1	27.02993	27.04158	TE31
0	2	29.97925	30.01745	TE02
4	0	29.97925	30.01745	TE40
1	2	30.90190	30.89727	TE12
2	2	33.51782	33.51086	TE22
4	1	33.51782	33.51086	TE41
3	2	37.47406	37.52181	TE32
5	0	37.47406	37.52181	TE50

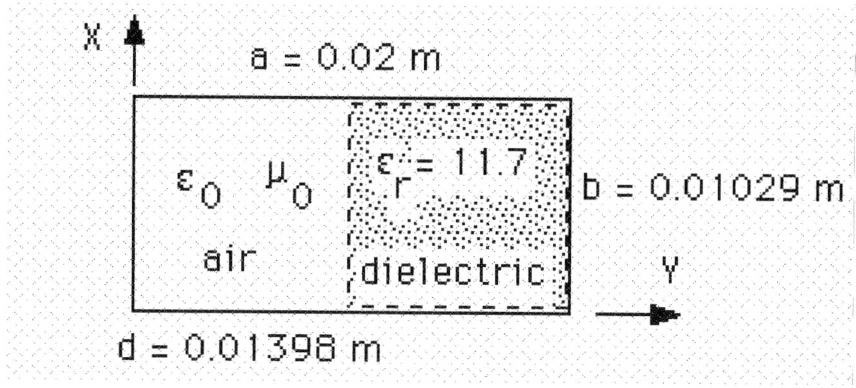

FIGURE 6.21
Partially filled rectangular waveguide.

given by Eq. (6.5.35). The mode cutoff frequencies for TE to z polarization using the analytical expression are also tabulated in Table 6.2. Again, there is excellent agreement between the data obtained by analytical expression and the FDTD results.

6.5.3 Dielectric Slab-Loaded Rectangular Waveguide

A rectangular waveguide which is partially loaded with a slab of dielectric material is discussed in this section. The dielectric slab is located vertically on the right side of rectangular waveguide, as shown in Fig. 6.21. The relative permittivity and permeability of the dielectric slab are $\epsilon_r = 11.7$ and $\mu_r = 1$, respectively. The side lengths for the rectangular waveguide along the x and y coordinates are $b = 0.01029$ m and $a = 0.02$ m, and the unfilled empty rectangular region is of width $d = 0.01398$ m.

For TE to z polarization, the interior region of the rectangular cross section is modeled by discretizing the region into structured square cells, $\Delta x = \Delta y = \Delta$, with $i = 54$ mesh points along the x-axis and $j = 104$ mesh points along the y-axis. The spatial increment is $\Delta = 0.0001942$ m and the corresponding time-stepping increment is $\Delta t = \Delta/\sqrt{2}c_{\text{max}}$, where c_{max} is the maximum wave velocity in the waveguide region given by 3×10^8 m/s. Following the case study of hollow rectangular waveguide, the boundary condition at the rectangular perfectly conducting walls is simulated by appropriately truncating cells at the transverse components of electric field, E_x and E_y. No explicit boundary conditions are required at the vertical interface separating the air medium and the dielectric medium. In the arrangement of lattice square cells, the parallel x components of the electric field are placed at the interface.

Similarly, for TM to z polarization, the rectangular cross section is modeled by $i = 54$ mesh points along the x-axis and $j = 104$ mesh points along the y-axis. The boundary condition at the rectangular perfectly conducting walls is simulated by appropriately truncating cells at the axial components of the electric field, E_z. Again, no explicit boundary conditions are enforced at the vertical interface separating the air medium and the dielectric medium, and the axial z components of the electric field are placed at the interface.

Initially, there is no excitation inside the waveguide region, and all the spatial magnetic and electric field components are assumed to be zero. The waveguide is excited by injecting one-fourth of a Gaussian pulse, given by Eq. (6.5.31), with its peak placed at the center cell of waveguide. For

Table 6.3 Mode Cutoff Frequencies of a Partially Dielectric Slab-Loaded Rectangular Waveguide

m	h	p	Analytical (GHz)	Numerical (GHz)	Mode	
0	82.935	283.681	3.95716	3.99801	TE	LSE
0	198.245	678.104	9.45909	9.46196	TE	LSE
0	256.629	877.806	12.24479	12.26058	TE	LSE
1	35.065	1014.991	14.79641	14.79265	TE	LSM
1	185.147	1190.332	17.15156	17.05819	TE	LSM
1	204.023	1225.903	17.63235	17.68442	TM	LSE
0	375.291	1283.693	17.90663	17.85780	TE	LSE
1	226.928	1359.946	19.45110	19.72354	TE	LSE
1	277.672	1384.886	19.79055	19.72354	TE	LSM
0	451.953	1545.918	21.56448	21.58928	TE	LSE
1	413.883	1737.820	24.61946	24.81098	TM	LSM
1	419.390	1753.201	24.83075	24.81098	TM	LSE
0	541.377	1851.795	25.83125	25.58729	TE	LSE
1	478.590	1922.419	27.15866	27.05452	TM	LSE
1	482.834	1934.793	27.32911	27.18649	TE	LSM
2	64.367	2027.747	29.56299	29.31877	TE	LSM
2	209.753	2139.636	31.05965	31.05124	TE	LSM

TE, axial z components of the magnetic field are initially excited, and for TM, axial z components of the electric field are excited.

Table 6.3 lists the numerically calculated mode cutoff frequencies for the partially loaded dielectric slab rectangular waveguide for the TE and TM polarizations. These results are obtained by first recording the temporal variation of fields at the cell ($i = 5,\ j = 5$) for $N = 16,384$ and then applying the FFT algorithm to convert the time domain data to the frequency domain.

To validate the results of cutoff frequencies, analytical expressions are utilized. The partially loaded dielectric slab rectangular waveguide can be analyzed in terms of normal modes of propagation based on the longitudinal-section electric (LSE) modes and the longitudinal-section magnetic (LSM) modes. Instead of using the eigenvalue equations for the field components directly, the axial and transverse components of the electric and magnetic field distribution are expressed in terms of either a magnetic-type Hertzian vector potential or an electric-type Hertzian vector potential. The vector potentials are selected in the transverse plane normal to the air and dielectric interface and are expanded in terms of a set of rectangular orthogonal functions. Enforcing the field boundary conditions on the perfectly conducting boundary and on the air/dielectric media interface, the following analytical expressions for the cutoff frequencies for the two types of normal modes can be obtained. For LSE modes, the magnetic-type Hertzian vector potential, $\mathbf{\Pi}_h$, is given by

$$\mathbf{\Pi}_h = \boldsymbol{a}_y \Psi_h(x, y) e^{-j\beta z} \tag{6.5.37}$$

$$\Psi_h(x, y) = A \sin(hy) \cos\left(\frac{m\pi x}{b}\right) \qquad 0 < y < d \tag{6.5.38}$$

$$\Psi_h(x, y) = B \sin[p(a - y)] \cos\left(\frac{m\pi x}{b}\right) \qquad d < y < a, \tag{6.5.39}$$

where m is an integer, and the parameters p and h are obtained from the following of transcendental equations:

$$h\tan[p(a-d)] = -p\tan(hd) \tag{6.5.40}$$

$$p^2 = h^2 + (\epsilon_r - 1)k^2. \tag{6.5.41}$$

Similarly for LSM modes, the electric-type Hertzian vector potential, $\mathbf{\Pi}_e$, is given by

$$\mathbf{\Pi}_e = \boldsymbol{a}_y \Psi_e(x, y) e^{-j\beta z} \tag{6.5.42}$$

$$\Psi_e(x, y) = A\cos(hy)\sin\left(\frac{m\pi x}{b}\right) \qquad 0 < y < d \tag{6.5.43}$$

$$\Psi_e(x, y) = B\cos[p(a-y)]\sin\left(\frac{m\pi x}{b}\right) \qquad d < y < a, \tag{6.5.44}$$

where m is an integer, and the parameters p and h are obtained from the following of transcendental equations:

$$-p\tan[p(a-d)] = h\epsilon_r\tan(hd) \tag{6.5.45}$$

$$p^2 = h^2 + (\epsilon_r - 1)k^2. \tag{6.5.46}$$

The LSE mode and LSM mode cutoff frequencies are given by

$$f_c = \frac{c_0}{2\pi}\left[h^2 + \left(\frac{m\pi}{b}\right)^2\right]^{1/2}. \tag{6.5.47}$$

The cutoff frequencies for the TM and TE polarizations are compared with the results obtained from the previous equations for LSE and LSM modes and are also shown in Table 6.3. Again, there is excellent agreement between the data obtained by analytical expression and the FDTD numerical results.

6.5.4 Shielded Microstrip Lines

The determination of mode cutoff frequencies based on the FDTD technique is extended to shielded microstrip lines. Figure 6.22 shows the geometry of a single microstrip line within a

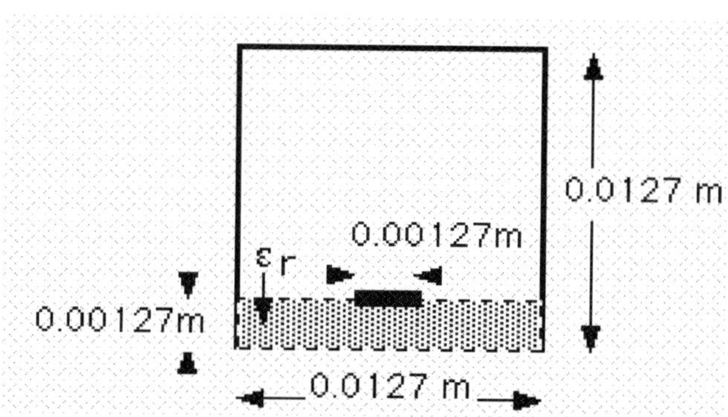

FIGURE 6.22
Shielded microstrip line.

Table 6.4 Mode Cutoff Frequencies of a Shielded Microstrip Line, TM and TE Polarizations

Mode	Numerical (GHz)	Polarization	Mode	Numerical (GHz)	Polarization
1	11.24738	TE	6	21.76123	TE
2	16.13754	TE	7	21.76123	TM
3	16.87106	TM	8	22.49475	TE
4	19.31614	TE	9	22.49475	TM
5	20.53868	TE	10	24.20631	TE

square cross-sectional waveguide. The thickness of the microstrip line is specified in terms of a single cell of the finite-difference grids. The dielectric slab is located horizontally at the bottom portion of the waveguide. The relative permittivity and permeability of the dielectric slab are $\epsilon_r = 8.875$ and $\mu_r = 1$, respectively. The side lengths for the close square waveguide along the x and y coordinates are $a = b = 0.0127$ m and the dielectric-filled region is of height $d = 0.00127$ m.

For TM polarization, the interior region of the square cross section is modeled by discretizing the region into many structured square cells, $\Delta x = \Delta y = \Delta$, with $i = 61$ mesh points along the x-axis and $j = 61$ mesh points along the y-axis. The spatial increment is $\Delta = 0.000212$ m and the corresponding time-stepping increment (in seconds), $\Delta t = \Delta/\sqrt{2}c_{\max}$, where $c_{\max} = 3 \times 10^8$ m/s is the maximum wave velocity in the unfilled waveguide region. For determining the mode cutoff frequencies, only the temporal variation of the axial component of the field at the cell $(i = 3,\ j = 33)$ needs to be recorded. To obtain the mode cutoff frequencies, the FFT algorithm is utilized to convert the time domain data, with the total number of time steps, $N = 16,384$. The mode cutoff frequencies for TM and TE to z polarizations are tabulated in Table 6.4.

In the numerical case studies presented so far, proper spatial resolution is required to represent the distribution of the various modes at the cutoff condition. In fact, the choice of the spatial discretization is completely dictated by the truncation of the frequency spectrum and the highest modal distribution to be recovered. For the two-dimensional TM or TE case studies, a minimum of at least 10 structured square cells per wavelength in a free-space medium yields accuracy acceptable for engineering applications.

6.6 FDTD FORMULATION FOR TWO-DIMENSIONAL OPEN-REGION PROBLEMS

An efficient time domain solution based on the FDTD technique for determining the mode cutoff frequencies of the closed-region structure was presented in the previous section. In fact, for modeling the electromagnetic scattering, coupling, and propagation characteristics of the close-type or the open-type geometry, a complete three-dimensional, full-wave analysis is required, and this is addressed later. Here, a detailed solution procedure based on the FDTD technique is presented for the analysis of arbitrarily-shaped and composition material structure with an external plane wave excitation. In general, all six scalar full-wave equations (Eqs. 6.5.1–6.5.6) must be considered in the rectangular coordinate system with fields given by electric field components $E_x(x, y, z, t)$, $E_y(x, y, z, t)$, and $E_z(x, y, z, t)$, and magnetic field components $H_x(x, y, z, t)$, $H_y(x, y, z, t)$, and $H_z(x, y, z, t)$. Equations (6.5.1)–(6.5.6) simplify dramatically only for the cutoff propagation condition or in the special case of normal plane wave external excitation. This introduces decoupling of the fields based on TM and TE polarizations of the open- or close-type structure.

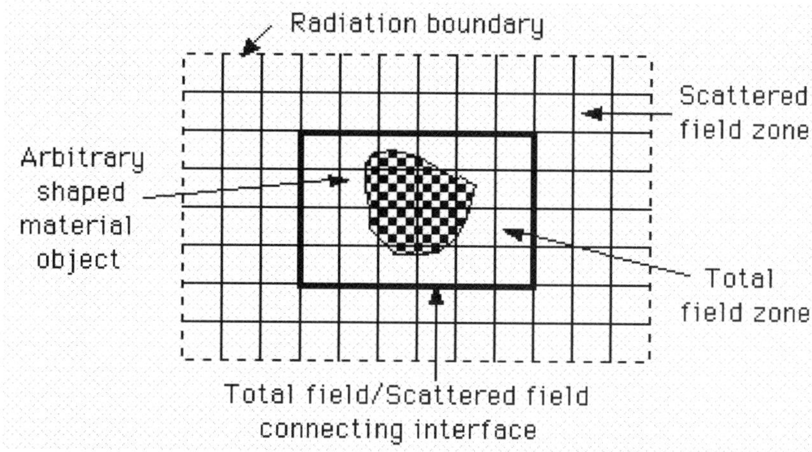

FIGURE 6.23
Two-dimensional FDTD modeling for scattering and coupling interaction.

Figure 6.23 shows that the FDTD field computation zone is divided into an interior total-field region and an exterior scattered-field region that are separated by a connecting interface. The arbitrarily shaped complex material structure is embedded in the total-field zone. The interior total-field region and the exterior free-space region are directly coupled, resulting in electromagnetic wave scattering and interaction in all directions. Here, we consider only a two-dimensional structure that is infinite in extent and has a uniform cross-sectional geometry along its longitudinal z-axis. For the special case of normal excitation with no propagation of the external incident field along the axis of structure, the three-dimensional electromagnetic full wave reduces to the two-dimensional FDTD with either TM to z polarization or TE to z polarization. Hence, the FDTD computational zone is the finite rectangular surface at $z = 0$ plane. The two-dimensional cross-sectional geometry that is embedded in the total-field zone can be conducting, homogeneous or inhomogeneous, and isotropic or anisotropic with dielectric and magnetic material properties. As discussed in Section 6.5, for the TM case, nonzero components of the field are given by the axial component of the electric field and the transverse components of the magnetic field, as indicated in Fig. 6.16 and Eqs. (6.5.7)–(6.5.9). Similarly, for the TE case, nonzero components of the field are given by the axial component of the magnetic field and the transverse components of the electric field, as indicated in Fig. 6.17 and Eqs. (6.5.24)–(6.5.26).

As shown in Fig. 6.23, when the embedded material structure is excited with an incident field, in addition to the electromagnetic field distribution in the material region, one has to consider the distribution of scattered fields in the complete surrounding medium. The two-dimensional FDTD modeling needs to simulate the complete material cross section at the $z = 0$ plane, including the complete surrounding planar region that is infinite in extent. However, the practical computational memory resource is limited. Thus, the FDTD numerical solution technique is limited to only mapping the infinite domain onto a bounded one by constructing an artificial boundary condition to simulate an infinite domain. This artificial boundary condition imposes the so-called absorbing radiation boundary condition on the various electric and magnetic field distributions at the four numerical truncation planes, which ideally require no reflection of the outgoing waves from the truncated boundary back into the computation domain of interest. Figures 6.16 and 6.17 show the two-dimensional FDTD spatial distribution of the various nonzero field components in an overall rectangular computational domain. The arrangement of the FDTD lattice cells and the location of field components shown in the figures are applicable both in the total-field zone and in the scattered-field zone. The iterative time-stepping numerical algorithms developed in Section 6.5 for the two polarizations are incorporated in Fig. 6.23 for both the total-field zone and the scattered-field zone independently in the finite $z = 0$ domain. At the connecting interface of the total-field and the scattered-field zones, the external plane wave incident field excitation is systematically simulated, similar to that done for the one-dimensional case, by appropriately resolving any inconsistency in the direct application of the time-stepping algorithm.

6.6.1 Absorbing Radiation Boundary Condition

In general, it is difficult to construct a perfect absorbing radiation boundary condition that gives no reflections at the truncation planes. Instead, an approximation to an ideal absorbing boundary condition is implemented under a specific numerical scheme to minimize the computational error due to spatial truncation. Therefore, the radiation boundary condition is enforced as an additional time-stepping algorithm to truncate the computational region, which is located few cells away from the connecting interface, and yet simulates a virtual extension of the computational domain to infinity.

Currently, there are two basic types of operators for simulating radiation boundary condition: mode annihilating and one-way wave approximation. The mode-annihilating operator is based on the idea of enforcing the radiation wave equation by killing most of the low-order modal wave terms from a convergent expansion series of the far-field scattered wave. The far-field scattered wave should satisfy the Sommerfeld radiation condition. Convergent series expansions of the far-field-type scattered wave are available for both two-dimensional and three-dimensional cases. In the two-dimensional case, a cylindrical coordinate system is generally used for the representation of far-field scattered wave, and the mode-annihilating operator normally eliminates the cylindrical outgoing waves. The mode-annihilating absorbing boundary operator can be constructed to any desired order by using boundary operators and their algorithms.

Depending on the accuracy requirement and the computational resource available, it is also possible to simulate an ideal perfectly matched layer at the numerical truncating planes. This approach is based on the one-dimensional lossy layered model discussed earlier. For each layer near the truncation plane, the four medium parameters are chosen appropriately by equating the ratio of electric conductivity and its permittivity to the ratio of magnetic conductivity and its permeability. This results in an ideal gradually attenuating outgoing plane wave with minimum or no reflection into the computational zone. In fact, the perfectly matched layers can also be simulated based on the one-dimensional inverse scattering model discussed in Section 6.3.

Another alternative, yet practically ideal method to enforce radiation boundary condition is based on the one-way propagation wave equation. For a homogeneous, isotropic, and time-independent two-dimensional domain, the well-known ideal one-way wave dispersion relation is given by

$$k_x = \frac{\omega}{c}\left[1 - \left(k_y \frac{c}{\omega}\right)^2\right]^{1/2}, \tag{6.6.1}$$

where c is the wave velocity, k_x and k_y are the wave numbers in x- and y-coordinate directions, respectively, and ω is the time-harmonic angular frequency. Generally, a partial differential equation can be constructed from the previous dispersion relation. However, Eq. (6.6.1) is not a rational function, and a numerically useful partial differential equation cannot be constructed. The radical in Eq. (6.6.1), however, can be replaced by a series approximation. The most popular absorbing boundary condition was proposed by using Pade approximation and provides a very good absorption for the incident angle normal to the boundary, but it deteriorates for angles away from the normal incidence. Many alternative approximations to the radical, such as Pade, interpolation using Chebyshev points, least squares, Chebyshev polynomials, Chebyshev on a subinterval interpolation using Newman points, and Chebyshev–Pade, have been proposed and discussed in the literature. These approximations, in fact, have better wide-angle behavior.

In the following, the partial differential equation for the absorbing boundary condition is utilized to develop a time-stepping algorithm at the truncation boundary planes. Either central difference or forward difference may be used for discretizing. The most popular second-order boundary operator is based on the central-difference technique, and the absorbing boundary condition is well-posed to guarantee the existence of the numerical solution in the domain of interest. Well-posedness refers to the existence of a unique solution, along the boundary of interest whose norm at $t = t_0$ can be estimated in terms of the initial data. In other words, a well-posed partial differential equation will render a numerical solution with bounded field magnitude. In fact, many seemingly reasonable approximations for one-way wave equation are ill-posed and hence useless in practice. It is highly desirable to be able to identify these problems in advance.

Many other basic types of absorbing boundary conditions have been proposed depending on specific applications and the numerical accuracy requirement. An error-correcting superabsorption algorithm can be incorporated to improve the accuracy of the current second-order absorbing boundary condition. In the following, a second-order boundary operator is developed systematically. The algorithm can be extended using higher order operators which are suitable for the time domain analysis of the open-type complex geometries.

6.6.2 Second-Order Radiation Boundary Condition

Among many absorbing boundary conditions available in the literature, the second-order Mur's radiation boundary condition is the most popular and is discussed here with modification to study electromagnetic scattering and coupling. The operator can be derived from the dispersion relation for the free-space wave equation. The two-dimensional wave equation in the rectangular system is

$$\frac{\partial^2 U}{\partial x^2} + \frac{\partial^2 U}{\partial y^2} - \frac{1}{c^2}\frac{\partial^2 U}{\partial t^2} = 0, \qquad (6.6.2)$$

where U is a scalar field component and c is the wave phase velocity. In rectangular coordinates, the plane wave time-harmonic solution to the previous wave equation is

$$U(x, y, t) = e^{j(\omega t - k_x x - k_y y)}. \qquad (6.6.3)$$

On substituting into the wave equation (Eq. 6.6.2), the following dispersion relation is obtained:

$$\frac{\omega^2}{c^2} = k_x^2 + k_y^2 = k^2. \qquad (6.6.4)$$

For a specific time-harmonic frequency ω, the solution of k_x and k_y constitutes a system of concentric circles in the $(k_x,\ k_y)$ plane, which suggests that the solution of Eq. (6.6.2) admits plane waves traveling in all directions. Solving for k_x in Eq. (6.6.4),

$$k_x = \pm k(1 - s^2)^{1/2} \quad \text{for } -1 \leq s \leq 1 \qquad (6.6.5)$$

$$s = \frac{k_y}{k}. \qquad (6.6.6)$$

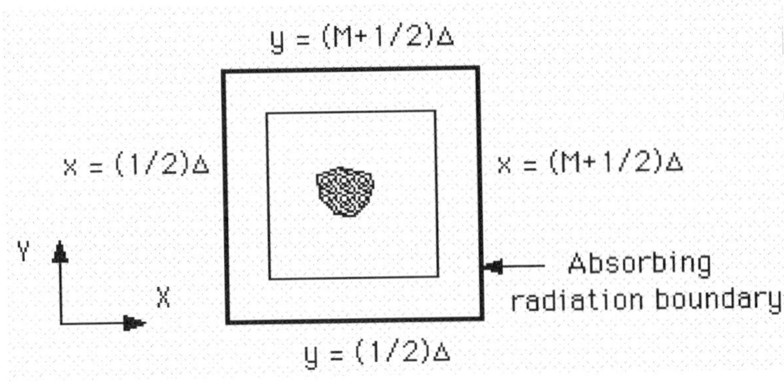

FIGURE 6.24
Two-dimensional region with truncation planes modeled with an artificial absorbing radiation boundary.

In Eq. (6.6.5), the $\pm$ sign corresponds to the dispersion relation of x-directed and $-x$-directed traveling waves, respectively. The dispersion relation cannot be used directly to construct a practical one-way wave equation because of the radical term. An approximation to the radical term in Eq. (6.6.5) is made to obtain a one-way wave equation by utilizing the first two terms in a Taylor series approximation:

$$(1 - s^2)^{1/2} \approx 1 - \frac{1}{2}s^2. \tag{6.6.7}$$

Using the previous two-term approximation, a partial differential equation which has the traveling-wave solution can be constructed. Referring to Fig. 6.24, corresponding to the four sides of the truncated absorbing boundary, the following relationships are obtained:

$$\frac{1}{c}\frac{\partial^2 U}{\partial x \partial t} = \frac{1}{c^2}\frac{\partial^2 U}{\partial t^2} - \frac{1}{2}\frac{\partial^2 U}{\partial y^2} \tag{6.6.8}$$

$$\frac{1}{c}\frac{\partial^2 U}{\partial x \partial t} = -\frac{1}{c^2}\frac{\partial^2 U}{\partial t^2} + \frac{1}{2}\frac{\partial^2 U}{\partial y^2} \tag{6.6.9}$$

$$\frac{1}{c}\frac{\partial^2 U}{\partial y \partial t} = \frac{1}{c^2}\frac{\partial^2 U}{\partial t^2} - \frac{1}{2}\frac{\partial^2 U}{\partial x^2} \tag{6.6.10}$$

$$\frac{1}{c}\frac{\partial^2 U}{\partial y \partial t} = -\frac{1}{c^2}\frac{\partial^2 U}{\partial t^2} + \frac{1}{2}\frac{\partial^2 U}{\partial x^2}, \tag{6.6.11}$$

at $x = 0$, $x = h$, $y = 0$, and $y = h$ boundaries, respectively.

Equation (6.6.8), which is valid at the $x = 0$ boundary, will allow a traveling wave solution only in the $-x$ direction. To apply Eqs. (6.6.8)–(6.6.11) numerically, a simple finite-difference scheme is introduced to derive a radiation boundary time-stepping algorithm (Fig. 6.25). This scheme involves implementing the partial derivatives in Eq. (6.6.8) as numerical central differences of the auxiliary U component, $U^n(1/2, j)$, located one-half space cell from the grid boundary at $(0, j)$. The mixed partial x and t derivatives on the left-hand side of Eq. (6.6.8) are

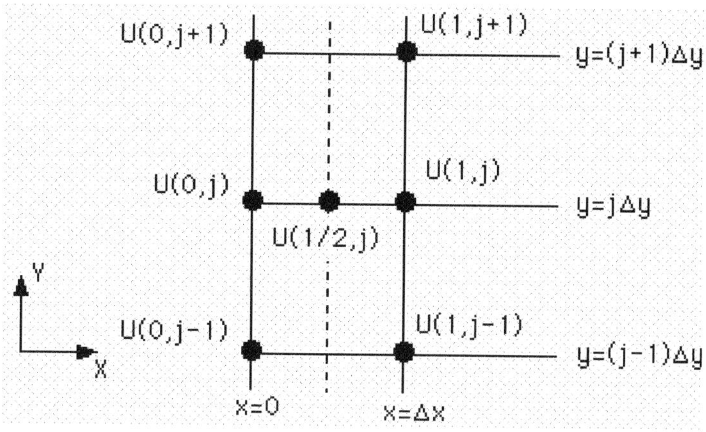

FIGURE 6.25
Position of field component near $x = 0$ truncation boundary.

implemented as

$$\frac{\partial^2 U^n(1/2, j)}{\partial x \partial t} = \frac{\left[\frac{\partial U^{n+1}(1/2,j)}{\partial x} - \frac{\partial U^{n-1}(1/2,j)}{\partial x}\right]}{2\Delta t} \tag{6.6.12}$$

$$= \frac{[U^{n+1}(1, j) - U^{n+1}(0, j)]}{2\Delta t \Delta x} - \frac{[U^{n-1}(1, j) - U^{n-1}(0, j)]}{2\Delta t \Delta x}. \tag{6.6.13}$$

The second-order partial derivative with respect to t on the right-hand side of Eq. (6.6.8) is implemented as an average of time derivatives at the adjacent points $(0, j)$ and $(1, j)$:

$$\frac{\partial^2 U^n(1/2, j)}{\partial t^2} = \frac{\left[\frac{\partial^2 U^n(0,j)}{\partial t^2} + \frac{\partial^2 U^n(1,j)}{\partial t^2}\right]}{2} \tag{6.6.14}$$

$$= \frac{[U^{n+1}(0, j) - 2U^n(0, j) + U^{n-1}(0, j)]}{2\Delta t^2} + \frac{[U^{n+1}(1, j) - 2U^n(1, j) + U^{n-1}(1, j)]}{2\Delta t^2}, \tag{6.6.15}$$

The second-order partial derivative with respect to y on the right-hand side of Eq. (6.6.8) is implemented as an average of y derivatives at the adjacent points $(0, j)$ and $(1, j)$:

$$\frac{\partial^2 U^n(1/2, j)}{\partial y^2} = \frac{\left[\frac{\partial^2 U^n(0,j)}{\partial y^2} + \frac{\partial^2 U^n(1,j)}{\partial y^2}\right]}{2} \tag{6.6.16}$$

$$= \frac{[U^n(0, j+1) - 2U^n(0, j) + U^n(0, j-1)]}{2\Delta y^2} + \frac{[U^n(1, j+1) - 2U^n(1, j) + U^n(1, j-1)]}{2\Delta y^2}. \tag{6.6.17}$$

On substituting Eqs. (6.6.13), (6.6.15), and (6.6.17) into Eq. (6.6.8) and solving for $U^{n+1}(0, j)$, the following time-stepping algorithm for the component U along the $x = 0$ grid boundary is obtained which implements the desired second-order radiation boundary condition:

$$\begin{aligned}U^{n+1}(0, j) = &-U^{n-1}(1, j) + \frac{c\Delta t - \Delta x}{c\Delta t + \Delta x}[U^{n+1}(1, j) + U^{n-1}(0, j)] \\ &+ \frac{2\Delta x}{c\Delta t + \Delta x}[U^{n}(0, j) + U^{n}(1, j)] + \frac{(c\Delta t)^2 \Delta x}{2\Delta y^2 (c\Delta t + \Delta x)} \\ &\times [U^{n}(0, j+1) - 2U^{n}(0, j) + U^{n}(0, j-1) \\ &+ U^{n}(1, j+1) - 2U^{n}(1, j) + U^{n}(1, j-1)].\end{aligned} \tag{6.6.18}$$

Using the same procedure, the radiation boundary conditions at the remaining grid truncation boundaries can be obtained by utilizing Eqs. (6.6.9)–(6.6.11). For the case of square grid boundaries $\Delta x = \Delta y = \Delta$, and the radiation boundary conditions can be summarized as

$$\begin{aligned}U^{n+1}(0, j) = &-U^{n-1}(1, j) + \frac{c\Delta t - \Delta}{c\Delta t + \Delta}[U^{n+1}(1, j) + U^{n-1}(0, j)] \\ &+ \frac{2\Delta}{c\Delta t + \Delta}[U^{n}(0, j) + U^{n}(1, j)] + \frac{(c\Delta t)^2}{2\Delta (c\Delta t + \Delta)} \\ &\times [U^{n}(0, j+1) - 2U^{n}(0, j) + U^{n}(0, j-1) \\ &+ U^{n}(1, j+1) - 2U^{n}(1, j) + U^{n}(1, j-1)]\end{aligned} \tag{6.6.19}$$

$$\begin{aligned}U^{n+1}(h, j) = &-U^{n-1}(h-1, j) + \frac{c\Delta t - \Delta}{c\Delta t + \Delta}[U^{n+1}(h-1, j) + U^{n-1}(h, j)] \\ &+ \frac{2\Delta}{c\Delta t + \Delta}[U^{n}(h, j) + U^{n}(h-1, j)] + \frac{(c\Delta t)^2}{2\Delta (c\Delta t + \Delta)} \\ &\times [U^{n}(h, j+1) - 2U^{n}(h, j) + U^{n}(h, j-1) \\ &+ U^{n}(h-1, j+1) - 2U^{n}(h-1, j) + U^{n}(h-1, j-1)]\end{aligned} \tag{6.6.20}$$

$$\begin{aligned}U^{n+1}(i, 0) = &-U^{n-1}(i, 1) + \frac{c\Delta t - \Delta}{c\Delta t + \Delta}[U^{n+1}(i, 1) + U^{n-1}(i, 0)] \\ &+ \frac{2\Delta}{c\Delta t + \Delta}[U^{n}(i, 0) + U^{n}(i, 1)] + \frac{(c\Delta t)^2}{2\Delta (c\Delta t + \Delta)} \\ &\times [U^{n}(i+1, 0) - 2U^{n}(i, 0) + U^{n}(i-1, 0) \\ &+ U^{n}(i+1, 1) - 2U^{n}(i, 1) + U^{n}(i-1, 1)]\end{aligned} \tag{6.6.21}$$

$$\begin{aligned}U^{n+1}(i, h) = &-U^{n-1}(i, h-1) + \frac{c\Delta t - \Delta}{c\Delta t + \Delta}[U^{n+1}(i, h-1) + U^{n-1}(i, h)] \\ &+ \frac{2\Delta}{c\Delta t + \Delta}[U^{n}(i, h) + U^{n}(i, h-1)] + \frac{(c\Delta t)^2}{2\Delta (c\Delta t + \Delta)} \\ &\times [U^{n}(i+1, h) - 2U^{n}(i, h) + U^{n}(i-1, h) + U^{n}(i+1, h-1) \\ &- 2U^{n}(i, h-1) + U^{n}(i-1, h-1)],\end{aligned} \tag{6.6.22}$$

at $x = 0$, $x = h$, $y = 0$, and $y = h$ boundaries, respectively.

From the previous equations, it is clear that the difference scheme cannot be implemented for field components located at the grid corners since some of the necessary data used are just outside of the grid. In fact, it is necessary to implement a special corner radiation boundary condition by utilizing available field data in the grid. The radiation boundary conditions, given by Eqs. (6.6.19)–(6.6.22), are extensively utilized in the study of two-dimensional electromagnetic propagation, scattering, and coupling with the time domain excitation. The radiation boundary conditions based on the second-order operator cannot absorb all outgoing waves completely when the number of time steps becomes very large. The current FDTD codes suppress spurious reflections of outward-propagating wave analogs by 95–99%. In order to improve numerical accuracy, a slight modification using the error-correcting superabsorption algorithm can be considered. This error-controlling algorithm is not restricted for any particular boundary absorbing operator. The algorithm based on the second-order boundary operator can be extended using higher order operators which are suitable for the time domain analysis of the open-type complex geometry.

6.7 PLANE WAVE SOURCE CONDITION

The two-dimensional numerical algorithms developed earlier have a linear dependence on the components of electric and magnetic fields. In fact, these algorithms can be applied with equal validity to the incident-field vector components, the scattered-field vector components, or the total field consisting of the sum of incident- and scattered-field vector components. The current FDTD codes utilize this property to zone the numerical space lattice into two distinct regions separated by a rectangular virtual planar interface to connect the fields in each region (Fig. 6.23).

This type of lattice zoning provides many algorithm features to enhance the computational flexibility and dynamic range. The connecting condition provided at the interface of the inner and outer regions simultaneously generates an arbitrary plane wave in the total-field region, having specified time-dependent waveform, angle of incidence, and angle of polarization. The connecting interface is transparent to the outgoing scattered-field waves. As discussed earlier, the arbitrarily shaped material structure is embedded in the total-field zone, in which the total fields are marched in time. The required continuity of the total tangential electric and magnetic fields across the interface of staggered unit cells separating dissimilar media parameters is satisfied automatically. This basically avoids the numerical simulation problem inherent in the pure scattered-field FDTD code, in which the enforcement of the continuity of total tangential fields is a separate simulation process requiring the incident field to be computed at all interfaces of the dissimilar media and then added to the values of the time-marched scattered fields at the interfaces.

Furthermore, the low levels of the total field in deep shadow regions or cavities of the embedded interacting material structure are computed directly in the total-field zone. In a pure scattered-field code, the low levels of total field are obtained by computing the incident field and these are added to the values of the time-marched scattered fields. Thus, the scattered-field code relies on near cancellation of the incident- and scattered-field components of the total field in deep shadow and cavity regions. This introduces undesirable subtraction noise and affects the computational dynamic range, which is the ratio of the maximum to the minimum accurately computed field level in the space lattice. The central-difference, finite-difference total-field and scattered-field algorithm yields a dynamic range more than 30 dB higher than that of the pure scattered-field algorithm.

Connecting Condition at the Interface of Total-Field and Scattered-Field Regions For two-dimensional TM and TE algorithms, the four rectangular faces comprising the interface of region 1 (total field) and region 2 (scattered field) contain the electric and magnetic field components, which require spatial finite difference of the various field components to advance

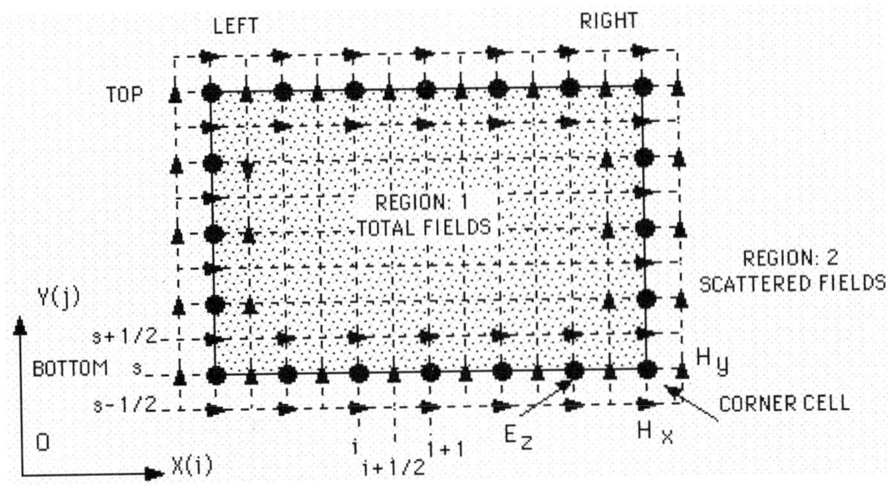

FIGURE 6.26
Calculation of field components at connecting interface (TE case).

by time stepping. As discussed in Section 6.2 and shown in Eqs. (6.2.36)–(6.2.38) for the one-dimensional plane wave simulation, there is a problem of inconsistency.

On the region 1 side of the interface, the fields to be used in the finite-difference expression are the total fields. Similarly, on the region 2 side of the interface, the fields to be used in the finite-difference expression are the scattered fields. The problem of inconsistency near the interface can be resolved by using the values of the incident electric and magnetic field components, $\boldsymbol{E}^{\mathrm{i}}$ and $\boldsymbol{H}^{\mathrm{i}}$, which are assumed to be known or calculable at each lattice point.

To illustrate the simulation of the plane wave source condition at the connecting interface, as shown in Fig. 6.26, consider the two-dimensional TM to z polarization along with Eqs. (6.5.15)–(6.5.17). The total electric and magnetic field components, E_z^{t} and H_y^{t}, lie exactly on the interface of region 1 and region 2 located at $j = s$. The time-stepping, finite-difference expressions (Eqs. 6.5.15–6.5.17) are slightly modified to achieve consistency when applied at $(i, j = s)$. Referring to Eq. (6.5.17), the calculation of the total z component of the electric field at the connecting interface (i, s) is given by

$$
\begin{aligned}
E_z^{\mathrm{s}\,n+1}(i,s) = {} & C_c(i,s)\left[E_z^{\mathrm{t}\,n}(i,s) - E_z^{\mathrm{i}\,n}(i,s)\right] \\
& + \frac{C_d(i,s)}{\Delta x}\left[H_y^{\mathrm{t}\,n+1/2}(i+1/2,s) - H_y^{\mathrm{i}\,n+1/2}(i+1/2,s)\right] \\
& - \frac{C_d(i,s)}{\Delta x}\left[H_y^{\mathrm{t}\,n+1/2}(i-1/2,s) - H_y^{\mathrm{i}\,n+1/2}(i-1/2,s)\right] \\
& + \frac{C_d(i,s)}{\Delta y}\left[H_x^{\mathrm{s}\,n+1/2}(i,s-1/2) - H_x^{\mathrm{s}\,n+1/2}(i,s+1/2)\right] \\
& + \frac{C_d(i,s)}{\Delta y}\left[H_x^{\mathrm{i}\,n+1/2}(i,s+1/2)\right],
\end{aligned}
\qquad (6.7.1)
$$

and the successive time stepping of the incident electric field only at the connecting interface

$(i,\ s)$ yields the total electric field

$$E_z^{\mathrm{t}\,n+1}(i,s) = E_z^{\mathrm{i}\,n+1}(i,s) + E_z^{\mathrm{s}\,n+1}(i,s). \tag{6.7.2}$$

Referring to Eq. (6.5.15), the calculation of the scattered x component of the magnetic field at the layer $(i,\ s-1/2)$ adjacent to the connecting interface in region 2 is accomplished by

$$\begin{aligned} H_x^{\mathrm{s}\,n+1/2}(i,s-1/2) = {} & C_a(i,s-1/2)\left[H_x^{s\,n-1/2}(i,s-1/2)\right] + \frac{C_b(i,s-1/2)}{\Delta y}\left[E_z^{\mathrm{s}\,n}(i,s-1)\right] \\ & - \frac{C_b(i,s-1/2)}{\Delta y}\left[E_z^{t\,n}(i,s)\right] + \frac{C_b(i,s-1/2)}{\Delta y}\left[E_z^{i\,n}(i,s)\right], \end{aligned} \tag{6.7.3}$$

and similarly, referring to Eq. (6.5.16), the calculation of the scattered y component of the magnetic field at the connecting interface $(i+1/2,\ s)$ is accomplished by

$$\begin{aligned} H_y^{\mathrm{t}\,n+1/2}(i+1/2,s) = {} & C_a(i+1/2,s)\left[H_y^{\mathrm{t}\,n-1/2}(i+1/2,s)\right] + \frac{C_b(i+1/2,s)}{\Delta x}\left[E_z^{\mathrm{t}\,n}(i+1,s)\right] \\ & - \frac{C_b(i+1/2,s)}{\Delta x}\left[E_z^{\mathrm{t}\,n}(i,s)\right]. \end{aligned} \tag{6.7.4}$$

The previous equations for the plane wave source simulation are valid at the bottom connecting interface $j = s$. In fact, proper treatment is required to consider the time stepping of the electric and magnetic field components at the left-end and right-end interface corner points to take into account the additional field component located in the scattered-field region 2. The previous equations can be repeated for the remaining three sides, namely, the top, left, and right connecting interfaces.

In summary, Eqs. (6.7.1)–(6.7.4) are the modified versions which are used in tandem for the original time-stepping expressions for the TM to z polarization. Referring to Eq. (6.7.2), it should be noted that plane wave simulation and time stepping for the TM case is applied only to the z component of the electric field that lies exactly on the region 1 and region 2 interface. These electric field components are indicated by dots in Fig. 6.26.

Calculation of Incident Field to Implement the Connecting Condition Referring to Fig. 6.26, the calculation of incident field data for E_z^{i}, H_x^{i}, and H_y^{i} is at or near the interface of regions 1 and 2 is required to implement the connecting condition for the TM case. A simple approach to calculate these data for arbitrary incident angles is based on a table lookup procedure which describes the space–time behavior of the incident wave. In fact, this is similar to the one-dimensional propagation model described in Section 6.2.

Consider the FDTD simulation of an incident plane wave propagating in the xy-coordinate plane. The incident plane wave electric field vector is polarized along the z-axis, and the incident magnetic field vector is polarized in a plane perpendicular to the z-axis. There is no propagation of the incident fields parallel to the z-axis, indicating that the incident excitation fields are uniform and completely independent of the z-coordinate variable. Hence,

$$\frac{\partial}{\partial z} = 0 \tag{6.7.5}$$

$$\boldsymbol{E}_0 = E_0\boldsymbol{a}_z \tag{6.7.6}$$

$$\boldsymbol{H}_0 = H_{0x}\boldsymbol{a}_x + H_{0y}\boldsymbol{a}_y. \tag{6.7.7}$$

The incident electric and magnetic fields based on the forward-traveling plane waves are given by

$$\boldsymbol{E}^{\text{i}}(\rho) = E_z^{\text{i}}(x, y)\boldsymbol{a}_z \tag{6.7.8}$$

$$E_z^{\text{i}}(x, y) = E_0 e^{-jk\, \boldsymbol{a}_k \cdot \boldsymbol{\rho}} \tag{6.7.9}$$

$$\boldsymbol{H}^{\text{i}}(\rho) = \boldsymbol{H}_0 e^{-jk\, \boldsymbol{a}_k \cdot \boldsymbol{\rho}} \tag{6.7.10}$$

$$\boldsymbol{a}_k = \cos(\phi^{\text{i}})\boldsymbol{a}_x + \sin(\phi^{\text{i}})\boldsymbol{a}_y, \tag{6.7.11}$$

where k is the propagation constant of the free-space medium and ϕ^{i} is the arbitrary angle of incidence of the TM plane wave.

The action of the connecting condition is to generate a plane wave originating from the bottom left corner of the region 1 and region 2 interface. This point is designated as the local origin $\bigcirc_1$ with grid coordinates (p, s). The location $\bigcirc_1$, in fact, is the first FDTD electric field grid point that is contacted by the wavefront of the incident field. All other grid points denoting the location of electric and magnetic vector field components in region 1 are contacted by the wavefront after a certain time retardation satisfied by the field causality condition. Referring to Figure 6.27, d is the distance in terms of the number of spatial grid cells along the wave propagation vector from the local origin $\bigcirc_1$ to a perpendicular dropped from the field vector component located on the connecting interface. It is given by

$$d = \boldsymbol{k} \cdot [(x - p)\boldsymbol{a}_x + (y - s)\boldsymbol{a}_y]. \tag{6.7.12}$$

The spatial grid coordinates of the field vector component can be given in terms of either integers or integers plus or minus half a grid cell since the field components are staggering alternatively by half a grid cell. The local origin $\bigcirc_1$ formed by the bottom left interface corner can simulate only arbitrary incident angles for the range $0 < \phi^{\text{i}} < \pi/2$. For other incident angles, the local origin is relocated at different interface corners $\bigcirc_2$, $\bigcirc_3$, and $\bigcirc_4$ as indicated in Fig. 6.27.

The incident fields $\boldsymbol{E}^{\text{i}}$ and $\boldsymbol{H}^{\text{i}}$ can be calculated at any location of a field vector component using a simple analytical expression once the retardation distance d is known. However, even for

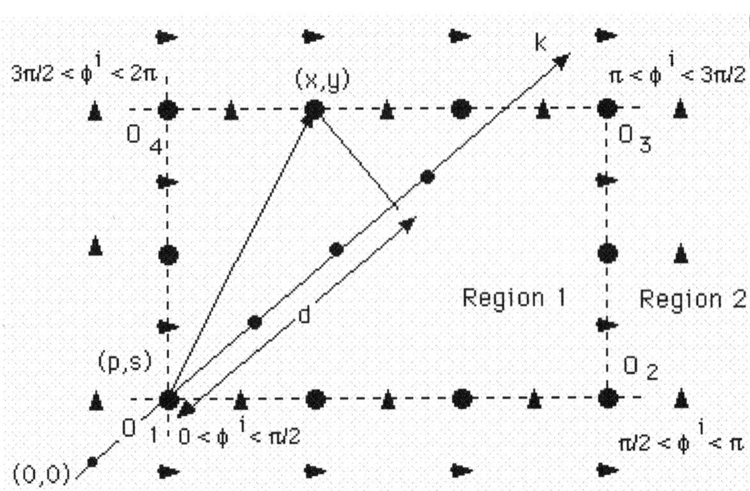

FIGURE 6.27
Coordinates for calculation of incident plane wave field.

the simplest incident field space–time behavior, for example, a single-frequency time-harmonic sinusoidal excitation required for steady-state analysis, the total amount of computational time to implement the connecting condition may be very long, especially if large two-dimensional or three-dimensional region 1 and region 2 interfaces are involved. To reduce the computational burden, an approach based on a lookup table can be employed. Referring to the discussion in Section 6.2, an auxiliary one-dimensional FDTD grid is placed along the wave propagation vector of the incident wave so that the local origin $\bigcirc_1$ of the region 1 and region 2 interface coincides with one of the z components of the electric field of the one-dimensional grid. The goal is to time step the one-dimensional auxiliary grid, using the same spatial discretization Δ, time discretization Δt, and time step number n as in the two-dimensional TM grid, to model the free-space propagation of the desired incident plane wave. Similar expressions can be developed for the TE to z polarization by interchanging the role of the electric field and the magnetic field components.

6.8 NEAR- TO FAR-FIELD TRANSFORMATION

Referring to Fig. 6.23, the FDTD technique yields both the region 1 total fields and the region 2 scattered fields. For many electromagnetic scattering and coupling applications, there is much interest in obtaining equivalent frequency domain data consisting of the magnitude and phase spectrum of the electric and magnetic fields both in the near-field and far-field regions with time-harmonic incident excitation. To obtain frequency domain data, an obvious procedure is to record the time history of electric and magnetic field components as a function of time step at each spatial lattice and apply the discrete Fourier transform (DFT) calculation at the excitation frequency using Eq. (6.5.32). The DFT procedure is computationally expensive, but it is ideal for broad-band, multiple-frequency studies obtained, for example, with an incident excitation consisting of a Gaussian time pulse superimposed with a carrier signal. The DFT calculation converts the time domain data of real electric and magnetic fields to data of the corresponding frequency domain phasor electric and magnetic fields having both the magnitude and phase information.

For the time-harmonic, single-frequency excitation, an alternative method is to avoid DFT calculation completely and study the variation of time domain response as a function of time step for each of the electric and magnetic field components. For example, corresponding to the single-frequency sinusoidal excitation which is applied continuously for a number of cycles, the time domain response of a given field component at a specific lattice cell consists of the transient solution plus the steady-state time-harmonic solution. The transient part of the field solution vanishes after a certain number of excitation cycles, which is approximately given by the quality factor Q of the electromagnetic material structure being analyzed. The steady-state solution follows the basic pattern of the time-harmonic sinusoidal excitation and its time history directly yields the frequency domain magnitude and phase information. In order to extract the magnitude and phase information of a field component at a lattice cell, the convergence of the steady-state solution must be traced by observing the maximum and minimum field response values being repeated cycle after cycle along with any constant shift in the response with respect to the time axis.

Accurate magnitude and phase information for sinusoidal steady state can be obtained by observing the peak positive and negative excursions of the fields over a complete cycle of the incident wave after having time stepped through two to five cycles of the transient period following the beginning of the time stepping. For some two-dimensional and three-dimensional problems, a constant offset of particular computed field components can be possible. Hence, to obtain correct magnitude and phase data, (i) the peak-to-peak value of the sinusoidal response at any point should be observed to eliminate the effects of any DC offset on the computation of the phasor magnitude, and (ii) the zero crossing of the field waveform may not be useful in determining the relative phase, and it is necessary to locate the zero-derivative points of the waveform for this purpose. With discrete time steps, the algorithm will have an uncertainty of one time step or less and operates by noting where the sign of the waveform time derivative changes from positive to negative or negative to positive. Thus, for every time step, an algorithm is written to test the desired field waveform for a change of sign of the time derivative. If a sign change is detected, the latest computed field value is stored in an array. By also testing when the half-cycle and full-cycle intervals have passed, the zero-to-peak field values can be determined. The desired relative phase of the field component can be obtained either as a phase lead or lag with respect to the phase center in the two-dimensional grid or any specific phase reference of a field component. The phase lead or lag is measured in terms of the number of time steps, and the phase is correspondingly converted into an equivalent radians within a cycle.

In principle, the electromagnetic scattering by an arbitrary material structure can be determined if the induced surface electric and magnetic currents are known. The induced currents can be integrated to calculate the near fields and the far fields. However, the body surface may have a complex shape or may be loaded with complex materials in some way to make implementation of the needed surface integrals difficult, and this is a unique problem for each type of scatterer. A useful alternative would be to obtain scattered-field information from off-surface near-field data rather than from the on-surface current data. The near-field information could be computed using the FDTD method, which can easily account for the complexities introduced by the scatterer shape and composition. Furthermore, the near-field data could be integrated along arbitrary, planar, virtual surfaces which completely surround the scatterer of interest.

A provision of a well-defined scattered-field region in the FDTD lattice permits the natural near- to far-field transformation, as illustrated in Fig. 6.28 for the two-dimensional case. The tangential scattered electric and magnetic fields are computed via the FDTD along a virtual close contour C located in the outer (scattered-field) region of the spatially discretized lattice. Since the electric and magnetic fields are staggered, for TM to z polarization the tangential electric field components are located on the virtual close contour. The tangential magnetic field components located half a cell on either side of the virtual close contour are averaged and placed on the contour. By utilizing the basic theory of electromagnetic equivalence, the tangential electric and magnetic field components are converted into the corresponding tangential equivalent magnetic current and

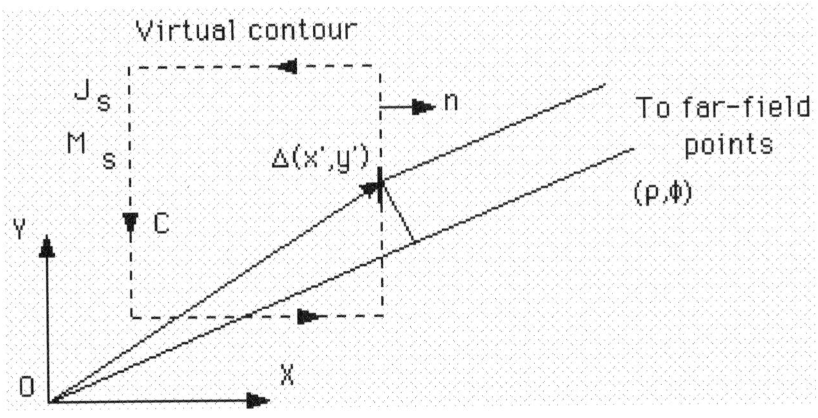

FIGURE 6.28
Near- to far-field integration contour in the scattered-field region.

equivalent electric current. These equivalent currents are weighted by the free-space Green's function and then integrated over the close contour C to obtain the far-field response. The current FDTD codes can provide complete magnitude and phase far-field scattered patterns, the monostatic and bistatic radar cross section for a particular sinusoidal steady state, and plane wave external illumination. The near-field virtual integration contour has a fixed rectangular shape which is independent of the shape of the material composition of the enclosed structure being modeled.

The electric and magnetic field distribution in the far-field region can be obtained by making use of the large argument approximation for the free-space Green's function. For TM to z polarization, in the far-field region there exist only the z component of the scattered electric field and the corresponding ϕ component of scattered magnetic field. In the far-field region, the scattered electric and magnetic fields behave as pure cylindrical waves propagating in the radial direction and are directly related through the medium intrinsic impedance.

The equivalent electric current and the equivalent magnetic current along the virtual close rectangular contour C are given by

$$\boldsymbol{a}_z J_s(\boldsymbol{\rho}, \omega) = \boldsymbol{a}_n \times \boldsymbol{a}_x H_x(\boldsymbol{\rho}) \quad \text{on top and bottom sides} \tag{6.8.1}$$

$$= \boldsymbol{a}_n \times \boldsymbol{a}_y H_y(\boldsymbol{\rho}) \quad \text{on left and right sides} \tag{6.8.2}$$

$$\boldsymbol{a}_s M_s(\boldsymbol{\rho}, \omega) = E_z(\boldsymbol{\rho}) \boldsymbol{a}_z \times \boldsymbol{a}_n \quad \text{on all four sides.} \tag{6.8.3}$$

Due to the equivalent electric and magnetic currents, the z component of scattered electric fields in the far-field region is given by

$$E_z^s(\boldsymbol{\rho}) = -j\omega A_z(\boldsymbol{\rho}) + j\omega\eta\, \boldsymbol{a}_\phi \cdot \boldsymbol{F}(\boldsymbol{\rho}), \tag{6.8.4}$$

where the magnetic vector potential $A_z(\boldsymbol{\rho})$, and the electric potential $\boldsymbol{F}(\boldsymbol{\rho})$ for the far-field region are given by

$$A_z(\boldsymbol{\rho}) = \frac{\mu}{4j} \int_C J_s(\boldsymbol{\rho}') H_0^{(2)}[k|\boldsymbol{\rho} - \boldsymbol{\rho}'|]\, dl' \tag{6.8.5}$$

$$\boldsymbol{F}(\boldsymbol{\rho}) = \frac{\epsilon}{4j} \int_C M_s(\boldsymbol{\rho}') \boldsymbol{a}_s' H_0^{(2)}[k|\boldsymbol{\rho} - \boldsymbol{\rho}'|]\, dl'. \tag{6.8.6}$$

In the large argument approximation, $|k\rho| \to \infty$, the Hankel function can be written as

$$H_0^{(2)}(k\rho) \sim \left(\frac{j2}{\pi k\rho}\right)^{1/2} e^{-jk\rho}. \tag{6.8.7}$$

The scattered electric field distribution in the far-field region reduces to

$$E_z^s(\phi) \sim \kappa \sum_{n=1}^{N} [\eta J_n + M_n \cos(\phi - \Omega_n)] e^{jk\rho_n \cos(\phi - \phi_n)} \Delta_n, \tag{6.8.8}$$

where Ω_n is the angle between the normal on C and the x-axis, and

$$\kappa = \frac{k}{\sqrt{8\pi k\rho}} e^{-jk\rho} e^{-j3\pi/4}. \tag{6.8.9}$$

The bistatic radar cross section (RCS) of the scatterer can be calculated using Eq. (6.8.8) and is

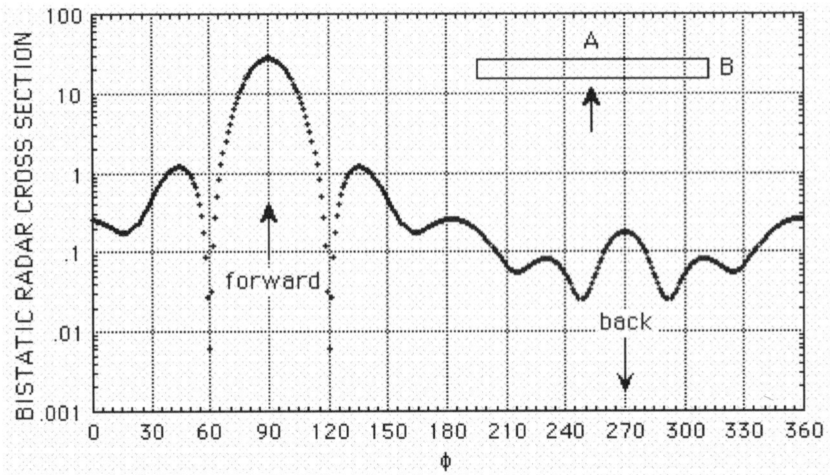

FIGURE 6.29
Bistatic radar cross section of a rectangular permeable dielectric scatterer (TM case).

given by the ratio of scattered-field power to the incident-field power:

$$\mathrm{RCS}(\phi) = \lim_{\rho\to\infty} 2\pi\rho \left| \frac{E_z^{\mathrm{s}}(\phi, \omega)}{E_z^{\mathrm{i}}(\phi, \omega)} \right|^2 . \tag{6.8.10}$$

Figure 6.29 shows a plot of the bistatic radar cross section of the lossless, homogeneous, permeable dielectric scatterer for the normal TM incident plane wave excitation with angle of incidence $\phi^{\mathrm{i}} = 90^\circ$. The geometry of rectangular slab scatterer is embedded in the total-field zone of the FDTD lattice (Fig. 6.23). The length and width of the scatterer along the x and y coordinates are $A = 2$ m and $B = 0.25$ m. The homogeneous material of the scatterer has relative permittivity $\epsilon_{\mathrm{r}} = 2$ and relative permeability $\mu_{\mathrm{r}} = 1.5$. The scatterer is excited by a plane wave with time-harmonic frequency $f = 300$ MHz. This rectangular cross section is modeled by discretizing the region into structured square cells, $\Delta x = \Delta y = \Delta$, with $i = 40$ cells along the x-axis and $j = 5$ cells along the y-axis. For this spatial discretization, the spatial increment is $\Delta = 0.05$ m or 20 cells per wavelength, and the corresponding time-stepping increment is $\Delta t = \Delta/\sqrt{2}c_{\max}$, where $c_{\max} = 3 \times 10^8$ m/s.

To determine the far field and the RCS, the incident field is time stepped for 15 cycles, and the phasor electric and magnetic field components along the close virtual contour C are recorded and applied in the near- to far-field transformation. The bistatic RCS data shown in Fig. 6.29 are also numerically validated using the MoM solution. In the two-dimensional frequency domain MoM technique, the coupled combined field integral equations (CFIEs) are solved to generate the equivalent electric current and magnetic current on the surface of the scatterer, and then the RCS results are obtained using the similar near- to far-field transformation. As can be seen, the bistatic data exhibit a symmetrical distribution with respect to the direction of excitation about which the scatterer also has geometrical symmetry. Since the bistatic RCS distribution is proportional to the scattering power, the far-field scattering pattern is directly obtained by taking the square root of the RCS distribution.

Similarly, Fig. 6.30 shows a plot of the bistatic RCS of the lossless, homogeneous, permeable dielectric scatterer for the normal TE incident plane wave excitation with an angle of incidence

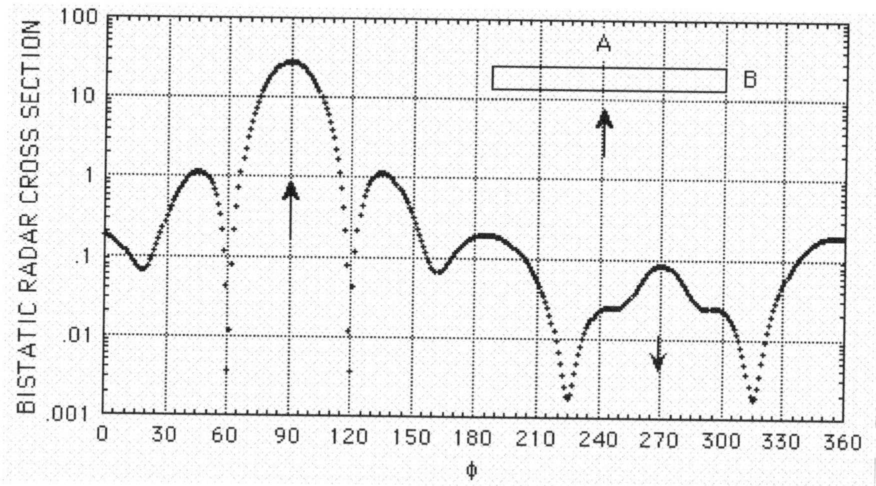

FIGURE 6.30
Bistatic radar cross section of the rectangular permeable dielectric scatterer (TE case).

$\phi^{\mathrm{i}} = 90^\circ$. The bistatic RCS data shown in Fig. 6.30 are also numerically validated using the MoM solution by solving the (CFIEs) for the TE case.

6.9 FDTD MODELING OF CURVED SURFACES

In this section, the FDTD technique is generalized to include an accurate modeling of the curved surfaces. The generalization is based on the contour path method that accurately models the electromagnetic effects associated with curved surfaces but retains the ability to model corners and edges of the arbitrarily shaped structure. In the following, the contour path method is discussed for the two-dimensional case involving electromagnetic wave scattering and coupling.

A significant limitation in the rectangular structured FDTD models of the material structure with smooth curved surface has been the stepped-edge approximation of the actual surface. Although structured FDTD models discussed earlier are ideal for modeling wave propagation, scattering, and penetration of low Q structures, the stepped-edge approximation of the curved surface and aperture wall can shift the center frequencies of resonant responses by 1 or 2% for Q factors and can even introduce spurious nulls. In the case of scattering and interaction by complex shaped objects, the use of stepped-surface approximation limits the application of the FDTD technique to modeling of the important target cases in which the surface roughness, exact curvature, and dielectric and permeable loading are important in determining the RCS. Depending on the computational resources available and modeling accuracy required, there are three types of FDTD-conformable surface models that can be used to treat arbitrary surface configuration, not including the hybrid numerical methods which aid the FDTD computation at selected regions by utilizing either the finite-element time domain or the integral equation time domain approaches. The following should be considered regarding the usefulness of each FDTD model based on the computer resources involved in the mesh generation: the severity of the numerical artifacts introduced by grid distortion, including numerical instability, dispersion, nonphysical wave reflection, and subtraction noise; the limitation of the near-field computational range due to the subtraction noise; and the comparative computer resources for

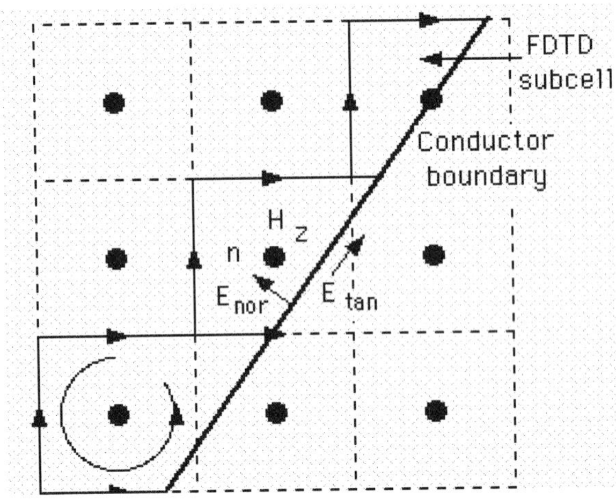

FIGURE 6.31
FDTD structure cells adjacent to nonconforming conducting surface.

running realistic models, especially for very large two-dimensional and three-dimensional target shapes.

1. Locally distorted grid model: This approach preserves the basic rectangular grid arrangement of the structured field components at all space cells except those immediately adjacent to the nonconforming structure surface. Only space cells adjacent to the material structure surface are appropriately deformed to conform with the surface locus. Slightly modified time-stepping expressions for the field components adjacent to the surface are obtained by applying either the modified finite volume technique or the contour path surface technique (Fig. 6.31).
2. Globally distorted grid model—body fitted: This approach employs the available numerical mesh-generation scheme to construct nonrectangular grids which are continuously and globally stretched to conform with the smoothly shaped structure. In effect, the rectangular grid is mapped to a numerically generated coordinate system wherein the structure surface contour occupies a locus of constant equivalent radius.
3. Globally distorted grid model—unstructured: This approach employs the available numerical mesh generation scheme to construct nonrectangular grids composed of an unstructured array of space-filling cells. The general shape of the structure surface features is appropriately fit into the unstructured grid, with local grid resolution and cell shape selected to provide the desired geometric modeling aspects.

6.9.1 Perfectly Conducting Object: The TE Case

The contour path modeling of the perfectly conducting object with TE to z polarization is considered here using the integral form of Maxwell's equations (Eqs. 6.1.1 and 6.1.2). Referring to Fig. 6.17, time-stepping (Eqs. 6.5.27–6.5.29) are applicable in the complete FDTD computational domain except near the nonconforming perfectly conducting surface. For the TE case, the normally rectangular Faraday contours surrounding each H_z component near the object are deformed so as to conform to its surface. Each H_z component is assumed to represent the average value of the magnetic field within the patch bounded by the distorted Faraday contour. The total tangential electric field, $E_{\tan}$, on the distorted contour at the perfectly conducting object surface is zero. Along the remaining straight portions of the bounded contour, E_x and E_y components of the electric field are assumed to have no variation along the respective contour segments. The basic assumptions discussed in Section 6.2 are introduced to convert the operator form to the corresponding difference form of structured time-stepping equations. Where possible, these electric field components are calculated using the rectangular Ampere contours from the adjacent H_z components. The Ampere contours are not deformed. Also, the calculation of Ampere contours which cross the media boundary is not used, necessitating that the E_x and E_y components of the electric field along the corresponding Faraday contour segments, if needed, are computed in an alternative procedure. The electric field components of the Faraday segments which intersect the object's surface, but are not tangent to it, are approximated according to the following two numerical procedures.

One possible method is to obtain the projection onto the Faraday contour segment of electric field value at the intersection of the segment and the object surface. The electric field at this intersection, E_{nor}, is normal to the object's surface and is calculated by setting up an auxiliary Ampere's law contour computation along the surface. The H_z magnetic field values needed for the auxiliary computation are interpolated from the H_z field components near the surface.

Referring to the nearest-neighbor approximation shown in Fig. 6.32, an alternative procedure is to use the nearest E_x or E_y electric field component that is collinear with the Faraday contour segment and on the same side of the media interface as the Faraday segment. Referring to Eq. (6.5.29), after applying Faraday's law for the illustrative contour of Fig. 6.32, the following special FDTD time-stepping relationship is obtained for the H_z magnetic field component immediately adjacent to the object surface:

$$\begin{aligned} H_z^{n+1}(i,j) = {} & C(i,j)H_z^n(i,j) + D(i,j)E_y^{n+1/2}(i-1/2,j)\Delta(i-1/2,j) \\ & - D(i,j)E_y^{n+1/2}(i+1/2,j)\Delta(i+1/2,j) \\ & + D(i,j)E_x^{n+1/2}(i,j+1/2)\Delta(i,j+1/2), \end{aligned} \tag{6.9.1}$$

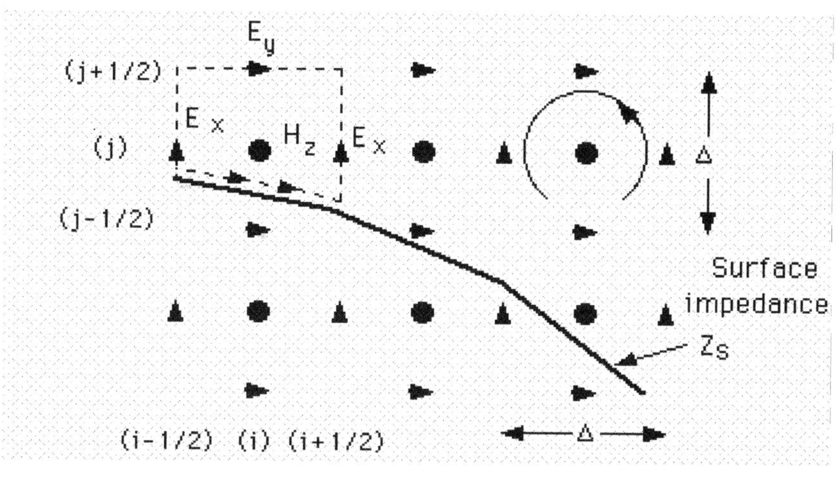

FIGURE 6.32
Contour path FDTD model for conducting surface (TE case).

where the lattice media coefficients for the subcell (i, j) are given by

$$R(i, j) = \frac{\mu_0 \mu_r(i, j) A(i, j)}{\Delta t}, \tag{6.9.2}$$

$$S(i, j) = 1/2[\sigma^{(m)}(i, j) A(i, j) + Z_s(i, j)\delta(i, j)], \tag{6.9.3}$$

$$C(i, j) = \frac{R(i, j) - S(i, j)}{R(i, j) + S(i, j)}, \tag{6.9.4}$$

$$D(i, j) = \frac{1}{R(i, j) + S(i, j)}, \tag{6.9.5}$$

where $A(i, j)$ is the area of the (i, j) subcell covered by the Faraday contour, $\delta(i, j)$ is the length of the conductor along the Faraday contour intersecting the (i, j) subcell, and $Z_s(i, j)$ is the surface impedance along the Faraday contour within the (i, j) subcell.

The time-stepping expression (Eq. 6.9.1) needs to be further modified to include other electric field components in the neighborhood of the conductor surface for the case of a standard stretched subcell or nonstandard subcell. In fact, this is accomplished by modifying the Faraday contour to include collinear electric field components. In Eq. (6.9.1), the only data needed to describe a distorted contour are the area of the patch within the contour, the intercept points of the object surface contour with the structured grid lines, the subtended arc length of the object surface, information on whether electric field components along the Faraday contour are calculable using the regular time-stepping algorithm, and the variation of the surface impedance with position along the object surface contour. Furthermore, for the TE case, no magnetic or electric field components in the FDTD space lattice other than the H_z components adjacent to the object surface require the modified time-stepping relationship.

The modeling discussion presented previously is also applicable to the standard stretched FDTD grid as shown in Fig. 6.33. In Eq. (6.9.1), the stretched segment length along the Faraday contour on the right side must be appropriately incorporated. Furthermore, in modeling the arbitrary surface based on the contour path method, invariably one has to account for the nonstandard

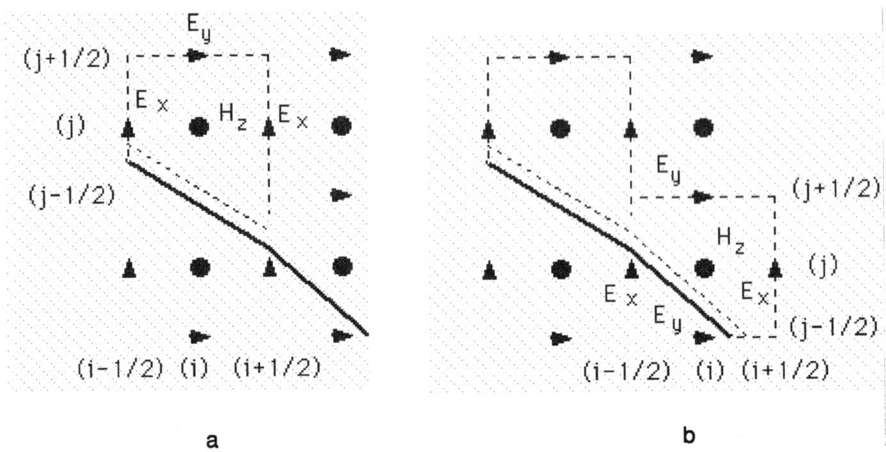

FIGURE 6.33
(a) Stretched standard grid and (b) nonstandard grid.

grid as shown in Fig. 6.33. This requires an additional electric field component to be included in Eq. (6.9.1).

Referring to Eqs. (6.5.29) and (6.9.1), after applying Faraday's law to the illustrative contour of Fig. 6.33, the following special FDTD time-stepping relationship is obtained for the H_z magnetic field component immediately adjacent to the object surface:

$$\begin{aligned} H_z^{n+1}(i, j) = C(i, j)H_z^n(i, j) &+ D(i, j)E_y^{n+1/2}(i - 1/2, j)\Delta(i - 1/2, j) \\ &- D(i, j)E_y^{n+1/2}(i + 1/2, j)\Delta(i + 1/2, j) \\ &+ D(i, j)E_x^{n+1/2}(i, j + 1/2)\Delta(i, j + 1/2) \\ &- D(i, j)E_x^{n+1/2}(i, j - 1/2)\Delta(i, j - 1/2), \end{aligned} \tag{6.9.6}$$

where the lattice media coefficients for the subcell (i, j) are given by Eqs. (6.9.2)–(6.9.5). The basic modeling concepts of the contour path method are now applied to analyze the near total electric and magnetic field distribution of a two-dimensional perfecting conducting circular cylindrical scatterer. The geometry of the circular scatterer is embedded completely in the total-field zone of the structured FDTD lattice, as shown in Fig. 6.23. The radius of the scatterer is $a = 0.796$ m and the surface impedance is $Z_s = 0$. The scatterer is excited by an incident field with time-harmonic frequency $f = 300$ MHz. This circular cross section is modeled by discretizing the scatterer region into structured square cells, $\Delta x = \Delta y = \Delta$, with $i = 80$ cells along the x-axis and $j = 80$ cells along the y-axis. For this spatial discretization, the spatial increment is $\Delta = 0.0199$ m and the corresponding time-stepping increment (in seconds) is $\Delta t = \Delta/c_{\max}$, where $c_{\max} = 3 \times 10^8$ m/s is the wave velocity in the air region. The fine sampling utilized in this study is not always required but has been selected to validate the basic numerical modeling approach of the contour path method and to provide additional data points to display the near-field distribution. Particularly, for determining the far-field distribution and the RCS, the circular cross section can be modeled by discretizing the scatterer region with $i = 20$ cells along the x-axis and $j = 20$ cells along the y-axis.

Figure 6.34 shows a plot of the total axial component of the magnetic field near the surface of the perfectly conducting circular scatterer for a normal TE polarized incident plane wave excitation

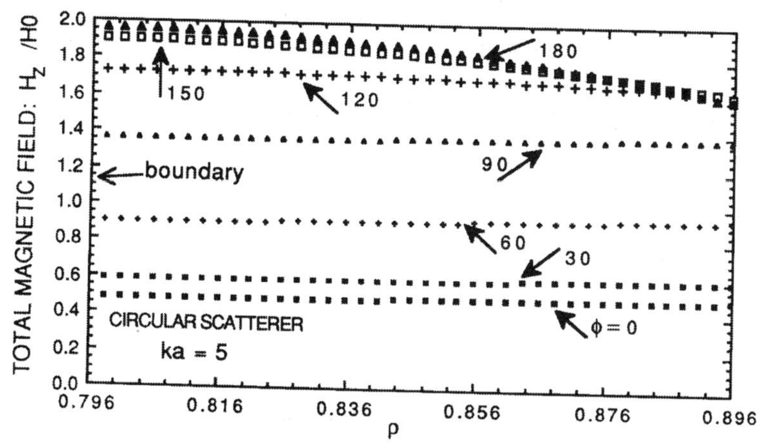

FIGURE 6.34
Distribution of the axial component of the total magnetic field.

with angle of incidence $\phi^{\mathrm{i}} = 0^\circ$. In this case study, the FDTD cells adjacent to the circular surface do not conform with the shape of the surface. For the TE analysis, Eqs. (6.5.27)–(6.5.29) are directly applied along with the special time-stepping contour path FDTD expression (Eq. 6.9.1) to calculate the H_z component of the total magnetic field at all cells adjacent to the circular surface having either the subcell or the stretched cell configuration. To determine the near total field, the incident field is time stepped for 12 cycles and the phasor electric and magnetic field components are recorded. The near total magnetic field result shown in Fig. 6.34 is also numerically validated using a series-type analytical solution by representing the scattered field in terms of outgoing cylindrical modes. The numerical results are normalized with respect to the amplitude of incident magnetic field H_0. The numerical data are displayed as a function of the radial coordinate variable from a point on the surface of the circular conducting scatterer for fixed observation angles. In fact, the total magnetic field data at the boundary surface directly represent the induced surface electric current density on the circular scatterer. The FDTD result and the series solution show excellent validation, with <0.1% error for the fine discretization selected in this case study.

6.9.2 Perfectly Conducting Object: The TM Case

The contour path modeling of the perfectly conducting object with TM to z polarization is considered in this section using the integral form of Maxwell's equations (Eqs. 6.1.1 and 6.1.2). Referring to Fig. 6.16, the time-stepping expressions (Eqs. 6.5.15–6.5.17) are applicable in the complete FDTD computational domain except near the nonconforming perfectly conducting surface (see Fig. 6.16).

For the TM case, the normally rectangular Faraday contours surrounding each H_x and H_y component near the object are deformed so as to conform to its surface. Each component of the magnetic field is assumed to represent the average value of the magnetic field within the patch bounded by the distorted Faraday contour. The total tangential electric field, E_{tan}, on the distorted contour at the perfectly conducting object surface is zero, assuming the surface impedance $Z_{\mathrm{s}} = 0$. Along the remaining straight portions of the bounded contour, E_z components of the electric field are assumed to have no variation along the respective contour segments. The basic assumptions discussed in Sections 6.2 and 6.9.1 are introduced to convert the operator form to the corresponding difference form of structured time-stepping equations. The Ampere contours are not deformed. Also, the calculation of Ampere contours which cross the media boundary is not used, necessitating that the E_z components of the electric field along the corresponding Faraday contour segments, if needed, are computed in an alternative procedure. Referring to Eqs. (6.5.15) and (6.5.16), after applying Faraday's law for the illustrative contour of Fig. 6.35, the following special FDTD time-stepping relationships are obtained for H_x and H_y magnetic field components immediately adjacent to the object surface:

$$H_x^{n+1/2}(i, j-1/2) = H_x^{n-1/2}(i, j-1/2) - C_1(i, j-1/2)E_z^n(i, j) \tag{6.9.7}$$

$$H_y^{n+1/2}(i+1/2, j) = H_y^{n-1/2}(i+1/2, j) - C_2(i+1/2, j)E_z^n(i, j), \tag{6.9.8}$$

where

$$C_1(i, j-1/2) = \frac{\Delta t}{\mu_0 \mu_{\mathrm{r}}(i, j-1/2)\Delta(i, j-1/2)}, \tag{6.9.9}$$

$$C_2(i+1/2, j) = \frac{\Delta t}{\mu_0 \mu_{\mathrm{r}}(i+1/2, j)\Delta(i+1/2, j)}, \tag{6.9.10}$$

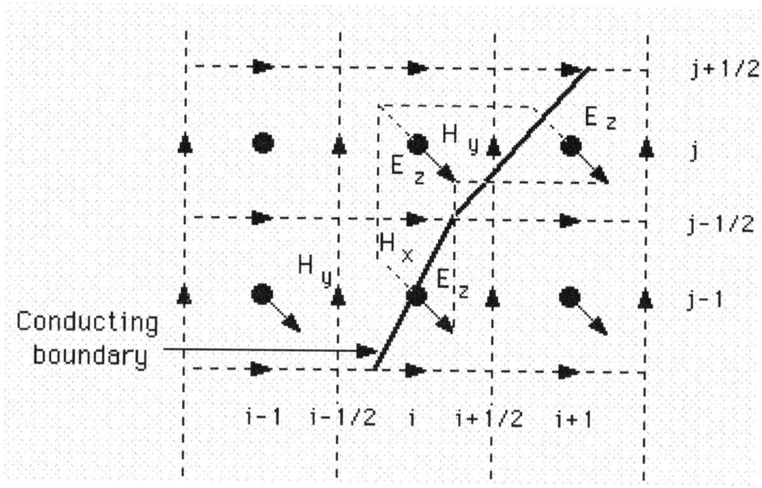

FIGURE 6.35
Contour path FDTD model for conducting surface (TM case).

where $\Delta(i, j-1/2)$ and $\Delta(i+1/2, j)$ are the lengths along the Faraday contour parallel to the conductor surface and intersecting the conductor surface, respectively.

The basic TM modeling concepts of the contour path method are now applied to analyze the near-scattered electric and magnetic field distribution of a two-dimensional perfectly conducting circular cylindrical scatterer. The geometry of the circular scatterer is embedded completely in the total-field zone of the structured FDTD lattice, as shown in Fig. 6.23. The radius of the scatterer is $a = 0.796$ m and the surface impedance is $Z_s = 0$. The scatterer is excited by an incident field with time-harmonic frequency $f = 300$ MHz. This circular cross section is modeled by discretizing the scatterer region into structured square cells, $\Delta x = \Delta y = \Delta$, with $i = 81$ cells along the x-axis and $j = 81$ cells along the y-axis. The fine sampling utilized in this study is not always required but has been selected to validate the basic numerical modeling approach of the contour path method and to provide additional data points to display the near-scattered field distribution.

Figure 6.36 shows a plot of the scattered axial component of the electric field near the surface of the perfectly conducting circular scatterer for a normal TM polarized incident plane wave excitation with angle of incidence $\phi^i = 0°$. The FDTD cells adjacent to the circular surface do not conform to the shape of the conducting surface. For the TM analysis, Eqs. (6.5.15)–(6.5.17) are directly applied along with the special time-stepping contour path FDTD expressions (Eqs. 6.9.7 and 6.9.8) to calculate H_x and H_y components of the total magnetic field at all cells adjacent to the circular surface having either the subcell or the stretched cell configuration. To determine the near total field, the incident field is time stepped for 12 cycles, and the phasor total electric and magnetic field components are recorded. To obtain the axial scattered electric field in the total-field zone, the axial incident component of the electric field is subtracted from the total axial electric field component. The near-scattered electric field result shown in Fig. 6.36 is also numerically validated using the series-type-analytical solution by representing the scattered field in terms of outgoing cylindrical modes. The numerical results are normalized with respect to the amplitude of the incident TM electric field, E_0. The numerical data are displayed as a function of the radial coordinate variable from a point on the surface of circular conducting scatterer for fixed observation angles. The FDTD result and the series solution show excellent validation, with $<0.1\%$ error for the fine discretization.

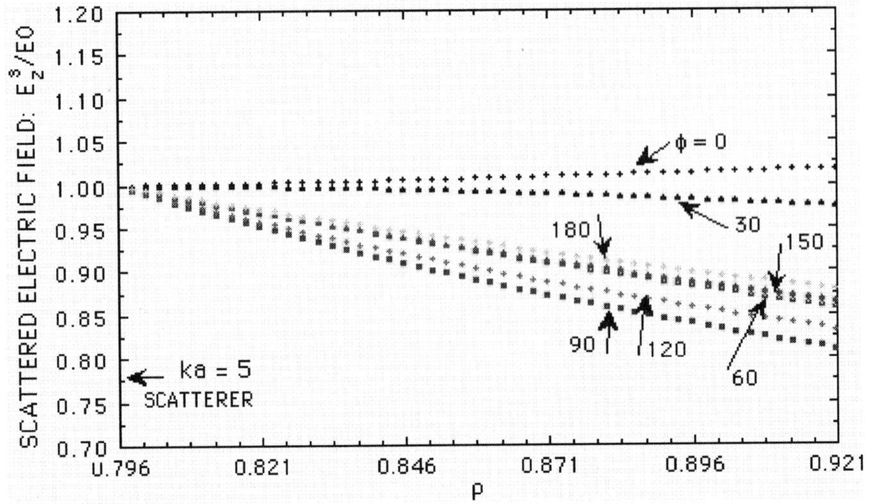

FIGURE 6.36
Distribution of the axial component of the scattered electric field.

6.9.3 Homogeneous Dielectric Object: The TE Case

The contour path modeling of the homogeneous dielectric object with TE to z polarization is considered here using the integral form of Maxwell's equations (Eqs. 6.1.1 and 6.1.2). Referring to Fig. 6.17, the time-stepping Eqs. (6.5.27)–(6.5.29) are applicable in the complete FDTD computational domain except near the nonconforming dielectric surface. For the TE case, the normally rectangular Faraday contours surrounding each H_z component near and on either side of the dielectric object are deformed so as to conform to its surface. Each H_z component is assumed to represent the average value of the magnetic field within the patch bounded by the distorted Faraday contour. The total tangential electric field, $E_{\tan}$, on the distorted contour at the dielectric object surface is continuous and, similarly, the total normal electric flux density, D_{nor}, on the distorted contour at the dielectric object surface is also continuous. Along the remaining straight portions of the bounded contour, E_x and E_y components of the electric field are assumed to have no variation along the respective contour segments. The basic assumptions discussed in Sections 6.2 and 6.9.1 are introduced to convert the operator form to the corresponding difference form of the structured time-stepping equations. Where possible, the electric field components are calculated using the rectangular Ampere contours from the adjacent H_z components. Also, the calculation of Ampere contours which cross the media boundary is not used, necessitating that the E_x and E_y components of the electric field along the corresponding Faraday contour segments, if needed, are computed in an alternative procedure using the nearest-neighbor approximation.

Referring to Eq. (6.5.29), after applying Faraday's law for the illustrative contour of Fig. 6.37, the following special FDTD time-stepping relationship is obtained for the H_z magnetic field component immediately adjacent to the dielectric object surface:

$$\begin{aligned} H_z^{n+1}(i,j) = {} & C_a(i,j)H_z^n(i,j) + C_b(i,j)E_t^{n+1/2}(p)L_p \\ & + C_b(i,j)E_x^{n+1/2}(i,j+1/2)\Delta(i,j+1/2) \\ & + C_b(i,j)E_y^{n+1/2}(i-1/2,j)\Delta(i-1/2,j) \\ & - C_b(i,j)E_x^{n+1/2}(i-1,j-1/2)\Delta(i,j-1/2), \end{aligned} \tag{6.9.11}$$

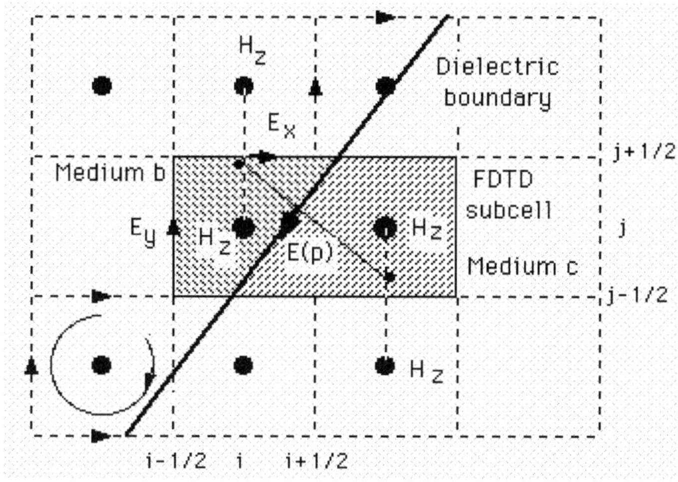

FIGURE 6.37
FDTD structure cells adjacent to nonconforming dielectric surface.

where

$$R_b(i, j) = \frac{\Delta t}{\mu_0 \mu_r(i, j) A(i, j)}, \tag{6.9.12}$$

$$C_a(i, j) = \frac{\left[1 - \frac{\sigma^{(m)}(i,j)\Delta t}{2\mu_0\mu_r(i,j)}\right]}{\left[1 + \frac{\sigma^{(m)}(i,j)\Delta t}{2\mu_0\mu_r(i,j)}\right]}, \tag{6.9.13}$$

$$C_b(i, j) = \frac{R_b(i, j)}{\left[1 + \frac{\sigma^{(m)}(i,j)\Delta t}{2\mu_0\mu_r(i,j)}\right]}, \tag{6.9.14}$$

$E_t^{n+1/2}(p)$ is the tangential component of the electric field at the pth segment along the interface, and L_p is the length of the pth segment along the interface.

Furthermore, the tangential component of the electric field at the pth segment along the interface is calculated using the following time-stepping algorithm obtained by using Ampere's contour as shown in Fig. 6.37:

$$E_t^{n+1/2}(p) = C_c(p) E_t^{n-1/2}(p) + C_d(p)\left[H_{zc}^n(p) - H_{zb}^n(p)\right], \tag{6.9.15}$$

where the lattice media coefficients for the pth segment along the interface connecting the two subcells (i, j) and $(i, j+1)$ are given by

$$R_d(p) = \frac{\Delta t}{\epsilon_0 \left[\epsilon_{rb}(p) L_b(p) + \epsilon_{rc}(p) L_c(p)\right]}, \tag{6.9.16}$$

$$C_{1c}(p) = 1 - \frac{\Delta t\left[\sigma_b^{(e)}(p) L_b(p) + \sigma_c^{(c)}(p) L_c(p)\right]}{\epsilon_0 \left[\epsilon_{rb}(p) L_b(p) + \epsilon_{rc}(p) L_c(p)\right]}, \tag{6.9.17}$$

$$C_{2c}(p) = 1 + \frac{\Delta t\left[\sigma_b^{(e)}(p)L_b(p) + \sigma_c^{(c)}(p)L_c(p)\right]}{\epsilon_0\left[\epsilon_{rb}(p)L_b(p) + \epsilon_{rc}(p)L_c(p)\right]}, \tag{6.9.18}$$

$$C_c(p) = \frac{C_{1c}(p)}{C_{2c}(p)}, \tag{6.9.19}$$

$$C_c(p) = \frac{R_d(p)}{C_{2c}(p)} \tag{6.9.20}$$

where the subscript b refers to the left-side medium subcell (i, j) and the subscript c refers to the right-side medium subcell $(i, j+1)$. The z components of the magnetic field in Eq. (6.9.15) are the appropriate interpolated values of the magnetic field components

$$H_{zc}^n(p) = D_1 H_z^n(i+1, j) + D_2 H_z^n(i+1, j-1) \tag{6.9.21}$$

$$H_{zb}^n(p) = D_3 H_z^n(i, j) + D_4 H_z^n(i, j+1), \tag{6.9.22}$$

and D_1–D_4 are corresponding linear interpolation constants.

Time-stepping Eqs. (6.9.11) and (6.9.15) must be further modified to include other electric field components in the neighborhood of the dielectric surface for the case of a standard stretched subcell or nonstandard subcell. In fact, this is accomplished by modifying the Faraday contour to include collinear electric field components. Furthermore, for the TM case duality can be invoked to interchange the role of the magnetic and electric field components in the FDTD space lattice adjacent to the dielectric object surface to determine the corresponding modified time-stepping relationships.

The basic modeling concepts of the contour path method are now applied to analyze the on-surface total electric and magnetic field distribution of a two-dimensional permeable and dielectric circular cylindrical scatterer. The geometry of the circular scatterer is embedded completely in the total-field zone of the structured FDTD lattice as shown in Fig. 6.23. The radius of the scatterer is $a = 0.796$ m. The scatterer is excited by a TE polarized incident field with time-harmonic frequency $f = 300$ MHz. This circular cross section is modeled by discretizing the scatterer region into structured square cells, $\Delta x = \Delta y = \Delta$, with $i = 80$ cells along the x-axis and $j = 80$ cells along the y-axis. For this spatial discretization, the spatial increment is $\Delta = 0.0199$ m and the corresponding time-stepping increment (in seconds) is $\Delta t = \Delta/\sqrt{2}c_{\max}$, where $c_{\max} = 3 \times 10^8$ m/s is the wave velocity in air. The fine sampling utilized in this study is not always required but has been selected to validate the basic numerical modeling approach of the contour path method and to provide additional data points to display the on-surface field distribution.

Figure 6.38 shows a plot of the axial equivalent magnetic current and the angular equivalent electric current on the surface of circular scatterer for a TE polarized normal incident plane wave excitation with angle of incidence $\phi^{\mathrm{i}} = 0°$. The FDTD cells adjacent to the circular surface do not conform to the shape of the surface. For the TE analysis, Eqs. (6.5.27)–(6.5.29) are directly applied along with special time-stepping contour path FDTD Eqs. (6.9.11) and (6.9.15). To obtain convergence of the near total-field data, the incident field is time stepped for 14 cycles, and the phasor electric and magnetic field components are recorded. The equivalent currents shown in Fig. 6.38 are also numerically validated using the series-type analytical solution by representing the axial scattered magnetic field in terms of outgoing cylindrical modes. The numerical results of angular electric current are normalized with respect to the amplitude of incident magnetic field H_0 and, similarly, the corresponding numerical results of axial magnetic current are normalized

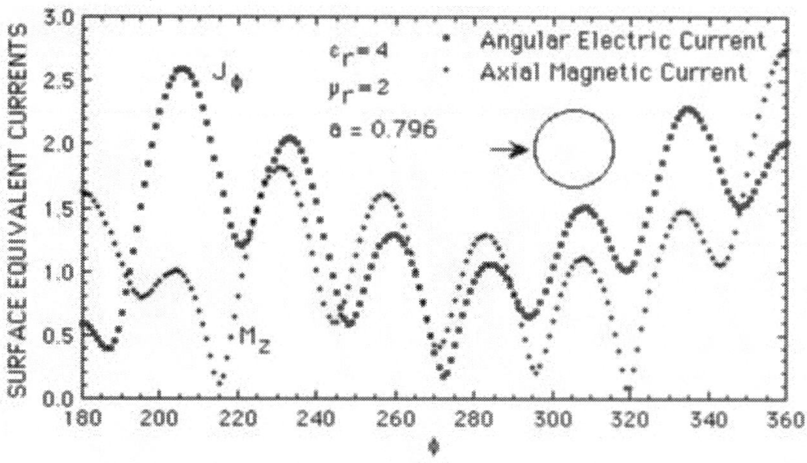

FIGURE 6.38
Magnitude distribution of surface equivalent currents on the permeable dielectric scatterer (TE case).

with respect to the amplitude of incident electric field E_0. The numerical data are displayed as a function of the angular coordinate variable on the surface of the circular scatterer. The FDTD result and the series solution show excellent validation, with <0.1% error for the fine discretization.

6.10 FDTD FORMULATION FOR THREE-DIMENSIONAL CLOSED-REGION PROBLEMS

The numerical solution based on the FDTD technique is investigated further for determining the scattering, penetration, and interaction phenomena associated with the general three-dimensional closed- or open-type arbitrarily shaped material structures. The structured three-dimensional FDTD models along with the modified contour path algorithms can also be used for analyzing the mode propagation dispersion characteristics of the closed- and open-type waveguides. As discussed in Section 6.5, to determine the mode propagation characteristics of the closed- or open-type waveguide geometry, a complete electromagnetic full-wave analysis is required. The full-wave equations decouple into two sets consisting of the TM and TE polarized field distributions. At the cutoff condition, the electric and magnetic modal fields degenerate to only three components for either the TE modes or the TM modes. Practically no power is propagated along the waveguide axis since there exists only the corresponding transverse distribution of either the electric field or the magnetic field. For all frequencies above the mode cutoff frequency, all components of the transverse electric and magnetic fields, E_x, E_y and H_x, H_y, exist, including the axial component of magnetic field H_z for the TE case and the axial component of electric field E_z for the TM case.

Unlike the method used for the mode cutoff frequency calculation, which is analyzed directly as a two-dimensional problem, a full-wave three-dimensional analysis is required to determine the propagating mode dispersion relationship. Referring to Fig. 6.15, the closed-type waveguide geometry is uniform along the axial z-coordinate direction and the propagation parameter β can be interpreted in terms of the axial time delay of the modal waves propagating along its axis. Hence, it is not always necessary to numerically model the complete three-dimensional closed-type waveguide volume, except for an incremental width along the axial coordinate direction. This modeling approach is generally referred to as the compact two-dimensional FDTD technique.

In the following sections, an efficient numerical approach is discussed based on the compact two-dimensional FDTD technique which retains most of the basic numerical modeling aspects discussed earlier for determining the mode cutoff frequencies. The compact two-dimensional FDTD technique is based on the idea that all adjacent field components in the axial propagation direction can be related by a gradual phase-shift term given by $e^{-j\beta z}$, where z is the axial direction of propagation. Compared with the traditional three-dimensional FDTD technique, the compact two-dimensional FDTD technique has advantages of less computer memory requirement and there is no need to use an absorbing termination condition. However, in the case of the compact two-dimensional FDTD technique, the propagation constant β must first be chosen as an input parameter to proceed with the time-stepping of the finite-difference process. This procedure is awkward in the general waveguide analysis and design compared with many other numerical methods. Especially in the case of the frequency-domain method, the time-harmonic excitation frequency is always chosen first and then the mode propagation parameter β is sought. However, this FDTD modeling approach does not impose any serious limitation since by randomly testing several values of β, the associated relationship between the mode propagation parameter β and the corresponding time-harmonic frequency ω can be established.

6.10.1 Three-Dimensional Full-Wave Analysis

The numerical method for the full-wave analysis based on the three-dimensional FDTD technique, which has a broad range of electromagnetic analysis and design applications, is now discussed. Similar to the two-dimensional cases, with this approach the continuous electromagnetic fields in a finite volume of space are sampled at distinct points in space and time, as shown in Fig. 6.14.

Figure 6.13 shows a structured unit cubic cell in the conventional space lattice in the Cartesian coordinate system. Each magnetic field component is surrounded by four circulating electric field components and, similarly, each electric field component is surrounded by four circulating magnetic field components. By applying the numerical procedure used earlier to derive the time-stepping expressions for the one-dimensional and two-dimensional cases, the three-dimensional time-stepping finite-difference equations for electric and magnetic field components can be obtained by analyzing all the six scalar-coupled partial differential equations (Eqs. 6.5.1–6.5.6). Using the rectangular coordinate system, let $E_x(x, y, z, t)$, $E_y(x, y, z, t)$, and $E_z(x, y, z, t)$ represent electric field components and $H_x(x, y, z, t)$, $H_y(x, y, z, t)$, and $H_z(x, y, z, t)$ represent the magnetic field components.

Referring to Fig. 6.13, let Δx, Δy, and Δz represent the lattice spatial increments along the x, y, and z coordinates, respectively. Δt represents the time increment. Furthermore, let (i, j, k) indices represent location $(i\Delta x, j\Delta y, k\Delta z)$. Lastly, $F^n(i, j, k) = F(i\Delta x, j\Delta y, k\Delta z, n\Delta t)$. It is assumed that the width of each spatial cell is very small, and for all practical purpose each cell can be assumed to be a linear and homogeneous region. For the first-order partial derivatives, a centered finite-difference approximation with second-order accuracy is introduced with respect to the x-, y-, z-coordinate spatial variables and the t time variable. The following finite-difference approximations for the first-order derivatives of the electric and magnetic field components are accurate up to the second power of their corresponding linear basis function discretization:

$$\frac{\partial F^n(i,j,k)}{\partial t} = \frac{F^{n+1/2}(i,j,k) - F^{n-1/2}(i,j,k)}{\Delta t} + \bigcirc(\Delta t^2) \tag{6.10.1}$$

$$\frac{\partial F^n(i,j,k)}{\partial x} = \frac{F^n(i+1/2,j,k) - F^n(i-1/2,j,k)}{\Delta x} + \bigcirc(\Delta x^2) \tag{6.10.2}$$

$$\frac{\partial F^n(i,j,k)}{\partial y} = \frac{F^n(i,j+1/2,k) - F^n(i,j-1/2,k)}{\Delta y} + \bigcirc(\Delta y^2) \tag{6.10.3}$$

$$\frac{\partial F^n(i,j,k)}{\partial z} = \frac{F^n(i,j,k+1/2) - F^n(i,j,k-1/2)}{\Delta z} + \bigcirc(\Delta z^2). \tag{6.10.4}$$

Using the basic integral form of Maxwell's equations (Eqs. 6.1.1 and 6.1.2) and the previous first-order finite-difference approximations for the various partial derivatives, the three-dimensional full-wave scalar partial differential equations (Eqs. 6.5.1–6.5.6) yield the following time-stepping algorithms within the closed- or open-type waveguide geometry. The magnetic field components are given by

$$\begin{aligned} H_x^{n+1/2}(i,j+1/2,k+1/2) = {} & C_a(i,j+1/2,k+1/2)H_x^{n-1/2}(i,j+1/2,k+1/2) \\ & + \frac{C_b(i,j+1/2,k+1/2)}{\Delta z}E_y^n(i,j+1/2,k+1) \\ & - \frac{C_b(i,j+1/2,k+1/2)}{\Delta z}E_y^n(i,j+1/2,k) \\ & - \frac{C_b(i,j+1/2,k+1/2)}{\Delta y}E_z^n(i,j+1,k+1/2) \\ & + \frac{C_b(i,j+1/2,k+1/2)}{\Delta y}E_z^n(i,j,k+1/2) \end{aligned} \tag{6.10.5}$$

$$\begin{aligned} H_y^{n+1/2}(i+1/2,j,k+1/2) = {} & C_a(i+1/2,j,k+1/2)H_y^{n-1/2}(i+1/2,j,k+1/2) \\ & + \frac{C_b(i+1/2,j,k+1/2)}{\Delta x}E_z^n(i+1,j,k+1/2) \\ & - \frac{C_b(i+1/2,j,k+1/2)}{\Delta x}E_z^n(i,j,k+1/2) \\ & - \frac{C_b(i+1/2,j,k+1/2)}{\Delta z}E_x^n(i+1/2,j,k+1) \\ & + \frac{C_b(i+1/2,j,k+1/2)}{\Delta z}E_x^n(i+1/2,j,k) \end{aligned} \tag{6.10.6}$$

$$\begin{aligned} H_z^{n+1/2}(i+1/2,j+1/2,k) = {} & C_a(i+1/2,j+1/2,k)H_z^{n-1/2}(i+1/2,j+1/2,k) \\ & + \frac{C_b(i+1/2,j+1/2,k)}{\Delta y}E_x^n(i+1/2,j+1,k) \\ & - \frac{C_b(i+1/2,j+1/2,k)}{\Delta y}E_x^n(i+1/2,j,k) \\ & - \frac{C_b(i+1/2,j+1/2,k)}{\Delta x}E_y^n(i+1,j+1/2,k) \\ & + \frac{C_b(i+1/2,j+1/2,k)}{\Delta x}E_y^n(i,j+1/2,k), \end{aligned} \tag{6.10.7}$$

and the electric field components are given by

$$E_x^{n+1}(i+1/2,j,k) = C_c(i+1/2,j,k)E_x^n(i+1/2,j,k) + \frac{C_d(i+1/2,j,k)}{\Delta y}H_z^{n+1/2}(i+1/2,j+1/2,k) - \frac{C_d(i+1/2,j,k)}{\Delta y}H_z^{n+1/2}(i+1/2,j-1/2,k) - \frac{C_d(i+1/2,j,k)}{\Delta z}H_y^{n+1/2}(i+1/2,j,k+1/2) + \frac{C_d(i+1/2,j,k)}{\Delta z}H_y^{n+1/2}(i+1/2,j,k-1/2) \quad (6.10.8)$$

$$E_y^{n+1}(i,j+1/2,k) = C_c(i,j+1/2,k)E_y^n(i,j+1/2,k) + \frac{C_d(i,j+1/2,k)}{\Delta z}H_x^{n+1/2}(i,j+1/2,k+1/2) - \frac{C_d(i,j+1/2,k)}{\Delta z}H_x^{n+1/2}(i,j+1/2,k-1/2) - \frac{C_d(i,j+1/2,k)}{\Delta x}H_z^{n+1/2}(i+1/2,j+1/2,k) + \frac{C_d(i,j+1/2,k)}{\Delta x}H_z^{n+1/2}(i-1/2,j+1/2,k) \quad (6.10.9)$$

$$E_z^{n+1}(i,j,k+1/2) = C_c(i,j,k+1/2)E_z^n(i,j,k+1/2) + \frac{C_d(i,j,k+1/2)}{\Delta x}H_y^{n+1/2}(i+1/2,j,k+1/2) - \frac{C_d(i,j,k+1/2)}{\Delta x}H_y^{n+1/2}(i-1/2,j,k+1/2) - \frac{C_d(i,j,k+1/2)}{\Delta y}H_x^{n+1/2}(i,j+1/2,k+1/2) + \frac{C_d(i,j,k+1/2)}{\Delta y}H_x^{n+1/2}(i,j-1/2,k+1/2), \quad (6.10.10)$$

where the lattice media coefficients are given by

$$R_b(i,j,k) = \frac{\Delta t}{\mu_0\mu_r(i,j,k)}, \quad (6.10.11)$$

$$C_a(i,j,k) = \frac{\left[1 - \frac{\sigma^{(m)}(i,j,k)\Delta t}{2\mu_0\mu_r(i,j,k)}\right]}{\left[1 + \frac{\sigma^{(m)}(i,j,k)\Delta t}{2\mu_0\mu_r(i,j,k)}\right]}, \quad (6.10.12)$$

$$C_b(i,j,k) = \frac{R_b(i,j,k)}{\left[1 + \frac{\sigma^{(m)}(i,j,k)\Delta t}{2\mu_0\mu_r(i,j,k)}\right]}, \quad (6.10.13)$$

and

$$R_d(i, j, k) = \frac{\Delta t}{\epsilon_0 \epsilon_r(i, j, k)}, \tag{6.10.14}$$

$$C_c(i, j, k) = \frac{\left[1 - \frac{\sigma^{(e)}(i,j,k)\Delta t}{2\epsilon_0\epsilon_r(i,j,k)}\right]}{\left[1 + \frac{\sigma^{(e)}(i,j,k)\Delta t}{2\epsilon_0\epsilon_r(i,j,k)}\right]}, \tag{6.10.15}$$

$$C_d(i, j, k) = \frac{R_d(i, j, k)}{\left[1 + \frac{\sigma^{(e)}(i,j,k)\Delta t}{2\epsilon_0\epsilon_r(i,j,k)}\right]}. \tag{6.10.16}$$

The various electric and magnetic field components in the three-dimensional, time-stepping finite-difference algorithms, given by Eqs. (6.10.11)–(6.10.16), are all real quantities and functions of the variables (x, y, z, t). With the previous iterative equations, the new value of a field vector component at any lattice point depends only on its previous value and on the previous values of the components of the other field vector at adjacent points. Therefore, at any given time step, the computation of a field vector can proceed either one point at a time or, if p parallel processors are employed concurrently, p points at a time.

To implement the previous iterative algorithms for a spatial region having a continuous or piecewise continuous variation of the material properties with spatial position, it is desirable to define and store the media lattice coefficients, Eqs. (6.10.11)–(6.10.16), for each vector field component. For a space region with a finite number of media having distinct electric and magnetic properties, the required computer storage can be drastically reduced.

The choice of space increments Δx, Δy, and Δz and time increment Δt is dictated by the reasons of accuracy and algorithm stability, respectively. To ensure the stability of the computed fields, Δt is chosen to satisfy the following inequality for the three-dimensional lattice model:

$$\Delta t \leq \frac{1}{c_{\max}} \left[\frac{1}{\Delta x^2} + \frac{1}{\Delta y^2} + \frac{1}{\Delta z^2}\right]^{-1/2}, \tag{6.10.17}$$

where $c_{\max}$ is the maximum wave velocity within the model.

The FDTD algorithms for Maxwell's equations derived previously cause numerical dispersion of the simulated wave modes in the computational lattice. In fact, the phase velocity of numerical wave modes in the FDTD grid can vary with modal wavelength, direction of propagation within the grid, and grid discretization. This numerical dispersion can lead to nonphysical results such as pulse distortion, artificial anisotropy, and pseudo-refraction. The numerical dispersion is a factor in the FDTD modeling that must be accounted for to understand the operation of the algorithm and its accuracy limits. In fact, the discretization introduces a numerical low-pass filtering effect wherein the wavelength of propagating numerical modes has a lower bound of two or three cells, depending on the propagation direction. The numerical dispersion relation for the conventional three-dimensional FDTD technique is given by

$$\left(\frac{v\Delta t}{\Delta x}\right)^2 \sin^2\left(\frac{k_x \Delta x}{2}\right) + \left(\frac{v\Delta t}{\Delta y}\right)^2 \sin^2\left(\frac{k_y \Delta y}{2}\right) + \left(\frac{v\Delta t}{\Delta z}\right)^2 \sin^2\left(\frac{k_z \Delta z}{2}\right) = \sin^2\left(\frac{\omega \Delta t}{2}\right), \tag{6.10.18}$$

where k_x, k_y, and k_z are the wave numbers along x, y, z coordinates, ω is the time-harmonic angular frequency, and the phase velocity $v = (\mu\epsilon)^{-1/2}$ in a homogeneous, isotropic medium. As a result, the FDTD modeling of time pulses having a finite duration (and thus infinite bandwidth) can result in progressive pulse distortion because the higher spatial frequency components propagate more slowly than the lower spatial frequency components, and the very high spatial frequency components with wavelengths less than 2 or 3 cells are rejected. This numerical dispersion causes broadening of finite-duration pulses and leaves a residue of high-frequency ringing on the trailing edges due to the relatively slowly propagating high-frequency components. Comparing the exact and numerical phase velocity as a function of FDTD grid resolution, the pulse distortion can be bounded by obtaining the Fourier spatial frequency spectrum of the desired pulse and selecting a grid cell size so that the principal spectral components are resolved with at least 10 cells per wavelength. This would limit the spread of numerical phase velocities of the principal spectral components to $<1\%$, regardless of wave propagation angle in the grid.

In addition to the numerical phase velocity anisotropy and pulse distortion effects, the numerical dispersion can lead to pseudo-refraction of propagating modes if the grid cell size is a function of position in the lattice. Such variable cell gridding would also vary the grid resolution of propagating modes and thereby perturb the modal phase velocity distribution. This would lead to nonphysical reflection and refraction of grid modes at interfaces between grid regions having different cell sizes, even if these interfaces were located in free-space regions, just as physical waves undergo reflection and refraction at interfaces of dielectric media having different indices of refraction. The degree of nonphysical refraction is dependent on the magnitude and abruptness of the change of modal phase velocity distribution and may be estimated using conventional theory for wave refraction at the dielectric interfaces.

6.10.2 Compact Two-Dimensional FDTD Algorithm

The numerical modeling of the closed waveguide structure based on the full-wave analysis requires complete modeling coverage along the axial propagation direction for several wavelengths. Thus, beside modeling the waveguide cross section, many layers of cells are needed along the axial propagation direction. Therefore, huge computer memory space and lengthy computational time are required, which are the main drawbacks of the traditional three-dimensional FDTD algorithm. Furthermore, the electric and magnetic field radiation termination conditions are needed at the end layer of boundary cells to simulate the infinite length of the uniform waveguide.

Particularly for the uniform waveguide geometries, one can take advantage of properties of the propagating modes. The electric and magnetic field variations along the z-coordinate axis of the closed waveguide can be written as

$$F(x, y, z) = F(x, y)e^{-j\beta z}, \tag{6.10.19}$$

where β is the axial propagation constant. Hence, a two-dimensional finite-difference mesh with a reduced grid size along the z-axis can be adopted for numerical modeling of the waveguide. This leads to an effective algorithm for determining the dispersion relationship of the closed waveguide structure. Based on Eq. (6.10.19), the adjacent electric and magnetic field components in the axial propagation direction are related as

$$E^n(i, j, k \pm 1) = E^n(i, j, k)e^{\mp j\beta\Delta z} \tag{6.10.20}$$

$$H^n(i, j, k \pm 1) = H^n(i, j, k)e^{\mp j\beta\Delta z}, \tag{6.10.21}$$

where the electric and magnetic field components are the phasor or complex spectral quantities.

The first-order partial derivatives with respect to z are replaced with $-j\beta$ and the corresponding discretized expressions are

$$\frac{E^n(i,j,k) - E^n(i,j,k+1)}{\Delta z} = j\beta E^n(i,j,k+1/2) \tag{6.10.22}$$

$$= j\beta E^n(i,j,k)e^{-j\beta\frac{\Delta z}{2}} \tag{6.10.23}$$

$$\frac{H^n(i,j,k) - H^n(i,j,k+1)}{\Delta z} = j\beta H^n(i,j,k+1/2) \tag{6.10.24}$$

$$= j\beta H^n(i,j,k)e^{-j\beta\frac{\Delta z}{2}}. \tag{6.10.25}$$

On substituting the previous simplification into the three-dimensional FDTD time-stepping algorithms, given by Eqs. (6.10.5)–(6.10.10), the following compact, two-dimensional, finite-difference, time-stepping expressions are obtained; for magentic field components,

$$\begin{aligned} H_x^{n+1/2}(i,j+1/2) = {} & C_a(i,j+1/2)H_x^{n-1/2}(i,j+1/2) \\ & - C_b(i,j+1/2)j\beta E_y^n(i,j+1/2)e^{-j\beta\frac{\Delta z}{2}} \\ & - \frac{C_b(i,j+1/2)}{\Delta y}E_z^n(i,j+1) + \frac{C_b(i,j+1/2)}{\Delta y}E_z^n(i,j) \end{aligned} \tag{6.10.26}$$

$$\begin{aligned} H_y^{n+1/2}(i+1/2,j) = {} & C_a(i+1/2,j)H_y^{n-1/2}(i+1/2,j) \\ & + \frac{C_b(i+1/2,j)}{\Delta x}E_z^n(i+1,j) - \frac{C_b(i+1/2,j)}{\Delta x}E_z^n(i,j) \\ & + C_b(i+1/2,j)\, j\beta E_x^n(i+1/2,j)e^{-j\beta\frac{\Delta z}{2}} \end{aligned} \tag{6.10.27}$$

$$\begin{aligned} H_z^{n+1/2}(i+1/2,j+1/2) = {} & C_a(i+1/2,j+1/2)H_z^{n-1/2}(i+1/2,j+1/2) \\ & + \frac{C_b(i+1/2,j+1/2)}{\Delta y}E_x^n(i+1/2,j+1) \\ & - \frac{C_b(i+1/2,j+1/2)}{\Delta y}E_x^n(i+1/2,j) \\ & - \frac{C_b(i+1/2,j+1/2)}{\Delta x}E_y^n(i+1,j+1/2) \\ & + \frac{C_b(i+1/2,j+1/2)}{\Delta x}E_y^n(i,j+1/2), \end{aligned} \tag{6.10.28}$$

and for electric field components,

$$\begin{aligned} E_x^{n+1}(i+1/2,j) = {} & C_c(i+1/2,j)E_x^n(i+1/2,j) \\ & + \frac{C_d(i+1/2,j)}{\Delta y}H_z^{n+1/2}(i+1/2,j+1/2) \\ & - \frac{C_d(i+1/2,j)}{\Delta y}H_z^{n+1/2}(i+1/2,j-1/2) \\ & + C_d(i+1/2,j)j\beta H_y^{n+1/2}(i+1/2,j)e^{j\beta\frac{\Delta z}{2}} \end{aligned} \tag{6.10.29}$$

$$E_y^{n+1}(i, j+1/2) = C_c(i, j+1/2)E_y^n(i, j+1/2) - C_d(i, j+1/2)j\beta H_x^{n+1/2}(i, j+1/2)e^{j\beta\frac{\Delta z}{2}} - \frac{C_d(i, j+1/2)}{\Delta x}H_z^{n+1/2}(i+1/2, j+1/2) + \frac{C_d(i, j+1/2)}{\Delta x}H_z^{n+1/2}(i-1/2, j+1/2) \quad (6.10.30)$$

$$E_z^{n+1}(i, j) = C_c(i, j)E_z^n(i, j) + \frac{C_d(i, j)}{\Delta x}H_y^{n+1/2}(i+1/2, j) - \frac{C_d(i, j)}{\Delta x}H_y^{n+1/2}(i-1/2, j) - \frac{C_d(i, j)}{\Delta y}H_x^{n+1/2}(i, j+1/2) + \frac{C_d(i, j)}{\Delta y}H_x^{n+1/2}(i, j-1/2). \quad (6.10.31)$$

The three-dimensional FDTD unit cell was shown in Fig. 6.13. It is now modified into a compact two-dimensional unit cell which is shown in Fig. 6.39. Thus, the compact FDTD numerical process is operated purely based on the two-dimensional cells, and the closed waveguide requires modeling only of its cross section. As shown in Fig. 6.39, the compact two-dimensional cells can be only half the thickness of a cell along the z-propagation axis. As discussed in Section 6.5, the numerical computation is restricted to a single layer of cells to model the closed waveguide cross section. Significant reduction in both computer memory and running time are achieved using this procedure. It should be noted that only the real field quantities are involved in the conventional three-dimensional time domain algorithm. However, the electric and magnetic field components involved in the previous compact time-stepping algorithm are complex spectral quantities having both magnitude and phase variation. Compact time-stepping Eqs. (6.10.26)–(6.10.31) are rewritten in the following with further simplification suitable for numerical computation:

$$E^n(i, j, k+1/2) - E^n(i, j, k-1/2) = \left[e^{-j\beta\frac{\Delta z}{2}} - e^{j\beta\frac{\Delta z}{2}}\right]E^n(i, j, k) = -2\sin\left(\frac{\beta\Delta z}{2}\right)e^{j\frac{(\pi-\beta\Delta z)}{2}}E^n(i, j, k-1/2) \quad (6.10.32)$$

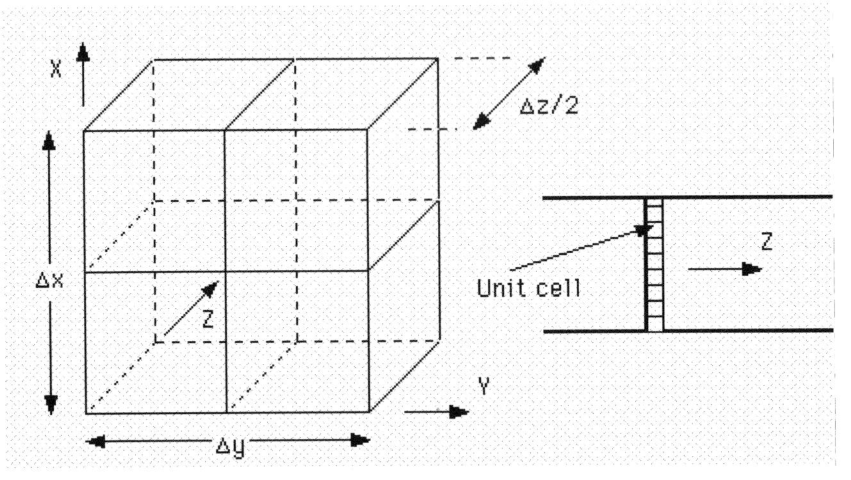

FIGURE 6.39
Modeling of waveguide cross section by compact FDTD cells.

$$H^n(i,j,k+1/2) - H^n(i,j,k-1/2) = \left[e^{-j\beta\frac{\Delta z}{2}} - e^{j\beta\frac{\Delta z}{2}}\right]H^n(i,j,k)$$
$$= -2\sin\left(\frac{\beta\Delta z}{2}\right)e^{j\frac{(\pi-\beta\Delta z)}{2}}H^n(i,j,k-1/2). \qquad (6.10.33)$$

On substituting the previous simplification into the compact FDTD time-stepping algorithm, given by Eqs. (6.10.26)–(6.10.31), the following alternative two-dimensional, finite-difference, time-stepping expressions are obtained; for magentic field components,

$$\begin{aligned} H_x^{n+1/2}(i,j+1/2) = {} & C_a(i,j+1/2)H_x^{n-1/2}(i,j+1/2) \\ & - \frac{C_b(i,j+1/2)}{\Delta z}2\sin\left(\frac{\beta\Delta z}{2}\right)e^{j\frac{(\pi-\beta\Delta z)}{2}}E_y^n(i,j+1/2) \\ & - \frac{C_b(i,j+1/2)}{\Delta y}E_z^n(i,j+1) + \frac{C_b(i,j+1/2)}{\Delta y}E_z^n(i,j) \end{aligned} \qquad (6.10.34)$$

$$\begin{aligned} H_y^{n+1/2}(i+1/2,j) = {} & C_a(i+1/2,j)H_y^{n-1/2}(i+1/2,j) \\ & + \frac{C_b(i+1/2,j)}{\Delta z}2\sin\left(\frac{\beta\Delta z}{2}\right)e^{j\frac{(\pi-\beta\Delta z)}{2}}E_x^n(i+1/2,j) \\ & + \frac{C_b(i+1/2,j)}{\Delta x}E_z^n(i+1,j) - \frac{C_b(i+1/2,j)}{\Delta x}E_z^n(i,j) \end{aligned} \qquad (6.10.35)$$

$$\begin{aligned} H_z^{n+1/2}(i+1/2,j+1/2) = {} & C_a(i+1/2,j+1/2)H_z^{n-1/2}(i+1/2,j+1/2) \\ & + \frac{C_b(i+1/2,j+1/2)}{\Delta y}E_x^n(i+1/2,j+1) \\ & - \frac{C_b(i+1/2,j+1/2)}{\Delta y}E_x^n(i+1/2,j) \\ & - \frac{C_b(i+1/2,j+1/2)}{\Delta x}E_y^n(i+1,j+1/2) \\ & + \frac{C_b(i+1/2,j+1/2)}{\Delta x}E_y^n(i,j+1/2), \end{aligned} \qquad (6.10.36)$$

and for electric field components,

$$\begin{aligned} E_x^{n+1}(i+1/2,j) = {} & C_c(i+1/2,j)E_x^n(i+1/2,j) \\ & + \frac{C_d(i+1/2,j)}{\Delta z}2\sin\left(\frac{\beta\Delta z}{2}\right)e^{j\frac{(\pi+\beta\Delta z)}{2}}H_y^{n+1/2}(i+1/2,j) \\ & + \frac{C_d(i+1/2,j)}{\Delta y}H_z^{n+1/2}(i+1/2,j+1/2) \\ & - \frac{C_d(i+1/2,j)}{\Delta y}H_z^{n+1/2}(i+1/2,j-1/2) \end{aligned} \qquad (6.10.37)$$

$$E_y^{n+1}(i, j+1/2) = C_c(i, j+1/2)E_y^n(i, j+1/2) - \frac{C_d(i, j+1/2)}{\Delta z} 2\sin\left(\frac{\beta \Delta z}{2}\right) e^{j\frac{(\pi+\beta\Delta z)}{2}} H_x^{n+1/2}(i, j+1/2) - \frac{C_d(i, j+1/2)}{\Delta x} H_z^{n+1/2}(i+1/2, j+1/2) + \frac{C_d(i, j+1/2)}{\Delta x} H_z^{n+1/2}(i-1/2, j+1/2) \quad (6.10.38)$$

$$E_z^{n+1}(i, j) = C_c(i, j)E_z^n(i, j) + \frac{C_d(i, j)}{\Delta x} H_y^{n+1/2}(i+1/2, j) - \frac{C_d(i, j)}{\Delta x} H_y^{n+1/2}(i-1/2, j) - \frac{C_d(i, j)}{\Delta y} H_x^{n+1/2}(i, j+1/2) + \frac{C_d(i, j)}{\Delta y} H_x^{n+1/2}(i, j-1/2). \quad (6.10.39)$$

Equations (6.10.34)–(6.10.39) are the compact, two-dimensional, time-stepping algorithms for the calculation of electric and magnetic field components. With the field solutions obtained based on the space–time difference equations, various field components for a given time step can be calculated iteratively. The new value of a field component at any cell point depends only on its value in the previous time step and the previous values of the components of other field at the adjacent spatial points. Hence, at any given time step, the computation of a field component will proceed one point at a time. The choice of space increments Δx, Δy, and Δz and time increment Δt is dictated by accuracy and algorithm stability, respectively. The development of the stability condition is similar to that of the conventional three-dimensional FDTD scheme for a homogeneous medium. On a rectangular lattice system with cell sizes Δx, Δy, and Δz and time increment Δt, the finite-difference algorithm is stable provided

$$\left(\frac{v\Delta t}{\Delta x}\right)^2 \sin^2\left(\frac{k_x \Delta x}{2}\right) + \left(\frac{v\Delta t}{\Delta y}\right)^2 \sin^2\left(\frac{k_y \Delta y}{2}\right) + \left(\frac{v\Delta t}{\Delta z}\right)^2 \sin^2\left(\frac{k_z \Delta z}{2}\right) \leq 1. \quad (6.10.40)$$

The wave numbers in the previous equation are unknown and their resultant maximum value is equal to 1. In the worst-case scenario, one can replace all the $\sin^2$ terms with unity. To ensure the stability of the computed electric and magnetic field components, Δt is chosen to satisfy the inequality for the three-dimensional model (Eq. 6.10.17).

For the compact two-dimensional FDTD technique, the derivative along the z-coordinate axis is calculated exactly, and $k_z = \beta$ is the preassigned value. Since the width of axial discretization Δz used in the computation is generally very small compared with the propagation wavelength, by taking the limit of $\Delta z \geq 0$ Eq. (6.10.40) becomes

$$\left(\frac{v\Delta t}{\Delta x}\right)^2 \sin^2\left(\frac{k_x \Delta x}{2}\right) + \left(\frac{v\Delta t}{\Delta y}\right)^2 \sin^2\left(\frac{k_y \Delta y}{2}\right) + \left(\frac{v\Delta t \beta}{2}\right)^2 \leq 1. \quad (6.10.41)$$

Then, the stability condition for the compact two-dimensional FDTD scheme is obtained as

$$\Delta t \leq \frac{1}{c_{\max}} \left[\frac{1}{\Delta x^2} + \frac{1}{\Delta y^2} + \frac{\beta^2}{4}\right]^{-1/2}. \quad (6.10.42)$$

Furthermore, referring to Eq. (6.10.18), the numerical dispersion relationship for the compact two-dimensional FDTD technique reduces to

$$\left(\frac{v\Delta t}{\Delta x}\right)^2 \sin^2\left(\frac{k_x \Delta x}{2}\right) + \left(\frac{v\Delta t}{\Delta y}\right)^2 \sin^2\left(\frac{k_y \Delta y}{2}\right) + \left(\frac{v\Delta t\beta}{2}\right)^2 = \sin^2\left(\frac{\omega \Delta t}{2}\right). \quad (6.10.43)$$

6.10.3 Evaluation of Dispersion Characteristics

Compared to the modeling of the waveguide geometry by the traditional three-dimensional FDTD technique, the modeling by the compact two-dimensional method is simplified. Only the cross section need to be considered in the modeling process as shown in Fig. 6.39. In addition, there is no need for an absorbing termination condition as is required with the traditional modeling. As in the conventional approach, the electric and magnetic field components used in computation are the real quantities. However, they are spectral complex field quantities in the compact method. For determining the mode dispersion relationship, a numerical procedure similar to the one followed for determining mode cutoff frequencies can be adopted. In fact, it is interesting to observe that the case of mode cutoff condition is a special case of the modeling discussed previously with the axial propagation constant $\beta = 0$. Numerical results of the mode dispersion characteristics for a hollow rectangular waveguide and a shielded microstrip line are calculated to demonstrate applicability of the compact algorithm.

Referring to Tables 6.1 and 6.2, let us consider the cross section of a hollow, air-filled, rectangular closed waveguide. The side lengths along the x- and y-coordinate variables are $b = 0.01$ m and $a = 0.02$ m. The rectangular cross section is modeled by discretizing the interior region into structured square cells, $\Delta x = \Delta y = \Delta$, with $i = 25$ cells along the x-axis and $j = 50$ cells along the y-axis. For this spatial discretization, the spatial increment is $\Delta = 0.0004$ m and the corresponding time-stepping increment (in seconds) is $\Delta t = \Delta/\sqrt{3}c_{\text{max}}$, where $c_{\text{max}} = 3 \times 10^8$ m/s is the wave velocity in the rectangular waveguide region with the waveguide region having free-space medium parameters.

Initially, there is no excitation inside the waveguide region and all the spatial magnetic and electric field components are assumed to be zero. Similar to the procedure discussed in Section 6.5 for determining mode cutoff frequencies, the rectangular waveguide is excited by injecting an impulse one-fourth Gaussian field, (Eq. 6.5.31). For determining the TE mode dispersion relationship, the initial arbitrary excitation field is assigned to the axial component of the magnetic field near the center of the lattice grids at time instant $t = 0$. Since the axial component of magnetic field H_z is excited initially, only the TE modes are triggered accordingly. Even though all six electric and magnetic field components are coupled, the axial component of the electric field is not excited at all inside the closed waveguide region. The axial propagation constant β is now chosen to be 235.095 as an input data, and time-stepping Eqs. (6.10.34)–(6.10.39) are implemented. As a function of time, the axial component of the magnetic field inside the rectangular waveguide at the grid point ($i = 5$, $j = 5$) is recorded for a total of $N = 8192$ time steps. Furthermore, based on Eq. (6.5.32), Fourier transform of the time domain data is performed with a frequency resolution Δf selected as $1/(N\Delta t)$. It is interesting to observe that the frequency spectrum of the axial component of the magnetic field has an isolated spike-like peak distribution; the spikes are in fact the mode propagation frequencies that the hollow rectangular waveguide can support. The mode cutoff frequencies for the hollow rectangular waveguide for $\beta = 0$ for the TE polarization are tabulated in Table 6.2. Corresponding to the propagation constant $\beta = 235.095$, the mode excitation frequency may be viewed as the frequency obtained due to a gradual shift of the mode cutoff frequency spectrum. In fact, the distribution of waveguide mode numbers

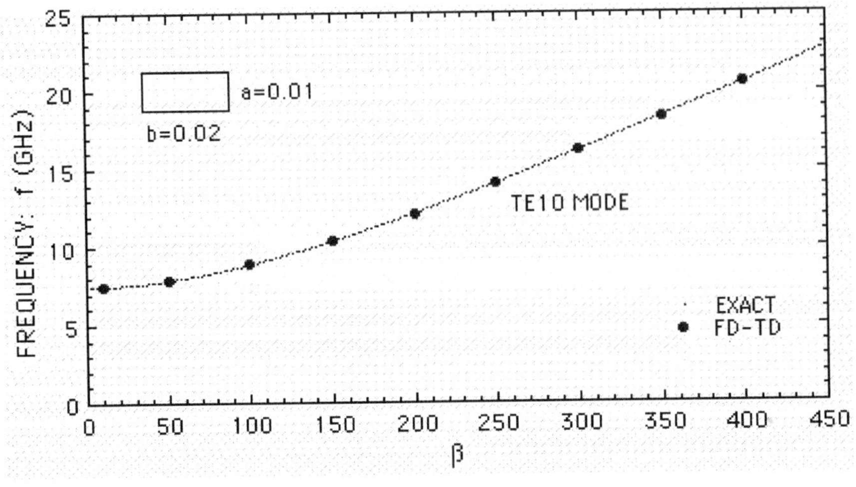

FIGURE 6.40
Mode dispersion characteristics of the hollow rectangular waveguide (TE10 mode).

is well preserved. This frequency shift can be read directly from the isolated spike-like peak distribution.

The previous numerical procedure is repeated by choosing different values of axial propagation constant β. The corresponding frequencies of excitation from the first peaks of the frequency spectrum are recorded to obtain the dispersion characteristics for the TE10 mode. Figure 6.40 shows the result of dispersion characteristics for the TE10 mode compared with the analytical result. The analytical dispersion relationship for the TE10 mode can be calculated using the expression

$$\beta_{10} = k\left[1 - \left(\frac{\pi}{ka}\right)^2\right]^{1/2} \tag{6.10.44}$$

$$k^2 = \omega^2 \mu_0 \epsilon_0. \tag{6.10.45}$$

There exists an excellent agreement between the data obtained by the analytical method and those from the FDTD technique. This procedure is now extended for the case of a shielded microstrip line. Table 6.4 shows the geometry of a single microstrip line within a square cross-section waveguide geometry. The thickness of the microstrip line is specified in terms of a single cell of the finite-difference grids. The dielectric slab is located horizontally at the bottom portion of the waveguide. The relative permittivity and relative permeability of the dielectric slab are $\epsilon_r = 8.875$ and $\mu_r = 1$. The side lengths for the closed square waveguide along the x and y coordinates are $a = b = 0.0127$ m, and the dielectric-filled region is of height $d = 0.00127$ m. For this case, various mode cutoff frequencies for TM and TE to z polarizations are shown in Table 6.4.

The mode dispersion characteristics of the first dominant propagating mode of the shielded single microstrip line are now investigated. The waveguide square cross section is modeled by discretizing the region into structured square cells, with $i = 40$ cells along the x-axis and $j = 40$ cells along the y-axis. For this discretization, the spatial increment is $\Delta = 0.0003175$ m and the corresponding time-stepping increment is $\Delta t = \Delta/\sqrt{3}c_{\max}$, where $c_{\max} = 3 \times 10^8$ m/s is the wave velocity in the rectangular waveguide region with the waveguide region having free-space medium parameters.

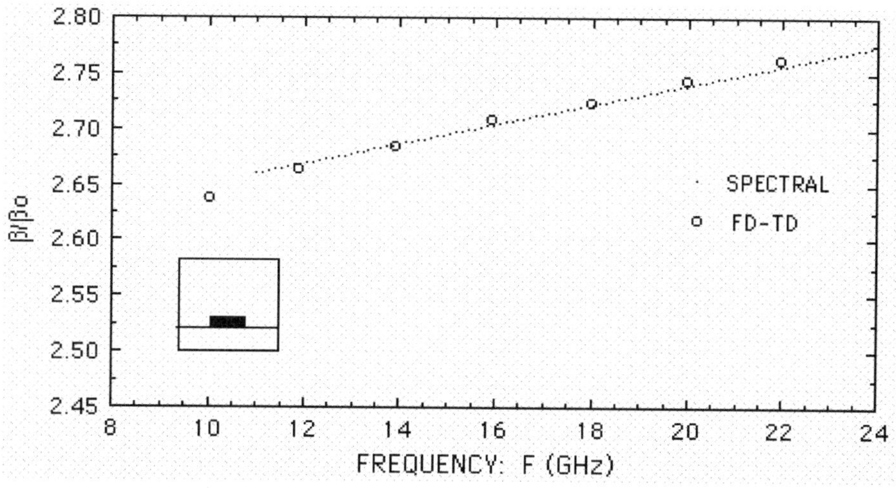

FIGURE 6.41
Mode dispersion characteristics of shielded single microstrip line (first dominant mode).

The Gaussian excitation field (Eq. 6.5.31) is again chosen as the initial excitation and is assigned to the axial electric field component E_z at $t = 0$ for the computation of the dominant mode dispersion relationship. The input value of the propagation constant β is varied in the range from 500 to 1400 in increments of 100 to determine the corresponding frequencies that can be supported by the shielded single microstrip line. As a function of the number of time steps, the transverse component of magnetic field H_x inside the waveguide at the spatial grid point $(i = 3, j = 18)$ is recorded for a total of $N = 4096$ time steps. The Fourier transform of the time domain data based on Eq. (6.5.32) is performed. Figure 6.41 shows the result of dispersion characteristics for the first dominant propagating mode of the shielded single microstrip line.

The previous two case studies correspond to the closed type waveguides. In fact, the compact two-dimensional FDTD algorithm is also applicable for open-type waveguide structures, such as dielectric and fiber optical waveguides. The compact two-dimensional algorithm requires additional treatment of its four computational truncation boundaries in terms of the absorbing radiation boundary conditions.

6.11 FDTD FORMULATION FOR THREE-DIMENSIONAL OPEN-REGION PROBLEMS

An efficient time domain solution based on the compact two-dimensional FDTD technique for determining the mode dispersion characteristics of the closed type structure was presented in the previous section. For modeling the electromagnetic scattering, coupling, and interaction characteristics of the open-type structure, a complete three-dimensional full-wave analysis is required. In the following, a detailed solution procedure based on the three-dimensional FDTD technique is presented for the analysis of arbitrary shape and composition material structure with an external plane wave excitation.

In general, the six scalar full-wave Maxwell's equations (Eqs. 6.5.1–6.5.6) must be considered. Using the rectangular coordinate system, the vector fields are given by the three electric field components E_x, E_y, and E_z and the three magnetic field components H_x, H_y, and H_z. Also, note that all the field components are functions of the space variables x, y, and z and the time varible t. Similar to that for the two-dimensional FDTD models (see Fig. 6.23), the three-dimensional FDTD field computation zone is divided into an interior total-field region and an exterior scattered-field region that are separated by a close connecting surface. The arbitrarily shaped complex material structure is embedded in the total-field zone. The interior total-field region and the exterior free-space region are directly coupled, resulting in electromagnetic wave scattering and interaction in all directions. The three-dimensional material geometry that is embedded in the total-field zone can be conducting, homogeneous or inhomogeneous, and isotropic or anisotropic with dielectric and magnetic material properties. The embedded material structure is excited with an external time-varying incident field. In addition to the electromagnetic field distribution in the material region, one must also consider the distribution of scattered fields in the complete surrounding medium. The three-dimensional FDTD modeling needs to simulate the complete volumetric material region, including the complete surrounding volumetric region that is infinite in extent. However, the practical computational memory resource is limited. Thus, the FDTD numerical solution technique is limited to the mapping the infinite domain onto a finite rectangular volumetric domain by constructing an artificial boundary condition to simulate an infinite domain. This artificial boundary condition imposes the so-called three-dimensional absorbing radiation boundary condition on the various electric and magnetic field components at the six numerical truncation surfaces, which ideally require no reflection of the outgoing waves from the truncated boundary planes back into the computation domain of interest. The iterative time-stepping numerical algorithms developed in Section 6.10 (Eqs. 6.10.5–6.10.10) are incorporated directly for both the total-field zone and the scattered-field zone independently in the finite rectangular volumetric domain. At the close connecting virtual surface of the total-field and the scattered-field zones, the external plane wave incident field excitation is systematically simulated similar to the one-dimensional and two-dimensional cases by appropriately resolving any inconsistency in the direct application of the FDTD time-stepping algorithms.

6.11.1 Second-Order Radiation Boundary Condition

The absorbing boundary condition discussed in Section 6.6.2 for the two-dimensional case can be easily extended with appropriate modification to study three-dimensional electromagnetic scattering and coupling. Again, the operator can be derived from the dispersion relationship for the free-space wave equation. The three-dimensional wave equation in rectangular system is

$$\frac{\partial^2 U}{\partial x^2} + \frac{\partial^2 U}{\partial y^2} + \frac{\partial^2 U}{\partial z^2} - \frac{1}{c^2}\frac{\partial^2 U}{\partial t^2} = 0, \tag{6.11.1}$$

where U is a scalar field component and c is the wave phase velocity. In rectangular coordinates, the plane wave time-harmonic solution to the previous wave equation is given by

$$U(x, y, z, t) = e^{j(\omega t - k_x x - k_y y - k_z z)}. \tag{6.11.2}$$

On substituting into the wave equation (Eq. 6.11.1), the following dispersion relation is obtained:

$$\frac{\omega^2}{c^2} = k_x^2 + k_y^2 + k_z^2. \tag{6.11.3}$$

For a specific time-harmonic frequency ω, solving for k_x in Eq. (6.11.3) yields

$$k_x = \pm k(1 - s^2)^{1/2} \quad \text{for } -1 \le s \le 1 \tag{6.11.4}$$

$$s^2 = \left(\frac{k_y}{k}\right)^2 + \left(\frac{k_z}{k}\right)^2. \tag{6.11.5}$$

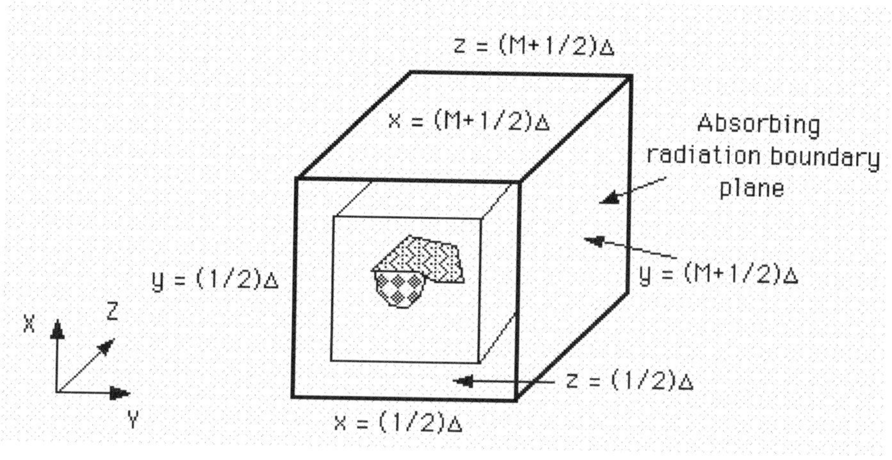

FIGURE 6.42
Three-dimensional region with six truncation planes modeled with an artificial absorbing radiation boundary.

In Eq. (6.11.4), the $\pm$ sign corresponds respectively to the dispersion relation of x- and $-x$-directed traveling waves, respectively. The dispersion relation cannot be used directly to construct a practical one-way wave equation because of the radical term. An approximation to the radical term of Eq. (6.11.4) is made to obtain a one-way wave equation by utilizing the first two terms of the Taylor series approximation:

$$(1 - s^2)^{1/2} \approx 1 - 1/2\, s^2. \tag{6.11.6}$$

Using the previous two-term approximation, a partial differential equation which has the traveling-wave solution can be constructed. Referring to Fig. 6.42, corresponding to the six surfaces of the truncated absorbing boundary, the following relationships are obtained:

$$\frac{1}{c}\frac{\partial^2 U}{\partial x \partial t} = \frac{1}{c^2}\frac{\partial^2 U}{\partial t^2} - \frac{1}{2}\frac{\partial^2 U}{\partial y^2} - \frac{1}{2}\frac{\partial^2 U}{\partial z^2} \tag{6.11.7}$$

$$\frac{1}{c}\frac{\partial^2 U}{\partial x \partial t} = -\frac{1}{c^2}\frac{\partial^2 U}{\partial t^2} + \frac{1}{2}\frac{\partial^2 U}{\partial y^2} + \frac{1}{2}\frac{\partial^2 U}{\partial z^2} \tag{6.11.8}$$

$$\frac{1}{c}\frac{\partial^2 U}{\partial y \partial t} = \frac{1}{c^2}\frac{\partial^2 U}{\partial t^2} - \frac{1}{2}\frac{\partial^2 U}{\partial z^2} - \frac{1}{2}\frac{\partial^2 U}{\partial x^2} \tag{6.11.9}$$

$$\frac{1}{c}\frac{\partial^2 U}{\partial y \partial t} = -\frac{1}{c^2}\frac{\partial^2 U}{\partial t^2} + \frac{1}{2}\frac{\partial^2 U}{\partial z^2} + \frac{1}{2}\frac{\partial^2 U}{\partial x^2} \tag{6.11.10}$$

$$\frac{1}{c}\frac{\partial^2 U}{\partial z \partial t} = \frac{1}{c^2}\frac{\partial^2 U}{\partial t^2} - \frac{1}{2}\frac{\partial^2 U}{\partial x^2} - \frac{1}{2}\frac{\partial^2 U}{\partial y^2} \tag{6.11.11}$$

$$\frac{1}{c}\frac{\partial^2 U}{\partial z \partial t} = -\frac{1}{c^2}\frac{\partial^2 U}{\partial t^2} + \frac{1}{2}\frac{\partial^2 U}{\partial x^2} + \frac{1}{2}\frac{\partial^2 U}{\partial y^2} \tag{6.11.12}$$

at $x = 0$, $x = h$, $y = 0$, $y = h$, $z = 0$, and $z = h$ boundaries, respectively.

For FDTD simulation of the vector Maxwell's equations, the radiation boundary conditions given by Eqs. (6.11.7)–(6.11.12) are applied to the individual Cartesian components of the vector electric and magnetic fields that are located tangential to the six lattice boundary planes.

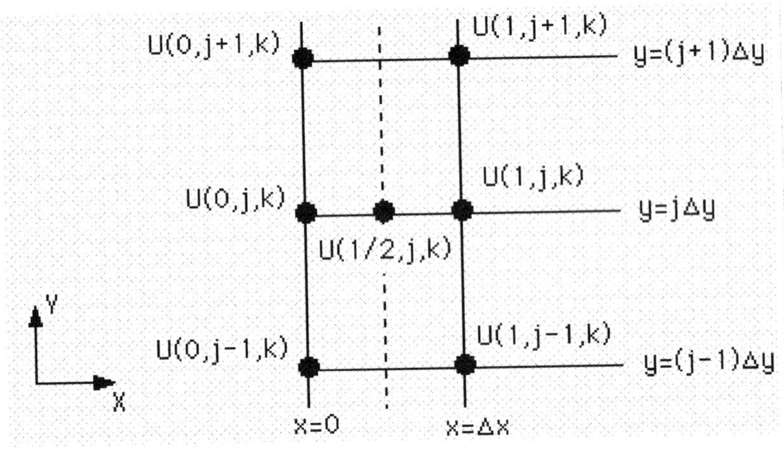

FIGURE 6.43
Position of field component near the $x = 0$ truncation boundary plane.

Equation (6.11.7), which is valid at $x = 0$ boundary, will allow traveling waves only in the $-x$ direction. To apply Eqs. (6.11.7)–(6.11.12) numerically, a simple finite-difference scheme is introduced to derive a radiation boundary time-stepping algorithm (Fig. 6.43). This scheme involves implementing the partial derivatives in Eq. (6.11.7) as numerical central differences of the auxiliary U component, $U^n(1/2, j, k)$, located one-half space cell from the grid boundary plane at $(0, j, k)$. The mixed partial x and t derivatives on the left-hand side of Eq. (6.11.7) are implemented as

$$\frac{\partial^2 U^n(1/2, j, k)}{\partial x \partial t} = \frac{\left[\frac{\partial U^{n+1}(1/2,j,k)}{\partial x} - \frac{\partial U^{n-1}(1/2,j,k)}{\partial x}\right]}{2\Delta t} \tag{6.11.13}$$

$$= \left[\frac{U^{n+1}(1, j, k) - U^{n+1}(0, j, k)}{2\Delta t \Delta x}\right] - \left[\frac{U^{n-1}(1, j, k) - U^{n-1}(0, j, k)}{2\Delta t \Delta x}\right]. \tag{6.11.14}$$

The second-order partial t derivatives on the right-hand side of Eq. (6.11.7) are implemented as an average of time derivatives at the adjacent points $(0, j, k)$ and $(1, j, k)$, given by

$$\frac{\partial^2 U^n(1/2, j, k)}{\partial t^2} = \frac{1}{2}\left[\frac{\partial^2 U^n(0, j, k)}{\partial t^2} + \frac{\partial^2 U^n(1, j, k)}{\partial t^2}\right] \tag{6.11.15}$$

$$= \left[\frac{U^{n+1}(0, j, k) - 2U^n(0, j, k) + U^{n-1}(0, j, k)}{2\Delta t^2}\right] + \left[\frac{U^{n+1}(1, j, k) - 2U^n(1, j, k) + U^{n-1}(1, j, k)}{2\Delta t^2}\right]. \tag{6.11.16}$$

Then, the second-order partial y derivatives on the right-hand side of Eq. (6.11.7) are implemented as an average of y derivatives at the adjacent points $(0, j, k)$ and $(1, j, k)$, given by

$$\frac{\partial^2 U^n(1/2, j, k)}{\partial y^2} = \frac{1}{2}\left[\frac{\partial^2 U^n(0, j, k)}{\partial y^2} + \frac{\partial^2 U^n(1, j, k)}{\partial y^2}\right] \tag{6.11.17}$$

$$= \left[\frac{U^n(0, j+1, k) - 2U^n(0, j, k) + U^n(0, j-1, k)}{2\Delta y^2}\right] + \left[\frac{U^n(1, j+1, k) - 2U^n(1, j, k) + U^n(1, j-1, k)}{2\Delta y^2}\right]. \tag{6.11.18}$$

Furthermore, the second-order partial z derivatives on the right-hand side of Eq. (6.11.7) are implemented as an average of z derivatives at the adjacent points $(0, j, k)$ and $(1, j, k)$, given by

$$\frac{\partial^2 U^n(1/2, j, k)}{\partial z^2} = \frac{1}{2}\left[\frac{\partial^2 U^n(0, j, k)}{\partial z^2} + \frac{\partial^2 U^n(1, j, k)}{\partial z^2}\right] \tag{6.11.19}$$

$$= \left[\frac{U^n(0, j, k+1) - 2U^n(0, j, k) + U^n(0, j, k-1)}{2\Delta z^2}\right] + \left[\frac{U^n(1, j, k+1) - 2U^n(1, j, k) + U^n(1, j, k-1)}{2\Delta z^2}\right]. \tag{6.11.20}$$

On substituting Eqs. (6.11.14), (6.11.16), (6.11.18), and (6.11.20) into Eq. (6.11.7) and solving for $U^{n+1}(0, j, k)$, the following time-stepping algorithm for the component U along the $x = 0$ grid boundary plane is obtained which implements the desired second-order radiation boundary condition:

$$\begin{aligned} U^{n+1}(0, j, k) = {} & -U^{n-1}(1, j, k) + \frac{c\Delta t - \Delta x}{c\Delta t + \Delta x}[U^{n+1}(1, j, k) + U^{n-1}(0, j, k)] \\ & + \frac{2\Delta x}{c\Delta t + \Delta x}[U^n(0, j, k) + U^n(1, j, k)] \\ & + \frac{(c\Delta t)^2 \Delta x}{2\Delta y^2(c\Delta t + \Delta x)}[U^n(0, j+1, k) - 2U^n(0, j, k) \\ & + U^n(0, j-1, k) + U^n(1, j+1, k) - 2U^n(1, j, k) + U^n(1, j-1, k)] \\ & + \frac{(c\Delta t)^2 \Delta x}{2\Delta z^2(c\Delta t + \Delta x)}[U^n(0, j, k+1) - 2U^n(0, j, k) \\ & + U^n(0, j, k-1) + U^n(1, j, k+1) - 2U^n(1, j, k) + U^n(1, j, k-1)]. \end{aligned} \tag{6.11.21}$$

Using the same procedure and following the details of two-dimensional development, the radiation boundary conditions at the remaining five grid truncation boundary planes can be obtained by utilizing Eqs. (6.11.8)–(6.11.12). From the previous equation, it is clear that the difference scheme cannot be implemented for the field components located at the grid corners since some of the necessary data used are just outside of the grid planes. In fact, it is necessary to implement a special corner radiation boundary condition by utilizing available field data in the grid planes. The previous radiation boundary conditions are, in fact, extensively utilized in the study of three-dimensional electromagnetic propagation, scattering, and coupling with the external time domain excitation. The radiation boundary conditions based on the second-order operator cannot absorb all outgoing waves completely for a very large number of time steps. The current FDTD codes suppress spurious reflections of outward-propagating wave analogs by 95–99%. In order to improve numerical accuracy at the truncated planes, a slight modification using the error-correcting superabsorption algorithm can be used. This error-controlling algorithm is not restricted for any particular boundary absorbing operator. The previous algorithm based on the second-order boundary operator can be extended, using higher order operators which are suitable for the time domain analysis of the open-type complex material geometry.

6.11.2 Three-Dimensional Plane Wave Source Condition

The three-dimensional numerical algorithms developed earlier have a linear dependence on the components of electric and magnetic fields. In fact, they can be applied with equal validity to the incident-field vector components, the scattered-field vector components, or the total field consisting of the sum of incident- and scattered-field vector components. Referring to Fig. 6.42, the current FDTD codes utilize this property to zone the volumetric numerical space lattice into two distinct regions separated by a rectangular virtual planar surface to connect the fields in each volumetric region.

Similar to the two-dimensional case, this type of lattice zoning provides a number of algorithm features to enhance the computational flexibility and dynamic range. The connecting condition provided at the virtual surface of the inner and outer volumetric regions simultaneously generates an arbitrary plane wave in the total-field region having specified time-dependent wave form, angle of incidence, and angle of polarization. The connecting surface is transparent to the outgoing scattered field waves. As discussed earlier, the arbitrarily shaped material structure is embedded in the total-field zone where the total fields are marched in time. The required continuity of the total tangential electric and magnetic field components across the interface of staggered unit cells separating dissimilar media parameters is satisfied automatically. Furthermore, the low levels of the total field in deep shadow regions or cavities of the embedded interacting material structure are computed directly in the total-field zone.

Connecting Condition at the Interface of Total-Field and Scattered-Field Regions In the case of three-dimensional FDTD modeling, the six rectangular surfaces comprising the virtual surface-type interface of region 1 (total field) and region 2 (scattered field) contain the electric and magnetic field components, which require spatial finite difference of the various field components to advance by time stepping. As discussed in Section 6.7, for the case of three-dimensional plane wave simulation there is a problem of inconsistency.

On the region 1 side of the interface, the fields to be used in the finite-difference expression are the total fields. Similarly, on the region 2 side of the interface, the fields to be used in the finite-difference expression are the scattered fields. The problem of inconsistency near the interface can be resolved by using the values of the incident electric and magnetic field components which are assumed to be known or calculable at each lattice point. To illustrate the simulation of the plane wave source condition at the connecting interface, consider the three-dimensional FDTD Eqs. (6.10.5)–(6.10.10). Referring to Fig. 6.42, the total electric field components given by E_x^{t} and E_y^{t} lie exactly on the front interface of region 1 and region 2. Time-stepping, finite-difference Eqs. (6.10.5)–(6.10.10) are correspondingly modified to achieve consistency when applied at the front interface.

The electric field components on the front interface can be calculated by

$$E_x^{n+1}(i = i_0 + 1/2, \cdots, i_1 - 1/2,\ j = j_0, \cdots, j_1,\ k = k_0)$$

$$\begin{aligned} E_x^{n+1}(i, j, k_0) = {} & C_c(i, j, k_0)E_x^n(i, j, k_0) \\ & + \frac{C_d(i, j, k_0)}{\Delta y}\left[H_z^{n+1/2}(i, j + 1/2, k_0) - H_z^{n+1/2}(i, j - 1/2, k_0)\right] \\ & - \frac{C_d(i, j, k_0)}{\Delta z}\left[H_y^{n+1/2}(i, j, k_0 + 1/2) - H_y^{n+1/2}(i, j, k_0 - 1/2)\right] \\ & + \frac{C_d(i, j, k_0)}{\Delta z}\left[H_y^{\mathrm{i}\,n+1/2}(i, j, k_0 - 1/2)\right] \end{aligned} \tag{6.11.22}$$

and

$$E_y^{n+1}(i = i_0, \cdots, i_1,\ j = j_0 + 1/2, \cdots, j_1 - 1/2,\ k = k_0)$$

$$\begin{aligned} E_y^{n+1}(i, j, k_0) = {} & C_c(i, j, k_0)E_y^n(i, j, k_0) \\ & + \frac{C_d(i, j, k_0)}{\Delta z}\left[H_x^{n+1/2}(i, j, k_0 + 1/2) - H_x^{n+1/2}(i, j, k_0 - 1/2)\right] \\ & - \frac{C_d(i, j, k_0)}{\Delta x}\left[H_z^{n+1/2}(i + 1/2, j, k_0) - H_z^{n+1/2}(i - 1/2, j, k_0)\right] \\ & - \frac{C_d(i, j, k_0)}{\Delta z}\left[H_x^{\mathrm{i}\,n+1/2}(i, j, k_0 - 1/2)\right]. \end{aligned} \tag{6.11.23}$$

The magnetic field components outside the front interface can be calculated by

$$H_x^{n+1/2}(i = i_0, \cdots, i_1,\ j = j_0 + 1/2, \cdots, j_1 - 1/2,\ k = k_0 - 1/2)$$

$$\begin{aligned} H_x^{n+1/2}(i, j, k_0 - 1/2) = {} & C_a(i, j, k_0 - 1/2)H_x^{n-1/2}(i, j, k_0 - 1/2) \\ & + \frac{C_b(i, j, k_0 - 1/2)}{\Delta z}\left[E_y^n(i, j, k_0) - E_y^n(i, j, k_0 - 1)\right] \\ & - \frac{C_b(i, j, k_0 - 1/2)}{\Delta y}\left[E_z^n(i, j + 1/2, k_0 - 1/2)\right. \\ & \left. - E_z^n(i, j - 1/2, k_0 - 1/2)\right] \\ & - \frac{C_b(i, j, k_0 - 1/2)}{\Delta z}\left[E_y^{\mathrm{i}\,n}(i, j, k_0 - 1)\right] \end{aligned} \tag{6.11.24}$$

and

$$H_y^{n+1/2}(i = i_0 + 1/2, \cdots, i_1 - 1/2, j = j_0, \cdots, j_1, k = k_0 - 1/2)$$

$$\begin{aligned} H_y^{n+1/2}(i, j, k_0 - 1/2) = {} & C_a(i, j, k_0 - 1/2) H_y^{n-1/2}(i, j, k_0 - 1/2) \\ & + \frac{C_b(i, j, k_0 - 1/2)}{\Delta x} \big[E_z^n(i + 1/2, j, k_0 - 1/2) \\ & - E_z^n(i - 1/2, j, k_0 - 1/2)\big] \\ & - \frac{C_b(i, j, k_0 - 1/2)}{\Delta z} \big[E_x^n(i, j, k_0) - E_x^n(i, j, k_0 - 1)\big] \\ & + \frac{C_b(i, j, k_0 - 1/2)}{\Delta z} \big[E_x^{in}(i, j, k_0 - 1)\big]. \end{aligned} \tag{6.11.25}$$

The previous equations for the plane wave source simulation are valid at the front connecting interface $k = k_0$. In fact, the time stepping of the electric and magnetic field components at the left-end and right-end interface edge points must take into account the additional field components located in the scattered-field region 2. The previous equations can be repeated at the remaining five interfaces, namely, the back, top, bottom, left, and right connecting interfaces.

Calculation of the Incident Field to Implement the Connecting Condition The calculation of incident electric and magnetic field data for $\boldsymbol{E}^{\mathrm{i}}$ and $\boldsymbol{H}^{\mathrm{i}}$ is required at or near the interface surfaces of regions 1 and 2 to implement the connecting condition for the three-dimensional case. A simple approach to calculate this data for arbitrary incident angles is based on a table look-up procedure which describes the space–time behavior of the incident wave. In fact, this is similar to the two-dimensional propagation model described in Section 6.7.

Consider the FDTD simulation of an incident plane wave propagating in a three-dimensional system. The incident plane wave electric field vector and the incident magnetic field vector are always perpendicular to each other and are polarized parallel in a plane which is always perpendicular to the direction of propagation. Let

$$\boldsymbol{E}_0 = E_{0x}\boldsymbol{a}_x + E_{0y}\boldsymbol{a}_y + E_{0z}\boldsymbol{a}_z \tag{6.11.26}$$

$$\boldsymbol{H}_0 = H_{0x}\boldsymbol{a}_x + H_{0y}\boldsymbol{a}_y + H_{0z}\boldsymbol{a}_z. \tag{6.11.27}$$

The incident electric and magnetic fields based on the forward-traveling plane waves are given by

$$\boldsymbol{E}^{\mathrm{i}}(\boldsymbol{r}) = \boldsymbol{E}_0 e^{j\boldsymbol{k}\cdot\boldsymbol{r}} \tag{6.11.28}$$

$$\boldsymbol{H}^{\mathrm{i}}(\boldsymbol{r}) = \boldsymbol{H}_0 e^{j\boldsymbol{k}\cdot\boldsymbol{r}}, \tag{6.11.29}$$

and the actual direction propagation of the plane wave field is given by

$$\boldsymbol{k} = k\boldsymbol{a}_k = k(\sin\theta^{\mathrm{i}}\cos\phi^{\mathrm{i}}\boldsymbol{a}_x + \sin\theta^{\mathrm{i}}\sin\phi^{\mathrm{i}}\boldsymbol{a}_y + \cos\theta^{\mathrm{i}}\boldsymbol{a}_z), \tag{6.11.30}$$

where $(\theta^{\mathrm{i}}, \phi^{\mathrm{i}})$ are the angles of incidence in the spherical coordinate system.

Following the discussion in Section 6.7, the action of the connecting condition is to generate a plane wave originating from the bottom left corner of the region 1 and region 2 interface.

This point is designated as the local origin $\bigcirc_1$ with grid coordinates (p, s, r) and is the first FDTD electric field grid point that is contacted by the wave front of the incident field. All other grid points denoting the location of electric and magnetic vector field components in region 1 are contacted by the wave front after a certain time retardation satisfied by the field causality condition. Referring to Fig. 6.27, d is the distance in terms of the number of spatial grid cells along the wave propagation vector from the local origin $\bigcirc_1$ to a perpendicular dropped from the field vector component located on the connecting interface. It is given by

$$d = \boldsymbol{a}_k \cdot [(x - p)\boldsymbol{a}_x + (y - s)\boldsymbol{a}_y + (z - r)\boldsymbol{a}_z]. \tag{6.11.31}$$

The incident fields $\boldsymbol{E}^{\text{i}}$ and $\boldsymbol{H}^{\text{i}}$ can be calculated at any location of a field vector component using a simple analytical expression once the retardation distance d is known. However, even for the simplest incident-field space–time behavior, for example a single-frequency time-harmonic sinusoidal excitation required for the steady-state analysis, the total amount of computer calculation to implement the connecting condition may be very large, especially if large three-dimensional region 1 and region 2 interfaces are involved. To reduce the computational burden, an approach based on a lookup table can be employed.

6.12 NEAR- TO FAR-FIELD TRANSFORMATION FOR THE THREE-DIMENSIONAL CASE

The electromagnetic scattering by an arbitrary material structure can be determined if the induced surface electric and magnetic currents are known. The induced currents can be integrated to calculate the near fields and the far fields. However, the body surface may have a complex shape or may be loaded with complex materials in some way so as to make implementation of the needed surface integrals difficult and, in fact, this is a unique problem for each type of scatterer. Similar to the two-dimensional case, a useful alternative would be to obtain scattered field information from off-surface near-field data rather than from the on-surface induced current data. The near-field phasor information could be computed using the FDTD method, which can easily account for the complexities introduced by the arbitrary scatterer shape and composition. Furthermore, the near-field data could be integrated on a close virtual surface which completely surrounds the scatterer of interest.

A provision of a well-defined scattered-field region in the three-dimensional FDTD lattice permits the natural near- to far-field transformation, as illustrated in Fig. 6.44. The tangential phasor scattered electric and magnetic fields are computed via the FDTD along a virtual close surface S located in the outer (scattered-field) region of the spatially discretized lattice. Since the electric and magnetic fields are staggered, the tangential electric field components are located on the virtual close surface. The tangential magnetic field components located half a cell on either side of the virtual close surface are averaged and placed on the contour. By utilizing the basic theory of electromagnetic equivalence principle, the tangential electric and magnetic field components are converted into the corresponding tangential equivalent magnetic current and equivalent electric current on all six virtual planes. These equivalent currents are weighted by the free-space Green's function and then integrated over the close surface S to obtain the far-field response. The current three-dimensional FDTD codes can provide the complete magnitude and phase far-field scattered patterns, monostatic and bistatic RCS for a particular sinusoidal steady state, and plane wave external illumination. The near-field virtual integration surface has a fixed rectangular shape, independent of the shape of the material composition of the enclosed structure being modeled.

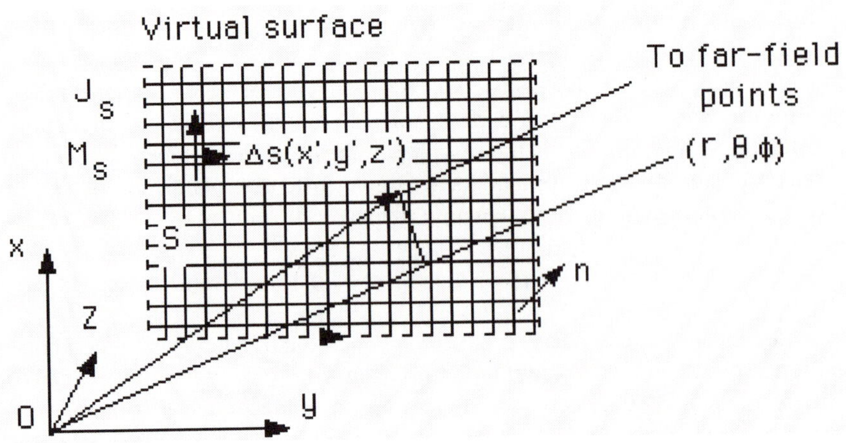

FIGURE 6.44
Near- to far-field integration surface in scattered-field region.

The electric and magnetic field distribution in the far-field region can be obtained by making use of the large argument approximation for the free-space Green's function. In the far-field region, using the spherical coordinate system, the scattered electric and magnetic fields behave as pure spherical waves propagating in the radial direction and are directly related through the medium intrinsic impedance.

The equivalent electric current and the equivalent magnetic current along the virtual close rectangular surface S are given by

$$\boldsymbol{J}_s(\boldsymbol{r}, \omega) = \boldsymbol{a}_n \times \boldsymbol{H}(\boldsymbol{r}) \tag{6.12.1}$$

$$\boldsymbol{M}_s(\boldsymbol{r}, \omega) = \boldsymbol{E}(\boldsymbol{r}) \times \boldsymbol{a}_n. \tag{6.12.2}$$

On the surface S the normal unit vector in Eqs. (6.12.1) and (6.12.2) is appropriately defined depending on which virtual integration surface is considered. Due to the equivalent electric current and the magnetic current, in the far-field region the total magnetic vector potential and the total electric vector potential are given by the surface integrals

$$\boldsymbol{A}(\boldsymbol{r}) = \frac{e^{-jkr}}{4\pi r} \int\int_S \boldsymbol{J}_s(\boldsymbol{r}') e^{jkr\cos\xi}\, ds' \tag{6.12.3}$$

$$\boldsymbol{F}(\boldsymbol{r}) = \frac{e^{-jkr}}{4\pi r} \int\int_S \boldsymbol{M}_s(\boldsymbol{r}') e^{jkr\cos\xi}\, ds' \tag{6.12.4}$$

$$r\cos\xi = \sin\theta(x'\cos\phi + y'\sin\phi) + z'\cos\theta. \tag{6.12.5}$$

For θ and ϕ polarizations, the following scattered far-field expressions are obtained for the three-dimensional case:

$$E_\theta(\theta, \phi) = -jk[\eta A_\theta + F_\phi] \tag{6.12.6}$$

$$E_\phi(\theta, \phi) = -jk[\eta A_\phi - F_\theta] \tag{6.12.7}$$

$$A_\theta = A_x \cos\theta\cos\phi + A_y \cos\theta\sin\phi - A_z \sin\theta \tag{6.12.8}$$

$$F_\theta = F_x \cos\theta \cos\phi + F_y \cos\theta \sin\phi - F_z \sin\theta \tag{6.12.9}$$

$$A_\phi = -A_x \sin\phi + A_y \cos\phi \tag{6.12.10}$$

$$F_\phi = -F_x \sin\phi + F_y \cos\phi. \tag{6.12.11}$$

The surface integrals are converted into sum over various FDTD surface patches on the virtual surface S. The calculation is carried out on all the six surfaces. The bistatic RCS of the three-dimension scatterer can be calculated using Eqs. (6.12.6) and (6.12.7) and is given by the ratio of scattered-field power to incident-field power:

$$\mathrm{RCS}(\theta,\phi) = \lim_{r\to\infty} 4\pi r^2 \left| \frac{E^{\mathrm{s}}_\theta(\theta,\phi) + E^{\mathrm{s}}_\phi(\theta,\phi)}{E^{\mathrm{i}}_\theta(\theta,\phi) + E^{\mathrm{i}}_\phi(\theta,\phi)} \right|^2. \tag{6.12.12}$$

6.12.1 RCS of a Flat-Plate Scatterer

The thick flat-plate scatterer discussed in this section is composed of a graphite-fiber/epoxy matrix composite material. To emphasize the material anisotropy, all graphite fibers are aligned parallel to each other. For the purposes of FDTD numerical modeling, in the direction parallel to the graphite fibers the relative permittivity ϵ_r is assumed to be 1.0 and the conductivity is assumed to be 3.7×10^7 S/m. In Fig. 6.45, the dimensions of the thick flat-plate scatterer are $10 \times 0.625 \times 30$ cm^3.

The thick flat-plate scatterer is externally excited by a time-harmonic plane wave with the incident electric field polarized along the z-coordinate direction. At the excitation frequency of 1 GHz, the free-space wavelength is $\lambda_0 = 30$ cm. For the FDTD three-dimensional model, a uniform cell size of 0.625 cm ($\lambda_0/48$) is selected. The flat plate is simulated by $16 \times 1 \times 48$ grid cells embedded in an overall grid lattice of $32 \times 18 \times 64$ cells with 221,184 unknown field components. The plane wave incident field is time stepped up to six cycles, equivalent to 576 time steps. Following the procedure discussed previously, the tangential phasor scattered electric and magnetic fields are computed via the FDTD along a virtual close surface S located in the

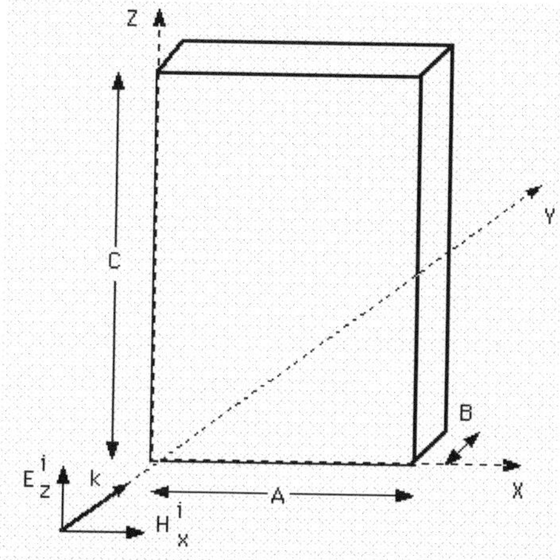

FIGURE 6.45
Geometry of a thick plate scatterer.

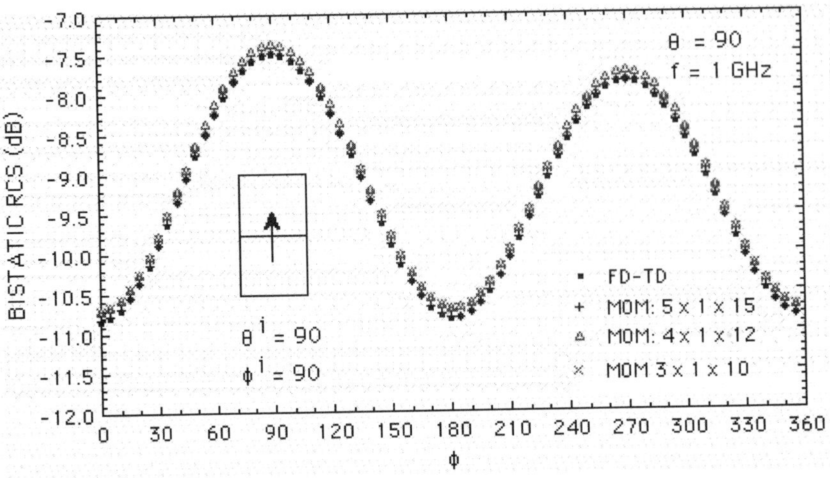

FIGURE 6.46
Bistatic radar cross section of a thick plate.

outer scattered-field region of the spatially discretized lattice. The near- to far-field transformation is applied on the rectangular virtual close surface located three cells from the truncated radiation boundary. Figure 6.46 shows a plot of the bistatic RCS for angles of incidence $\theta^{\mathrm{i}} = 90^{\circ}$, $\phi^{\mathrm{i}} = 90^{\circ}$. The FDTD results are now compared with respect to the solution obtained by MoM surface patching analysis of the electric field integral equation. The flat plate is spanned by dividing the scatterer into $3 \times 1 \times 10$ divisions, resulting in a total of 172 uniform triangular surface patches and a 258×258 complex valued system matrix. The unknown induced electric currents are determined first and the bistatic RCS is then calculated. The results obtained from the FDTD and MoM solution compare very well for all observation angles in the horizontal plane.

6.13 COMPUTER RESOURCES AND MODELING IMPLICATIONS

The FDTD places very strong demands on the computing system in regard to memory space, bandwidth, and floating point speed. It is desirable to understand current trends in computer technology to deduce whether the FDTD technique can take advantage of these trends to provide practical software for the engineering electromagnetic analysis and design.

The floating point operation (flop) rate of single-processor computers has steadily increased during the past 10 years. However, none of the current technologies provide much more than a 10-fold increase in the single-processor floating point rates beyond those currently available. System problems such as circuit packing and cooling, combined with propagation delays due to the finite velocity of information transmission along any path, may combine to limit the single-processor computation rate regardless of the speed of the logic elements employed. Therefore, attention has increasingly turning to the use of vector processing and parallel processing to surmount this practical engineering boundary.

Vector processing is applicable to those electromagnetic modeling problems which can be structured to have explicit mathematical operations that are repetitively performed on large sets of numbers. The FDTD technique is one example of a computer technique which inherently qualifies for vector processing. In the limit, one processor could be assigned to each field component or unit cell of the space lattice, and all would then work in parallel under ideal conditions.

Large electromagnetic models increasingly require more computer memory than is available in the central processor. However, the auxiliary memory currently available, normally a magnetic disk, is not very satisfactory for two reasons. First, access times for new data may be milliseconds for a disk versus microseconds for semiconductor central memory. Second, data flow rates, such as reading or writing to memory, may be several hundred thousand words per second for a high-speed disk versus more than 10 million words per second for semiconductor memory. As a result, models which require substantial disk memory resources may devote most of the program execution time to accessing various points in the disk and transferring data to and from the disk. Thus, the models are often constrained by the input/output machine boundary. Thus, even a superfast central processor unit (CPU) or an array of CPUs can be reduced in efficiency if either insufficient high-speed central memory is provided or the input/output bandwidth to auxiliary memory is inadequate.

Currently, it appears that the suitable algorithms for the electromagnetic analysis and design of electrically large structures are all of the iterative type and avoid the need for the generation and inversion of large-scale matrices. Iterative methods can usually be structured with fully explicit algorithms that are suitable for vector processing. The impact of direct modeling at the 50- to 100-wavelength scale in three dimension will be profound. This capability will permit a detailed, optimized design of the electromagnetic wave scattering and penetration characteristics of structures of realistic size and complexity. Wave interaction phenomenology will be observable in the time domain and in the frequency domain and in the near field and interior of the structures as well as in the far field and exterior of the structures. Large antennas of complex shapes and material composition will be the subject of direct modeling as will the dielectric, magnetic, and optics structures, including integrated-optics structures. The available information about near- and far-field regions in the time domain and frequency domain will permit numerical testing of competing schemes for modeling inverse scattering and target identification problems.

Indeed, the objects that can be modeled will be so complex (spanning hundreds of millions of three-dimensional geometry points) that the advanced computer-aided design and manufacturing techniques will be needed to permit effective geometry generation of internal and external structural details. Furthermore, the output data will be so voluminous for both near and far fields that innovative data display techniques, such as motion picture or television picture generation directly from the computed output, will be needed to display the wave motion near-critical interaction region, diffraction, near-zone magnitude and phase, and far-zone magnitude and phase.

For current and future research, the FDTD technique is ideal for the latest generations of vector-processing and parallel-processing supercomputers and related workstations. A highly promising research area involves development of an optimized FDTD software for such machines along with practically efficient geometry-generation algorithms. In the next 10 years, such optimized FDTD software should permit modeling of the three-dimensional structures spanning 50–100 wavelengths or more with computer program running times measured in the minutes per look angle. Application of the direct FDTD modeling to large, realistic problems involving RCS, microwave penetration, antennas, integrated optics, and inverse scattering would then be possible.

6.14 CONCLUDING REMARKS

In this chapter, I have presented a detailed overview of the FDTD method to solve electromagentic scattering/interaction problems in both closed and open regions. Although the treatment presented in this work is of a tutorial nature with emphasis on understanding the technique, in reality the method has been applied to extremely complex situtations. As a result, this method continues to receive much attention from the scientific and the engineering community.

ACKNOWLEDGMENTS

The author expresses sincere thanks to his professional colleagues Dr. Shen Chen of Harris Corporation, Prof. Gregory Kriegsmann of New Jersey Institute of Technology, Prof. Allen Taflove of Northwestern University, and Prof. S. M. Rao of Auburn University for their many interesting discussions in the research programs and contributions dealing with the development and successive validation of the finite-difference time domain technique.

BIBLIOGRAPHY

[1] T. Itoh (Ed.), *Numerical Techniques for Microwave and Millimeter-Wave Passive Structures*, Wiley, New York, 1989.
[2] K. Umashankar and A. Taflove, *Computational Electromagnetics—Integral Equation Method*, Artech House, Boston, 1993.
[3] R. Sorrentino (Ed.), *Numerical Methods for Passive Microwave and Millimeter Wave Structures*, Selected Preprint Series, IEEE Press, New York, 1989.
[4] B. E. A. Saleh and M. C. Teich, *Fundamentals of Photonics*, Wiley, New York, 1991.
[5] R. Mittra and S. W. Lee, *Analytical Techniques in the Theory of Guided Waves*, Macmillan, New York, 1971.
[6] C. E. Baum, "On the Singularity Expansion Method for the Solution of Electromagnetic Interaction Problems," *Interaction Notes*, Note 88, KAFB, Albuquerque, NM, 1971.
[7] K. Umashankar and A. Taflove, "A Novel Method to Analyze Electromagnetic Scattering of Complex Objects," *IEEE Trans. Electromag. Compat.*, vol. 24, pp. 397–405, 1982.
[8] A. Taflove and K. Umashankar, "Radar Cross Section of General Three-Dimensional Scatterers," *IEEE Trans. Electromag. Compat.*, vol. 25, pp. 433–440, 1983.
[9] K. Umashankar and A. Taflove, *Analytical Models for Electromagnetic Scattering*, Electromag. Sci. Div., Rome Air Development Center, Hanscom AFB, Final Rep. No. RADC-TR-85-87, Contract F19268-82-C-0140, Boston, 1985.
[10] A. Taflove and K. Umashankar, "Review of FDTD Numerical Modeling of Electromagnetic Wave Scattering and Radar Cross Section," *Proc. IEEE*, vol. 77, pp. 682–699, 1989.
[11] T. G. Jurgens, A. Taflove, K. Umashankar, and T. G. Moore, "Finite Difference Time Domain Modeling of Curved Surfaces," *IEEE Trans. Antennas Propagat.*, vol. 40, pp. 357–366, 1992.
[12] K. Umashankar, S. Chaudhari, and A. Taflove, "Finite Difference Time Domain Formulation of an Inverse Scattering Scheme for Remote Sensing of One Dimensional Inhomogeneous Lossy Layered Media, Part I—One Dimensional Case," *J. Electromag. Waves Appl.*, vol. 8, pp. 489–508, 1994.
[13] M. Fusco, "FDTD Algorithm in Curvilinear Coordinates," *IEEE Trans. Antennas Propagat.*, vol. 38, pp. 76–89, 1990.
[14] I. Kim and W. J. R. Hoeffer, "A Local Mesh Refinement Algorithm for the Time Domain Finite Difference Method Using Maxwell's Curl Equations," *IEEE Trans. Microwave Theory Tech.*, vol. 38, pp. 812–815, 1990.
[15] R. Holland, "Finite Difference Solution of Maxwell's Equation in Generalized Non-Orthogonal Coordinates," *IEEE Trans. Nucl. Sci.*, vol. 30, pp. 4589–4591, 1983.
[16] G. Mur, "Absorbing Boundary Conditions for the Finite-Difference Approximation of the Time Domain Electromagnetic Field Equations," *IEEE Trans. Electromag. Compat.*, vol. 23, pp. 377–382, 1981.
[17] L. N. Trefethen and L. Halpern, "Well-Posedness of One-Wave Wave Equations and Absorbing Boundary Conditions," *Math. Comput.*, vol. 47, pp. 421–435, 1986.
[18] K. K. Mei and J. Fang, "Superabsorption—A Method to Improve Absorbing Boundary Conditions," *IEEE Trans. Antennas Propagat.*, vol. 40, pp. 1001–1010, 1992.

[19] S. M. Rao, *Electromagnetic Scattering and Radiation of Arbitrarily Shaped Surfaces by Triangular Patch Modeling*, PhD dissertation, University of Mississippi, Oxford, 1980.

[20] K. Umashankar, "Numerical Analysis of Electromagnetic Scattering and Interaction Based on Integral Equation and Method of Moments Technique," *Wave Motion*, pp. 1–33, 1989.

[21] B. Beker and K. Umashankar, "Analysis of Electromagnetic Scattering by Arbitrary Shaped Anisotropic Objects Using Combined Field Surface Integral Equation Formulation," *J. Electromag.*, vol. 9, pp. 215–230, 1989.

[22] S. W. Chen, *Characterization of Open and Close Type Waveguide Structures Based on Finite Difference Time Domain Method*, PhD dissertation, Department of Electrical Engineering and Computer Science, University of Illinois at Chicago, Chicago, 1994.

[23] X. Zhang, J. Fang, K. K. Mei, and Y. Liu, "Calculations of the Dispersive Characteristics of Microstrips by the Time-Domain Finite Difference Method," *IEEE Trans. Microwave Theory Tech.*, vol. 36, pp. 263–267, 1988.

[24] J. B. Davis and C. A. Muilwyk, "Numerical Solution of Uniform Hollow Waveguides with Boundaries of Arbitrary Shapes," *Proc. IEEE*, vol. 113, pp. 277–284, 1966.

[25] K. Bierwirth, N. Schulz, and F. Arndt, "Finite Difference Analysis of Rectangular Dielectric Waveguide Structure," *IEEE Trans. Microwave Theory Tech.*, vol. 34, pp. 1104–1114, 1986.

[26] A. Christ and H. Hartnagel, "Three-Dimensional Finite Difference Method for the Analysis of Microwave Device Embedding," *IEEE Trans. Microwave Theory Tech.*, vol. 35, pp. 688–696, 1987.

[27] N. Morita, "A Method Extending the Boundary Condition for Analyzing Guided Modes of Dielectric Waveguides of Arbitrary Cross-Sectional Shape," *IEEE Trans. Microwave Theory Tech.*, vol. 30, pp. 6–11, 1982.

[28] S. Xiao, R. Vahldieck, and H. Jin, "Full-Wave Analysis of Guided Wave Structures Using a Novel 2-D FDTD," *IEEE Microwave Guided Wave Lett.*, vol. 2, pp. 165–167, 1992.

[29] A. C. Cangellaris, "Numerical Stability and Numerical Dispersion of a Compact 2-D/FDTD Method Used for the Dispersion Analysis of Waveguides," *IEEE Microwave Guided Wave Lett.*, vol. 3, pp. 3–5, 1993.

[30] J. C. Olivier, "On the Synthesis of Exact Free Space Absorbing Boundary Conditions for the Finite-Difference Time Domain Method," *IEEE Trans. Antennas Propagat.*, vol. 40, pp. 456–460, 1992.

[31] F. Moglie, T. Rozzi, P. Marcozzi, and A. Schiavoni, "A New Termination Condition for the Application of FDTD Techniques to Discontinuity Problem in Close Homogeneous Waveguide," *IEEE Microwave Guided Wave Lett.*, vol. 2, pp. 475–477, 1992.

[32] Z. Bi, K. Wu, C. Wu, and J. Litva, "A Dispersive Boundary Condition for Microstrip Component Analysis Using the FDTD Method," *IEEE Trans. Microwave Theory Tech.*, vol. 40, pp. 774–777, 1992.

CHAPTER 7

Transmission Line Modeling Method

G. K. GOTHARD
Harris Corporation

F. J. GERMAN
Raytheon Systems Company

The relationship between electromagnetic field quantities and voltage and current on transmission lines is well-known. As early as 1944, Kron [1] developed an equivalent transmission line circuit to model free space. This relationship has been used in the solution of electromagnetic (EM) problems [2]. In 1971, Johns and Beurle [3] used these principles along with the theory of pulse propagation on transmission lines to develop a numerical technique compatible with computational techniques for the solution of two-dimensional electromagnetic problems. This method was called the transmission line matrix method and is now called the transmission line modeling (TLM) method. This method has been used for many applications such as the analysis of various waveguides [4, 5], calculation of induced surface currents on objects due to an incident electromagnetic pulse [6], and radar cross sections due to plane wave incidence [7].

The TLM method provides one with a reliable numerical procedure with which to analyze a wide variety of electromagnetic problems. The TLM method allows one to model arbitrary geometries which may be composed of anisotropic inhomogeneous materials. This method is not limited by the frequency of interest but only by the available computer resources. The TLM method allows one to define a problem as a set of boundary conditions and material constituents in a two- or three-dimensional Cartesian mesh. An initial excitation is defined and impulses are propagated throughout the mesh using the scattering theory of simple TEM transmission lines. The method makes available all the spatial field components at the center of each node in the mesh. Since the TLM is a transient analysis, all the necessary information about the frequency response for frequencies from 0 Hz up to some upper limit imposed by the TLM mesh is available. In this chapter, we focus on using the TLM to solve electromagnetic problems. However, the TLM had been used in other applications, such as diffusion [8] and acoustics [9] problems.

In this chapter, we present the TLM starting at a basic level to outline the fundamental TLM concepts. Next, we present more advanced topics and list numerous references for readers desiring more in-depth detail. The goal of this chapter is to enable someone with a basic understanding of field theory and transmission line concepts to be able not only to understand the TLM method but also to develop a working TLM algorithm and work physical problems. Additionally, through the use of extensive references, those experienced with the TLM method may be able to make use of this work for more advanced topics.

7.1 THE TWO-DIMENSIONAL TLM

Two-dimensional TLM calculations are typically carried out on a Cartesian grid, or mesh, composed of individual rectangular cells, or nodes, linked together. Each TLM node contains two shunt or series-connected ideal transmission lines, and an iterative technique is employed to generate a solution to the problems modeled. In this section, the rationale behind utilizing transmission line networks to solve EM problems is presented along with the methodology.

First, the time domain wave equation defining EM wave propagation in two-dimensional space for the transverse magnetic (TM) to z case is outlined, followed by the time domain wave equation defining propagation on two-dimensional shunt (parallel)-connected ideal transmission lines. Next, we show that both wave equations are identical provided certain equalities are enforced. Scattering matrix theory is outlined and applied to the TLM shunt node, and mesh generation, excitation, and truncation is highlighted. After outlining the rules governing spatial discretization, we show how to obtain field output from the TLM pulses. Finally, the series TLM node and an outline for generating a two-dimensional TLM algorithm are presented.

7.1.1 Time Domain Wave Equation

Consider Maxwells' equations in a source-free region for nondispersive media. First, we examine Faradays' law, given by

$$\nabla \times \boldsymbol{E} = \frac{-\partial \boldsymbol{B}}{\partial t} = -\mu \frac{\partial \boldsymbol{H}}{\partial t}, \tag{7.1.1}$$

where $\boldsymbol{E}$ is the electric field, $\boldsymbol{B}$ is the magnetic flux density, μ is the permeability, and $\boldsymbol{H}$ is the magnetic field. Expanding in Cartesian coordinates, we have

$$\begin{aligned}
(\nabla \times E)_x &= \frac{\partial E_z}{\partial y} - \frac{\partial E_y}{\partial z} = -\mu \frac{\partial H_x}{\partial t} \\
(\nabla \times E)_y &= \frac{\partial E_x}{\partial z} - \frac{\partial E_z}{\partial x} = -\mu \frac{\partial H_y}{\partial t} \\
(\nabla \times E)_z &= \frac{\partial E_y}{\partial x} - \frac{\partial E_x}{\partial y} = -\mu \frac{\partial H_z}{\partial t}.
\end{aligned} \tag{7.1.2}$$

Examining Ampere's law under the same conditions,

$$\nabla \times \boldsymbol{H} = \frac{\partial \boldsymbol{D}}{\partial t} = \epsilon \frac{\partial \boldsymbol{E}}{\partial t}, \tag{7.1.3}$$

where $\mathbf{D}$ is the electric flux density and ϵ is the permittivity. Again, expanding in Cartesian coordinates we have

$$\begin{aligned}
(\nabla \times H)_x &= \frac{\partial H_z}{\partial y} - \frac{\partial H_y}{\partial z} = \epsilon \frac{\partial E_x}{\partial t} \\
(\nabla \times H)_y &= \frac{\partial H_x}{\partial z} - \frac{\partial H_z}{\partial x} = \epsilon \frac{\partial E_y}{\partial t} \\
(\nabla \times H)_z &= \frac{\partial H_y}{\partial x} - \frac{\partial H_x}{\partial y} = \epsilon \frac{\partial E_z}{\partial t}.
\end{aligned} \tag{7.1.4}$$

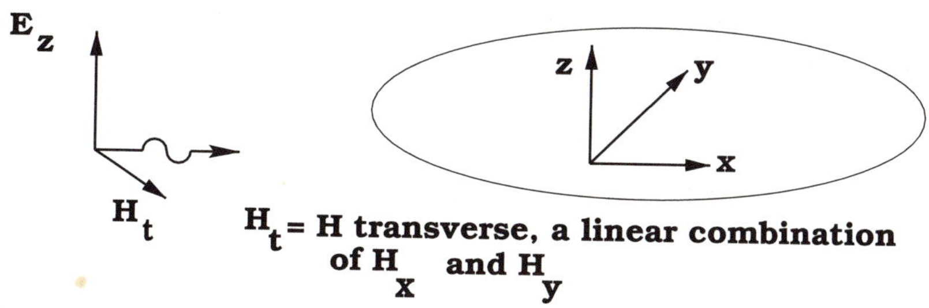

FIGURE 7.1
A section of two-dimensional space for an arbitrary two-dimensional problem involving the TM to z case. For this example, a field E_z, H_t, ($H_t = H_{\text{transverse}}$, a linear combination of H_x and H_y and problem dependent) is incident upon an infinite cylinder.

Equations (7.1.2) and (7.1.4) are the three-dimensional expansions of Eq. (7.1.1) and (7.1.3), respectively. We wish to simplify them to the two-dimensional transverse magnetic (TM) to z case (Fig. 7.1). For the TM to z case, the following conditions are true:

$$E_x = E_y = H_z = 0$$
$$\frac{\partial}{\partial z}(E, H) = 0.$$

Applying the previous conditions to Eqs. (7.1.2) and (7.1.4) yields, after simplifying,

$$\frac{\partial E_z}{\partial y} = -\mu \frac{\partial H_x}{\partial t} \tag{7.1.5}$$

$$\frac{\partial E_z}{\partial x} = \mu \frac{\partial H_y}{\partial t} \tag{7.1.6}$$

$$\frac{-\partial H_x}{\partial y} + \frac{\partial H_y}{\partial x} = \epsilon \frac{\partial E_z}{\partial t}. \tag{7.1.7}$$

Taking the derivative of Eq. (7.1.5) with respect to y, the derivative of Eq. (7.1.6) with respect to x, and the derivative of Eq. (7.1.7) with respect to t and combining yields, after simplifying,

$$\frac{\partial^2 E_z}{\partial x^2} + \frac{\partial^2 E_z}{\partial y^2} = \mu\epsilon \frac{\partial^2 E_z}{\partial t^2}, \tag{7.1.8}$$

where Eq. (7.1.8) is the two-dimensional TM to z free-space wave equation. In the next section, the time-domain wave equation defining wave propagation on ideal shunt-connected transmission lines is derived.

7.1.2 Time Domain Transmission Line Equation

Consider a section of transmission line of length Δl with an inductance L in Henrys (H) per meter and a capacitance C in Farads (F) per meter as shown in Fig. 7.2. Assume the transmission line is a lossless, nondispersive (Z_o does not vary with frequency), ideal transmission line and set the values of C and L such that $Z_o = (L/C)^{1/2} = 377\ \Omega$. The "T" equivalent circuit of this transmission line is shown in Fig. 7.3, in which the distributed inductance and capacitance of the transmission line are represented as lumped elements. In order to obtain a two-dimensional

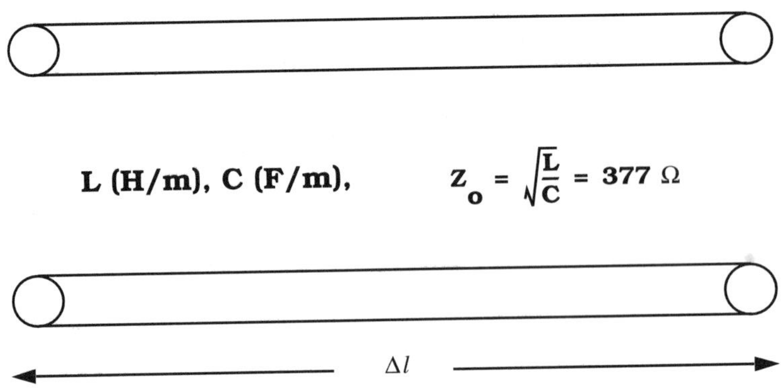

FIGURE 7.2
A section of ideal transmission line of length Δl with an inductance L(H/m) and capacitance C(F/m).

circuit, we will use the shunt or parallel combination of Fig. 7.3 as is illustrated in Fig. 7.4, which shows the equivalent T circuit of the shunt T-line model. Note that the currents I_1, I_2, I_3, and I_4 may be defined as

$$I_1 = I_y\left(y - \frac{\Delta l}{2}, t\right), \qquad I_3 = -I_y\left(y + \frac{\Delta l}{2}, t\right) \tag{7.1.9}$$

$$I_2 = I_x\left(x - \frac{\Delta l}{2}, t\right), \qquad I_4 = -I_x\left(x + \frac{\Delta l}{2}, t\right) \tag{7.1.10}$$

Figure 7.5 shows the x-directed branch of the equivalent circuit shown in Fig. 7.4. Applying Kirchoff's voltage law (KVL) about the loop of Fig. 7.5 yields

$$-V_z\left(x - \frac{\Delta l}{2}, t\right) + L\frac{\Delta l}{2}\frac{\partial I_x}{\partial t} + L\frac{\Delta l}{2}\frac{\partial I_x}{\partial t} + V_z\left(x + \frac{\Delta l}{2}, t\right) = 0, \tag{7.1.11}$$

which simplifies to

$$\frac{V_z(x + (\Delta l/2), t) - V_z(x - (\Delta l/2), t)}{\Delta l} = -L\frac{\partial I_x}{\partial t}. \tag{7.1.12}$$

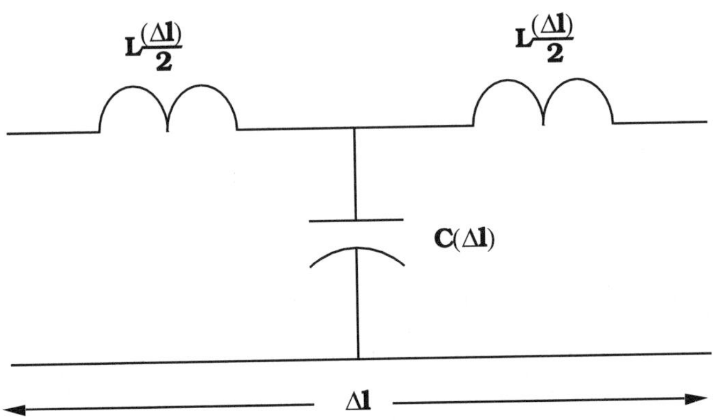

FIGURE 7.3
The T equivalent circuit model of the transmission line in Fig. 7.2. Notice that L and C are per unit length parameters and therefore, to derive lumped circuit elements, they are multiplied by the length of the lines, Δl.

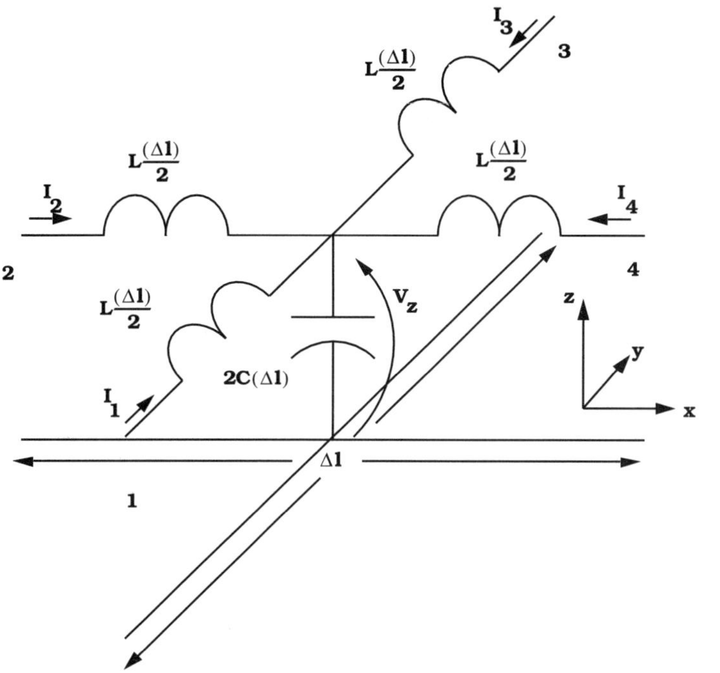

FIGURE 7.4
The T equivalent circuit model of the shunt transmission lines of Fig. 7.3, with current I_1, I_2, I_3, and voltage V_z defined in the figure.

Allowing $\Delta l = \Delta x$, in the limit as $\Delta l \to 0$, Eq. (7.1.12) becomes

$$\frac{\partial V_z}{\partial x} = -L\frac{\partial I_x}{\partial t}. \tag{7.1.13}$$

Similarly, allowing $\Delta l = \Delta y$ and applying KVL about the y-directed loop of the circuit shown

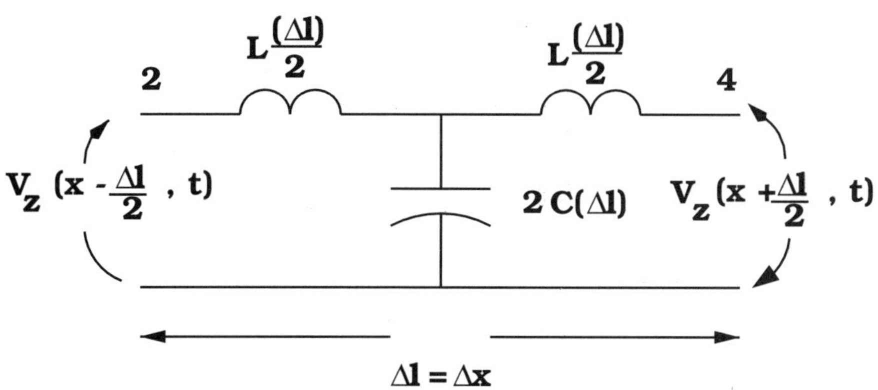

FIGURE 7.5
The x-directed branch of the "T" equivalent circuit model shown in Fig. 7.4.

in Fig. 7.4 leads to

$$\frac{\partial V_z}{\partial y} = -L\frac{\partial I_y}{\partial t}. \tag{7.1.14}$$

Applying Kirchoffs' current law at the center of the circuit representation in Fig. 7.4 yields

$$\begin{aligned} I_x\left(x+\frac{\Delta l}{2},t\right) - I_x\left(x-\frac{\Delta l}{2},t\right) + I_y\left(y+\frac{\Delta l}{2},t\right) \\ -I_y\left(y-\frac{\Delta l}{2},t\right) + 2C\Delta l\frac{\partial V_z}{\partial t} = 0. \end{aligned} \tag{7.1.15}$$

Rewriting Eq. (7.1.15) yields

$$\frac{I_x(x+(\Delta l/2),t) - I_x(x-(\Delta l/2),t)}{\Delta l} + \frac{I_y(y+(\Delta l/2),t) - I_y(y-(\Delta l/2),t)}{\Delta l} = -2C\frac{\partial V_z}{\partial t}. \tag{7.1.16}$$

In the limit as $\Delta l \to 0$ (and $\Delta l = \Delta x = \Delta y$), Eq. (7.1.16) becomes

$$\frac{\partial I_x}{\partial x} + \frac{\partial I_y}{\partial y} = -2C\frac{\partial V_z}{\partial t}. \tag{7.1.17}$$

Differentiating Eq. (7.1.13) with respect to x, Eq. (7.1.14) with respect to y, and Eq. (7.1.17) with respect to t yields, after combining and simplifying,

$$\frac{\partial^2 V_z}{\partial x^2} + \frac{\partial^2 V_z}{\partial y^2} = 2CL\frac{\partial^2 V_z}{\partial t^2}, \tag{7.1.18}$$

where Eq. (7.1.18) is the shunt transmission line wave equation. Next, we discuss the equivalences between the free-space and transmission line wave equations.

7.1.3 Equating Maxwell's and the Circuit Equations

Equations (7.1.5)–(7.1.7), and (7.1.8) are compared with Eqs. (7.1.13), (7.1.14), (7.1.17), and (7.1.18) as follows:

$$\begin{aligned}
\frac{\partial E_z}{\partial y} &= -\mu\frac{\partial H_x}{\partial t} & \frac{\partial V_z}{\partial x} &= -L\frac{\partial I_x}{\partial t} \\
\frac{\partial E_z}{\partial x} &= \mu\frac{\partial H_y}{\partial t} & \frac{\partial V_z}{\partial y} &= -L\frac{\partial I_y}{\partial t} \\
\frac{-\partial H_x}{\partial y} + \frac{\partial H_y}{\partial x} &= \epsilon\frac{\partial E_z}{\partial t} & \frac{\partial I_x}{\partial x} + \frac{\partial I_y}{\partial y} &= -2C\frac{\partial V_z}{\partial t} \\
\frac{\partial^2 E_z}{\partial x^2} + \frac{\partial^2 E_z}{\partial y^2} &= \mu\epsilon\frac{\partial^2 E_z}{\partial t^2} & \frac{\partial^2 V_z}{\partial x^2} + \frac{\partial^2 V_z}{\partial y^2} &= 2CL\frac{\partial^2 V_z}{\partial t^2}.
\end{aligned}$$

These equations are identical provided the following equalities are enforced.

$$E_z = V_z,$$
$$H_y = -I_x,$$
$$H_x = I_y,$$
$$\mu = L,$$
$$\epsilon = 2C.$$

Therefore, solving the equivalent circuit model equations and applying the previous equalities results in a solution to Maxwell's equations. However, since the transmission line equations are just as complex as the Maxwell's equations, solving them directly is just as difficult. For obvious reasons, we would not want to actually build the transmission line network or the equivalent circuit to solve the problem. Therefore, in order to take advantage of the circuit to wave relationship, we use scattering theory and the properties of pulse propagation on ideal transmission lines to characterize the transmission lines and use this model to solve the Maxwell's equations. In the following section, we discuss scattering theory.

7.1.4 General Scattering Matrix Theory

The scattering matrix describing a microwave network system provides a complete description of the network as seen at its ports [10]. For example, consider the four-port network in Fig. 7.6. Given a knowledge of the scattering matrix associated with the network, it is unnecessary to know what components comprise the interior of the network. The scattering matrix provides the

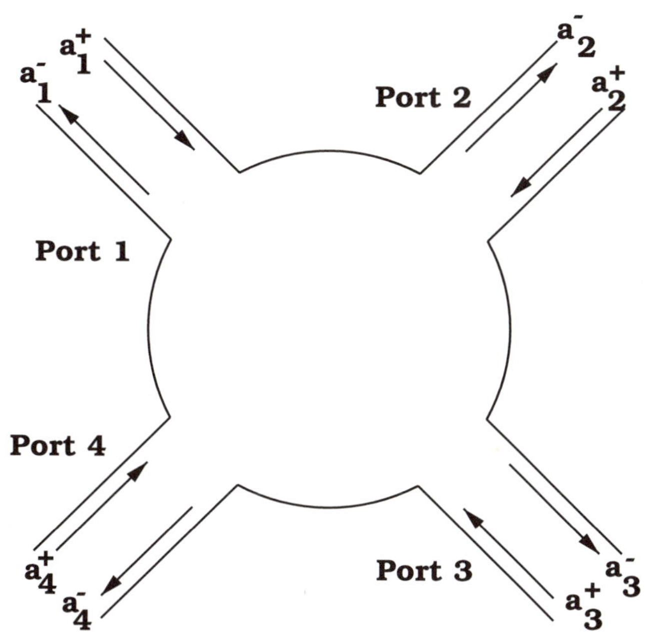

FIGURE 7.6
A general four-port microwave network, where forward (or incident) and backward (or reflected) waves are defined as $a^{+}_{1\text{–}4}$ and $a^{-}_{1\text{–}4}$, respectively.

information necessary to determine the output at all four ports given any input. This makes a generalized scattering matrix very convenient to use, especially with complex systems.

The scattering matrix which describes the microwave network shown in Fig. 7.6 is

$$\begin{bmatrix} a_1^- \\ a_2^- \\ a_3^- \\ a_4^- \end{bmatrix} = \begin{bmatrix} S_{11} & S_{12} & S_{13} & S_{14} \\ S_{21} & S_{22} & S_{23} & S_{24} \\ S_{31} & S_{32} & S_{33} & S_{34} \\ S_{41} & S_{42} & S_{43} & S_{44} \end{bmatrix} \begin{bmatrix} a_1^+ \\ a_2^+ \\ a_3^+ \\ a_4^+ \end{bmatrix}, \tag{7.1.19}$$

where $a_{1\text{–}4}^-$ are the backward-traveling waves and $a_{1\text{–}4}^+$ are the forward-traveling waves associated with ports 1–4, and S_{lm}, where $l = 1\text{–}4$ and $m = 1\text{–}4$, are the scattering parameters.

To determine the scattering parameters, a signal is injected into one port (incident or forward wave) with all other inputs at all other ports set to zero. This implies that all the ports with no signals injected are match terminated so that no reflections occur. If this procedure is followed at port 1 in Fig. 7.6, we have the following: ($a_2^+ = a_3^+ = a_4^+ = 0$), and ports 2–4 are match terminated. Then, applying the matrix of Eq. (7.1.19):

$$\begin{bmatrix} a_1^- \\ a_2^- \\ a_3^- \\ a_4^- \end{bmatrix} = \begin{bmatrix} S_{11}a_1^+ \\ S_{21}a_1^+ \\ S_{31}a_1^+ \\ S_{41}a_1^+ \end{bmatrix}, \tag{7.1.20}$$

which yields

$$S_{11} = \frac{a_1^-}{a_1^+} \quad S_{21} = \frac{a_2^-}{a_1^+} \quad S_{31} = \frac{a_3^-}{a_1^+} \quad S_{41} = \frac{a_4^-}{a_1^+}. \tag{7.1.21}$$

If this procedure is followed for ports 2–4, the scattering matrix is complete. Notice that the self terms (diagonal elements of the scattering matrix) are the reflection coefficients (quantities used to determine the amount of the incident signal reflected back into the originating port) associated with each port, and the off-diagonal terms are the transmission coefficients (quantities used to determine the amount of the incident signal transmitted from the incident port into the other ports). In Section 7.1.5, the scattering matrix theory will be applied to the shunt node of Fig. 7.4.

7.1.5 Applying Scattering Theory to the Free-Space Shunt T-Line

Consider the shunt-connected transmission line of Fig. 7.7 (the top view of Fig. 7.4). Allowing $V = a$ in Eq. (7.1.19),

$$\begin{bmatrix} V_1^- \\ V_2^- \\ V_3^- \\ V_4^- \end{bmatrix} = \begin{bmatrix} S_{11} & S_{12} & S_{13} & S_{14} \\ S_{21} & S_{22} & S_{23} & S_{24} \\ S_{31} & S_{32} & S_{33} & S_{34} \\ S_{41} & S_{42} & S_{43} & S_{44} \end{bmatrix} \begin{bmatrix} V_1^+ \\ V_2^+ \\ V_3^+ \\ V_4^+ \end{bmatrix} \tag{7.1.22}$$

or

$$[V]^- = [S][V]^+, \tag{7.1.23}$$

where $V_{1\text{–}4}^-$ are the reflected voltage pulses, $V_{1\text{–}4}^+$ are the incident voltage pulses, and S is the scattering matrix.

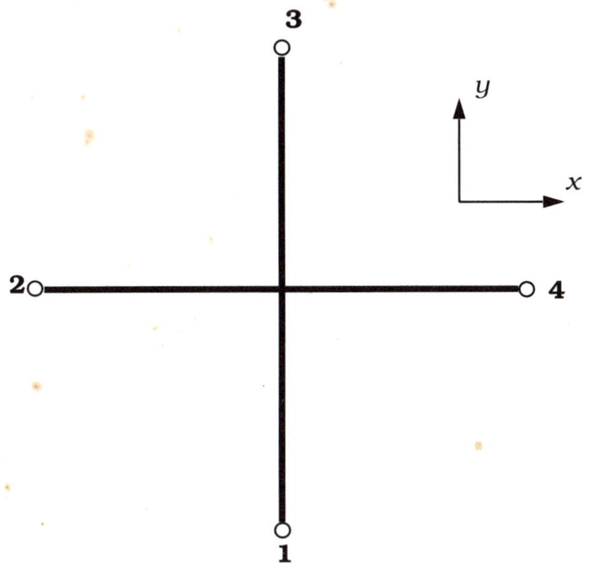

FIGURE 7.7
The shunt transmission lines of Fig. 7.4 (top view). The cell size is $\Delta l \times \Delta l$.

We know that the self terms are reflection coefficients and the rest are transmission coefficients,

$$S_{ij} = \begin{cases} i = j \rightarrow \text{reflection coefficient} \\ i \neq j \rightarrow \text{transmission coefficient,} \end{cases}$$

and therefore, from the transmission line network of Fig. 7.7, we can determine the voltage reflection and transmission coefficients from the expressions

$$S_{ij} = \begin{cases} \frac{Z_{\text{to}} - Z_{\text{from}}}{Z_{\text{to}} + Z_{\text{from}}} & i = j \\ \frac{2Z_{\text{to}}}{Z_{\text{to}} + Z_{\text{from}}} & i \neq j, \end{cases}$$

where Z_{to} is the impedance that the port in question "sees" (where a signal is going to), and Z_{from} is the impedance of the shunt link line (where the signal is coming from). For example, apply a signal to port 1 of Fig. 7.7 and match terminate ports 2–4. Since we have enforced $Z = Z_o$ on the four legs of Fig. 7.7, solving for the transmission and reflection coefficients yields

$$Z_{\text{to}} = Z_o//Z_o//Z_o = \frac{1}{\frac{1}{Z_o} + \frac{1}{Z_o} + \frac{1}{Z_o}} = \frac{Z_o}{3}$$

$$Z_{\text{from}} = Z_o$$

$$S_{11} = \frac{\frac{Z_o}{3} - Z_o}{\frac{Z_o}{3} + Z_o} = \frac{-1}{2}$$

$$S_{21} = S_{31} = S_{41} = \frac{1}{2}.$$

By symmetry, since ports 2–4 "look" like port 1, we get the following equation:

$$\begin{bmatrix} V_1 \\ V_2 \\ V_3 \\ V_4 \end{bmatrix}_{n+1}^{\mathrm{r}} = \frac{1}{2} \begin{bmatrix} -1 & 1 & 1 & 1 \\ 1 & -1 & 1 & 1 \\ 1 & 1 & -1 & 1 \\ 1 & 1 & 1 & -1 \end{bmatrix} \begin{bmatrix} V_1 \\ V_2 \\ V_3 \\ V_4 \end{bmatrix}_{n}^{\mathrm{i}}, \tag{7.1.24}$$

where the superscripts i and r refer to incident and reflected fields, respectively, and the subscript n represent time increments. Applying Eq. (7.1.24) to the TLM shunt node of Fig. 7.7 results in known reflected and transmitted voltages at a time Δt, where Δt is the time increment between applications of the scattering matrix, after any incident voltage is applied to the ports. This scattering matrix of Eq. (7.1.24) forms the basis for the TLM computational algorithm. Pulses injected into the mesh at time 0 can be propagated throughout the TLM mesh by repeated application of Eq. (7.1.24) to each node in the mesh. Once Eq. (7.1.24) is applied to a node and incident impulses have become reflected impulses, the reflected impulses become incident impulses on adjacent nodes through a connection process (described in detail in Section 7.1.7). Since appropriate application of the scattering matrix defined in Eq. (7.1.24) results in a solution to the transmission line equations from Section 7.1.2, applying the equalities of Section 7.1.5 results in a solution to Maxwell's equations. Therefore, solutions to physical electromagnetic interaction problems can be obtained through this technique.

At this point, the scattering matrix defined in Eq. (7.1.24) can only be used for homogeneous problems. Since this is limiting, it is desirable to modify the matrix to allow the modeling of inhomogeneous media. This will be addressed in the next section.

7.1.6 Modeling Inhomogeneous Lossy Media

To model lossy, inhomogeneous media with the TLM shunt node model, it is necessary to add two additional transmission line stubs to the model: one to model the relative permittivity, ϵ_{r}, by increasing the capacitance C at each node and one to model loss by absorbing power at each node.

To model the additional permittivity in the shunt TLM node, an additional open-circuit stub of length $\Delta l/2$ and variable characteristic admittance Y_{o} is added to the shunt node (as shown in Fig. 7.8). The stub adds to each node an additional lumped capacitance $CY_{\mathrm{o}}\Delta l/2$ [11–12] so that at each node

$$C_{\mathrm{tot}} = 2C\Delta l + CY_{\mathrm{o}}\frac{\Delta l}{2} = 2C\Delta l\left(1 + \frac{Y_{\mathrm{o}}}{4}\right). \tag{7.1.25}$$

The capacitive stub is chosen to be $\Delta l/2$ to maintain the time step synchronization. In other words, the reflected and transmitted pulses on the node legs as well as on the capacitive stub must arrive and leave the node center at the same time. This is because our TLM algorithm is a discrete, not continuous, algorithm, and the scattering process can only occur at discrete intervals.

To model loss in the shunt TLM node, an additional match-terminated stub is added to the shunt node. This stub adds to each node an additional characteristic admittance G_{o} (as shown in Fig. 7.8) normalized to Z_{o}. Since the stub used to model loss absorbs energy and returns none to the node, the length of the G_{o} stub can be thought of as either infinite or match terminated.

Repeating the steps of Sections 7.1.2 and 7.1.3 while including the conductivity σ_{e} in Maxwell's equations, and the additional admittance from the circuit model of the shunt transmission line,

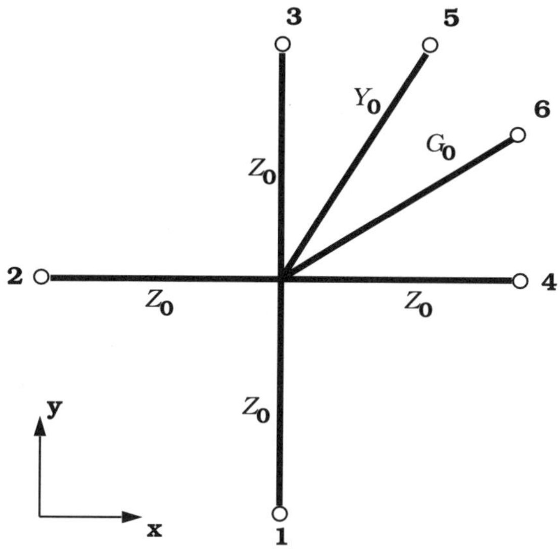

FIGURE 7.8
The TLM shunt node including additional transmission line stubs with characteristic admittances Y_o and G_o. These stubs will be used to include the permittivity ϵ_r and the conductivity σ_e. The link lines are all $\Delta l/2$.

leads to the following relationships:

$$Y_o = 4(\epsilon_r - 1) \tag{7.1.26}$$

$$G_o = \sigma_e \Delta l Z_{\text{t-line}}, \tag{7.1.27}$$

where $Z_{\text{t-line}} = Z_o$, and Z_o is the impedance of the transmission link lines. Thus, using the relationships in Eqs. (7.1.26) and (7.1.27) and the additional stubs discussed previously, permittivity and loss can be modeled at each TLM node. In order to take advantage of the new node, a new scattering matrix describing the shunt node of Fig. 7.8 needs to be derived.

Applying the scattering matrix theory of Section 7.1.4 to this new node yields the following relationship between the incident and reflected pulses at the node:

$$\begin{bmatrix} V_1 \\ V_2 \\ V_3 \\ V_4 \\ V_5 \end{bmatrix}_{n+1}^{r} = \left(\frac{1}{4+Y}\right) \times \begin{bmatrix} (-2-Y) & 2 & 2 & 2 & 2Y_o \\ 2 & (-2-Y) & 2 & 2 & 2Y_o \\ 2 & 2 & (-2-Y) & 2 & 2Y_o \\ 2 & 2 & 2 & (-2-Y) & 2Y_o \\ 2 & 2 & 2 & 2 & (Y-4) \end{bmatrix} \begin{bmatrix} V_1 \\ V_2 \\ V_3 \\ V_4 \\ V_5 \end{bmatrix}_{n}^{i}, \tag{7.1.28}$$

where $Y = (Y_o + G_o)$. Since the impulses scattered into the loss stub represent lost energy, there is no need to compute it explicitly. This new scattering matrix can now be used to model single and multiple regions of differing constitutive parameters, including lossy media, and inhomogeneous media by allowing the user to define an impedance at each node in the TLM mesh.

7.1.7 Excitation of the TLM Mesh and Metallic Boundaries

Typically, the two-dimensional TLM mesh is excited with an ideal delta impulse, or series of impulses, beginning at some time $n\Delta t$, $n = 0, 1, 2, \cdots$. The TLM mesh can be excited at any location or locations with the physical problem to be solved determining these locations. The excitation can be defined by

$$N_{\text{in}}(t) = \sum_{n=1}^{N} a_n \delta(t - n\Delta t), \tag{7.1.29}$$

where $N_{\text{in}}(t)$ is the input excitation of the TLM mesh as a function of time, a_n are weighting coefficients, the excitation is impulses at $\delta(t - n\Delta t)$, and Δt is the time increment between applications of the scattering matrix. Additionally, the initial TLM mesh excitation begins at $n = 0$.

As an example of the TLM mesh excitation, consider a section of two-dimensional space discretized for a TLM model, as shown in Fig. 7.9. A point source is excited by applying unity impulses incident at node (i, j) from all four ports, at a time $t = 0$ as shown in Fig. 7.9a. In Eq. (7.1.29), this is equivalent to allowing the weighting function a_n to be

$$a_n = \begin{cases} 1.0 & n = 0 \\ 0.0 & n > 0. \end{cases}$$

Next, the scattering matrix is applied at all nodes in the TLM mesh (however, at $t = 0$, node (i, j) is the only node with nonzero incident pulses).

After applying the scattering matrix at time $t = \Delta t$, we have reflected pulses on (i, j), as shown in Fig. 7.9b, due to the application of the scattering matrix defined in Eq. (7.1.24) as follows:

$$\begin{bmatrix} V_1 \\ V_2 \\ V_3 \\ V_4 \end{bmatrix}_1^{\mathrm{r}} = \frac{1}{2} \begin{bmatrix} -1 & 1 & 1 & 1 \\ 1 & -1 & 1 & 1 \\ 1 & 1 & -1 & 1 \\ 1 & 1 & 1 & -1 \end{bmatrix} \begin{bmatrix} 1 \\ 1 \\ 1 \\ 1 \end{bmatrix}_{\mathrm{o}}^{\mathrm{i}}.$$

Next, we must connect the TLM shunt nodes to each other by passing the reflected pulses onto the neighboring nodes at time $t = \Delta t^+$. This is called the connection process, as illustrated in Fig. 7.9c, which can be represented in closed form by

$$\begin{aligned} {}^{n+1}V_3^{\mathrm{i}}(i, j-1) &= {}^{n}V_1^{\mathrm{r}}(i, j) \\ {}^{n+1}V_4^{\mathrm{i}}(i-1, j) &= {}^{n}V_2^{\mathrm{r}}(i, j) \\ {}^{n+1}V_1^{\mathrm{i}}(i, j+1) &= {}^{n}V_3^{\mathrm{r}}(i, j) \\ {}^{n+1}V_2^{\mathrm{i}}(i+1, j) &= {}^{n}V_4^{\mathrm{r}}(i, j). \end{aligned} \tag{7.1.30}$$

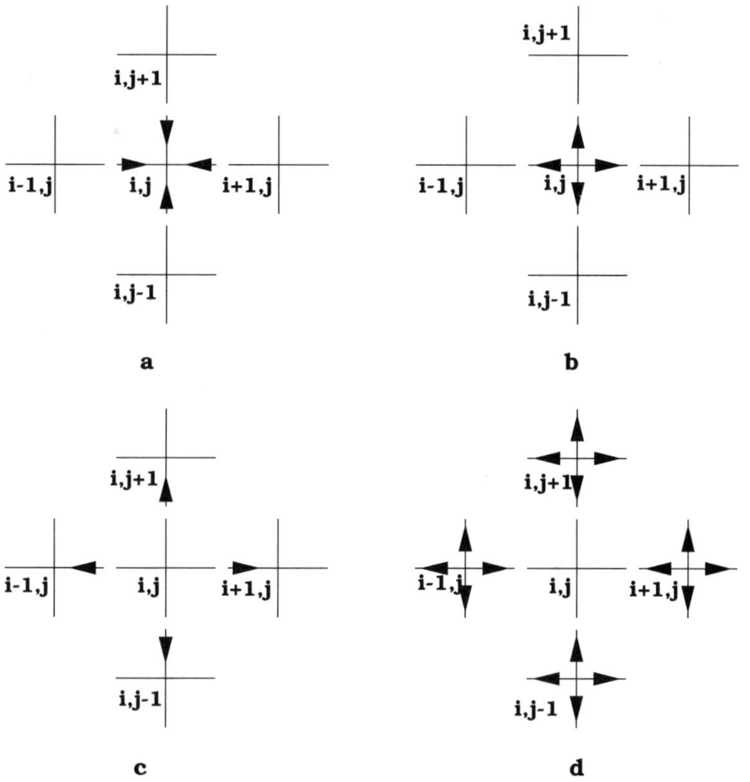

FIGURE 7.9
Excitation and propagation on a TLM mesh. (a) Incident pulses defined as initial excitation at time $t = 0$ and node (i, j). (b) Reflected pulses at time $t = \Delta t$. (c) Connection routine where reflected pulses become incident pulses on adjacent nodes at time $t = \Delta t^{+}$. (d) Reflected pulses at time $t = 2\Delta t$.

After the connection process has been completed for each node in the mesh, the reflected pulses become the incident pulses on adjacent nodes, and the scattering matrix is applied again at time $t = 2\Delta t$ (Fig. 7.9d). This process is repeated N times, where N is determined from the physical problem being modeled. By repeated application of this process, the energy initially applied at node $(i, j), n = 0$, is propagated outward on the TLM mesh in an approximately cylindrical pattern (since the mesh is Cartesian, it is only an approximate cylindrical pattern). Much like ripples in water after a pebble is tossed in, the incident impulses spread outward on the TLM mesh and so the incident impulses interact with the model of the physical problem, generating a solution (in this case, for a two-dimensional point source in free space). In order to model physical problems, one requirement is the ability to model metallic conductors.

Figure 7.10 illustrates the method used to simulate a metallic boundary in TLM. Following the methodology in Fig. 7.9, when a metallic boundary is involved the incident impulses are handled as usual at each node in the mesh. However, after the scattering matrix has been applied and the incident voltage impulses have become reflected voltage impulses, the voltage incident upon a metallic boundary is not connected with a neighboring node in the connection routine (in this example, there is no neighbor). The voltage pulse is returned to the node from which it came, having been inverted in sign as shown in Fig. 7.10b.

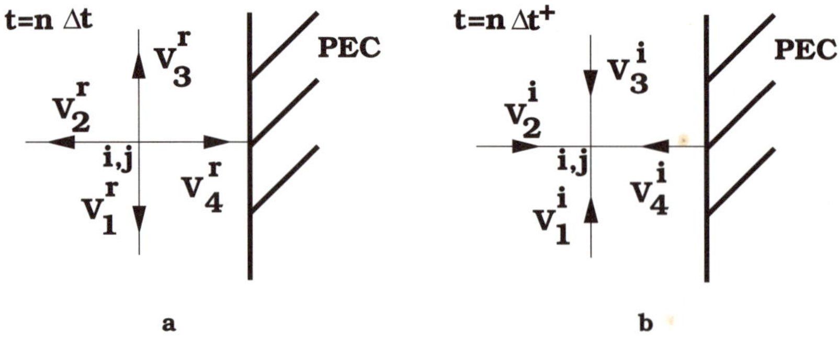

FIGURE 7.10
This figure illustrates the method used in TLM to simulate a metallic boundary, or perfect electrical conductor (PEC). (a) At $t = n\Delta t$, all the pulses are reflected; (b) at $t = n\Delta t^+$, the pulse incident on the metallic boundary is reflected and returned with a negative value ($V_4^i(n\Delta t^+) = -V_4^r(n\Delta t)$), becoming incident upon the node (i, j) center.

In other words, pulses incident upon metallic boundaries "see" a reflection coefficient of $\Gamma = -1$ and are therefore reflected with an opposite sign and can be characterized by

$$^{n+1}V_l^{\,i}(i, j) = {}^{n}V_l^{\,r}(i, j)\Gamma; \qquad \Gamma = -1.0, \tag{7.1.31}$$

where l represents the node port. Additionally, when using a regular Cartesian grid, all boundaries are defined at a spacing of $\Delta l/2$ away from node centers for the same reason that the capacitive stub used to simulate relative permittivity was chosen to be $\Delta l/2$ long; to preserve time synchronization. As a result, the voltage pulse reflected from the metal boundary returns to the node center at the same time as the other voltage pulses.

Using this technique, it is possible to define resistive sheets, or boundaries, by allowing the reflection coefficient Γ in Eq. (7.1.31) to lie between $-1 \leq \Gamma \leq 1$. Also, it is possible to utilize the concept of a reflection coefficient to terminate the TLM mesh at artificially imposed boundaries, such as mesh truncation to free space. This will be discussed, along with other mesh truncation techniques, in the next section.

7.1.8 TLM Mesh Truncation Conditions

Because any TLM mesh must be composed of a finite number of nodes, some means must be introduced for artificially terminating the computational space at the mesh extents. For closed-region problems such as a resonator, the mesh is naturally terminated by the boundaries of the structure. For problems involving radiation of waves in an open region, or for those problems that require a matched termination of a guided wave structure, an artificial mesh truncation condition must be introduced. This absorbing boundary condition (ABC) must permit fields propagating outward on the mesh to do so without introducing any significant nonphysical reflections.

There are many types of ABCs available for use in computational field models [13]. These have been applied with varying degrees of success in many numerical methods such as finite difference, finite element, and TLM. Here, we limit our discussion to only those techniques which have been successfully applied to the TLM method.

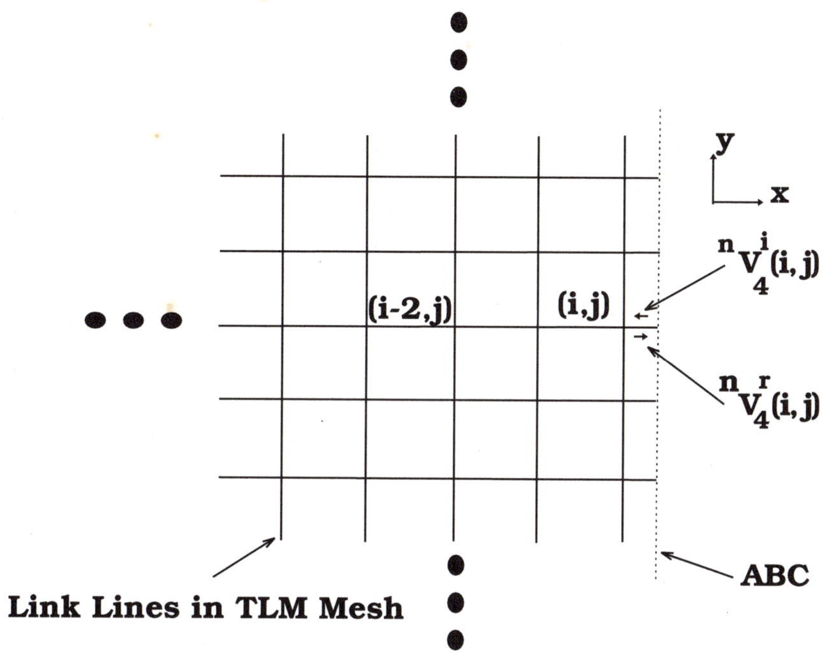

FIGURE 7.11
A two-dimensional mesh section terminated to infinite space.

In order to understand why special conditions are required to absorb waves at the mesh edges, consider the two-dimensional TLM mesh section shown in Fig. 7.11. At the right-hand edge of the mesh we wish to use an ABC to simulate an infinite space. The incoming voltage pulse ${}^nV_4^{\mathrm{i}}(i, j)$ would normally be obtained during the connection procedure from ${}^{n-1}V_2^{\mathrm{r}}(i+1, j)$. However, the node at $(i+1, j)$ does not exist since it falls outside of the computational mesh, and so some means must be devised to compute ${}^nV_4^{\mathrm{i}}(i, j)$ from known pulses within the mesh. ABCs provide methods for calculating these quantities from known fields/voltages within the computational domain.

Perhaps the most obvious approach would be to introduce a reflection coefficient, Γ_{ABC}, at the boundary such that the incident pulses at the boundary nodes can be computed as

$$ {}^nV_4^{\mathrm{i}} = \Gamma_{\mathrm{ABC}}\,{}^{n-1}V_4^{\mathrm{r}}. \tag{7.1.32} $$

Note that the superscripts i and r refer to incident and reflected voltages at the node inside the ABC and not at the plane of the boundary.

In order to compute Γ_{ABC} to simulate a reflectionless ABC, recall that for the two-dimensional shunt node the center of the node has an inductance L and a capacitance $2C$ which models a medium with impedance $(L/2C)^{1/2}$, and this leads to

$$ \Gamma_{\mathrm{ABC}} = \frac{1-\sqrt{2}}{1+\sqrt{2}} = -0.172. \tag{7.1.33} $$

This reflection coefficient provides a reflectionless termination only for waves that are normally incident on the boundary of a stubless and cubic two-dimensional shunt mesh. For a general discussion of computing Γ_{ABC} for a general graded and inhomogeneous TLM mesh, the reader is referred to German *et al.* [7].

The matched termination is very easy to implement but its applications are limited since it provides perfect absorption only for a single-incident angle. This leads to a very narrow-band ABC when used to terminate many types of problems, such as waveguides.

An alternative ABC that provides very good wideband absorption is the Johns's matrix. This technique, which is based on time domain diakoptics [14], computes the unknown voltage pulses at the boundary via a time domain convolution with an analytically derived or numerically computed time domain Green's function, i.e., the Johns's matrix [15]. For general structures this method is computationally prohibitive. However, for the case of terminating waveguide structures for which the spatial behavior of the fields at the waveguide ports is analytically known, the method simplifies considerably and provides a very good wideband ABC for waveguide discontinuity problems. The Johns's matrix has also been used to segment large problems into a number of smaller ones that are solved separately and later reconnected via a convolution process [16].

The most generally applicable and computationally feasible ABCs to be implemented in the TLM method to date are those based on one-way wave equations. The most popular of these used in the TLM method is Higdon's ABC [17–20]. The general formulation of Higdon's ABC allows it to be adjusted by a proper choice of parameters to provide optimal absorption on many types of problems. An nth-order Higdon ABC allows for perfect absorption at n incident angles which can be specified by the analyst. Typically, a second-order Higdon ABC is sufficient. While a third-order ABC may improve results in some instances, higher order Higdon ABCs are rarely worth the additional complexity.

The Higdon ABC may be applied directly to the TLM voltage pulses since they are linear combinations of the actual electric and magnetic field quantities. The voltage pulse to be injected back into the mesh at the nth time step (${}^nV_4^{\mathrm{i}}$ in Fig. 7.11) is computed from voltage pulses at neighboring nodes in the mesh at the current and previous time steps. The number of voltage pulses that must be stored at the boundary nodes from previous time steps depends on the order of the ABC being used.

For example, consider the following equation for predicting ${}^nV_4^{\mathrm{i}}(i, j)$ in Fig. 7.11 using a second-order Higdon ABC:

$$\begin{aligned}{}^nV_4^{\mathrm{i}}(i,j) = {} & \left[A_{01}\,{}^{n-1}V_4^{\mathrm{i}}(i,j) + A_{02}\,{}^{n-2}V_4^{\mathrm{i}}(i,j)\right] \\ & + \left[A_{10}\,{}^{n}V_4^{\mathrm{i}}(i-1,j) + A_{11}\,{}^{n-1}V_4^{\mathrm{i}}(i-1,j) + A_{12}\,{}^{n-2}V_4^{\mathrm{i}}(i-1,j)\right] \\ & + \left[A_{20}\,{}^{n}V_4^{\mathrm{i}}(i-2,j) + A_{12}\,{}^{n-1}V_4^{\mathrm{i}}(i-2,j) + A_{22}\,{}^{n-2}V_4^{\mathrm{i}}(i-2,j)\right]. \qquad (7.1.34)\end{aligned}$$

Examination of Eq. (7.1.34) shows that the pulse incident on branch 4 of the node at (i, j) at the nth time step can be computed using the same voltage pulse from the previous two time steps (first bracketed term) and the corresponding voltage pulses at the two nodes immediately to the left of the boundary node at the current (known from the standard scatter/connect procedure within the mesh) and previous two time steps. The local (in both space and time) nature of this class of ABCs makes them relatively easy to implement and computationally efficient.

The coefficients A_{lm} are calculated as follows [18]:

$$\begin{aligned}
A_{01} &= (\alpha_1 + \alpha_2)\\
A_{02} &= -\alpha_1\alpha_2\\
A_{10} &= (\beta_1 + \beta_2)\\
A_{11} &= (\gamma_1 + \gamma_2 - \alpha_1\beta_2 - \beta_1\alpha_2)\\
A_{12} &= -(\alpha_1\gamma_2 + \gamma_1\alpha_2)\\
A_{20} &= -\beta_1\beta_2\\
A_{21} &= -(\beta_1\gamma_2 + \gamma_1\beta_2)\\
A_{22} &= -(\gamma_1\gamma_2)\\
\alpha_{\mathrm{i}} &= \frac{a - g_{\mathrm{i}}(1-b)}{(a - 1 - g_{\mathrm{i}}(1-b) - \epsilon_{\mathrm{i}}\Delta l)}\\
\beta_{\mathrm{i}} &= \frac{(a - 1 + g_{\mathrm{i}}b)}{(a - 1 - g_{\mathrm{i}}(1-b) - \epsilon_{\mathrm{i}}\,\Delta l)}\\
\gamma_i &= \frac{(-a - bg_{\mathrm{i}})}{(a - 1 - g_{\mathrm{i}}(1-b) - \epsilon_{\mathrm{i}}\Delta l)}\\
g_{\mathrm{i}} &= \frac{\cos\theta_{\mathrm{i}}}{c}\frac{\Delta l}{\Delta t} = \frac{\cos\theta_{\mathrm{i}}}{c}\frac{\Delta l 2c}{\Delta l} = 2\cos\theta_{\mathrm{i}}.
\end{aligned}$$

These coefficients depend on the incidence angles chosen for perfect absorption (θ_{i}), the mesh size (Δl), time step (Δt), and various space–time averaging and damping factors (a, b, and ϵ_{i}). These parameters can be chosen to provide optimal absorption for specific problems [18].

The Higdon ABC provides a very versatile ABC for use in the TLM algorithms. It has been implemented in two- and three-dimensional meshes. Its usefulness has been demonstrated for open-region scattering and radiation problems as well as for a variety of guided wave structures.

Another ABC that has received recent attention is the so-called perfectly matched layer (PML) of Berenger [21]. This method was originally developed for use in a staggered finite-difference time domain (FDTD) mesh. It consists of adding a region of material (the PML) meeting certain criteria at the periphery of the mesh and mathematically splitting the field components into non-physical components for finite-difference updating. This method yields an ABC that is superior to that obtainable with one-way wave equation ABCs such as Higdon's. Efforts to implement the PML ABC into TLM algorithms to date have involved interfacing the TLM mesh to a FDTD section modeling the PML region [22–23]. Good absorption has been demonstrated but the interface between the meshes can be cumbersome to implement. Further work is needed to incorporate the benefits of the PML directly into the TLM scatter/connect procedure without resorting to an interface to a FDTD mesh.

7.1.9 Discretization of the TLM Spatial Grid

Since we cannot discretize a spatial grid into an infinite number of segments, we must approximate the physical problem modeled with the TLM by using a mesh with finite segments. This approximation results in some potential error for each problem being analyzed, and so a methodology must be found to determine the number of TLM cells needed for a given problem to minimize this discretization error. Notice that up until this point, no mention has been made of what size Δl must be nor the size of the time increment Δt between TLM scattering operations.

In order to determine some bounds for the TLM grid scheme, we first examine part of the formulation used in deriving the TLM algorithm. An approximation was made regarding the TLM cell size Δl which allowed us to make use of the fundamental theorem of calculus as shown in the following:

$$\frac{V_z\,(x+(\Delta l/2),t)-V_z\,(x-(\Delta l/2),t)}{\Delta l}=-L\frac{\partial I_x}{\partial t}$$

$$\text{In the limit as } \Delta l \to 0;\qquad \frac{\partial V_z}{\partial x}=-L\frac{\partial I_x}{\partial t}.$$

For this approximation to be valid, as a "rule of thumb" it is usually sufficient given by

$$\frac{\Delta l}{\lambda_{\min}} \leq \frac{1}{10} \text{ where } f_{\max} \propto \frac{1}{\lambda_{\min}},$$

$f_{\max}$ is the highest frequency of interest, and $\lambda_{\min}$ is the smallest wavelength of $f_{\max}$ in the material with the highest constitutive parameters. While this gives us a rule of thumb in determining the TLM grid scheme, it does not put any bounds on the errors which may result in using a finite grid. Therefore, a more rigorous bound must be obtained.

Since the TLM mesh has a finite grid scheme, it stands to reason that there is a definite upper limit on accurate frequency information available for a given mesh. In fact, as that limit is pushed, it is reasonable to assume that different frequency components on any wideband field simulated on the TLM grid will travel at different velocities, resulting in dispersion.

In order to determine the effects of the dispersion in the TLM mesh, first we define a field propagating in free space represented by

$$E_z(x,y)=E_0e^{-j(\omega t+\beta x\cos(\phi)+\beta y\sin(\phi))}, \tag{7.1.35}$$

where $E_z(x,y)$ is a two-dimensional z-polarized electric field propagating in the $\boldsymbol{a}_\rho$ direction at an angle ϕ with respect to the x-axis, E_0 is the magnitude, $\beta=\omega/c$, and c is the speed of light in free space.

We allow the field represented by Eq. (7.1.35) to be approximated on a TLM mesh by defining the time-harmonic voltage at the node center as

$${}^{n}V_z(x,y)=V_0e^{-j(\omega n\Delta t+\beta_n x\cos(\phi)+\beta_n y\sin(\phi))}, \tag{7.1.36}$$

where ${}^{n}V_z(x,y)$ is the voltage representation of $E_z(x,y)$ of Eq. (7.1.35), $\beta_n=\omega/v_n$, and v_n is the wave speed in the TLM mesh. The nodal relationship for lossless media can be shown to be [24]

$$\left[4-8\sin^2\left(\frac{\pi\,\Delta l}{\lambda}\right)\right]V_z(\omega)=(V_1+V_2+V_3+V_4), \tag{7.1.37}$$

where we have set $\Delta t = \Delta l/c$. With node (i, j) referenced to $(0, 0)$ we know that

$$\begin{aligned} V_1 &= V_z\left(0, -\frac{\Delta l}{2}\right) \\ V_2 &= V_z\left(-\frac{\Delta l}{2}, 0\right) \\ V_3 &= V_z\left(0, \frac{\Delta l}{2}\right) \\ V_4 &= V_z\left(\frac{\Delta l}{2}, 0\right). \end{aligned} \tag{7.1.38}$$

Substituting Eq. (7.1.38) into Eq. (7.1.37) and using the identity $\cos(2A) = (1 - 2\sin^2(A))$ gives

$$\sin^2\left(\frac{\pi\,\Delta l c}{\lambda v_n}\sin(\phi)\right) + \sin^2\left(\frac{\pi\,\Delta l c}{\lambda v_n}\cos(\phi)\right) = 2\sin^2\left(\frac{\pi\,\Delta l}{\lambda}\right). \tag{7.1.39}$$

As $\Delta l/\lambda \to 0$ and using the approximation $\sin(A) = A$ as $A \to 0$, Eq. (7.1.39) becomes

$$\left(\frac{\pi\,\Delta l c}{\lambda v_n}\sin(\phi)\right)^2 + \left(\frac{\pi\,\Delta l c}{\lambda v_n}\cos(\phi)\right)^2 = 2\left(\frac{\pi\,\Delta l}{\lambda}\right)^2$$

and solving this for v_n/c yields

$$\frac{v_n}{c} = \frac{1}{\sqrt{2}}. \tag{7.1.40}$$

The previous result implies that for small values of $\Delta l/\lambda$ and $\Delta t = \Delta l/c$, the wave speed on the TLM mesh is $c/\sqrt{2}$.

Additionally, consider the following cases. If we allow $\phi = 45^\circ$ in Eq. (7.1.39) and solve for v_n/c, this leads to the same result as that of Eq. (7.1.40) but in this case for all values of $\Delta l/\lambda$. This implies that the velocity of the TLM mesh is a constant at an angle of 45° to the mesh structure. Next, if we allow $\phi = 0^\circ$ in Eq. (7.1.39) and solve for v_n/c, we get

$$\frac{v_n}{c} = \frac{\pi\,\Delta l}{\lambda}\frac{1}{\sin^{-1}[\sqrt{2}\sin(\pi\,\Delta l/\lambda)]}. \tag{7.1.41}$$

Since the argument of the arcsine cannot be >1, this implies that the velocity on the TLM mesh is undefined for $\phi = 0$ and $\Delta l/\lambda < 0.25$, and at $\Delta l/\lambda = 0.25$ the first mesh cutoff frequency occurs at which the mesh velocity is $c/\sqrt{2}$. Additionally, if we allow $\Delta l/\lambda = 0.1$, the mesh velocity for $\phi = 0^\circ$ is $0.695c$, or only 1.8% less than the maximum of $c/\sqrt{2}$. What the dispersion analysis suggests is that allowing $\Delta l/\lambda$ to be much greater that 0.1 will lead to mesh dispersion which will distort the solution to the physical problem. Figure 7.12 shows a plot of the normalized transmission line velocity v_n/c verses $\Delta l/\lambda$ for the $\phi = 0^\circ$ and the $\phi = 45^\circ$ cases.

An alternate means of determining the maximum velocity of propagation on the TLM mesh is simply to consider the lumped capacitance and inductance at the TLM shunt node center.

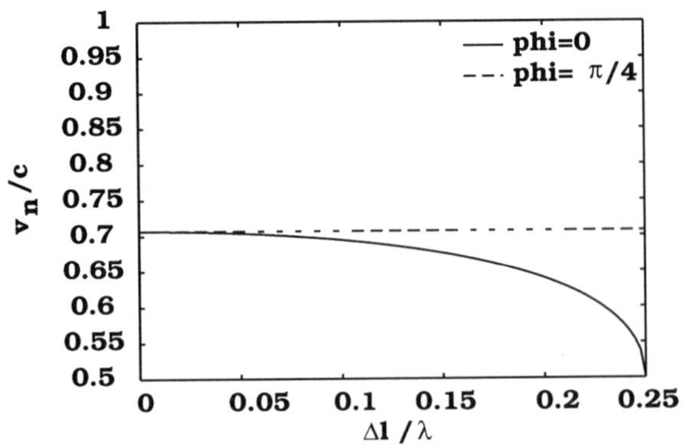

FIGURE 7.12
For the 2D TLM shunt node, the normalized velocity v_n/c versus the electrical cell size $\Delta l/\lambda$ for the $\phi = 0°$ and $\phi = 45°$ cases.

Remember from Section 7.1.3 that we set

$$L = \mu \qquad 2C = \epsilon$$

in Eq. (7.1.17). This implies that the velocity of TLM simulated fields propagating on the mesh is

$$v_n = \frac{1}{\sqrt{L(2C)}} = \frac{1}{\sqrt{2}\sqrt{LC}} = \frac{c}{\sqrt{2}},$$

which is the same upper limit obtained in the dispersion analysis.

For the dispersion analysis we chose $\Delta t = \Delta l/c$. This resulted in a velocity of propagation on the TLM mesh to be less than that of free space. In fact, the TLM mesh appears to be in a medium corresponding to $\epsilon_r = 2$, $\mu_r = 1$. We can correct this by fixing the time step Δt at

$$\Delta t = \frac{\Delta l}{\sqrt{2}c}. \tag{7.1.42}$$

Using the value for Δt in Eq. (7.1.42) and repeating the dispersion analysis results in a TLM mesh velocity of c, or that of free space. This does not eliminate dispersion on the TLM mesh but only allows the TLM mesh to appear to be free space, and thus simplifies postprocessing of the TLM output.

7.1.10 TLM Output

At any time step $n\Delta t$, the voltage pulses in the shunt TLM mesh can be related to the corresponding field quantities by the following:

$$^nE_z(i,j) = A(^nV_1(i,j) + {^nV_2}(i,j) + {^nV_3}(i,j) + {^nV_4}(i,j) + {^nV_5}(i,j)Y_o) \tag{7.1.43}$$

$$^nH_x(i,j) = B(^nV_4(i,j) - {^nV_2}(i,j)) \tag{7.1.44}$$

$$^nH_y(i,j) = B(^nV_3(i,j) - {^nV_1}(i,j)) \tag{7.1.45}$$

$$A = \frac{2}{\Delta l(4 + Y_o + Z_o)}$$

$$B = \frac{1}{\Delta l Z_{\text{t-line}}},$$

where (i, j) represents the node location, V_{1-5} are the impulses at node (i, j) on the four ports at time n, and $Z_{\text{t-line}} = Z_o$ is the impedance of the TLM link lines. Since the TLM uses voltage impulses, the output will be of the form

$$h_{\text{out}}(t) = \sum_{n=0}^{N} b_n \delta(t - n\Delta t), \tag{7.1.46}$$

where $h_{\text{out}}(t)$ is the output from the TLM mesh as a function of time, N is the number of time samples, and b_n is a weighting coefficient generated by the TLM algorithm and b_n weights a string of impulse functions over time. Thus, the TLM output will be the electric and magnetic field quantities over the time frame from $0 \to N\Delta t$. The output can be taken at any or all nodes (i, j) in the TLM mesh, with the physical problem determining these locations.

The form of the output is greatly determined by the form of the TLM excitation because the TLM excitation determines, along with the TLM cell size, the available frequency content contained in the output. In Section 7.1.7, the TLM mesh excitation was defined as

$$N_{\text{in}}(t) = \sum_{n=0}^{N} a_n \delta(t - n\Delta t),$$

and this excitation can be applied at any or all nodes (i, j).

For example, we can represent a_n in the previous equation as

$$a_n = \sin(\omega n \Delta t), \tag{7.1.47}$$

where the weighting coefficients are now representing a sinusoidal function with a radian frequency ω. The TLM output would then be the transient response of a sinusoid of radian frequency ω and would contain only very narrowband frequency information centered at ω. This is typically not the excitation of choice because this removes one of the principle advantages of time domain analysis, i.e., wideband data from a single computation.

As another example, the TLM mesh can be excited as it was in Section 7.1.7, with a_n being

$$a_n = \begin{cases} 1.0 & n = 0 \\ 0.0 & n > 0. \end{cases}$$

In this case, the delta distribution excitation has a theoretical bandwidth of infinity. This results in the TLM output containing useful frequency information from 0 Hz up to the frequency corresponding to $\Delta l/\lambda = 0.1$ of the TLM mesh. Exciting the TLM mesh with a delta distribution and generating an output is equivalent in network theory to determining the impulse response of a system. Thus, the output can be used to determine the response of the system to any general excitation $f_{\text{in}}(t)$ via a time domain convolution,

$$f_{\text{out}}(t) = \int_{-\tau}^{+\tau} f_{\text{in}}(t) h_{\text{out}}(t - \tau)\, d\tau, \tag{7.1.48}$$

where $f_{\text{out}}(t)$ is the time domain response of the system due to $f_{\text{in}}(t)$. Also, we may take the Fourier transform to get the frequency response, or transfer function, for the system by

$$H_{\text{out}}(\omega) = \int_{-\infty}^{+\infty} h_{\text{out}}(t) e^{-j\omega t}\, dt \approx \sum_{n=0}^{N} a_n e^{-j\omega n \Delta t}, \tag{7.1.49}$$

where $H_{\text{out}}(\omega)$ is the frequency response of the system and can be obtained using a discrete Fourier transform as shown in Eq. (7.1.49).

It is important to note that when using a delta function excitation for the TLM mesh, the unprocessed output appears very noisy. This is because the excitation contains frequency content outside the useful mesh bandwidth. Thus, the output must be processed using Eq. (7.1.48) and /or Eq. (7.1.49) or some other method to extract the useful data.

Additionally, to minimize the discretization error associated with this type of excitation, it is possible to use an excitation with limited frequency content, such as a Gaussian distribution, to excite the TLM mesh. If the Gaussian distribution is chosen with the proper frequency content for the mesh, much or all of the dispersive noise can be eliminated. While generally not necessary, some ABCs are sensitive to discretization error and break down without a band-limiting excitation.

A final problem is to determine the number of time samples N to be used. This is problem dependent, involving electrical problem size, resonance conditions, and the Q of the system. However, one simple method of determining the size of N is to do convergence tests. In other words, keep increasing the value of N until the desired results stop varying. However, this is not always reliable and must be approached with care. Some systems, especially those with low Q resonances, can appear to have convergent results when in fact that is not the case.

Until now, we have confined the analysis to TM problems. Next, we examine transverse electric (TE)-type problems by introducing the series TLM node and duality.

7.1.11 The Series Node and Duality

Shown in Fig. 7.13 is the series TLM node. Through a similar process as was used to obtain the two-dimensional shunt node, the two-dimensional series node can be derived by considering the dual of the TM to z case—the TE to z case. The scattering matrix defining the series TLM node has scattering properties similar to the shunt scattering matrix and is shown in Eq. (7.1.50) for inhomogeneous lossy media:

$$\begin{bmatrix} V_1 \\ V_2 \\ V_3 \\ V_4 \\ V_5 \end{bmatrix}_{n+1}^{\text{r}} = \frac{2}{4 + Z_{\text{s}} + R_{\text{s}}} \begin{bmatrix} C & 2 & 2 & -2 & -2 \\ 2 & C & -2 & 2 & 2 \\ 2 & -2 & C & 2 & 2 \\ -2 & 2 & 2 & C & -2 \\ -2Z_{\text{s}} & 2Z_{\text{s}} & 2Z_{\text{s}} & -2Z_{\text{s}} & D \end{bmatrix} \begin{bmatrix} V_1 \\ V_2 \\ V_3 \\ V_4 \\ V_5 \end{bmatrix}_{n}^{\text{i}}, \tag{7.1.50}$$

where

$$C = 2 + Z_{\text{s}} + R_{\text{s}},$$
$$D = 4 - Z_{\text{s}} + R_{\text{s}},$$

and Z_{s} is the stub impedance used to model the permeability μ_{r}, and R_{s} is the stub resistance used to model magnetic conductivity.

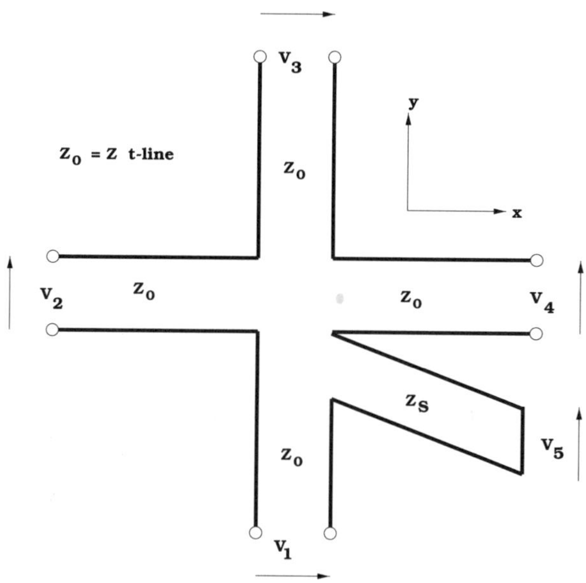

FIGURE 7.13
The series TLM node, with stub for permeability (μ_r) included.

The relationships between the permeability and conductivity to the transmission line parameters Z_s and R_s are

$$Z_s = 4(\mu_r - 1)$$

$$R_s = \frac{\sigma_m Z_{\text{t-line}}}{\Delta l},$$

where σ_m is the magnetic conductivity. To relate the voltage impulses to the field quantities, we have

$$^{n}H_z(i,j) = E\left({}^{n}V_1(i,j) - {}^{n}V_2(i,j) - {}^{n}V_3(i,j) + {}^{n}V_4(i,j) + {}^{n}V_5(i,j)\right) \quad (7.1.51)$$

$$^{n}E_x(i,j) = \frac{1}{\Delta l}\left({}^{n}V_1(i,j) - {}^{n}V_3(i,j)\right) \quad (7.1.52)$$

$$^{n}E_y(i,j) = -\frac{1}{\Delta l}\left({}^{n}V_2(i,j) - {}^{n}V_4(i,j)\right), \quad (7.1.53)$$

where

$$E = \frac{2}{\Delta l(4Z_{\text{t-line}} + Z_s + R_s)}.$$

The implementation of the TLM series node is nearly identical to that of the shunt node, and an algorithm written for one is easily modified for the other. In fact, it is possible, while limiting oneself to the two-dimensional TLM, to use the shunt or series nodes for both TE and TM problems using the concept of duality [25]. Using this concept, the following conditions, given by

$$H_z \equiv -E_z$$
$$E_x \equiv H_x$$
$$E_y \equiv H_y$$
$$\eta \equiv \frac{1}{\eta}$$
$$\mu \equiv \epsilon,$$

allow a single two-dimensional code to be used for both TE and TM problems [26].

7.1.12 Outline of the Algorithm for Two-Dimensional TLM Code

There are many ways to generate the TLM algorithm. Presented here is one possible organization scheme for a two-dimensional TLM program as presented in the following:

1. Define parameters: Specify the size of the scattering matrix (x, y), number of iterations, etc.
2. Declare variables: Specify all variables needed for the algorithm. It is usually good programming practice to define all variables, even if the compiler may not require it.
3. Read matrix size, number of boundaries, number of excitation points, number of output points, number of iterations, and the number of media boundaries: Some means of determining the size of these quantities is needed.
4. Read the boundary points, the type of boundary, and the reflection coefficient associated with the boundary: Specify the outer and any internal boundaries. Using a flagging system, the flags can be used in order to define metallic boundaries, edge of mesh boundaries, etc.
5. Read the excitation points and the type of excitation: Define the excitation (E_z, H_x, H_y for shunt case) and the excitation locations.
6. Read in output data locations: In the TLM, output points can be taken anywhere. Here, we specify the desired output locations and type of output desired.
7. Read in media blocks: If some media besides free space is considered, there must be some means to define the regions of differing media.
8. Clear the scattering matrix: Prepare to begin the TLM algorithm by clearing the matrix memory which stores the scattering matrix.
9. Initialize media blocks: "Load" the stubs which model ϵ_r or μ_r with the appropriate values.
10. Initialize excitation: Start the impulsive excitation of the TLM mesh (location and excitation) at time $t = 0$.
11. Begin iterations: Start the iterative process.
12. Calculate output: Determine the output as desired.
13. Perform the scattering operation: Apply the matrix operation $[V]^{\mathrm{r}}_{k+1} = [S]\,[V]^{\mathrm{i}}_{k}$ to each node in the TLM mesh. To save computation time, it is possible to skip this process provided the pulses on all ports of a node are zero.
14. Begin the switching operation: Let the reflected pulses on each node become incident on the neighboring nodes.

15. Switching is continued for the outer boundaries: Take the special cases into account (the metallic, outer, etc. boundaries).
16. Continue until the total number of iterations have occurred: After this step, continue the algorithm to calculate the quantities of interest.

7.2 THREE-DIMENSIONAL TLM

The discussion thus far has concentrated on the TLM method in two-dimensional space in order to present the basic concepts of the technique as simply as possible. However, the TLM method is fully extendible to three dimensions and there are various three-dimensional TLM formulations. Soon after its initial presentation in two dimensions, the TLM method was extended to a fully three-dimensional field model by Akhtarzad and Johns [4]. This first three-dimensional formulation, now referred to as the expanded node TLM method, was derived by connecting two-dimensional series and shunt nodes in a staggered grid configuration similar to the Yee grid in FDTD. It was later shown to be mathematically equivalent to the FDTD algorithm under certain restrictive conditions [27].

Further refinements resulted in the so-called asymmetrical condensed node TLM method which was derived independently by Amer [28] and Saguet and Pic [29]. While having the advantage of allowing all six field components in each cell to be located at the same spatial point, the node is asymmetrical. This asymmetry has the effect of causing waves traveling in opposite coordinate directions to propagate at slightly different velocities (e.g., waves traveling the the $+x$ and $-x$ directions). Despite this effect, the formulation has been used quite successfully for many electromagnetic interaction problems. In 1987, Johns and coworkers [30, 31] developed the symmetrical condensed node (SCN), which was a very significant advance in the TLM technique. Unlike any of the previous formulations, an equivalent lumped element circuit for this node was not used to derive the scattering matrix. The scattering properties for this node were derived directly from the Maxwell equations instead of from an equivalent transmission line network. While this field derivation eliminated the necessity of the voltage/current–electric/magnetic field equivalences, they were retained for conceptual reasons. The basic SCN is shown in Fig. 7.14, in which the original port numbering scheme used by Johns is shown. Each arm of the node is composed of two orthogonal link lines that carry two field polarizations. The method uses the same scatter/connect procedure described for the two-dimensional nodes in the previous section. The 12×12 scattering matrix for the SCN in a cubic mesh modeling free space is given by

$$\underline{\underline{S}} = \frac{1}{2}\begin{bmatrix} 0 & 1 & 1 & 0 & 0 & 0 & 0 & 0 & 1 & 0 & -1 & 0 \\ 1 & 0 & 0 & 0 & 0 & 1 & 0 & 0 & 0 & -1 & 0 & 1 \\ 1 & 0 & 0 & 1 & 0 & 0 & 0 & 1 & 0 & 0 & 0 & -1 \\ 0 & 0 & 1 & 0 & 1 & 0 & -1 & 0 & 0 & 0 & 1 & 0 \\ 0 & 0 & 0 & 1 & 0 & 1 & 0 & -1 & 0 & 1 & 0 & 0 \\ 0 & 1 & 0 & 0 & 1 & 0 & 1 & 0 & -1 & 0 & 0 & 0 \\ 0 & 0 & 0 & -1 & 0 & 1 & 0 & 1 & 0 & 1 & 0 & 0 \\ 0 & 0 & 1 & 0 & -1 & 0 & 1 & 0 & 0 & 0 & 1 & 0 \\ 1 & 0 & 0 & 0 & 0 & -1 & 0 & 0 & 0 & 1 & 0 & 1 \\ 0 & -1 & 0 & 0 & 1 & 0 & 1 & 0 & 1 & 0 & 0 & 0 \\ -1 & 0 & 0 & 1 & 0 & 0 & 0 & 1 & 0 & 0 & 0 & 1 \\ 0 & 1 & -1 & 0 & 0 & 0 & 0 & 0 & 1 & 0 & 1 & 0 \end{bmatrix}. \tag{7.2.1}$$

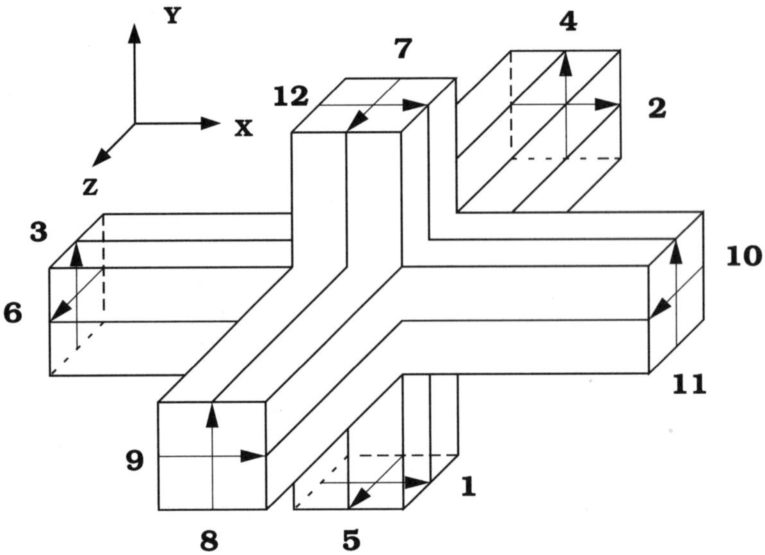

FIGURE 7.14
The original symmetrical condensed node with all port voltages shown.

The fields at the node center are obtained from the following equations:

$$E_x = -\frac{1}{2\Delta l}\left(V_1^{\mathrm{i}} + V_2^{\mathrm{i}} + V_9^{\mathrm{i}} + V_{12}^{\mathrm{i}}\right)$$
$$E_y = -\frac{1}{2\Delta l}\left(V_3^{\mathrm{i}} + V_4^{\mathrm{i}} + V_8^{\mathrm{i}} + V_{11}^{\mathrm{i}}\right) \quad (7.2.2)$$
$$E_z = -\frac{1}{2\Delta l}\left(V_5^{\mathrm{i}} + V_6^{\mathrm{i}} + V_7^{\mathrm{i}} + V_{10}^{\mathrm{i}}\right)$$

$$H_x = \frac{1}{2\Delta l Z_0}\left(V_4^{\mathrm{i}} - V_5^{\mathrm{i}} + V_7^{\mathrm{i}} - V_8^{\mathrm{i}}\right)$$
$$H_y = \frac{1}{2\Delta l Z_0}\left(-V_2^{\mathrm{i}} + V_6^{\mathrm{i}} + V_9^{\mathrm{i}} - V_{10}^{\mathrm{i}}\right) \quad (7.2.3)$$
$$H_z = \frac{1}{2\Delta l Z_0}\left(-V_3^{\mathrm{i}} + V_1^{\mathrm{i}} + V_{11}^{\mathrm{i}} - V_{12}^{\mathrm{i}}\right),$$

where Z_0 is the characteristic impedance of the link lines and Δl is the cell dimension.

The node can be modified by the addition of stubs to model inhomogeneous media in a graded TLM mesh. Loss stubs can also be included to model both electric and magnetic conductivities [32, 33]. For a node modeling a block of space of dimensions (u, v, w) in the (x, y, z) directions, and with relative permittivity ϵ_r, relative permeability μ_r, electric conductivity σ_e, and magnetic conductivity σ_m, the 18×18 scattering matrix is given in Table 7.1,

Table 7.1 The TLM Stub-Loaded SCN Matrix

Column No.			1	2	3	4	5	6	7	8	9	10	11	12	13	14	15	16	17	18
Short circuit stub (Y)			x	x	y	y	z	z	z	y	x	z	y	x	x	y	z			
Open-circuit stub (Z)			z	y	z	x	x	y	x	x	y	y	z	z				x	y	z
1	x	z	a	b	d	0	0	0	0	0	b	0	$-d$	c	g	0	0	0	0	i
2	x	y	b	a	0	0	0	d	0	0	c	$-d$	0	b	g	0	0	0	$-i$	0
3	y	z	d	0	a	b	0	0	0	b	0	0	c	$-d$	0	g	0	0	0	$-i$
4	y	x	0	0	b	a	d	0	$-d$	c	0	0	b	0	0	g	0	0	0	0
5	z	x	0	0	0	d	a	b	c	$-d$	0	b	0	0	0	0	g	$-i$	0	0
6	z	y	0	d	0	0	b	a	b	0	$-d$	c	0	0	0	0	g	0	i	0
7	z	x	0	0	0	$-d$	c	b	a	d	0	b	0	0	0	0	g	i	0	0
8	y	x	0	0	b	c	$-d$	0	d	a	0	0	b	0	0	g	0	$-i$	0	0
9	x	y	b	0	0	0	0	$-d$	0	0	a	d	0	b	g	0	0	0	i	0
10	z	y	0	$-d$	0	0	b	c	b	0	d	a	0	0	0	0	g	0	$-i$	0
11	y	z	$-d$	0	c	b	0	0	0	b	0	0	a	d	0	g	0	0	0	i
12	x	z	c	b	$-d$	0	0	0	0	0	b	0	d	a	g	0	0	0	0	$-i$
13	x		e	e	0	0	0	0	0	0	e	0	0	e	h	0	0	0	0	0
14	y		0	0	e	e	0	0	0	e	0	0	e	0	0	h	0	0	0	0
15	z		0	0	0	0	e	e	e	0	0	e	0	0	0	0	h	0	0	0
16		x	0	0	0	f	f	0	f	$-f$	0	0	0	0	0	0	0	j	0	0
17		y	0	$-f$	0	0	0	f	0	0	f	$-f$	0	0	0	0	0	0	j	0
18		z	f	0	$-f$	0	0	0	0	0	0	0	f	$-f$	0	0	0	0	0	j

where

$$
\begin{aligned}
a &= -\frac{(Y+G)}{2(Y+G+4)} + \frac{(Z+R)}{2(Z+R+4)},\\
b &= -\frac{4}{2(Y+G+4)},\\
c &= -\frac{(Y+G)}{2(Y+G+4)} - \frac{(Z+R)}{2(Z+R+4)},\\
d &= \frac{4}{2(Z+R+4)},\\
e &= b,\\
f &= zd,\\
g &= Yb,\\
h &= \frac{Y-G-4}{Y+G+4},\\
i &= d,\\
j &= \frac{4-R-Z}{4+Z+R},
\end{aligned}
\tag{7.2.4}
$$

the subscripts appropriate for G, R, Y, and Z are found from Table 7.1, and the normalized stub impedances and admittances are given by [28]

$$
\begin{aligned}
Y_x &= 2\left[\frac{vw}{u}\frac{\epsilon_r}{c\Delta t} - 2\right] & Y_y &= 2\left[\frac{uw}{v}\frac{\epsilon_r}{c\Delta t} - 2\right] & Y_z &= 2\left[\frac{vu}{w}\frac{\epsilon_r}{c\Delta t} - 2\right]\\
Z_x &= 2\left[\frac{vw}{u}\frac{\mu_r}{c\Delta t} - 2\right] & Z_y &= 2\left[\frac{uw}{v}\frac{\mu_r}{c\Delta t} - 2\right] & Z_z &= 2\left[\frac{vu}{w}\frac{\mu_r}{c\Delta t} - 2\right]\\
G_x &= \frac{\sigma_{\mathrm{ex}}(vw)}{u}Z_0 & G_y &= \frac{\sigma_{\mathrm{ey}}(uw)}{v}Z_0 & G_z &= \frac{\sigma_{\mathrm{ez}}(uv)}{w}Z_0\\
R_x &= \frac{\sigma_{\mathrm{mx}}(vw)}{Z_0 u} & R_y &= \frac{\sigma_{\mathrm{my}}(uw)}{Z_0 v} & R_z &= \frac{\sigma_{\mathrm{mz}}(uv)}{Z_0 w}.
\end{aligned}
\tag{7.2.5}
$$

The E-fields at the node center are calculated as

$$
\begin{aligned}
E_x &= -\frac{2}{u(4+Y_x+G_x)}(V_1+V_2+V_3+V_{12}+Y_xV_{13})\\
E_y &= -\frac{2}{v(4+Y_y+G_y)}(V_3+V_4+V_8+V_{11}+Y_yV_{14})\\
E_z &= -\frac{2}{w(4+Y_z+G_z)}(V_5+V_6+V_7+V_{10}+Y_zV_{15}),
\end{aligned}
\tag{7.2.6}
$$

and the H-fields are calculated as

$$
\begin{aligned}
H_x &= \frac{2}{Z_0u(4+Z_x+R_x)}(V_4-V_5+V_7-V_8-V_{16})\\
H_y &= \frac{2}{Z_0v(4+Z_y+R_y)}(-V_2+V_6+V_9-V_{10}-V_{17})\\
H_z &= \frac{2}{Z_0w(4+Z_z+R_z)}(V_1-V_3+V_{11}-V_{12}-V_{18}).
\end{aligned}
\tag{7.2.7}
$$

Since the power scattered into the loss stubs is not returned to the node at the next time step, there is no need to calculate or store the voltage pulses associated with the loss stubs. Because stubs are used to model all the cell properties (shape and material parameters), this TLM formulation is called the stub-loaded SCN. Other variations on the SCN will be discussed later.

Like all discrete field models, waves propagating through a SCN mesh experience numerical dispersion. The dispersion associated with the SCN has been examined by various researchers [34–36]. In general, the dispersion associated with the SCN TLM schemes is less than that for other TLM nodes as well as FDTD field models. The degree of dispersion depends on the number of stubs at the node and the characteristic impedance of the link lines. In theory, the superior dispersion characteristics of the SCN node allow fewer nodes to be used to model propagation in a fixed-size space.

In addition to dispersion, the SCN propagation equations lead to nonphysical propagating solutions called spurious modes [37]. Theoretically, these spurious modes can pollute simulation results and lead to instabilities in ABCs. In practice, there are few reports of spurious modes causing difficulties in practical field simulations, and they can often be suppressed with the use of an appropriate band-limited excitation [38].

Since Johns's original derivation of the SCN, many researchers have examined the node very closely from differing mathematical perspectives. Mlakar [39] developed an equivalent circuit for the SCN that yielded the same scattering matrix as that derived by Johns. While interesting from a theoretical point of view, the circuit was very complicated and of no real practical use. Other researchers have presented rigorous mathematical approaches to derive the scattering matrix for the SCN. Hein [40] presented a derivation of the SCN properties based on a propagator integral approach. Jin and Vahldieck [41] derived the SCN in terms of finite differencing and averaging using a combined time–space coordinate system. Krumpholz and Russer [42] presented a rigorous integral equation derivation based on the method of moments. These various mathematical approaches can often provide insight into the physical propagation processes modeled by the SCN even though they all lead to the same final result—the scattering matrix for the SCN.

A less rigorous approach to obtaining the scattering matrix for various SCNs has been presented [43]. It is based on combining the equivalent transmission line networks for the two-dimensional series and shunt nodes to determine the full three-dimensional SCN scattering matrix. Their equations, while arrived at heuristically, can also be obtained from a rigorous finite-difference averaging approach. In their most general form they provide a simple procedure for obtaining the scattering matrix for any SCN.

Although the SCN TLM method can be implemented directly in terms of equations that update the electric and magnetic field quantities explicitly, most computer codes maintain the efficiency and simplicity of the scattering matrix approach. Trenkic [44] studied the most efficient way to implement the algorithm for SCNs in order to minimize the computational operation count for the scatter/connect procedure.

As discussed previously, open- and short circuit stubs can be added to increase the total inductance and capacitance at the node. Alternatively, the inductance and capacitance of a cell can be adjusted by varying the impedance of the link lines. Any combination of stubs and link line impedances can be used provided the desired total inductance and time synchronization within the entire mesh is maintained. This was indeed the concept behind the hybrid node that was presented in the context of the SCN by Scaramuzza and Lowery [45], who used only capacitive stubs for adding capacitance to the node and adjusting the inductance by varying the link line impedances. Other variations were examined by Berini and Wu [46]. The advantage is that by reducing the number of stubs at each node, the number of voltage pulses that must be stored in memory is reduced as is the number of operations required to update the node at each time step.

Trenkic *et al.* [47] developed the super symmetrical condensed node (SSCN) by totally eliminating all the stubs and obtaining all the required nodal inductance and capacitance by adjusting the impedance of the link lines. Thus, a SCN capable of modeling arbitrary media with a variable mesh was obtained that required the storage of only 12 voltage pulses per node, the same number as that for the original SCN for homogeneous cubic cells. It should be noted that there is some computational overhead associated with calculating and storing the coefficients for the more complicated SSCN scattering matrix.

Trenkic *et al.* [48] recently presented the general theory of the SCN. With this general symmetrical condensed node, any combination of stubs and link line impedances can obtain the required inductance and capacitance at the node. In this way, the nodes can be optimized, for example, to minimize the numerical dispersion associated with the discrete TLM mesh [49]. There is certainly a trade-off associated with an increased number of stubs versus reduced dispersion error in a given TLM model. More work is needed to determine the optimum balance of these parameters, which is likely to be problem dependent.

7.3 SPECIAL FEATURES IN TLM

Previous sections have dealt with "standard" TLM modeling; that is, modeling of arbitrary structures and inhomogeneous medium using rectangular cells with a Cartesian TLM mesh. In this section, we present several modifications to the standard TLM method. These modifications extend the range of application of the method and improve the computational efficiency for certain classes of problems.

7.3.1 Frequency-Dependent Material

In time domain techniques, such as TLM, materials that display properties which vary as a function of frequency present a modeling challenge. Certainly, the TLM model can be excited with a sinusoidal waveform with the material parameters set for that frequency, and the simulation can be rerun at other frequencies of interest with the corresponding constitutive parameters. As was mentioned earlier, this is quite inefficient and totally negates the advantages of an impulsive time domain excitation for obtaining wideband data. Fortunately, several researchers have proposed solutions to this problem.

For cases in which reflection from the surface of a frequency-dependent medium must be modeled, but the actual penetration and propagation into the medium is not of interest, a reflection coefficient approach for the TLM voltage pulses at the material interface can be used [50]. For a voltage pulse V_i, where the i subscript refers to incidence on the boundary as shown in Fig. 7.15a, the reflected pulse in the frequency domain is given by

$$V_r(f) = V_i(f)\Gamma(f), \tag{7.3.1}$$

where $\Gamma(f)$ is the known frequency-dependent reflection coefficient at the media interface. For time domain computations the inverse Fourier transform is applied to Eq. (7.3.1) and the multiplication in the frequency domain becomes a convolution:

$$V_r(t) = V_i(t) \times \Gamma(t). \tag{7.3.2}$$

The time domain pulse incident on the boundary is convolved with a time domain reflection coefficient obtained by inverse Fourier transform of the frequency domain reflection coefficient, which

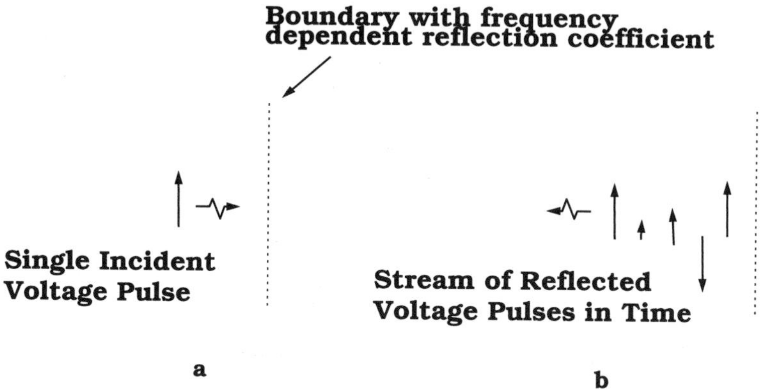

FIGURE 7.15
Reflection from a frequency-dependent boundary. (a) Single-incident voltage pulse results in (b) a stream of reflected pulses in time.

is obtained from the frequency-dependent material parameters. Thus, every voltage pulse incident on the boundary leads to a stream of reflected pulses in time as demonstrated in Fig. 7.15b. In this method, the string of time samples representing the reflection coefficient must be precomputed and stored. This is the same concept used in the Johns matrix approach. Another method that has been proposed for modeling the frequency-dependent reflection from ferrite tiles is given in Refs. [51, 52]. In this case, the reflection coefficient of the tiles in the frequency domain is implemented in the time domain by the use of digital filters. This approach does not require the storage of a complete time domain sequence for the reflection coefficient but does require the calculations associated with the filter. The method can be adapted for materials other than ferrite tiles [53]. In cases in which it is required to model the actual penetration and propagation in the frequency-dependent medium, modifications are required to the TLM SCN node instead of just at the boundary surface. The most general formulation is presented in [54] in which the required modifications are relatively easy to implement. The process consists of using equivalent sources at the TLM nodes to model the material parameters. These sources vary in complexity depending on the characteristics of the material being modeled. Another approach that treats anisotropic media via the derivation of a new SCN scattering matrix is presented in [55].

7.3.2 Alternative Meshing Schemes

The discussion thus far has treated TLM nodes based on Cartesian grids using rectangular or cubic cells. While applicable to a vast range of problems, there are cases in which a fixed Cartesian grid can lead to inaccuracies and/or excessive run times. An example is computing the scattering from a cylindrical surface or the resonances of a spherical cavity. In both these cases forcing the geometries to conform to a Cartesian grid requires a stair-step approximation to the actual surface. The coarseness of the stair stepping determines the accuracy of the results and is very problem dependent, such that it is difficult to derive common rules of thumb to use for different structures.

For problem geometries that naturally conform to orthogonal coordinate surfaces, such as the cylinder and sphere mentioned previously, it is possible to derive TLM models in these coordinate systems. A cylindrical grid SCN has been used successfully to model magnetron cavities [56]. A general formulation for curvilinear orthogonal TLM meshes in two dimensions was presented in [57] but little work has been done to extend this to the three-dimensional SCN.

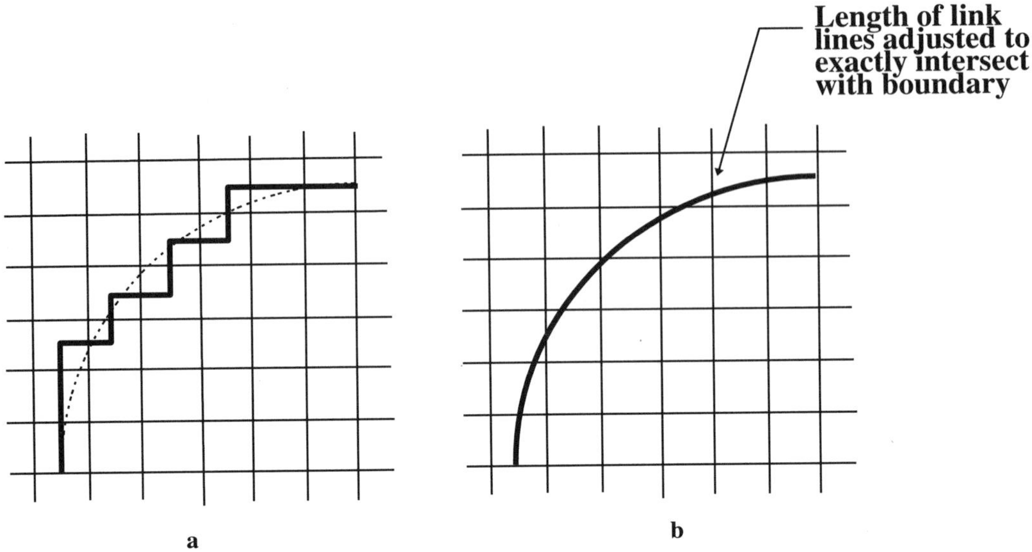

FIGURE 7.16
A section of circular boundary modeled using (a) a stair-step approximation to the true shape and (b) infinitesimally adjustable link line lengths to exactly intersect the boundary.

Another approach is to use a Cartesian grid and locally distort the cells that must conform to a geometry. This can be accomplished in the TLM by permitting link line lengths that intersect the boundary to be infinitesimally adjustable in length. Figure 7.16 compares a stair-cased approximation for a section of a cylinder modeled using adjustable boundaries. The derivation of the scattering matrix for such an adjustable SCN is given by German [58]. It has been demonstrated that, utilizing this technique, a coarser grid achieves the same level of accuracy as that of a stair-casing approximation. In addition to the modeling of curved or slanted boundaries, this technique also allows boundaries to be located at any point between nodes instead of just at the midway point as with the conventional SCN. The advantage of this technique is that only the affected cells require modification and the standard SCN can be used in the rest of the problem space.

Another situation that can lead to difficulties in meshing is the modeling of small features in a large computational space. The mesh must be fine enough in the region of interest to resolve the smallest geometric details, but using this cell size throughout the entire problem space can place a huge demand on computing resources. Use of a variable Cartesian mesh eases the computational burden since the mesh can be varied to be fine in specific regions. However, the variable, or graded mesh, still requires the cell size in any coordinate direction to extend across the entire mesh in the other two directions. An attractive alternative solution is to use a multigrid technique.

For multigrid approaches one or more finer TLM meshes are embedded in a coarser mesh. In this way, a fine mesh can be used around fine features, whereas the coarser mesh is used in the remainder of the problem space. Several formulations for multigrid TLM methods have been presented in the literature [59, 60]. These meshing alternatives are illustrated for a two-dimensional slice of a TLM mesh in Fig. 7.17. So called "smart cell" methods are also extremely useful for accurately modeling certain features that are smaller than the TLM cell size. Such subcell models formulate the interaction physics of the structure directly into the TLM algorithm such that the fine feature need not be actually discretized. Models such as this, including thin wires, thin slots in conducting planes, and thin dispersive material sheets, have been introduced [61–64].

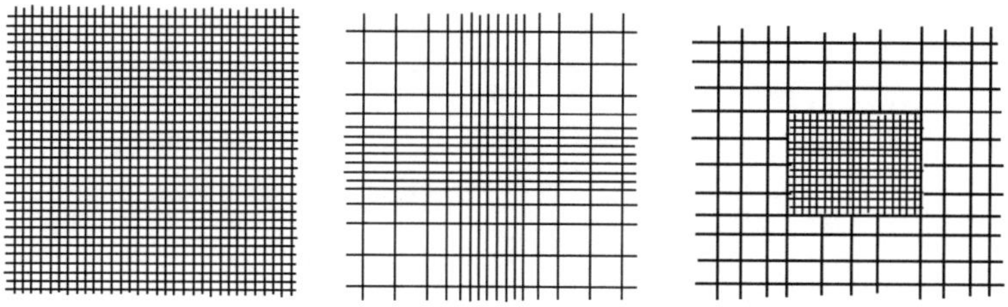

a b c

FIGURE 7.17
Various meshing techniques for modeling a small object in the center of the mesh: (a) fine mesh throughout; (b) variable, or graded, mesh; and (c) multigrid mesh.

These permit the introduction of such subcell-sized structures into a coarse TLM mesh without the need to use extremely fine cells. Smart cell models to accurately model thin conductive edges such as microstrip transmission lines have also been presented [65]. These methods generally involve formulating equivalent circuit parameters which model the physics of the structure (wire, slot, etc.) and then coupling this circuit directly to the TLM mesh via a modified scattering matrix and/or connection procedure.

It is also possible to incorporate circuit device models directly into a TLM mesh. Passive and nonlinear active device models have been successfully modeled in this way [66–68]. By including device models in full three-dimensional TLM field models, coupling and radiation effects can be analyzed. These effects are often difficult if not impossible to include in standard circuit analysis programs. With the trend to higher frequency circuits and ever-increasing computer switching speeds, it is becoming imperative to include the electromagnetic interaction effects directly into the circuit design cycle in order to avoid costly surprises after the circuit is built.

This section provided an overview of advanced formulations of the TLM method which can lead to faster and more accurate modeling of complex electromagnetic interaction problems. While the concentration of this book is on time domain techniques, it should be mentioned that the TLM method has been formulated as a steady-state frequency domain technique. In the frequency domain, pulses traveling between nodes are operated on by a complex propagation coefficient instead of a time delay. The waves reflected from all nodes are formulated as a function on the incident voltage sources in a sparse matrix equation. For details of this procedure, see Refs. [69, 70].

7.4 NUMERICAL EXAMPLES

In this section we present two application examples of the three-dimensional TLM algorithm. Both examples use a computer program based on the hybrid symmetrical condensed node (HSCN) as described in Section 7.2.

7.4.1 Antenna Array

The first example involves the analysis of an array of 16 flared-notch antennas. The general construction of a flared-notch antenna is shown in Fig. 7.18. The antenna element consists of a stripline feed line embedded between two dielectric substrates. The radiating element is a flared

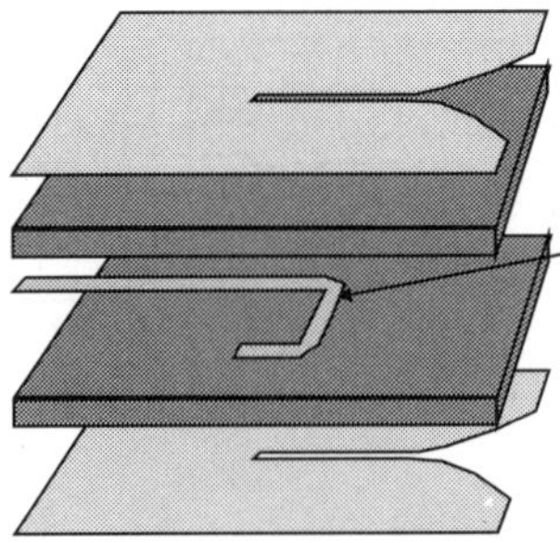

FIGURE 7.18
Construction details of a flared-notch antenna element showing the various layers.

slotline which is etched from the outer conductor layers, which also forms the stripline ground plane. These elements are attractive for phased array applications due to their relatively simple printed circuit construction and the wide bandwidth achievable. The element as shown in Fig. 7.18 consists of the stripline feed, the stripline-to-slotline transition which couples the energy from the feedline to the radiator, and the actual flared slotline radiator. Figure 7.19 shows a side view of the actual element used in the current example. The stripline feed is shown by the dashed line in the illustration and contains a step discontinuity and bend. The stripline has two step discontinuities before the flared radiator begins. These discontinuities in both the stripline and the slotline are for matching the high-impedance slotline radiator to the low-impedance stripline feed. Sixteen of these radiators were arranged in a square 4×4 element array. Figure 7.20 shows a front view of the array. In this figure radiation occurs out of the page. The metal test fixture was used to hold the elements in place. The total cross section of the array was 5.08×4.13 cm and the depth was 2.54 cm.

Modeling of the array presents a true modeling challenge due to the large aspect ratios involved. The TLM mesh must be fine enough to resolve the small-scale details of the feed line and transition and at the same time the problem space must be large enough to contain the entire array structure and enough space between the array and ABCs to minimize unwanted reflections. In order to achieve satisfactory discretization in the feed region a uniform TLM cell size of 169.33 μm was used throughout the problem space. Due to symmetry, a quarter of the array was included in the TLM model and appropriate boundary conditions were imposed at the symmetry planes. The total mesh size in the three coordinate directions was $200 \times 170 \times 130$ for a total of 7,820,000 HSCN

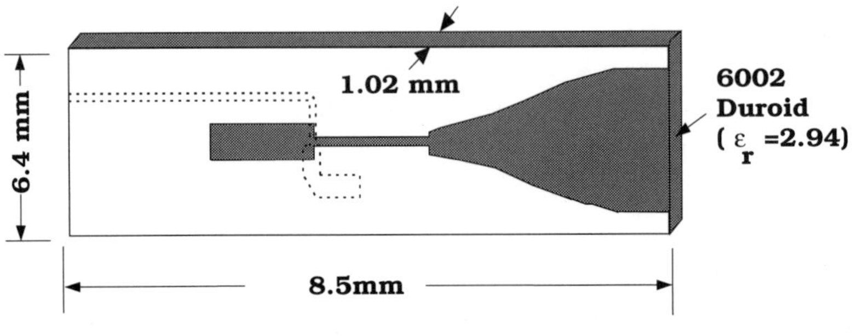

FIGURE 7.19
Side view of flared-notch antenna element used in the array example.

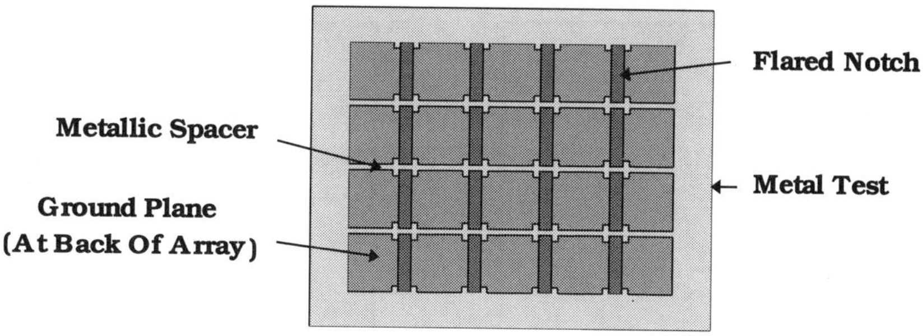

FIGURE 7.20
Front view of the 4 × 4 element flared notch array with test fixture.

cells. First order matched boundaries were found to provide an adequate ABC for this problem with approximately a quarter wavelength between the sides and front of the array and outer mesh boundaries at the lowest frequency of interest. Since very little radiation occurs to the back of the array, only five cells separated it from the ABC.

The simulation was run and the tangential electric and magnetic fields were saved on a virtual surface surrounding the entire array. These field quantities were then used to calculate the radiation pattern of the array using a near- to far-field transform [71]. The directivity, D, of the array was also computed as follows:

$$D(\theta, \phi) = 4\pi \frac{|E_F(\theta, \phi)|^2}{\eta P_{rad}}, \tag{7.4.1}$$

where $E_F(\theta, \phi)$ is the electric field at observation angles θ and ϕ in the far field and P_{rad} is the radiated power obtained by the following integration:

$$P_{rad} = \frac{1}{\eta} \int\int \left| E_F^2(\theta, \phi) \right| d\omega. \tag{7.4.2}$$

The free-space impedance is given by η and is 377 ohms.

A 4 × 4 test array was constructed and measured. The antenna patterns were measured over the entire hemisphere in front of the array. The integration shown for previously P_{rad} was performed over this hemisphere for both the measured and TLM data at 6, 12, and 18 GHz. Table 7.2 compares the measured and calculated values of the maximum directivity for the array. All elements of the array were excited with equal power and in phase. The agreement is quite good, with the largest discrepancy being 0.7 dB between measured and calculated values at 6 GHz.

Principal plane cuts of the normalized antenna patterns were also compared at the same three frequencies and are shown in Fig. 7.21. The E-plane is the vertical scan plane and the H-plane

Table 7.1 The TLM vs Measured Directivities

Frequency (GHz)	Directivity (dB) measured	Directivity (dB) TLM
6	9.7	9.0
12	13.6	14.0
18	17.2	16.8

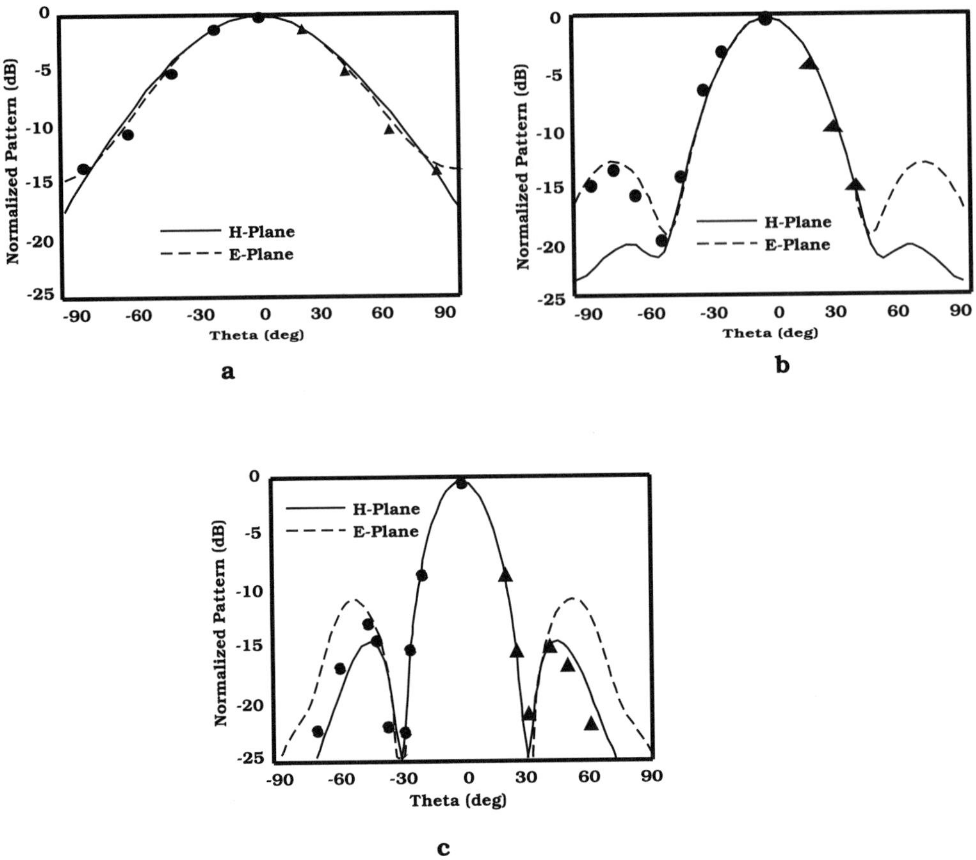

FIGURE 7.21
Principal plane radiation patterns at (a) 6, (b) 12, and (c) 18 GHz.

the horizontal. The continuous and dashed lines are the TLM computed patterns. The measured data are shown by circles (E-plane) and triangles (H-plane). Since the patterns are symmetric, the measured data are only included for negative observation angles in the E-plane and for positive observation angles in the H-plane. Thus, the circles (measured) should be compared to the dashed line (TLM) for negative angles, and the triangles (measured) should be compared to the solid line (TLM) for positive angles.

The agreement between the measured and computed data is quite good, especially considering the size and complexity of the TLM model. It is worth pointing out again that the TLM model used here encompasses the entire array structure including the feedline, radiator, and measurement fixture and thus incorporates the physics associated with the microwave feed network, radiation of the antenna elements, coupling between the antenna elements, and interaction with the test fixture holding the elements in place.

7.4.2 Electromagnetic Scattering

The previous example was concerned with the radiation produced by a source feeding an antenna. In this example, we consider how energy is scattered from a stationary object when illuminated with a plane wave at different incident angles. The geometry is shown in Fig. 7.22 and consists

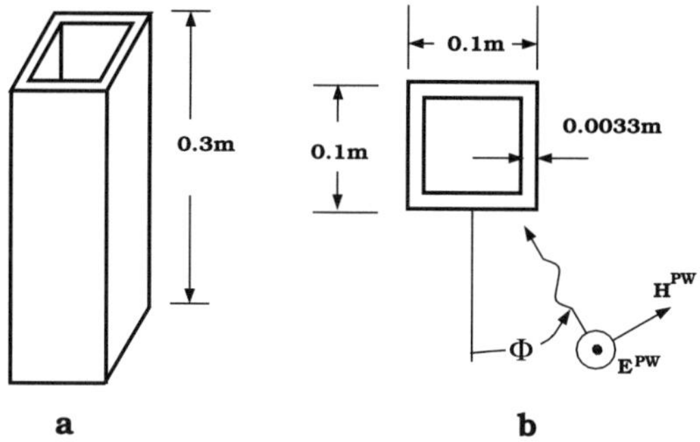

FIGURE 7.22
Geometry of scatterer and plane wave polarization: (a) oblique view and (b) top view.

of a perfectly conducting hollow tube of square cross section. The TLM model used a uniform mesh with a 3.3-mm cell size such that a single cell was used to model the metal thickness. The incident plane wave orientation is also shown.

In order to implement the source condition that produces the plane wave in the TLM model, the problem space is separated into a total- and scattered-field region as shown in Fig. 7.23 for an arbitrary scatterer. The plane wave source condition is implemented on the boundary between the two regions and results in the total field (incident and scattered) being present inside the boundary and only the scattered field (propagating outward on the mesh) outside of the boundary. In this way, the ABCs need only operate on the outward-propagating scattered field and the near- to far-field transform is easily performed in the scattered-field region. Details of implementing this procedure for the SCN TLM method are given by German [71].

The monostatic radar cross section as a function of illumination angle ϕ is shown in Fig. 7.24. Comparison is made to measured data given in [72]. The error bars represent the range of measured data at each look angle. The temporal profile of the excitation pulse used here was a Gaussian pulse with a relatively low-frequency content in order to minimize errors associated with mesh dispersion and the connection between the total- and scattered-field regions.

7.5 CONCLUDING REMARKS

In this chapter, we developed the TLM method for use in the solution of electromagnetic interaction problems in bounded and unbounded spaces. The method is general in nature and can be applied

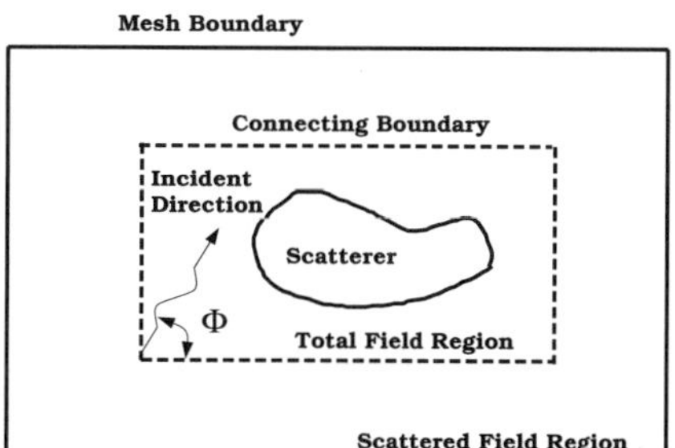

FIGURE 7.23
Separation of the TLM mesh into total- and scattered-field regions.

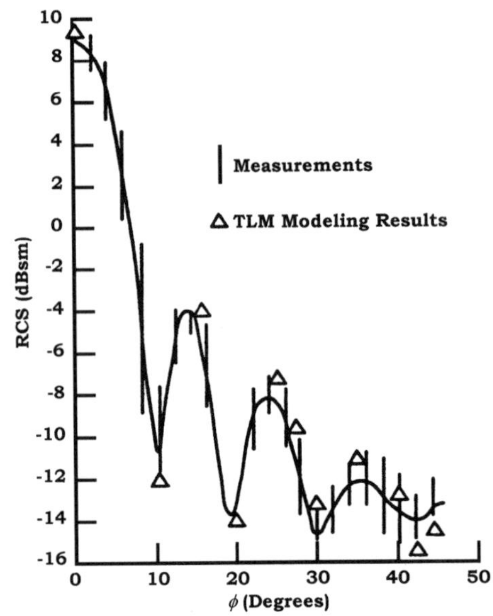

FIGURE 7.24
Monostatic radar cross section (RCS).

to a variety of problems such as analysis of microstrip lines, coating of radar-absorbing materials, and scattering by inhomogeneous material bodies.

BIBLIOGRAPHY

[1] G. Kron, "Equivalent Circuit of the Field Equations of Maxwell — I," *Proc. IRE*, vol. 32, pp. 289–299, 1944.

[2] J. R. Whinnery and S. Ramo, "A New Approach to the Solution of High-Frequency Field Problems," *Proc. IRE*, vol. 32, pp. 384–288, 1944.

[3] P. B. Johns and R. L. Beurle, "Numerical Solution of 2-Dimensional Scattering Problems Using a Transmission-Line Matrix," *Proc. IEEE*, vol. 118, pp. 1203–1209, 1971.

[4] S. Akhtarzad and P. B. Johns, "Three-Dimensional Transmission-Line Matrix Computer Analysis of Microstrip Resonators," *IEEE Trans. Microwave Theory Tech.*, vol. 23, pp. 990–998, 1975.

[5] P. Saguet and W. J. R. Hoefer, "The Modeling of Multiaxial Discontinuities in Quasi-Planar Structures with the Modified TLM Method," *Int. J. Numerical Modeling Elec. Networks Devices Fields I*, vol. 1, pp. 7–17, 1988.

[6] P. B. Johns, K. Akhtarzad, and Y. Rahhal, "Numerical Simulation of Interference in Aircraft Due to Electromagnetic Fields," *IERE International Conference on Electromagnetic Compatibility*, England, pp. 295–299, 1980.

[7] F. J. German, G. K. Gothard, L. S. Riggs, and P. M. Goggans, "The Calculation of Radar Cross Section (RCS) Using the TLM Method," *Int. J. Numerical Modeling Elec. Networks Devices Fields*, vol. 2, pp. 1797–1812, 1990.

[8] P. B. Johns, "A Simple Explicit and Unconditionally Stable Numerical Routine for the Solution of the Diffusion Equation," *Int. J. Numerical Methods Eng.*, vol. 11, pp. 1307–1328, 1977.

[9] A. H. M. Saleh and P. P. Blanchfield, "Analysis of Acoustic Radiation Patterns of Array Transducers Using the TLM Method," *Int. J. Numerical Modeling*, vol. 3, pp. 39–56, 1990.

[10] D. M. Pozar, *Microwave Engineering,* Addison-Wesley, Reading, MA, 1990.
[11] M. N. O. Sadiku, *Numerical Techniques in Electromagnetics*, CRC Press, Boca Raton, FL, 1992.
[12] C. Christopoulos, *The Transmission Line Modeling Method—TLM*. Oxford Univ. Press, Oxford, 1995.
[13] T. G. Moore, J. G. Blaschal, A. Taflove, and G. A. Kriegsmann, "Theory and Application of Radiation Boundary Operators," *IEEE Trans. Antennas Propagat.*, vol. 36, pp. 1797–1812, 1988.
[14] P. B. Johns and K. Akhtarzad, "The Use of Time Domain Diakoptics in Time Discrete Models of Fields," *Int. J. Numerical Methods Eng.*, vol. 17, pp. 1–14, 1981.
[15] W. J. R. Hoefer, "The Discrete Time Domain Green's Function or Johns Matrix—A New Powerful Concept in Transmission Line Modeling (TLM)," *Int. J. Numerical Modeling*, vol. 2, pp. 215–225, 1989.
[16] M. Righi, J. L. Herring, and W. J. R. Hoefer, "Efficient Hybrid TLM/Mode-Matching Analysis of Packaged Components," *IEEE Trans. Microwave Theory Tech.*, vol. 45, pp. 1715–1724, 1997.
[17] R. L. Higdon, "Numerical Absorbing Boundary Conditions for the Wave Equations," *Math. Comp.*, vol. 49, pp. 65–91, 1987.
[18] Z. Chen, M. M. Ney, and W. J. R. Hoefer, "Absorbing and Connecting Boundary Conditions for the TLM Method," *IEEE Trans. Microwave Theory Tech.*, vol. 41, pp. 2015–2024, 1993.
[19] J. A. Morente, J. A. Porti, and M. Khalladi, "Absorbing Boundary Conditions for the TLM Method," *IEEE Trans. Microwave Theory Tech.*, vol. 40, pp. 2095–2099, 1992.
[20] C. Eswarappa and W. J. R. Hoefer, "One-Way Equation Absorbing Boundary Conditions for the 3-D TLM Analysis of Planar and Quasi-Planar Structures," *IEEE Trans. Microwave Theory Tech.*, vol. 42, pp. 1669–1677, 1994.
[21] J. P. Berenger, "A Perfectly Matched Layer for the Absorption of Electromagnetic Waves," *J. Comp. Phys.*, vol. 114, pp. 110–117, 1994.
[22] C. Eswarappa and W. J. R. Hoefer, "Implementation of Berenger's Absorbing Boundary Conditions in TLM by Interfacing FDTD Perfectly Matched Layers," *Electron. Lett.*, vol. 31, pp. 1264–1266, 1995.
[23] N. Pena and M. M. Ney, "Absorbing-Boundary Conditions Using Perfectly Matched-Layer (PML) Techniques for Three-Dimensional TLM Simulations," *IEEE Trans. Microwave Theory Tech.*, vol. 45, pp. 1749–1755, 1997.
[24] C. R. Brewitt-Taylor, "On the Construction and Numerical Solution of Transmission-Line and Lumped Network Models of Maxwell's Equations," *Int. J. Numerical Methods Eng.*, vol. 15, pp. 13–30, 1980.
[25] R. F. Harrington, *Time-Harmonic Electromagnetic Fields*, McGraw-Hill, New York, 1961.
[26] W. J. R. Hoefer and P. P. M. So, *The EM Wave Simulator*, Wiley, New York, 1991.
[27] P. B. Johns, "On the Relationship between TLM and Finite-Difference Methods for Maxwell's Equations," *IEEE Trans. Microwave Theory Tech.*, vol. 35, pp. 60–61, 1987.
[28] A. Amer, "The Condensed Node TLM Method and Its Application to Transmission in Power Systems," PhD thesis, Nottingham University, Nottingham, UK, 1980.
[29] P. Saguet and E. Pic, "Utilization d'un Nouveau Type de Noeud dans la Methode TLM en 3 Dimensions," *Electron. Lett.*, vol. 18, pp. 478–480, 1982.
[30] P. B. Johns, "A Symmetrical Condensed Node for the TLM Method," *IEEE Trans. Microwave Theory Tech.*, vol. 35, pp. 370–377, 1987.
[31] R. Allen, A. Mallik, and P. B. Johns, "Numerical Results for the Symmetrical Condensed TLM Node," *IEEE Trans. Microwave Theory Tech.*, vol. 35, pp. 378–382, 1987.
[32] P. Naylor and R. A. Desai, "New Three Dimensional Symmetrical Condensed Lossy Node for Solution of Electromagnetic Wave Problems by TLM," *Electron. Lett.*, vol. 26, pp. 492–494, 1990.
[33] F. J. German, G. K. Gothard, and L. S. Riggs, "Modeling of Materials with Electric and Magnetic Losses with the Symmetrical Condensed Node TLM Method," *Electron. Lett.*, vol. 26, pp. 1307–1308, 1990.
[34] J. S. Nielsen and W. J. R. Hoefer, "A Complete Dispersion Analysis of the Condensed Node TLM Mesh," *IEEE Trans. Magn.*, vol. 27, pp. 1375–1384, 1991.
[35] M. Krumpholz and P. Russer, "On the Dispersion in TLM and FDTD," *IEEE Trans. Microwave Theory Tech.*, vol. 42, pp. 1275–1279, 1994.
[36] V. Trenkic, C. Christopoulos, and T. Benson, "Analytical Expansion of the Dispersion Relation for TLM Condensed Nodes," *IEEE Trans. Microwave Theory Tech.*, vol. 44, pp. 2223–2230, 1996.

[37] J. S. Nielsen and W. J. R. Hoefer, "Generalized Dispersion Analysis and Spurious Modes of 2-D TLM and 3-D TLM Formulations," *IEEE Trans. Microwave Theory Tech.*, vol. 41, pp. 1375–1384, 1993.

[38] S. Lindenmeier, B. Isele, R. Weigel, and P. Russer, "On the Suppression of Spurious Modes in Stub-Loaded 3D-Symmetrical Condensed TLM-Nodes," *First International Workshop on Transmission Line Matrix (TLM) Modeling*, Victoria, Canada, pp. 41–44, 1995.

[39] J. Mlakar, "Circuit Model for a Symmetrical Condensed TLM Node," *Int. J. Numerical Modeling*, vol. 6, pp. 183–193, 1993.

[40] S. Hein, "Consistent Finite Difference Modeling of Maxwell's Equations with Lossy Symmetrical Condensed TLM Node," *Int. J. Numerical Modeling*, vol. 6, pp. 207–220, 1993.

[41] H. Jin and R. Vahldieck, "Direct Derivations of TLM Symmetrical Condensed Node and Hybrid Symmetrical Condensed Node Form Maxwell's Equations Using Centered Differencing and Averaging," *IEEE Trans. Microwave Theory Tech.*, vol. 42, pp. 2554–2561, 1994.

[42] M. Krumpholz and P. Russer, "A Field Theoretical Derivation of TLM," *IEEE Trans. Microwave Theory Tech.*, vol. 42, pp. 1660–1668, 1994.

[43] P. Naylor and R. Ait-Sadi, "Simple Method for Determining 3D TLM Nodal Scattering in Nonscalar Problems," *Electon. Lett.*, vol. 28, pp. 2353–2354, 1992.

[44] V. Trenkic, C. Christopoulos, and T. Benson, "Efficient Computational Algorithms for TLM," *First International Workshop on Transmission Line Matrix (TLM) Modeling*, Victoria, Canada, pp. 77–80, 1995.

[45] R. A. Scaramuzza and A. J. Lowery, "Hybrid Symmetrical Condensed Node for TLM Method," *Electron. Lett.*, vol. 26, pp. 1947–1949, 1990.

[46] P. Berini and K. Wu, "A Pair of Hybrid Symmetrical Condensed TLM Nodes," *IEEE Microwave Guided Wave Lett.*, vol. 4, pp. 244–246, 1994.

[47] V. Trenkic, C. Christopoulos, and T. Benson, "Theory of the Symmetrical Super-Condensed Node for the TLM Method," *IEEE Trans. Microwave Theory Tech.*, vol. 43, pp. 1342–1348, 1995.

[48] V. Trenkic, C. Christopoulos, and T. Benson, "Development of a General Symmetrical Condensed Node for the TLM Method," *IEEE Trans. Microwave Theory Tech.*, vol. 44, pp. 2129–2135, 1996.

[49] V. Trenkic, C. Christopoulos, and T. Benson, "Optimization of TLM Schemes Based on the General Symmetrical Condensed Node," *IEEE Trans. Antennas Propagat.*, vol. 45, pp. 457–465, 1997.

[50] A. Mallik and C. P. Loller, "The Modeling of EM Leakage into Advanced Composite Enclosures Using the TLM Technique," *Int. J. Numerical Modeling*, vol. 2, pp. 241–248, 1989.

[51] J. F. Dawson, "Representing Ferrite Absorbing Tiles as Frequency Dependent Boundaries in TLM," *Electron. Lett.*, vol. 29, pp. 791–792, 1993.

[52] V. Trenkic, J. Paul, I. Argyri, and C. Christopoulos, "Modeling of Ferrite Tiles as Frequency Dependent Boundaries in General Time Domain TLM Schemes," *13th Annual Review of Applied Computer Electromagnetics*, Monterey, CA, pp. 630–637, 1997.

[53] J. A. Cole, J. F. Dawson, and S. J. Porter, "A Digital Filter Technique for Electromagnetic Modeling of Thin Composite Layers in TLM," *13th Annual Review of Applied Computational Electromagnetics*, Monterey, CA, pp. 686–693, 1997.

[54] J. Huang and K. Wu, "A Unified TLM Model for Wave Propagation of Electrical and Optical Structures Considering Permittivity and Permeability Tensors," *IEEE Trans. Microwave Theory Tech.*, vol. 43, pp. 2472–2477, 1995.

[55] L. R. A. X. de Menezes and W. J. R. Hoefer, "Modeling of General Constitutive Relationships in SCN TLM," *IEEE Trans. Microwave Theory Tech.*, vol. 44, pp. 854–861, 1996.

[56] R. Allen and M. J. Clark, "Application of the Symmetrized Transmission Line Matrix Method to the Cold Modeling of Magnetrons," *Int. J. Numerical Modeling*, vol. 1, pp. 61–70, 1988.

[57] H. Meliani, D. De Cogan, and P. B. Johns, "The Use of Orthogonal Curvilinear Meshes in TLM Models," *Int. J. Numerical Modeling*, vol. 1, pp. 221–238, 1988.

[58] F. J. German, "Infinitesimally Adjustable Boundaries in Symmetrical Condensed Node TLM Simulations," *9th Annual Review of Applied Computer Electromagnetics*, Monterey, CA, pp. 482–490, 1993.

[59] J. L. Herring and C. Christopoulos, "Solving Electromagnetic Field Problems Using a Multiple-Grid Transmission Line Modeling Method," *IEEE Trans. Antennas Propagat.*, vol. 42, pp. 1654–1658, 1994.

[60] A. J. Wlodarczyk, "New Multigrid Interface for the TLM Method," *Electron. Lett.*, vol. 32, pp. 1111–1112, 1996.

[61] A. J. Wlodarczyk and D. P. Johns, "New Wire Interface for Graded 3-D TLM," *Electron. Lett.*, vol. 28, pp. 728–729, 1992.

[62] J. A. Porti, J. A. Morente, M. Khalladi, and A. Callego, "Comparison of Thin Wire Models for TLM Method," *Electron. Lett.*, vol. 28, pp. 1910–1911, 1992.

[63] A. Mallik, D. P. Johns, and A. J. Wlodarczyk, "TLM Modeling of Wires and Slots," *10th Int. Zurich Symp. Electromag. Compatibility*, pp. 515–520, 1993.

[64] D. P. Johns, A. J. Wlodarczyk, and A. Mallik, "New TLM Models for Thin Structures," *Proc. Int. Conf. Comp. Electromag.*, pp. 335–338, 1991.

[65] L. Cascio, G. Tardioli, T. Rozzi, and W. J. R. Hoefer, "A Quasi-Static Modification of TLM at Knife Edge and 90° Wedge Singularities," *IEEE MTT-S Symp.*, Orlando, FL, pp. 23–26, 1995.

[66] J. Nielsen and W. J. R. Hoefer, "Modeling of Non-Linear Elements in a 3D Condensed Node TLM Mesh," *Int. J. Microwave Millimeter Wave Comput. Aided Design*, vol. 3, pp. 61–66, 1993.

[67] R. H. Voelker and R. J. Lomax, "A Finite-Difference Transmission Line Matrix Method Incorporating a Nonlinear Device Model," *IEEE Trans. Microwave Theory Tech.*, vol. 38, pp. 302–312, 1990.

[68] P. Russer, P. So, and W. J. R. Hoefer, "Modeling of Nonlinear Active Regions in TLM," *IEEE Microwave Guided Wave Lett.*, vol. 1, pp. 8–10, 1991.

[69] D. P. Johns, A. J. Wlodarczyk, and A. Mallik, "New TLM Technique for Steady-State Field Solutions in Three-Dimensions," *Electron. Lett.*, vol. 28, pp. 1692–1694, 1992.

[70] H. Jin and R. Vahldieck, "The Frequency Domain Transmission-Line Matrix Method—A New Concept," *IEEE Trans. Microwave Theory Tech.*, vol. 40, pp. 2207–2218, 1992.

[71] F. J. German, "General Electromagnetic Scattering Analysis by TLM Method," *Electron. Lett.*, vol. 30, pp. 689–690, 1994.

[72] A. Taflove, K. R. Umashankar, and T. G. Jurgens, "Validation of FD-TD Modeling of the Radar Cross Section of Three Dimensional Structures Spanning up to Nine Wavelengths," *IEEE Trans. Electromag. Compatibility*, vol. 25, pp. 440–443, 1983.

CHAPTER 8

Finite-Element Time Domain Method

T. K. SARKAR and T. ROY
Department of Electrical Engineering and Computer Science, Syracuse University

M. SALAZAR-PALMA
Universidad Politécnica de Madrid, ETSI de Telecomunicación, Ciudad Universitaria

A. R. DJORDJEVIC
Department of Electrical Engineering, University of Belgrade

8.1 INTRODUCTION

In this chapter, another differential equation-based time domain method, the finite-element time domain (FETD) method, is developed. Unlike the finite-difference time domain (FDTD) and transmission line modeling methods dealt with in Chapters 6 and 7, the FETD can be used with unstructured grids which improves the modeling capabilities of arbitrary geometries. In the following, we discuss the method in detail.

The finite-element method was originally applied for the approximate solution of boundary value problems. It was a natural step in developing similar methods for initial value problems. Much of the original stimulus in the development of finite-element methods came from energy or variational principles. Many of the finite-element methods implemented for initial value problems have been motivated by variational principles but very few are strictly variational. Most of them either use variational principles in the space variables only, introducing the time variable at later stages, or use variational principles in space and time as a guide to the derivation of a functional that is not, in fact, stationary in the sense of the classical calculus of variations. These quasi-variational principles are well-known in other contexts such as heat conduction and thermodynamics. Normally, finite-element methods are based on extreme principles, stationary principles that are not extreme [1], quasi-variational principles, and the method of weighted residuals. The extreme principles exist only for restricted classes of equations. The advantage of

Pages 282–294: Portions reprinted, with permission, from *IEEE Transactions on Antennas and Propagation*, Vol. 46, No. 10 (October 1998): 1471–1477. © 1998 IEEE.

the method of weighted residuals is that it can be applied to general equations but it is not always clear how to choose the optimum weight factors. Stationary principles and quasi-variational principles help us to choose suitable weighting functions. These considerations are to some extent independent of the fact that finite-element methods usually involve weighting functions that are nonzero in only parts of the range of interest, which tends to produce stable numerical procedures.

Because there is a lack of finite-element methods based on extreme principle, we will use quasi-variational techniques in our formulation. In this analysis part of the undetermined parameters are assumed to be functions of time, whereas the other set of parameters are assumed to depend on spatial coordinates. This leads to two stages for the solution, both of which employ approximate methods. In the solution of time-dependent problems, the spatial approximation is considered first and the time approximation next. This semidiscrete variational approximation in space results in a set of ordinary differential equations in time, which must be further approximated to obtain a set of algebraic equations. The spatial approximation of time-dependent problems leads to a matrix differential equation of the form

$$[\boldsymbol{A}]\left\{\frac{\partial^2\Phi}{\partial t^2}\right\}+[\boldsymbol{B}]\{\Phi\}=\{\boldsymbol{P}\}. \tag{8.1.1}$$

There are several approximation schemes available for time derivatives. The most commonly used is the Newmark direct-integration method. In this method parameters can be chosen to control the accuracy and the stability of the scheme. It is known that with appropriate values of these parameters, the Newmark method yields an unconditionally stable scheme in linear problems [2].

Finite-element methods have been implemented using both implicit and explicit methods. For an implicit method to be competitive with an explicit method, the number of time iterations required to converge to a final solution must be significantly less since each time iteration requires a solution of a linear system of equations. The Newmark method applied to a wave equation is an implicit method. Researchers in mechanical engineering have used this method extensively in dynamic analysis of beams, trusses, etc. For computational electromagnetics Barkeshli *et al.* [2] used this method coupled with boundary integrals to solve transient electromagnetic field coupling for a metallic enclosure. The fields outside the computational domain are coupled through the H-field integral equation, and the coupled system of equations are solved by a leapfrogging technique. Gedney and Navsariwala [3] applied the Newmark algorithm to the time-dependent vector wave equation. They performed a stability analysis of the same wave equation and demonstrated that it leads to an unconditionally stable solution for certain choices of a few particular parameters. They applied their technique to a cavity resonator problem.

In our analysis, we have used a quasi-variational formulation for the two-dimensional wave equation. Using space discretization, we arrive at the second-order system of differential equations with respect to the time variable. Next, the Newmark integration scheme is used to transform the second-order equation into a set of simple algebraic equations, which are solved at each time step in a leapfrogging manner. In this approach, we used the Green's function to terminate the finite-element mesh, and because the fields are causal, at the terminating surface they can be calculated through the Green's function integral using induced surface currents computed from earlier time steps. If the source and the field point are separated by a distance R, the induced current at a particular time instant will have an effect on the field at a time R/c later, where c is the velocity of propagation in the medium considered. Hence, the boundary condition at the terminating surface at a particular time instant can be found exactly from earlier induced surface currents. The induced surface current on a particular subsection of the body is also determined from the boundary condition on the body. The time-stepping procedure continues to a time for which the data for a particular problem are desired.

8.1.1 Incident Field

In the time domain analysis we are required to turn on the excitation in a "numerically gentle" fashion. This is necessary to avoid introducing excitation frequency components that exceed the highest frequency for which the model is valid. Also, if we wish to derive specific characteristics of an object for particular frequencies, the spectral content of the excitation determines whether that frequency range is actually covered.

Instead of an ideal impulse, we use an approximate smoothed impulse which has a Gaussian temporal variation. This type of excitation is time and band limited and very well suited for numerical modeling. It decays very rapidly in both the time and frequency domains. The Gaussian plane wave in this work is given by

$$\boldsymbol{E}^{\mathrm{i}}(\boldsymbol{r}, t) = \boldsymbol{E}_{\mathrm{o}} \frac{4.0}{\sqrt{\pi} T} e^{-\gamma^2}, \tag{8.1.2}$$

where

$$\gamma = \frac{4.0}{T} \left[ct - ct_{\mathrm{o}} - \boldsymbol{r} \cdot \boldsymbol{k} \right], \tag{8.1.3}$$

$\boldsymbol{E}_{\mathrm{o}}$ denotes the amplitude, T denotes the width of the pulse, $\boldsymbol{k}$ is the unit vector along the direction of propagation, t_{o} is the time delay which denotes the time when the pulse attains its maximum value at the origin, and $\boldsymbol{r}$ is the position vector with respect to the origin. The pulse width T is defined such that the magnitude of the incident field falls to 2% of its peak value. In our work we have assumed $T = 2$ LM and $ct_{\mathrm{o}} = 3$ LM. Taking the Fourier transform of Eq. (8.1.2) and assuming $\boldsymbol{r} \cdot \boldsymbol{k} = 0$, we have

$$\mathcal{F}(\boldsymbol{E}^{\mathrm{i}}) = \frac{\boldsymbol{E}_{\mathrm{o}}}{c} e^{-j2\pi f t_{\mathrm{o}}} e^{-\left(\frac{\pi f T}{4c}\right)^2}. \tag{8.1.4}$$

The temporal and spectrum plots of two pulses in the time and frequency domain are shown in Figs. 8.1 and 8.2, respectively. As can be seen, by changing the width of the pulse one can either seek narrowband or broadband information.

In the following sections, the time domain solution of scattering by two-dimensional conducting bodies is considered in detail. In Sections 8.2 and 8.3, we consider the transverse magnetic (TM)

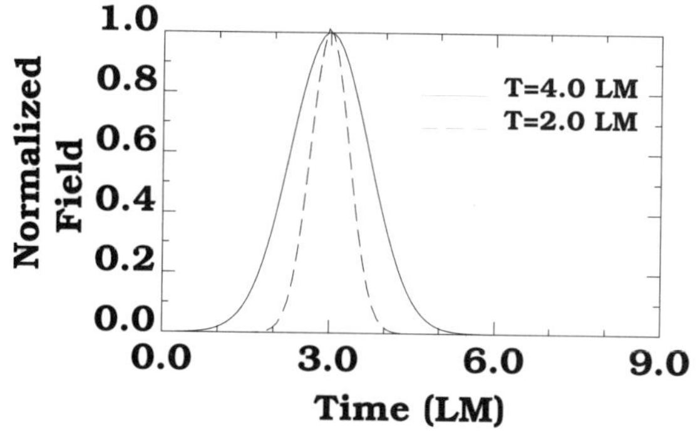

FIGURE 8.1
Gaussian pulse in the time domain.

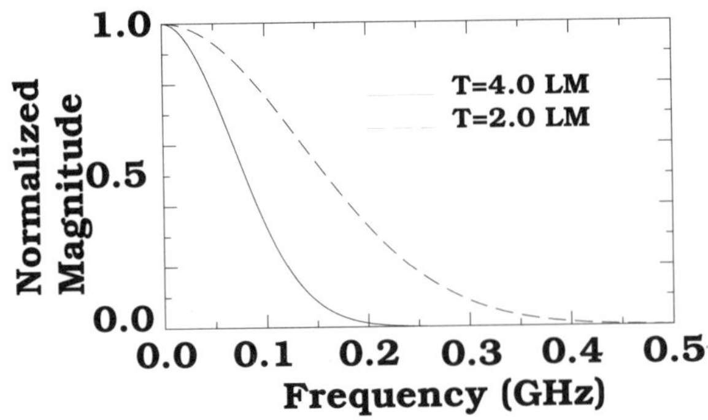

FIGURE 8.2
Fourier transform of the Gaussian pulse.

and the transverse electric (TE) cases, respectively. In each case, we provide a detailed description of the formulation and the numerical solution procedure, including the application of boundary conditions for the terminating surface. Smoothed impulse responses for selected objects are presented, discussed, and compared with responses of other computational methods. Finally, we present a discussion of the FETD method along with its merits and demerits.

8.2 TRANSVERSE MAGNETIC CASE

Previously, electromagnetic scattering problems were traditionally tackled by frequency domain techniques in which time-harmonic behavior was assumed, the main reason being the simplicity of the numerical approach for the frequency domain formulation. However, with the advent of fast and powerful computers there is a steady growth of time domain modeling. Initially, the research was mainly confined to integral equation formulations, but in recent years differential equation modeling has attracted wide attention.

The direct time domain solution has advantages when the solutions of transient electromagnetic scattering problems are desired. The computations are carried out by illuminating the object with a smoothed impulse waveform. The width of the pulse determines the band of frequency information to be derived for the particular object. The impulse response of an object provides useful information and can be treated as a characteristic signature of a target. First, all the electromagnetic information of scattering properties of a body is contained in the impulse response. The frequency response can be derived by the Fourier transform of the impulse response. Also, any response of a target can be computed by performing a convolution of the impulse response with the desired input waveform pulse.

As mentioned in Chapter 1, the integral equation technique to solve time domain problems received wide acceptance in the early stages of time domain research. However, differential equation techniques such as finite-difference and finite-element procedures have slowly begun to be used for time domain analysis. When solving an open-region problem using a differential equation procedure, one needs to discretize the entire space, which from a computational viewpoint is highly unrealistic. Hence, the question of solving the problem with a finite portion of the space domain arises. The introduction of a new surface which bounds the computational domain also poses a few problems. For a unique solution of the open problem, one needs to apply appropriate boundary conditions on that fictitious surface. There are several techniques used to apply boundary conditions on the fictitious surface based on the local or nonlocal nature of the formulation as discussed in Chapter 1, Section 1.5.

In [4] and [5], a hybrid method to terminate the finite-element meshes for frequency domain scattering problems has been developed. In that analysis the authors used a scalar Helmholtz equation for solving the fields and enforced the radiation condition on the terminating surface from the induced currents on the illuminated cylinder.

Also in the time domain analysis, one needs to determine appropriate boundary conditions for "exact" termination of the finite-element meshes. Because this procedure introduces an implicit time-stepping scheme, the computational domain should be kept small to speed up the computation. However, in this process, if the boundary is too close to the object it may introduce unwanted reflections. Hence, judicious placement of the terminating surface from the target results in reduced computational cost and proper termination of the mesh boundary.

Consider a cylinder illuminated by a Gaussian pulse in time whose electric field is parallel to the cylinder axis. This incident field produces axially directed induced surface currents on the conductor, a scattered magnetic field in the transverse direction from that induced current, and an electric field along the axis of the cylinder.

In this section, we solve the two-dimensional wave equation using the finite-element technique combined with the integral equation approach. For TM scattering problems, since $E_z(x, y, t)$ is the axial component of the electric field, the wave equation can be written as

$$\nabla^2 E_z(x, y, t) - \frac{1}{c^2}\frac{\partial^2 E_z(x, y, t)}{\partial t^2} = 0$$

$$\Rightarrow \frac{\partial^2 E_z(x, y, t)}{\partial x^2} + \frac{\partial^2 E_z(x, y, t)}{\partial y^2} - \frac{1}{c^2}\frac{\partial^2 E_z(x, y, t)}{\partial t^2} = 0$$

$$x, y \in \Re; \quad t \in (0, \infty), \qquad (8.2.1)$$

where z is the axial coordinate, $\Re$ is the cross section of the computational domain, E_z does not depend on z, and c is the velocity of the wave in the medium considered. Furthermore, we note that E_z may represent either the scattered electric field component or the total electric field component, which is the sum of the incident and the scattered fields. In this work, triangular finite elements are used to discretize the computational domain. The terminating surface will be four layers away from the boundary. It has been observed that keeping the boundary further away from the object improves the accuracy of this method. We have used conformal and regular meshes for the space domain; however, it is not necessary. The finite-element technique is node based and the field quantities will be computed at each node as opposed to staggered in space as is done in FDTD.

8.2.1 Formulation

In this analysis it is convenient to use a time domain Gaussian pulse because its frequency content can be controlled by changing its pulse width. The electric field of the incident Gaussian pulse, as shown in Fig. 8.1, is given by

$$\boldsymbol{E}^{\mathrm{i}}(\rho, t) = E_{\mathrm{o}}\frac{4.0}{\sqrt{\pi}T}e^{-\gamma^2}\boldsymbol{a}_z, \qquad (8.2.2)$$

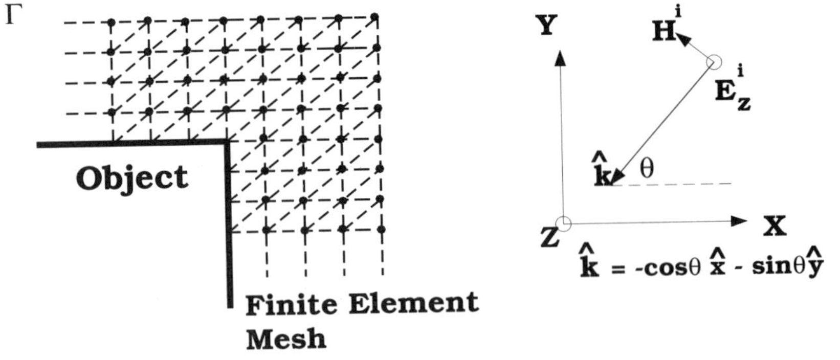

FIGURE 8.3
Finite-element mesh for the TM case.

where $\boldsymbol{a}_z$ is the unit vector along the z direction and T is the width of the Gaussian pulse. From Fig. 8.3, γ can be written as

$$\gamma = \frac{4.0}{T}\left[ct - ct_0 + x\cos\theta + y\sin\theta\right]. \tag{8.2.3}$$

Next, we define the total field $\boldsymbol{E}^{\mathrm{t}}(\boldsymbol{r}, t)$ as

$$\boldsymbol{E}^{\mathrm{t}}(\boldsymbol{r}, t) = \boldsymbol{E}^{\mathrm{i}}(\boldsymbol{r}, t) + \boldsymbol{E}^{\mathrm{s}}(\boldsymbol{r}, t), \tag{8.2.4}$$

where the scattered field $\boldsymbol{E}^{\mathrm{s}}(\boldsymbol{r}, t)$ is given by

$$\boldsymbol{E}^{\mathrm{s}}(\boldsymbol{r}, t) = -\frac{\partial \boldsymbol{A}}{\partial t}, \tag{8.2.5}$$

and $\boldsymbol{A}(\boldsymbol{r}, t)$ is the magnetic vector potential given by

$$\boldsymbol{A}(\boldsymbol{r}, t) = \frac{\mu}{4\pi}\int_C \int_{t'=-\infty}^{t-R/c} \boldsymbol{J}_{\mathrm{s}}(\boldsymbol{r}', t') G(\boldsymbol{r}, \boldsymbol{r}', t, t')\, dt'\, dl'. \tag{8.2.6}$$

In Eq. (8.2.6), $R = |\boldsymbol{r} - \boldsymbol{r}'|$ and $\boldsymbol{J}_{\mathrm{s}}$ is the induced surface current density which is given by $\boldsymbol{J}_{\mathrm{s}} = J_{\mathrm{s}z}\boldsymbol{a}_z$, dl' is the differential element of the contour C bounding the conductor cross section, and the Green's function for this problem is given by [6]

$$G(\boldsymbol{r}, \boldsymbol{r}', t, t') = \begin{cases} \frac{2c}{\sqrt{c^2(t-t')^2 - R^2}} & \text{for } c(t - t') > R \\ 0 & \text{for } c(t - t') < R. \end{cases} \tag{8.2.7}$$

Since $J_{\mathrm{s}z}$ is independent of the spatial coordinate z, there are no charges associated with this current; hence, the electric scalar potential Φ is zero everywhere. Using a backward difference for Eq. (8.2.5) at time instant t_n, we have

$$\boldsymbol{E}^{\mathrm{s}}(\boldsymbol{r}, t_n) = -\frac{1.5\boldsymbol{A}(\boldsymbol{r}, t_n) - 2\boldsymbol{A}(\boldsymbol{r}, t_{n-1}) + 0.5\boldsymbol{A}(\boldsymbol{r}, t_{n-2})}{\Delta t}, \tag{8.2.8}$$

where Δt is the finite time step used for the solution in a leapfrogging manner. Notice that the backward difference used in this case is a higher order approximation for the first derivative than that used in Chapters 3 and 4 for the implicit solution of the time domain integral equations. This higher order approximation is necessary in order to develop an accurate solution using the finite-element procedure. Due to the causal nature of the induced surface current, the magnetic vector potential at time instant t_n is given as

$$\boldsymbol{A}(\boldsymbol{r}, t_n) = \frac{\mu c}{2\pi} \int_C \int_{t'=0}^{t_n - R/c} \frac{\boldsymbol{J}_s(\boldsymbol{r}', t')}{\sqrt{c^2(t_n - t')^2 - R^2}} \, dt' \, dl'. \tag{8.2.9}$$

Because the conductor surface is divided into a number of subsections in the process of finite-element discretization, the integral over C in Eq. (8.2.9) can be broken into a finite sum over each subsection, which can be given as

$$\boldsymbol{A}(\boldsymbol{r}, t_n) = \frac{\mu c}{2\pi} \sum_i \int_{\Delta l_i} \int_{t'=0}^{t_n - R/c} \frac{\boldsymbol{J}_{\Delta l_i}(\boldsymbol{r}', t')}{\sqrt{c^2(t_n - t')^2 - R^2}} \, dt' \, dl', \tag{8.2.10}$$

where Δl_i is the cross-sectional length of the subsection. The transverse components of the electric and magnetic fields are

$$\boldsymbol{E}_t = 0 \tag{8.2.11}$$

$$\frac{\partial \boldsymbol{H}_t}{\partial t} = \frac{1}{\mu} \boldsymbol{a}_z \times \nabla_t E_z, \tag{8.2.12}$$

respectively. The transverse del operator ∇_t is given by $[\boldsymbol{a}_x(\partial/\partial x) + \boldsymbol{a}_y(\partial/\partial y)]$. The transverse components of the electric fields are $E_x = E_y = 0$, and from Eq. (8.2.12) we get

$$\frac{\partial H_x(\boldsymbol{r}, t)}{\partial t} = \frac{-1}{\mu} \frac{\partial E_z(\boldsymbol{r}, t)}{\partial y} \tag{8.2.13}$$

$$\frac{\partial H_y(\boldsymbol{r}, t)}{\partial t} = \frac{1}{\mu} \frac{\partial E_z(\boldsymbol{r}, t)}{\partial x}. \tag{8.2.14}$$

Since the cylinder is perfectly conducting, the boundary condition requires the tangential component of the total electric field to be zero on the conductor. Hence,

$$E_z(\boldsymbol{r}, t) = 0 \tag{8.2.15}$$

on the conductor surface. We can express the surface current density on the conductor in terms of the tangential magnetic field just outside (the superscript +) the conductor. Hence,

$$\boldsymbol{J}_s(\boldsymbol{r}, t) = \boldsymbol{a}_n \times \boldsymbol{H}_t^+(\boldsymbol{r}, t), \tag{8.2.16}$$

where $\boldsymbol{a}_n$ is the unit normal on the conductor surface in the transverse plane. Now computing the transverse magnetic fields at a time instant t_n and, using a first-order approximation for the time derivatives in Eqs. (8.2.13) and (8.2.14), we have

$$H_x(\boldsymbol{r}, t_n) \approx H_x(\boldsymbol{r}, t_{n-1}) - \frac{\Delta t}{\mu} \frac{\partial E_z(\boldsymbol{r}, t_n)}{\partial y} \tag{8.2.17}$$

$$H_y(\boldsymbol{r}, t_n) \approx H_y(\boldsymbol{r}, t_{n-1}) + \frac{\Delta t}{\mu} \frac{\partial E_z(\boldsymbol{r}, t_n)}{\partial x}. \tag{8.2.18}$$

Similarly, using Eq. (8.2.16), the induced current on the conductor surface at time instant t_n may be given as

$$\boldsymbol{J}_s(\boldsymbol{r}, t_n) \approx \boldsymbol{J}_s(\boldsymbol{r}, t_{n-1}) + \frac{\Delta t}{\mu}\frac{dE_z(\boldsymbol{r}, t_n)}{dn}, \tag{8.2.19}$$

where dn represents the normal derivative defined with respect to the contour C.

We note that using the finite-element procedure, as discussed in the following sections, we solve for E_z everywhere in the computational domain at time instant t_n. Next, we compute the magnetic fields using Eqs. (8.2.17) and (8.2.18), and the induced current on the structure using Eq. (8.2.19), as a postprocess.

8.2.2 Finite-Element Procedure

In this section, we use the variational formulation to transform the two-dimensional wave equation into a matrix equation. The open region surrounding the body is discretized into nonoverlapping finite elements. The domain of discretization is finite, and a fictitious boundary is introduced to bound the computational domain. In our solution, the fictitious boundary is only four finite-element layers away from the object as shown in Fig. 8.3. The region between the body surface and the artificial boundary is divided into triangular finite elements. The choice of triangular elements provides an easy simulation of any arbitrary boundary. The meshing around the structure need not be conformal to the body, but the numerical procedure is more efficient for the case of body-fitting meshes.

We start the mathematical formulation by using Ψ instead of E_z in Eq. (8.2.1) for the sake of notational convenience. To find Ψ for the two-dimensional solution region we approximate the function Ψ^p within the pth element and then interrelate the functions in various elements such that Ψ is continuous across interelement boundaries [7]. Thus,

$$\Psi(x, y, t) = \sum_{p=1}^{N} \Psi^p(x, y, t), \tag{8.2.20}$$

where N is the number of triangular elements into which the bounded domain is divided and Ψ^p represents the spatial and temporal variation in the pth triangular element. A semidiscrete finite-element model is assumed for discretization purposes, and we use a linear polynomial function $g_j^p(x, y)$ for each node $j = 1, 2$, and 3, and the space variables x and y over an element p. Thus,

$$\Psi^p = \sum_{j=1}^{3} f_j^p(t) g_j^p(x, y), \tag{8.2.21}$$

where f_j^p represents the unknown coefficient to be solved.

Here, we seek a solution space $V = \{v \in H^1(\Re)\}$, where H^1 denotes the Hilbert space for this problem. Now, multiplying Eq. (8.2.1) with v and integrating over the space yields

$$\int_\Re v\frac{\partial^2\Psi}{\partial x^2}\,ds + \int_\Re v\frac{\partial^2\Psi}{\partial y^2}\,ds - \frac{1}{c^2}\int_\Re v\frac{\partial^2\Psi}{\partial t^2}\,ds = 0 \quad v \in V, \tag{8.2.22}$$

where ds is the infinitesimal area of integration. Now using the divergence theorem, we get

$$\int_{\Gamma} v \frac{\partial \Psi}{\partial n}\, d\Gamma - \int_{\Re} \nabla v \cdot \nabla \Psi \, ds - \frac{1}{c^2} \int_{\Re} v \frac{\partial^2 \Psi}{\partial t^2}\, ds = 0, \tag{8.2.23}$$

where Γ denotes the boundary of the computational domain. The boundary integral term in Eq. (8.2.23) may be split into two terms. The first is extended to the part of the boundary with Dirichlet boundary conditions, Γ_D, where, following the conventional finite-element procedure, the weighting functions v are chosen to be identically zero. Thus, this first term is identically zero. The second term is extended to the part of the boundary with Neumann boundary conditions, Γ_N, where $\partial \Psi / \partial n$ has a known value. Thus,

$$\int_{\Gamma} v \frac{\partial \Psi}{\partial n}\, d\Gamma = \int_{\Gamma_N} v \left(\frac{\partial \Psi}{\partial x} n_1 + \frac{\partial \Psi}{\partial y} n_2 \right) d\Gamma, \tag{8.2.24}$$

where n_1 and n_2 are the direction cosines along the x and y directions, respectively.

For the problem considered in this section, Γ_D is composed of the contour of the conductor C plus the contour of the terminating surface, whereas Γ_N is composed of the symmetry walls, if any. We note that the evaluation of Eq. (8.2.24) depends on the type of formulation employed. If the total-field formulation is used, then $\partial \Psi / \partial n = 0$ over symmetry walls. On the other hand, if the scattered-field formulation is used, then

$$\frac{\partial \Psi}{\partial n} = - \left(\frac{\partial E_z^{\mathrm{i}}}{\partial x} n_1 + \frac{\partial E_z^{\mathrm{i}}}{\partial y} n_2 \right),$$

where E_z^{i} is given by Eq. (8.2.2). Thus, for the total-field formulation Eq. (8.2.24) may be set to zero, whereas for the scattered-field formulation, a nonzero value for Eq. (8.2.24) should be retained.

Using the approximation in Eq. (8.2.21) and using $v = g_i^p$, we apply the finite-element procedure for Eq. (8.2.23) as follows: for the first term in Eq. (8.2.23),

$$\int_{\Gamma_p \subset \Gamma_N} v \frac{\partial \Psi}{\partial n}\, d\Gamma = W^p \tag{8.2.25}$$

where

$$W^p = \begin{cases} 0 & \text{for the total-field formulation} \\ -\int_{\Gamma_p \subset \Gamma_N} g_i^p \left(\frac{\partial E_z^{\mathrm{i}}}{\partial x} n_1 + \frac{\partial E_z^{\mathrm{i}}}{\partial y} n_2 \right) d\Gamma & \text{for the scattered-field formulation.} \end{cases} \tag{8.2.26}$$

In Eqs. (8.2.25) and (8.2.26), Γ_p denotes the boundary of the pth element. Furthermore, the integral in Eq. (8.2.26) is extended to that part of Γ_p that coincides with Γ_N.

The second term in Eq. (8.2.23) can be written, for the pth element, as

$$\int_{\Re_p} \nabla v \cdot \nabla \Psi \, ds = \sum_{j=1}^{3} f_j^p(t) \int_{\Re_p} \left(\frac{\partial g_j^p}{\partial x} \frac{\partial g_i^p}{\partial x} + \frac{\partial g_j^p}{\partial y} \frac{\partial g_i^p}{\partial y} \right) ds \quad \text{for } i = 1, 2, 3, \tag{8.2.27}$$

where $\Re_p$ represents the region corresponding to element p.

Lastly, the third term of Eq. (8.2.23) can be written as

$$\frac{1}{c^2}\int_{\Re_p} v\frac{\partial^2\Psi}{\partial t^2}\,ds = \frac{1}{c^2}\sum_{j=1}^{3}\frac{d^2 f_j^p(t)}{dt^2}\int_{\Re_p} g_i g_j\,ds \quad \text{for } i = 1, 2, 3. \tag{8.2.28}$$

Now, defining

$$C_{ij} = \int_{\Re_p}\left(\frac{\partial g_j^p}{\partial x}\frac{\partial g_i^p}{\partial x} + \frac{\partial g_j^p}{\partial y}\frac{\partial g_i^p}{\partial y}\right)ds \tag{8.2.29}$$

$$T_{ij} = \int_{\Re_p} g_i^p g_j^p\,ds \tag{8.2.30}$$

and using Eqs. (8.2.25), (8.2.27), and (8.2.28), we can write the left-hand side of Eq. (8.2.23) for the pth element as

$$[\boldsymbol{W}^p] - [\boldsymbol{C}^p][f^p] - \frac{1}{c^2}[\boldsymbol{T}^p]\left[\frac{d^2 f^p}{dt^2}\right]. \tag{8.2.31}$$

Now assembling all such elements in the solution region, the ordinary differential equation of the assemblage is given by

$$\frac{1}{c^2}[\boldsymbol{T}]\left[\frac{d^2 f}{dt^2}\right] = [\boldsymbol{W}] - [\boldsymbol{C}][f]. \tag{8.2.32}$$

The matrices $[\boldsymbol{C}]$, $[\boldsymbol{T}]$, and $[\boldsymbol{W}]$ are the assemblage of individual coefficient matrices $[\boldsymbol{C}^p]$, $[\boldsymbol{T}^p]$, and $[\boldsymbol{W}^p]$, respectively. It is evident that $[\boldsymbol{C}]$, $[\boldsymbol{T}]$, and $[\boldsymbol{W}]$ are time-independent matrices. Furthermore, by replacing $1/c^2[\boldsymbol{T}]$ and $-[\boldsymbol{C}]$ with matrices $[\boldsymbol{B}]$ and $[\boldsymbol{A}]$, respectively, and denoting the time derivatives by dots, we have

$$[\boldsymbol{B}][\ddot{f}] = [\boldsymbol{W}] + [\boldsymbol{A}][f]. \tag{8.2.33}$$

Note that in Eq. (8.2.34), the column matrix $[f]$ represents E_z at the corresponding node. Furthermore, in this work, we did not consider any symmetry walls which implies that $[\boldsymbol{W}] = 0$. Thus, Eq. (8.2.33) reduces to

$$[\boldsymbol{B}][\ddot{f}] = [\boldsymbol{A}][f]. \tag{8.2.34}$$

8.2.3 Time-Stepping Procedure

Equation (8.2.34) is a second-order differential equation, which will be solved by Newmark's method [2, 3]. The Newmark method is unconditionally stable and introduces the best truncation error for a particular choice of certain parameters.

By denoting $[f]_n$ as the function value at time instant $t = t_n$, we start the algorithm by expanding $[\dot{f}]$ as

$$[\dot{f}]_{n+1} = [\dot{f}]_n + \{(1-\delta)[\ddot{f}]_n + \delta[\ddot{f}]_{n+1}\}\Delta t \tag{8.2.35}$$

and

$$[f]_{n+1} = [f]_n + (\Delta t)[\dot{f}]_n + \{(0.5 - \alpha)[\ddot{f}]_n + \alpha[\ddot{f}]_{n+1}\}\Delta t^2, \tag{8.2.36}$$

where $\alpha \geq 0.25(0.5+\delta)^2$ and δ are parameters that can be chosen to obtain accuracy and stability. Typically, the values for α and δ are 0.25 and 0.5, respectively, to guarantee the unconditional convergence.

Next, we multiply Eq. (8.2.35) by $[\boldsymbol{B}]$ and use Eq. (8.2.34) to obtain

$$[\boldsymbol{B}][\dot{f}]_{n+1} = [\boldsymbol{B}][\dot{f}]_n + \{(1-\delta)[\boldsymbol{A}][f]_n + \delta[\boldsymbol{A}][f]_{n+1}\}\,\Delta t. \tag{8.2.37}$$

Similarly, we multiply Eq. (8.2.36) by $[\boldsymbol{B}]$ and use Eq. (8.2.34) to obtain

$$\begin{aligned}[\boldsymbol{B}][f]_{n+1} = {} & [\boldsymbol{B}][f]_n + \Delta t[\boldsymbol{B}][\dot{f}]_n \\ & + \{(0.5-\alpha)[\boldsymbol{A}][f]_n + \alpha[\boldsymbol{A}][f]_{n+1}\}\,\Delta t^2.\end{aligned} \tag{8.2.38}$$

Writing Eq. (8.2.38) for time instant t_{n+1} and subtracting Eq. (8.2.38) from it and substituting for $[\dot{f}]_{n+1} - [\dot{f}]_n$ from Eq. (8.2.37) results in a two-step recurrence relation for the wave equation, which is given as

$$\begin{aligned}&\{[\boldsymbol{B}] - \Delta t^2\alpha[\boldsymbol{A}]\}[f]_{n+2} + \{-2[\boldsymbol{B}] - \Delta t^2(0.5+\delta-2\alpha)[\boldsymbol{A}]\}[f]_{n+1} \\ &\quad + \{[\boldsymbol{B}] - \Delta t^2(0.5-\delta+\alpha)[\boldsymbol{A}]\}[f]_n = 0.\end{aligned} \tag{8.2.39}$$

Now, replacing matrix $([\boldsymbol{B}] - \Delta t^2\alpha[\boldsymbol{A}])$ by $[\boldsymbol{L}]$, $-2[\boldsymbol{B}] - \Delta t^2(0.5+\delta-2\alpha)[\boldsymbol{A}]$ by $[\boldsymbol{M}]$, and $[\boldsymbol{B}] - \Delta t^2(0.5-\delta+\alpha)[\boldsymbol{A}]$ by $[\boldsymbol{N}]$, we can rewrite Eq. (8.2.39) in a compact form as

$$[\boldsymbol{L}][f]_{n+2} + [\boldsymbol{M}][f]_{n+1} + [\boldsymbol{N}][f]_n = 0. \tag{8.2.40}$$

Here, we note that the matrices $[\boldsymbol{L}]$, $[\boldsymbol{M}]$, and $[\boldsymbol{N}]$ in Eq. (8.2.40) do not change with respect to time, but they do depend on the length of the time step chosen for the marching on in time procedure. Replacing n by $n-2$ in Eq. (8.2.40) results in

$$[\boldsymbol{L}][f]_n = -[\boldsymbol{M}][f]_{n-1} - [\boldsymbol{N}][f]_{n-2}. \tag{8.2.41}$$

Furthermore, the matrix $[\boldsymbol{L}]$ can be split into two parts for two types of nodes. The first set of nodes are called free nodes, on which the electric field needs to be solved, and the second set are called prescribed nodes, on which $E_z(\boldsymbol{r}, t)$ are known. If all the free nodes are numbered first and the prescribed nodes last, we can rewrite Eq. (8.2.41) as

$$\begin{bmatrix}\boldsymbol{L}_{\mathrm{ff}} & \boldsymbol{L}_{\mathrm{fp}} \\ \boldsymbol{L}_{\mathrm{pf}} & \boldsymbol{L}_{\mathrm{pp}}\end{bmatrix}\begin{bmatrix} f_{\mathrm{f}} \\ f_{\mathrm{p}}\end{bmatrix}_n = -[\boldsymbol{M}][f]_{n-1} - [\boldsymbol{N}][f]_{n-2}, \tag{8.2.42}$$

where subscripts f and p refer to free and prescribed nodes, respectively. Note that Eq. (8.2.42) is a recurrence relation which can be used to compute $[f]_n$ for all values of n knowing $[f]_{n-1}$ and $[f]_{n-2}$.

In the finite-element method terminology, the nodes which are residing on the terminating surface and on the conductor boundary are called prescribed nodes. The electric fields for the prescribed nodes residing on the conductor are known from the boundary condition on the conducting surface, i.e., from Eq. (8.2.15), for all time steps. Furthermore, the electric fields at any time step t_n for the nodes residing on the terminating surface can be computed by using Eqs. (8.2.8) and (8.2.10) because we already know the induced current on the conductor for earlier times t_{n-1}, t_{n-2}, and so on. Thus,

$$A(\boldsymbol{r}, t_n) = \frac{\mu c}{2\pi}\Delta t \sum_i \sum_k \int_{\Delta l_i} \boldsymbol{J}_{\Delta l_i}(l_i, k\Delta t)\frac{dl'}{\sqrt{c^2(t_n - k\Delta t)^2 - R^2}}$$

$$0 \le k\Delta t \le t_n - R/c. \quad (8.2.43)$$

In Eq. (8.2.43), it should be noted that the summation for k indices is essentially the integration over time coordinates. Even though it is written in a simple manner for time integration, we have employed Simpson's rule to achieve higher order accuracy in the integration.

As mentioned previously, Eq. (8.2.1) can be solved by either the scattered-field $\boldsymbol{E}^{\mathrm{s}}$ or the total electric field, $\boldsymbol{E}^{\mathrm{i}} + \boldsymbol{E}^{\mathrm{s}}$. The boundary condition on the conductor surface and the radiation boundary condition on the terminating surface depend on the kind of formulation employed for a particular problem. Even though both of them produce accurate results, there may be numerical differences due to the error introduced in the numerical differentiation and integration procedure. If we employ a total-field formulation then the boundary condition on the conductor is satisfied by Eq. (8.2.15), whereas fields on the terminating surface will be given by $\boldsymbol{E}^{\mathrm{i}} + \boldsymbol{E}^{\mathrm{s}}$ through Eqs. (8.2.2) and (8.2.8). Similarly, for the scattered-field formulation, one needs to employ $-\boldsymbol{E}^{\mathrm{i}}$ on the conductor and use Eq. (8.2.8) to compute the radiation condition on the terminating surface.

The proposed method can be regarded as a hybrid of differential and integral equation techniques because a finite-element approach is applied for the electric field within the domain and a boundary condition expressed in terms of an integral (containing the unknown source distribution under the integral) formulated for the terminating surface.

8.2.4 Numerical Results

Square Cylinder We consider the case of a square cylinder with a side of 1.0 m, centered about the origin and illuminated by a Gaussian pulse of unit amplitude ($E_{\mathrm{o}} = 1$ V/m). The incident wave is traveling along the negative x-axis with $\theta = 0°$. Part of the finite-element grid of this problem is shown in Fig. 8.3. The finite-element mesh is chosen such that it is conformal to the scatterer. In the time domain problem we have used four layers of finite elements from the surface of the body. In this example, the fictitious surface is only 0.2 m away from the body. The terminating surface is composed of 80 nodes. Furthermore, there are 48 prescribed nodes on the conductor boundary. The inner layers between the conductor and the terminating surface consist of 192 free nodes. For the total-field formulation, the tangential electric field on the conductor is zero. Lastly, for this example, we employ the total-field formulation.

The pulse width of the Gaussian pulse is taken to be $T = 2$ LM. The delay of the pulse is $ct_{\mathrm{o}} =$ 3 LM. This means that the Gaussian pulse will reach its maximum at the origin at $ct_{\mathrm{o}} = 3$ LM. The time step (Δt) for this example is 1/21 LM. To start the time-stepping procedure, we turn on the Gaussian pulse at $t = 0$. We compute the fields for the free nodes and the fields on the terminating surface using Eq. (8.2.42). Because the fields are known for that particular instant, the induced surface currents are computed through Eq. (8.2.19) for that instant. Hence, the radiation condition on the terminating surface is applied through the Green's function integral. Now the scattered fields on the terminating surface are computed through Eq. (8.2.8). These values are

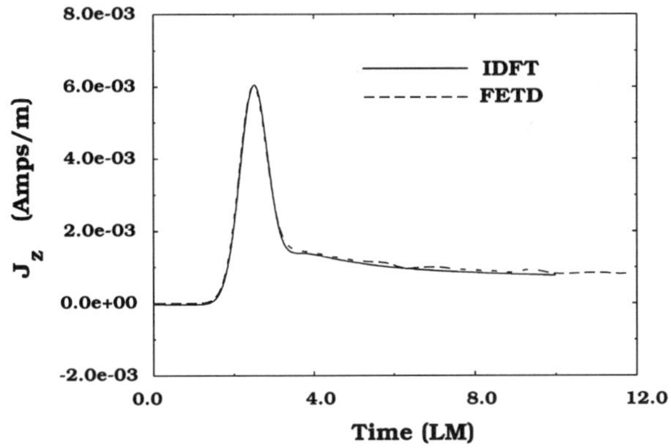

FIGURE 8.4
Induced current on the square conductor.

replaced in Eq. (8.2.42) and the time-stepping procedure continues to compute fields for the next time step. In our analysis, to obtain integration accuracy and numerical stability, we have used values of 0.25 and 0.5 for α and δ, respectively.

The induced surface current on the conductor for a point in the x-axis (0.5, 0.0) is plotted in Fig. 8.4 with respect to time. Because the peak of the incident pulse arrives at the front face of the square cylinder at 2.5 LM, the induced surface current reaches a maximum during that time. With the progression of time the incident pulse moves toward the back of the conductor; hence, the induced current at the middle of the front face decreases. After the pulse crosses it completely, the current starts decaying slowly.

The results from this method are compared with the inverse Fourier transformed frequency domain method of moments (MoM) results. We used 48 subsections to calculate the induced current on the conductor with the incident pulse being Gaussian but frequency transformed. The pulse basis functions and point-matching testing procedure are used to evaluate the current on the conductor in the frequency domain. We have chosen 256 equally spaced frequency points between 0 and 0.5 GHz. These current values in frequency domain are inverse transformed to obtain the time domain results. As can be seen, the results of this FETD method agree very well with inverse Fourier transformed frequency domain MoM results.

Now we discuss how the induced current on the surface of the conductor behaves in late time. Here, we have computed the induced current at the middle of the front face of the conductor, illuminated by Gaussian pulse. As can be seen from Fig. 8.5, the induced current at late time does not show any behavior of instability or any kind of oscillations. Even though we can observe small ripples, they die down slowly as time progresses. The total electric field on a node residing on the terminating surface is shown in Fig. 8.6 with respect to time. This node lies on the x-axis but just opposite to a node at the middle of the conductor. We can see that the pulse passes through this point and reaches its maximum at approximately 2.3 LM. However, the reflected pulse from the front face of the conductor, which is negative of the incident pulse, slowly arrives at the terminating surface. The negative pulse reaches its peak at approximately 3 LM. As can be seen, there are small oscillations in the late time, but it never increases as it progresses along the time axis.

Circular Cylinder We now consider the case of a circular cylinder with a diameter of 1.0 m. A four-layer finite-element mesh has been used for this problem. The meshing scheme used is similar to that shown in Fig. 8.3 and it is conformal to the object. In this example, the terminating

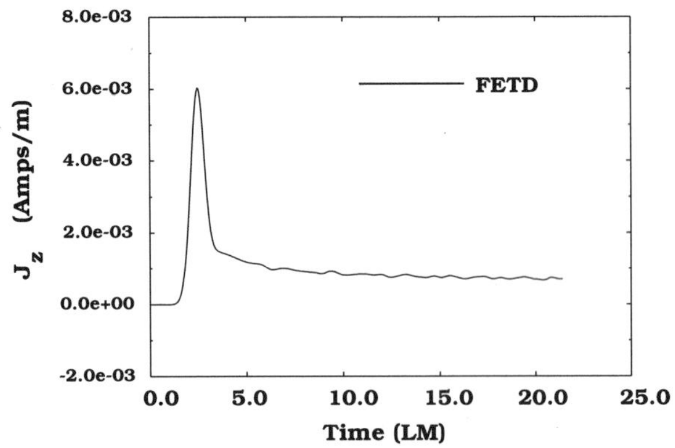

FIGURE 8.5
Induced current on the square conductor (late-time behavior).

surface is 0.2 m away from the conducting surface. In the finite-element grid scheme, there are 44 nodes on the conductor surface, 44 nodes on the outer boundary, and 132 nodes on the layers in between the conductor and the terminating surface.

The incident field, as before, is a Gaussian pulse with $T = 2$ LM and $ct_o = 3$ LM. The wave is of unit amplitude and traveling in the negative x direction. The coordinate axis is centered at the center of the circular cylinder. In this case, we also use the total-field formulation. Hence, the time-stepping procedure starts by assuming $\boldsymbol{E}^{\mathrm{i}}$ on the outside boundary and zero field for the conductor surface for all times. As before, we switch on the incident field at time $t = 0$ all over the computational space. Then, Eq. (8.2.42) is solved at every time step to compute electric fields at those nodes. With these known values of fields in the computational domain, we compute the normal derivative of the electric field to the conductor, for that time instant, to compute the induced current through Eq. (8.2.19). Now the scattered electric field is computed at the terminating surface at that instant using Eq. (8.2.8) and the known induced current on the conductor. After replacing the total electric field for the nodes at the terminating surface, we employ the recurrence relation to find electric fields for free nodes and compute the induced

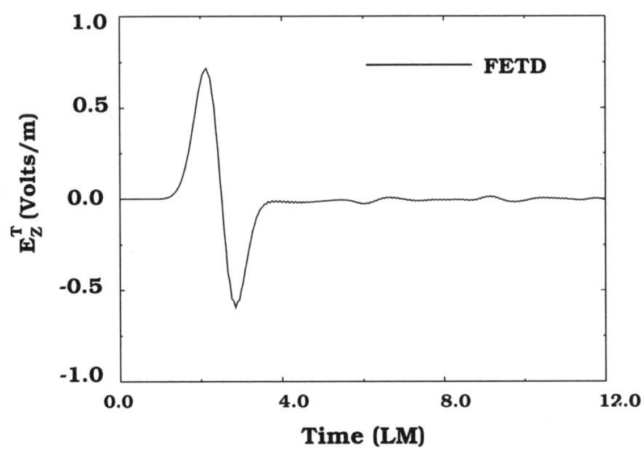

FIGURE 8.6
Total electric field on the terminating surface at point (0.7,0.0).

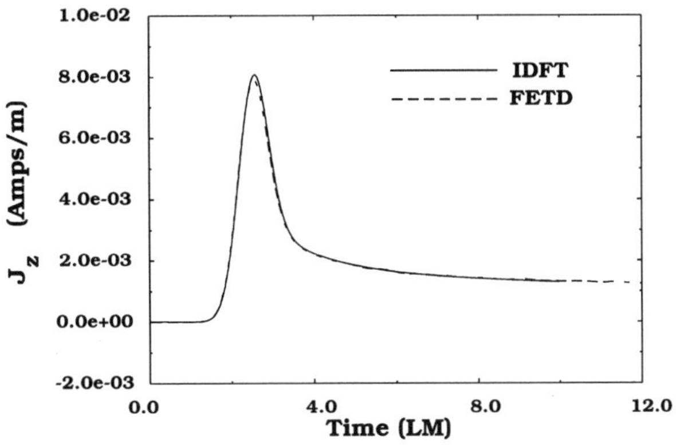

FIGURE 8.7
Induced current on the circular cylinder at point (0.0,0.5). Outer boundary is 0.2 m away.

current distribution on the conductor for the next time step. In this context, we should mention that for the initial few time steps, the total electric field on the terminating surface is essentially the incident field because the scattered field generated by the induced current requires a finite time to propagate to the field point.

The induced surface current on a node residing on the x-axis is plotted in Fig. 8.7 with respect to time. In this example, we have chosen a time step of 1/21 LM. As can be observed, the induced surface current reaches maximum near about 2.5 LM as the front face of the cylinder is illuminated by the peak of the Gaussian pulse. Again, the results are compared with frequency domain inverse discrete Fourier transformed (IDFT) MoM results. We used 256 equally spaced frequency points between 0 and 0.5 GHz and inverse Fourier transformed the data to produce the time response. As before, we used a pulse basis and point-matching testing procedure for the frequency domain results. For the MoM solution, 40 subsections are chosen on the conducting surface. It is evident from Fig. 8.7 that both solutions agree very well.

The total electric field for a node residing on the terminating surface has been plotted in Fig. 8.8. We can clearly observe that the total field reaches a maximum at approximately 2.1 LM and, after

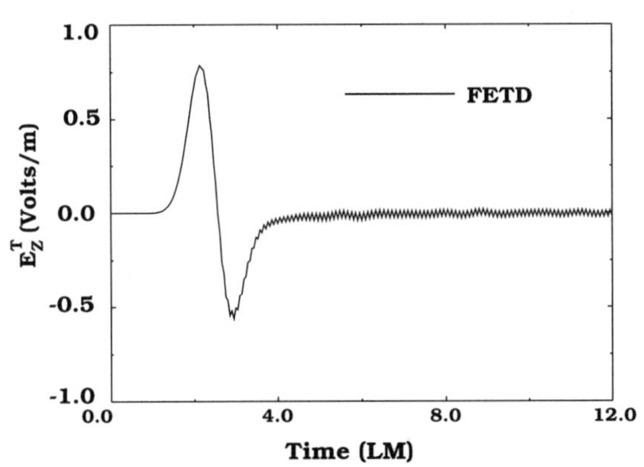

FIGURE 8.8
Total electric field on the terminating surface for the circular cylinder at point (0.0,0.7).

reflection from the conducting surface, reaches its negative maximum at approximately 3 LM before returning to zero.

Half Cylinder As a last example, we consider the scattering from a closed, half cylinder with a diameter of 1.0 m. The finite-element meshing of this structure is similar to those used in the previous examples. We employed four layers of finite elements to discretize the space between the fictitious surface and the scatterer. The terminating surface is conformal to the body and is only 0.2 m away from the surface of the conductor. The incident field in this case is coming from $\theta = 90^\circ$. We have used a total-field formulation for this problem. Electric field values are found for 180 free nodes at every time step. The tangential electric field on the scatterer, formed by 50 prescribed nodes, is zero for the total-field formulation. As before, the time-stepping procedure starts by assuming $\boldsymbol{E}^{\mathrm{i}}$ on the outside boundary, which is composed of 60 prescribed nodes. The normal derivative of the electric field is computed through a postprocess which gives the current distribution over a particular segment for a time instant through Eq. (8.2.19). As before, the finite-element matrix (Eq. 8.2.42) is solved for the unknown free nodes. These values are used to compute induced surface currents on the conductor. With these known currents at every time step, the radiation condition on the outside boundary is obtained through the Green's function integral equation (Eq. 8.2.8). We used the same width and delay of the Gaussian pulse as employed in the previous examples. The time step chosen is $1/21$ LM.

The induced surface current for a node residing on the y-axis with time progression is plotted in Fig. 8.9. As expected, the induced current reaches its peak when the peak of the incident wave hits the particular node. We compared frequency domain MoM results with the results from this method (FETD). We used 256 equally spaced frequency samples for the inverse Fourier transform and 52 subsections were chosen to calculate the induced current using MoM. As before, pulse basis functions and a point-matching testing procedure were used to evaluate the current on the conductor. The results agree well with the MoM values of the induced current. Even though there are no late-time oscillations, small ripples can be observed. The total electric field on a node residing on the terminating surface is plotted in Fig. 8.10. Similar conclusions about the incident and reflected Gaussian pulse from the conducting surface can be derived by examining the electric fields with time progression.

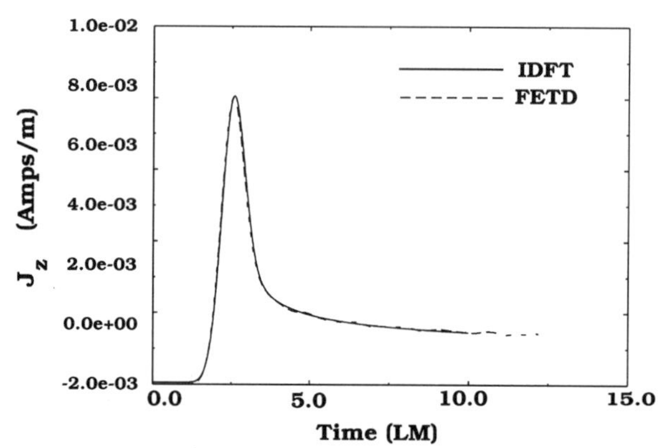

FIGURE 8.9
Induced current on the half cylinder at point (0.0,0.5). Outer boundary is 0.2 m away.

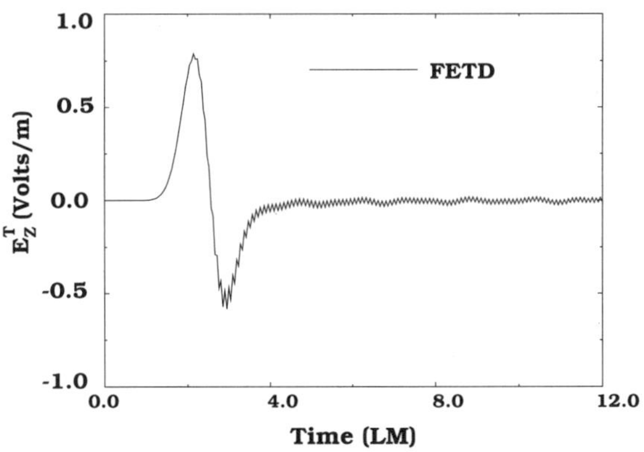

FIGURE 8.10
Total electric field on the terminating surface for the half cylinder at point (0.7,0.0).

8.3 TRANSVERSE ELECTRIC CASE

Previously, we demonstrated the application of the time domain finite-element method for the TM scattering problem. In that analysis, we used the wave equation for E_z fields and obtained the induced currents on the illuminated cylinder at every time step. We also noted that the total-field formulation is simpler compared to the scattered-field formulation.

In this section, we describe the procedure to solve the wave equation for the TE scattering problem for conducting bodies. Furthermore, we describe only the total-field formulation because this is also simpler for the TE case. However, one can develop a scattered-field formulation, if desired, with little extra computation.

As before, the incident wave is a smooth Gaussian pulse chosen to produce information for a band of frequencies. The magnetic field of the incident pulse is parallel to the cylinder axis. This incident field produces transverse-directed induced surface currents on the conductor, an axially directed scattered magnetic field from that induced current, whereas the electric field is purely transverse to the cylinder axis. For the TE scattering problem, since H_z is the axial component of the magnetic field, the wave equation can be written as

$$\nabla^2 H_z(x, y, t) - \frac{1}{c^2}\frac{\partial^2 H_z(x, y, t)}{\partial t^2} = 0$$

$$\Rightarrow \frac{\partial^2 H_z(x, y, t)}{\partial x^2} + \frac{\partial^2 H_z(x, y, t)}{\partial y^2} - \frac{1}{c^2}\frac{\partial^2 H_z(x, y, t)}{\partial t^2} = 0$$

$$x, y \in \Re; \quad t \in (0, \infty), \quad (8.3.1)$$

where z is the axial coordinate, H_z is independent of z for the TE case, and c is the velocity of the wave in the medium considered. As before, triangular finite elements are used to discretize the computational domain.

As noted earlier, for mesh termination to be "exact" one needs to find an appropriate radiation condition on the fictitious surface. It has been shown that the scattered field depends on earlier currents retarded by the distance between the source and the field point. Hence, computationally it is expensive to compute scattered fields unless the computational domain is kept very small, which implies that the terminating boundary must be close to the scatterer. However, keeping the fictitious boundary very close to the scatterer can produce undesired effects of abrupt termination. On the one hand, it is desirable to keep the number of unknowns in the finite-element matrix small so that a quick solution of the problem is possible; on the other hand, keeping the terminating surface at a sufficient distance from the structure avoids undesired reflection from the boundary. Keeping these conflicting requirements in mind, we note that introducing four layers between the scatterer and the fictitious surface vastly improves the solution with little added cost to the computational time. Even though we have used conformal and regular meshes for the space domain, it is not necessary to do so. The finite-element technique is node based, and the field quantities will be computed at each node.

8.3.1 Formulation

The incident waveform is a Gaussian plane wave as used previously. The pulse width of the Gaussian waveform is chosen in such a fashion that its highest frequency component does not invalidate the basic underlying assumption of space discretization. The magnetic field of the incident Gaussian pulse is given as

$$\boldsymbol{H}^{\mathrm{i}}(\boldsymbol{r}, t) = H_0 \frac{4.0}{\sqrt{\pi} T \eta} e^{-\gamma^2} \boldsymbol{a}_z, \tag{8.3.2}$$

where $\boldsymbol{a}_z$ is the unit vector along the z direction, and η is the free-space impedance.

Equation (8.3.1) is satisfied by the scattered magnetic field $\boldsymbol{H}^{\mathrm{s}}$ produced by the currents induced on the conductor as well as by the total magnetic field, which is given as

$$\boldsymbol{H}(\boldsymbol{r}, t) = \boldsymbol{H}^{\mathrm{i}}(\boldsymbol{r}, t) + \boldsymbol{H}^{\mathrm{s}}(\boldsymbol{r}, t). \tag{8.3.3}$$

The scattered field $\boldsymbol{H}^{\mathrm{s}}$ is given by

$$\boldsymbol{H}^{\mathrm{s}}(\boldsymbol{r}, t) = \frac{1}{\mu} \nabla \times \boldsymbol{A}, \tag{8.3.4}$$

where $\boldsymbol{A}$ is the magnetic vector potential, given by

$$\boldsymbol{A}(\boldsymbol{r}, t) = \frac{\mu}{4\pi} \int_C \int_{t'=-\infty}^{t-R/c} \boldsymbol{J}_{\mathrm{s}}(\boldsymbol{r}', t') G(\boldsymbol{r}, \boldsymbol{r}', t, t')\, dt'\, dl', \tag{8.3.5}$$

and $\boldsymbol{J}_{\mathrm{s}}$ is the induced surface current density given by

$$\boldsymbol{J}_{\mathrm{s}}(\boldsymbol{r}, t) = J_{\mathrm{s}}(\boldsymbol{r}, t) \boldsymbol{a}_{\tau}. \tag{8.3.6}$$

In Eqs. (8.3.5) and (8.3.6), $\boldsymbol{a}_\tau$ denotes the tangential unit vector, whereas dl' is the source point on the contour C bounding the conductor cross section. The Green's function for the TE case is the same as that for the TM case (Eq. 8.2.7) and is given by

$$G(\boldsymbol{r}, \boldsymbol{r}', t, t') = \begin{cases} \frac{2c}{\sqrt{c^2(t-t')^2 - R^2}} & \text{for } c(t-t') > R \\ 0 & \text{for } c(t-t') < R, \end{cases} \tag{8.3.7}$$

where $R = |\boldsymbol{r} - \boldsymbol{r}'|$. Now, replacing $\boldsymbol{A}(\boldsymbol{r}, t)$ in Eq. (8.3.3), the following expression for the scattered magnetic field is derived:

$$\begin{aligned}\boldsymbol{H}^{\mathrm{s}}(\boldsymbol{r},t) &= \frac{c}{2\pi}\nabla\times\int_C\int_{t'=-\infty}^{t-R/c}\frac{\boldsymbol{J}_{\mathrm{s}}(\boldsymbol{r}',t')}{\sqrt{c^2(t-t')^2-R^2}}\,dt'\,dl' \\ &= \frac{c}{2\pi}\oint_C \boldsymbol{dl}'\times\frac{\boldsymbol{a}_R}{R}\int_{t'=-\infty}^{t-R/c}\frac{(t-t')\frac{\partial}{\partial t'}J_{\mathrm{s}}(\boldsymbol{r}',t')}{\sqrt{c^2(t-t')^2-R^2}}\,dt', \end{aligned} \tag{8.3.8}$$

where $\boldsymbol{a}_R$ is the unit vector from the source point to the field point and $\boldsymbol{dl}'$ is the elemental vector along the contour of the conductor in an counter clockwise sense. Note that $\boldsymbol{H}^{\mathrm{s}}(\boldsymbol{r},t)$ is along the axial direction given by $\boldsymbol{H}^{\mathrm{s}}(\boldsymbol{r},t) = H^{\mathrm{s}}(\boldsymbol{r},t)\boldsymbol{a}_z$.

Due to the causal nature of the induced surface current, the magnetic vector potential at time instant t_n is given as

$$\boldsymbol{A}(\boldsymbol{r},t_n) = \frac{\mu c}{2\pi}\oint_C\int_{t'=0}^{t_n-R/c}\frac{\boldsymbol{J}_{\mathrm{s}}(\boldsymbol{r}',t')}{\sqrt{c^2(t_n-t')^2-R^2}}\,dt'\,dl'. \tag{8.3.9}$$

Because the conductor surface is divided into many subsections in the process of finite-element discretization, the integral over the conducting surface can be broken into finite sums, which can be given as

$$\boldsymbol{A}(\boldsymbol{r},t_n) = \frac{\mu c}{2\pi}\sum_i\int_{\Delta l_i}\int_{t'=0}^{t_n-R/c}\frac{\boldsymbol{J}_{\Delta l_i}(\boldsymbol{r}',t')}{\sqrt{c^2(t_n-t')^2-R^2}}\,dt'\,dl'. \tag{8.3.10}$$

Now, substituting Eq. (8.3.10) into Eq. (8.3.8) we get the scattered field at time instant t_n, which is

$$\boldsymbol{H}^{\mathrm{s}}(\boldsymbol{r},t_n) = \frac{c}{2\pi}\sum_i\int_{\Delta l_i}\boldsymbol{dl}'\times\frac{\boldsymbol{a}_R}{R}\int_{t'=0}^{t_n-R/c}\frac{(t_n-t')\frac{\partial}{\partial t'}\boldsymbol{J}_{\mathrm{s}}(\boldsymbol{r}',t')}{\sqrt{c^2(t_n-t')^2-R^2}}\,dt'. \tag{8.3.11}$$

Next, we note that $H_x = H_y = 0$ and, using Maxwell's equation, $\nabla\times\boldsymbol{H} = \epsilon(\partial\boldsymbol{E}/\partial t)$, we get

$$\frac{\partial E_x(\boldsymbol{r},t)}{\partial t} = \frac{1}{\epsilon}\frac{\partial H_z(\boldsymbol{r},t)}{\partial y} \tag{8.3.12}$$

$$\frac{\partial E_y(\boldsymbol{r},t)}{\partial t} = \frac{-1}{\epsilon}\frac{\partial H_z(\boldsymbol{r},t)}{\partial x}. \tag{8.3.13}$$

Since the cylinder is perfectly conducting, the boundary condition requires the tangential component of the total electric field to be zero on the conductor for all times. Hence,

$$\boldsymbol{E}\cdot\boldsymbol{a}_\tau = 0 \tag{8.3.14}$$

on the conductor surface.

We note that Eq. (8.3.14) is equivalent to setting the normal derivative of the total magnetic field on the conductor surface to zero. In other words, on the contour of the conductor surface the unknown H_z has Neumann boundary conditions equal to zero for the total-field formulation. However, if we use a scattered-field formulation, then the boundary condition is equal to the negative of $((\partial H_z^{\mathrm{i}}/\partial x)n_1 + (\partial H_z^{\mathrm{i}}/\partial y)n_2)$.

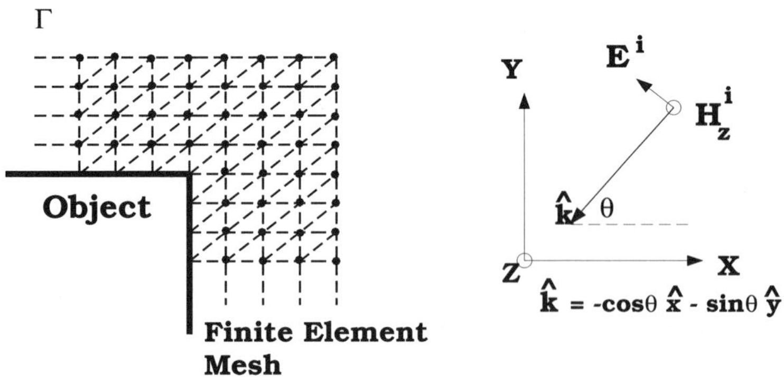

FIGURE 8.11
Finite-element mesh for the TE case.

We can express the surface current density on the conductor in terms of the tangential magnetic field just outside (the superscript +) the conductor by

$$\boldsymbol{J}_{\mathrm{s}}(\boldsymbol{r}, t_n) = \boldsymbol{a}_{\mathrm{n}} \times H_z^{+}(\boldsymbol{r}, t_n), \tag{8.3.15}$$

where $\boldsymbol{a}_{\mathrm{n}}$ is the unit normal on the conductor surface.

8.3.2 Finite-Element Procedure

In this section, the variational formulation is used to transform the two-dimensional wave equation into a matrix equation. As before, we place a fictitious boundary four layers away from the scatterer and discretize the space into triangular finite elements as shown in Fig. 8.11. Even though a uniform meshing is shown in Fig. 8.11, the formulation is general enough to handle any type of meshing. Furthermore, in this work we derive only the expressions for the total-field formulation.

As in the TM case, we start the mathematical formulation by using Ψ instead of H_z in Eq. (8.3.1) for the sake of notational convenience. To find Ψ for the two-dimensional solution region we approximate the function Ψ^p within an element p and then interrelate the functions in various elements such that the Ψ is continuous across interelement boundaries [7]. Thus,

$$\Psi(x, y, t) = \sum_{p=1}^{N} \Psi^p(x, y, t), \tag{8.3.16}$$

where N is the number of triangular elements into which the bounded domain is divided and Ψ^p represents the spatial and temporal variation in the pth triangular element. A semidiscrete finite-element model is assumed for discretization purposes, and we use a linear polynomial function for the space variables x and y over an element p. Thus,

$$\Psi^p = \sum_{j=1}^{3} f_j^p(t) g_j^p(x, y). \tag{8.3.17}$$

As in the TM case, we seek a solution space $V = \{v \in H^1(\Re)\}$, where H^1 denotes the Hilbert space for this problem. Multiplying Eq. (8.3.1) with v and integrating over the space yields

$$\int_{\Re} v \frac{\partial^2 \Psi}{\partial x^2}\, ds + \int_{\Re} v \frac{\partial^2 \Psi}{\partial y^2}\, ds - \frac{1}{c^2} \int_{\Re} v \frac{\partial^2 \Psi}{\partial t^2}\, ds = 0 \quad v \in V, \tag{8.3.18}$$

where ds is the infinitesimal area of integration. Now, following the mathematical procedures of Section 8.2.2 and using Eqs. (8.2.22)–(8.2.30), for the left-hand side of Eq. (8.3.18), we derive

$$-\sum_{j=1}^{3} f_j^p \int_{\Re_p} \left(\frac{\partial g_j^p}{\partial x} \frac{\partial g_i^p}{\partial x} + \frac{\partial g_j^p}{\partial y} \frac{\partial g_i^p}{\partial y} \right) ds + \int_{\Gamma_p \subset \Gamma_N} g_i^p \left(\frac{\partial \Psi}{\partial x} n_1 + \frac{\partial \Psi}{\partial y} n_2 \right) d\Gamma$$
$$-\frac{1}{c^2} \sum_{j=1}^{3} \ddot{f}_j^p \int_{\Re_p} g_i g_j\, ds \quad \text{for } i = 1, 2, 3, \tag{8.3.19}$$

where g's are the element shape functions and Γ_p is the part of the boundary of the pth element that coincides with the Neumann-type boundary of the computational domain.

For the problem considered in this section, the Neumann boundary is the contour of the conductor C. Furthermore, the Dirichlet boundaries are the terminating surface plus the symmetry walls, if any. Since we are considering the total-field formulation, the boundary integral of Eq. (8.3.19) is zero. A suitable modification, as described for the TM case, is required if the scattered-field formulation is considered.

Using the notations from Section 8.2.2, Eq. (8.3.19) can be written as

$$-[\boldsymbol{C}^p][f^p] - \frac{1}{c^2}[\boldsymbol{T}^p][\ddot{f}^p]. \tag{8.3.20}$$

Now, assembling all such elements in the solution region, the ordinary differential equation of the assemblage is given by

$$\frac{1}{c^2}[\boldsymbol{T}][\ddot{f}] = -[\boldsymbol{C}][f]. \tag{8.3.21}$$

The matrices $[\boldsymbol{C}]$ and $[\boldsymbol{T}]$ are the assemblage of individual coefficient matrices $[\boldsymbol{C}^p]$ and $[\boldsymbol{T}^p]$, respectively. It is evident that $[\boldsymbol{C}]$ and $[\boldsymbol{T}]$ are time-independent matrices. Furthermore, by replacing $1/c^2[\boldsymbol{T}]$ and $-[\boldsymbol{C}]$ with matrices $[\boldsymbol{B}]$ and $[\boldsymbol{A}]$, respectively, we have

$$[\boldsymbol{B}][\ddot{f}] = [\boldsymbol{A}][f]. \tag{8.3.22}$$

Note that in Eq. (8.3.22), the column matrix $[f]$ represents $H_z(\boldsymbol{r}, t)$ at the corresponding node.

8.3.3 Time-Stepping Procedure

The system of second-order differential equations given by Eq. (8.3.22) is similar to that of Eq. (8.2.34). Thus, once again, we follow the same steps as in Eqs. (8.2.35)–(8.2.39) to obtain

$$\{[\boldsymbol{B}] - \Delta t^2 \alpha [\boldsymbol{A}]\}[f]_{n+2} + \{-2[\boldsymbol{B}] - \Delta t^2 (0.5 + \delta - 2\alpha)[\boldsymbol{A}]\}[f]_{n+1}$$
$$+ \{[\boldsymbol{B}] - \Delta t^2 (0.5 - \delta + \alpha)[\boldsymbol{A}]\}[f]_n = 0, \tag{8.3.23}$$

which in a compact form yields

$$[\boldsymbol{L}][f]_{n+2} + [\boldsymbol{M}][f]_{n+1} + [\boldsymbol{N}][f]_n = 0. \tag{8.3.24}$$

Here, we note that proper choices for α and δ are important to obtain unconditionally stable results using Newmark's method. Typically, the values for α and δ are 0.25 and 0.5, respectively.

Next, we note that the matrices $[\boldsymbol{L}]$, $[\boldsymbol{M}]$ and $[\boldsymbol{N}]$ in Eq. (8.3.24) do not change with respect to time, but they do depend on the length of the time step chosen for the marching on in time procedure. Replacing n by $n-2$ in Eq. (8.3.24) results in

$$[\boldsymbol{L}][f]_n = -[\boldsymbol{M}][f]_{n-1} - [\boldsymbol{N}][f]_{n-2}. \tag{8.3.25}$$

Furthermore, the matrix $[\boldsymbol{L}]$ can be split into two parts for two types of nodes. The first set of nodes are called free nodes, on which the magnetic field needs to be solved, and the second set are called prescribed nodes, on which $H_z(\boldsymbol{r}, t)$ are known or can be computed easily. If all free nodes are numbered first and the prescribed nodes last, we can rewrite Eq. (8.3.25) as

$$\begin{bmatrix} \boldsymbol{L}_{\mathrm{ff}} & \boldsymbol{L}_{\mathrm{fp}} \\ \boldsymbol{L}_{\mathrm{pf}} & \boldsymbol{L}_{\mathrm{pp}} \end{bmatrix} \begin{bmatrix} f_{\mathrm{f}} \\ f_{\mathrm{p}} \end{bmatrix}_n = -[\boldsymbol{M}][f]_{n-1} - [\boldsymbol{N}][f]_{n-2}, \tag{8.3.26}$$

where subscripts f and p refer to free and prescribed nodes, respectively. Note that Eq. (8.3.26) is a recurrence relation which can be used to compute $[f]_n$ for all values of n knowing $[f]_{n-1}$ and $[f]_{n-2}$.

For the TE case, the nodes residing on the terminating surface are the prescribed nodes. The scattered magnetic fields at time step t_n for the nodes residing on the terminating surface are given by Eq. (8.3.8) in terms of the induced current on the conductor for earlier times. Nodes other than the prescribed nodes are called free nodes, for which the magnetic fields are determined by Eq. (8.3.26).

As mentioned for the TM case, Eq. (8.3.1) can be solved for either the total-field $\boldsymbol{H}(\boldsymbol{r}, t)$ or the scattered-field $\boldsymbol{H}^{\mathrm{s}}(\boldsymbol{r}, t)$. However, the boundary condition for the conducting surface and the radiation condition for the terminating surface depend on the formulation employed (total or scattered) for a particular problem. Even though they are equivalent conditions, numerical differences may arise due to discretization and not employing exact arithmetic for the field computation.

8.3.4 Numerical Results

Square Cylinder Consider the case of a square cylinder with a side of 1.0 m illuminated by a Gaussian pulse of unit amplitude ($E_{\mathrm{o}} = 1$ V/m) with as angle of arrival $\theta = 0°$. We have used a similar meshing scheme as that discussed previously. The terminating surface is only four layers away (0.2 m) and it is conformal to the scatterer. The terminating surface and the conducting surface are composed of 80 and 48 nodes, respectively. The inner layer between the conductor and the terminating surface consists of 192 free nodes.

The pulse width of the incident Gaussian pulse is taken to be $T = 2$ LM and the delay (ct_{o}) to be 3 LM. The time step (Δt) for this example is chosen to be $1/25c$. To derive the initial conditions for the time-stepping procedure, we follow the same procedure as outlined before. We turn on the Gaussian pulse at $t = 0$ over the whole computational space. Now, using Eq. (8.3.26), we compute the fields for the free nodes and the fields on the terminating surface. Hence, from the knowledge of the fields induced, the surface current can be computed by Eq. (8.3.15) at

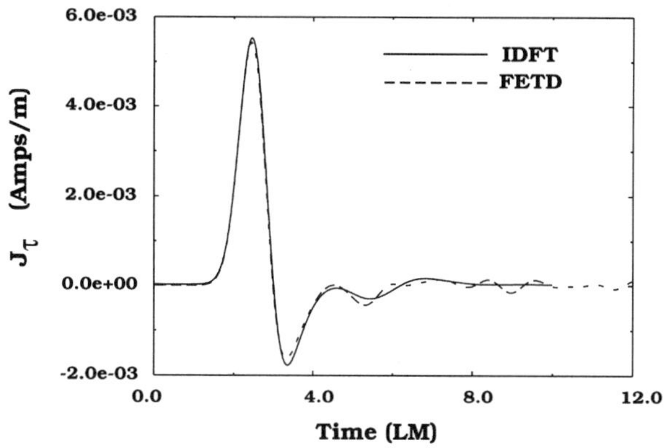

FIGURE 8.12
Induced current on the square conductor at point (0.5,0.0). Outer boundary is 0.2 m away.

time instant t_n. By applying the radiation condition on the terminating surface through the Green's function integral, the scattered fields on the terminating surface may be calculated from Eq. (8.3.11). These values are utilized in Eq. (8.3.26) and the time-stepping procedure continues to compute fields for the next time step. Values of 0.25 and 0.5 were chosen for α and δ, respectively.

To validate the accuracy of the method, the induced surface current for a node, located at $(0.5, 0.0)$ on the conductor, is plotted in Fig. 8.12 as a function of time. From the boundary condition for a perfect conductor, the reflected magnetic pulse is of the same order of magnitude as that of the incident pulse. As the peak of the incident pulse arrives at the front face of the square cylinder at 2.5 LM, the induced surface current is a maximum during this time and reaches double the value of the incident pulse. With the progression of time, the incident pulse moves toward the back of the conductor. These are other transient effects on the current before it returns to zero.

The induced current from this method is compared with currents obtained from the frequency domain MoM. We used 48 subsections to discretize the conductor surface. The pulse basis functions and point-matching testing procedure are used to evaluate the current on the conductor. We have chosen 256 equally spaced frequency points between 0 and 0.5 GHz to produce the frequency domain currents. These values are inverse Fourier transformed to derive the time domain results. As can be seen, the results from the FETD method agree very well with MoM results.

The total magnetic field on a node residing on the terminating surface (0.7,0.0) is shown in Fig. 8.13 with time progression. The total field is very similar to the induced current as should be the case. The pulse passes through this point and reaches its maximum at approximately 2.4 LM, as the reflected pulse from the conductor arrives at the same time. There are small oscillations in late time, but they never increase as solution progresses along the time axis.

Circular Cylinder Consider the case of a circular cylinder with a diameter of 1.0 m. A similar meshing scheme has been used as in the TM scattering case. The computational domain consists of four layers and the terminating surface is 0.2 m away from the conductor surface. There are 60 free nodes on the conducting surface and 60 prescribed nodes on the outside boundary or the terminating surface. Another 180 free nodes reside between the conductor and the terminating surface which, in reality, are elements of the unknown vector.

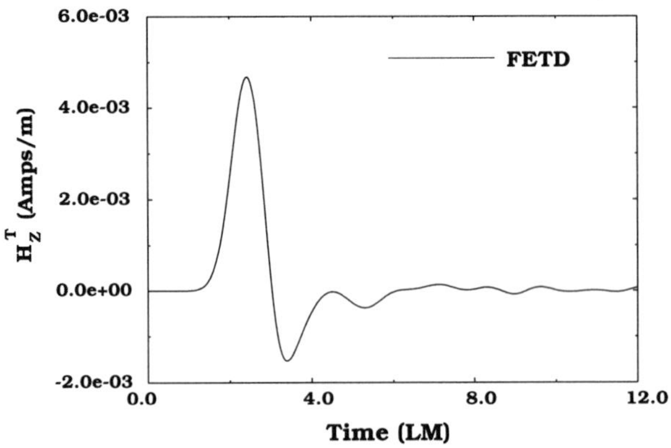

FIGURE 8.13
Total magnetic field on the terminating surface for the square cylinder at point (0.7,0.0).

The incident field, as before, is a Gaussian pulse with $T = 2$ LM, $ct_o = 3$ LM, $E_o = 1$, and $\theta = 0°$. The origin of the coordinate axis coincides with the center of the circular cylinder. Equation (8.3.26) is solved at every time step for the free nodes to compute the magnetic fields at those nodes. Because the magnetic fields are known all over the computational space for a particular instant, we use Eq. (8.3.15) to compute the current on the conductor. Next, the scattered magnetic field is computed at the terminating surface at that instant using Eq. (8.3.11). Now the recurrence relation is continued to compute the fields and currents for the next time step. In this context, for the initial few time steps the total magnetic field on the terminating surface is essentially the incident field because the field produced by the currents takes a finite time to travel from the source point to the field point.

The induced surface current on a node residing on the conductor surface is plotted in Fig. 8.14 with respect to time. In this example, we have chosen a time step of 0.05 LM. As can be observed, the induced surface current reaches maximum near 2.5 LM as the reflected pulse and incident pulse reach their maximum. The results in this case are compared with frequency domain MoM

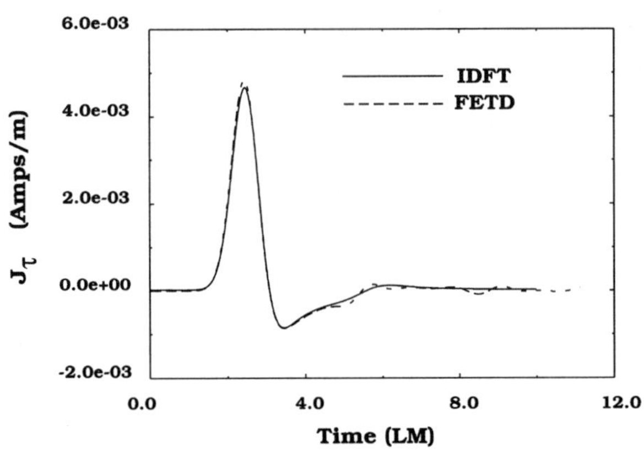

FIGURE 8.14
Induced current on the circular cylinder at point (0.5,0.0). Outer boundary is 0.2 m away.

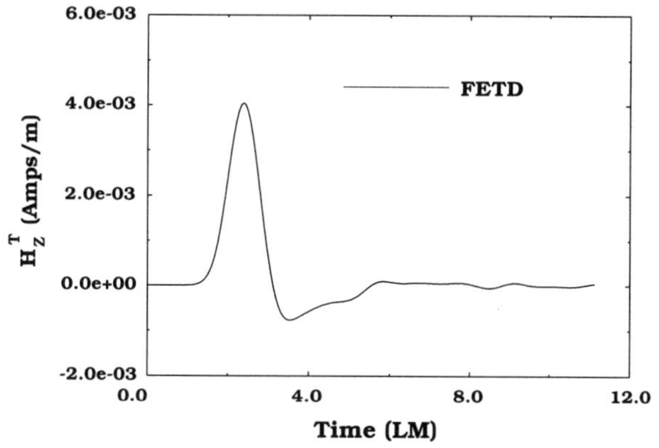

FIGURE 8.15
Total magnetic field on the terminating surface for the circular cylinder at point (0.7,0.0).

results. We used 256 equally spaced frequency points between 0 and 0.5 GHz and inverse Fourier transform technique to obtain the time domain result. As before, we used a pulse basis and point-matching testing procedure for the frequency domain results. For the MoM solution, 40 subsections were chosen on the conducting surface. Even though there is very good agreement in early times, as the incident wave passes over the object there are small oscillations in late time. However, these oscillations do not grow to produce late-time instabilities.

The total magnetic field for a node residing on the terminating surface is plotted in Fig. 8.15. We can clearly observe that the incident wave passes through the node and after reflection from the conducting surface is doubled before returning to zero.

Semicircular Cylinder As a last example, we consider the scattering from a semicircular, closed cylinder with a diameter of 1.0 m. We employed a similar meshing scheme as that described for the previous example. The terminating surface is conformal to the scatterer and only 0.16 m away from the conductor. The incident field in this case is coming from $\theta = 90^\circ$ and has the same parameters as those in the previous example. We used the total-field formulation for this problem. The magnetic field values are determined for 210 free nodes, located between the terminating surface and the conductor, at every time step. Furthermore, there are 60 free nodes on the conductor surface and 70 prescribed nodes on the terminating surface. As before, the time-stepping procedure starts by assuming $\boldsymbol{H}^{\mathrm{i}}$ on the outside boundary. Also, the finite-element matrix equation (Eq. 8.3.26) is solved for the unknown free nodes. The field values for the whole computational domain are used to compute induced current on the conductor. With these known currents for every time step, the radiation condition on the outside boundary is employed through the Green's function integral given by Eq. (8.3.11). The time step chosen is 1/32 LM.

The induced surface current for a node residing on the y-axis with time progression is plotted in Fig. 8.16. As expected, the induced current reaches its peak when the peak of the incident wave hits the particular node. We compared the IDFT (MoM) results with the results from this method (FETD). In the MoM solution, 52 subsections were used to calculate the induced current. The results agree well with the MoM values for the induced current. Even though there are no late-time oscillations, small ripples can be observed. The total magnetic field on a node residing on the terminating surface is plotted in Fig. 8.17. Similar conclusions about the incident and the reflected Gaussian pulse from the conductor surface can be made by examining the magnetic fields with time.

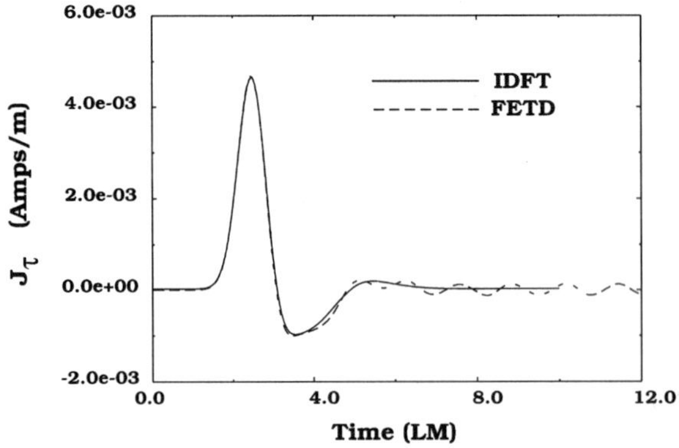

FIGURE 8.16
Total magnetic field on the terminating surface for the half cylinder at point (0.0,0.66).

8.4 CONCLUDING REMARKS

The time domain analysis for TM and TE scattering from perfectly conducting cylinders has been presented utilizing a finite-element method. The radiation condition in this procedure is enforced exactly through the Green's functions on a fictitious boundary which is placed only four layers away from the conductor surface. The space between the boundary and the scatterer is discretized by utilizing triangular elements. The wave equation is discretized in space to derive a set of ordinary differential equations in time. An implicit integration scheme known as Newmark's method is applied to the differential equations to derive a set of algebraic equations. The two-step recurrence relation is used to compute the field quantities in a leapfrog manner. Once the field quantities for the computational space for a particular time instant are known, the radiation condition is applied at the terminating surface through the Green's function. Scattering from a circular, square, and semicircular cylinder has been computed and compared with other available data such as that from the inverse Fourier transform of frequency domain MoM.

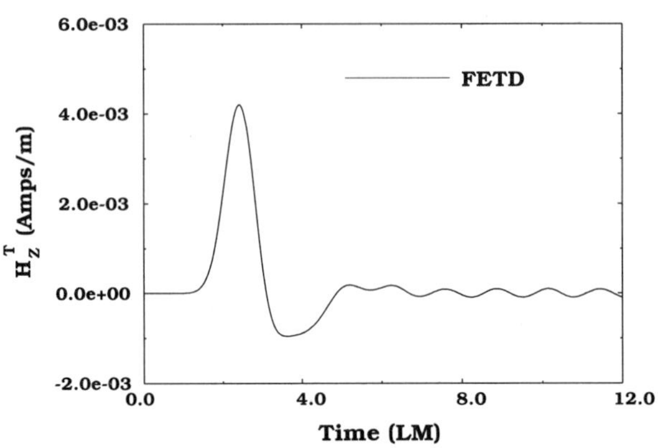

FIGURE 8.17
Total magnetic field on the terminating surface for the half cylinder at point (0.0,0.66).

The implicit integration scheme of Newmark makes the finite-element matrix equations unconditionally stable. Typically, the values of α and δ are prescribed across all the problems, resulting in reasonable accuracy. It should be noted that this method is particularly free of any late-time oscillations or spurious solutions. In this context, placing the boundary too close to the body or keeping it far away from the body introduces different kinds of problems. When it was placed too close, we observed small ripples in the late-time solution even though they do not produce any kind of instability. The exact reason for these types of ripples is unknown. They may be due to integration inaccuracy while computing the scattered field. Results in the late time improve after employing higher order approximation polynomials for the currents and with higher order integration rules. We observed that by moving the terminating surface away from the body the magnitude of the ripples decreases significantly. However, by keeping it very far away from the body and using very few layers between the body and the terminating surface, we introduce error while computing the normal derivative of the electric field, resulting in a loss of accuracy in currents at every time step. This in turn provides erroneous scattered fields at the terminating surface. Hence, it produces inappropriate radiation conditions on the surface. Next, it may be noted that the field values at the sharp corners are difficult to model with node based elements. Because the field is singular near the sharp edges, a denser mesh is necessary to model that region for accurate results at the expense of increasing the computational domain. This method can be used very effectively in solving electromagnetic scattering problems.

Lastly, this method has been applied to two-dimensional conducting scatterers only. However, it is important that this method be validated for more complex structures, such as three-dimensional bodies, material bodies, and bodies with apertures. Efforts are under way to extend this technique to such cases.

BIBLIOGRAPHY

[1] B. Noble, "Variational Finite Element Methods for Initial Value Problems," in *The Mathematics of Finite Elements and Applications*, J. R. Whiteman (Ed.), pp. 143–151, John Wiley, NY, 1993.

[2] S. Barkeshli, H. A. Sabbagh, D. J. Radecki, and M. Melton "A Novel Implicit Time Domain Boundary-Integral/Finite-Element Algorithm for Computing Transient Electromagnetic Field Coupling to a Metallic Enclosure," *IEEE Trans. Antennas Propagat.*, vol. 40, pp. 1155–1164, October 1992.

[3] S. Gedney and U. Navsariwala, "An Unconditionally Stable Finite Element Time Domain Solution of the Vector Wave Equation," *IEEE Microwave Guided Wave Lett.*, vol. 5, pp. 332–334, October 1995.

[4] T. Roy, T. K. Sarkar, A. R. Djordjevic, and M. Salazar, "A Hybrid Method for Terminating the Finite Element Mesh (Electrostatic Case)," *Microwave Optical Technol. Lett.*, vol. 8, pp. 282–287, April 1995.

[5] T. Roy, T. K. Sarkar, A. R. Djordjevic, and M. Salazar, "A Hybrid Method for Solution of Scattering by Conducting Cylinders (TM Case)," *IEEE Trans. Microwave Theory Tech.*, vol. 44, pp. 2145–2151, December 1996.

[6] P. M. Morse and H. Feshback, *Methods of Theoretical Physics*, McGraw-Hill, New York, 1953.

[7] C. Johnson, *Numerical Solution of Partial Differential Equations by the Finite Element Method*, Cambridge Univ. Press, Cambridge, UK, 1994.

CHAPTER 9

Finite-Volume Time Domain Method

P. BONNET, X. FERRIERES, B. L. MICHIELSEN, and P. KLOTZ
ONERA, France

J. L. ROUMIGUIÈRES
LASMEA, France

In this chapter, we present the finite-volume time domain (FVTD) method for the solution of Maxwell's equations. Initially, the FVTD method was developed as a numerical technique to solve the partial differential equations of hyperbolic conservation laws. The technique has been very successful in, for example, computational fluid dynamics, i.e., for solving the Euler and Navier–Stokes equations [1].

It is only quite recently that finite-volume methods have been applied to the numerical solution of the Maxwell equations. The method was used in electromagnetics by Shankar and Hall [2] and, at the same time, by Madsen and Ziolkowski [3], who proposed a modified finite-volume method based on a generalization of Yee's scheme in the finite-difference time domain (FDTD) method [4]. In these approaches, the authors were motivated by the desire to approximate the boundaries of irregular domains more accurately than through the use of the "staircase" approximations of the structured grids of the classical finite-difference methods and thus to improve the accuracy of the numerical schemes. Since that time, this method has been applied in several domains of electromagnetism, such as antenna problems, radar cross-section computations, and electromagnetic compatibility problems [5–13].

In the first section of this chapter, the background of the finite-volume method, relevant to electromagnetic applications, is given. The essential ingredients of the method are introduced through the presentation of an elementary finite-volume method for a pure initial value problem in free space.

In the second section, the finite-volume discretization of Maxwell's equations is elaborated in detail. The implementation of the boundary conditions on material obstacles is given, which makes the application of the method to actual electromagnetic problems possible. This section also presents a discussion of the stability of the method.

The third section extends the applicability of the finite-volume method in various ways. First, it is shown how one can incorporate thin-wire models in the method. Second, a hybrid FVTD–FDTD method is presented which improves the performances of both the FVTD and the FDTD methods. Finally, another approach to the FVTD method is presented which can be viewed as a mixture of a conventional FVTD method and a generalization of the FDTD method. In the last section, examples computed with the techniques presented in this chapter are shown and their results are compared to results obtained with other numerical methods or by measurement.

9.1 MAXWELL'S EQUATIONS AS A HYPERBOLIC CONSERVATIVE SYSTEM

In this section, we develop the fundamentals of the FVTD method for Maxwell's equations. The finite-volume method is based on the integral form of Maxwell's equations and the principal idea of the construction relies on the hyperbolic nature of Maxwell's equations, which implies that the field at space–time point (x, t) depends on initial data at time t_0 on a neighborhood of x with radius $c(t - t_0)$ (c is the speed of light). This enables one to construct the evolution of the solution in time using only local information. The integral form and the hyperbolic nature of Maxwell's equations will be discussed in greater detail in the following subsections.

We first present a short discussion of Maxwell's equations which allows us to introduce some definitions and terminology as we use them in the development of the method in the rest of this chapter. The fundamental equations modeling the electromagnetic field in a domain with dielectric, magnetizable, and conducting material is the system of Maxwell's equations:

$$\begin{cases} \nabla \times \boldsymbol{H} - \varepsilon \frac{\partial \boldsymbol{E}}{\partial t} - \sigma \boldsymbol{E} = \boldsymbol{J} \\ \nabla \times \boldsymbol{E} + \mu \frac{\partial \boldsymbol{H}}{\partial t} = \boldsymbol{K}. \end{cases} \tag{9.1.1}$$

The fields $\boldsymbol{E}, \boldsymbol{H}$ are R^3 vector-valued functions on space–time, with regularity properties to be defined later. The spatial domain is denoted $\Omega \subset R^3$ (possibly unbounded) and, generally, we consider a finite time interval $\mathcal{T} = (0, T) \subset R_+$. The constitutive parameters are supposed to be piecewise smooth functions of space and independent of time. The independent, prescribed source densities, $\boldsymbol{J}$ and $\boldsymbol{K}$, are supposed to have bounded spatial support, $\text{supp}(\boldsymbol{J}(t), \boldsymbol{K}(t)) \subset\subset R^3$ and $\boldsymbol{J}(x, t), \boldsymbol{K}(x, t) \equiv 0$ for $t < 0$. The initial–boundary value problem we are interested in here is to find the functions $\boldsymbol{E}$ and $\boldsymbol{H}$ for $t \in \mathcal{T}$ given that $(\forall x \in \Omega)\,[\lim_{t\to 0} \boldsymbol{E}(x, t) = \lim_{t\to 0} \boldsymbol{H}(x, t) = 0]$.

The partial differential equations are completed with continuity conditions on surfaces of discontinuity in the constitutive parameters

$$[\boldsymbol{E}_{\text{tg}}, \boldsymbol{H}_{\text{tg}}]_-^+ = 0$$
$$[\varepsilon \boldsymbol{a}_{\text{n}} \cdot \boldsymbol{E}, \mu \boldsymbol{a}_{\text{n}} \cdot \boldsymbol{H}]_-^+ = 0,$$

where $\boldsymbol{a}_{\text{n}}$ is a unit normal on the surface of discontinuity and $[f]_-^+$ denotes the difference between the limits on the surface from the positive side of the surface and the negative side of the surface.

On the boundary of the spatial domain of definition of the fields, which may coincide with the boundaries of perfectly conducting or perfectly magnetizable media, we can specify one of the tangential projections of the fields,

$$\begin{cases} \boldsymbol{E}_{\text{tg}}|_{\partial\Omega} = \boldsymbol{e} \\ \boldsymbol{H}_{\text{tg}}|_{\partial\Omega} = \boldsymbol{h}. \end{cases} \tag{9.1.2}$$

In the special cases of impenetrable obstacles, one would take $\boldsymbol{e} \equiv 0$ on perfect conductors and $\boldsymbol{h} \equiv 0$ on perfectly magnetizable media, respectively.

On thin conducting obstacles one often does not want to deal with the internal fields and replaces the obstacle by a set of continuity conditions. Let Σ be the surface, oriented by a normal $\boldsymbol{a}_{\mathrm{n}}$, representing the thin obstacle. Then, the following conditions are required:

$$\begin{cases} \boldsymbol{a}_{\mathrm{n}} \times [\boldsymbol{H}_{\mathrm{tg}}]_{-}^{+} - \sigma_{\Sigma}\{\boldsymbol{E}_{\mathrm{tg}}\}_{-}^{+} = \boldsymbol{J}_{\Sigma} \\ \boldsymbol{a}_{\mathrm{n}} \times [\boldsymbol{E}_{\mathrm{tg}}]_{-}^{+} = \boldsymbol{K}_{\Sigma}, \end{cases} \tag{9.1.3}$$

where $\{\boldsymbol{E}_{\mathrm{tg}}\}_{-}^{+}$ is the average of the two limits from either side of the surface Σ, σ_{Σ} is the sheet conductivity (the bulk conductivity of the thin obstacle times its thickness), and $\boldsymbol{J}_{\Sigma}$ and $\boldsymbol{K}_{\Sigma}$ are surface sources representing the internal sources in the thin obstacle.

The formulation given thus far is sometimes referred to as the total-field formulation. When treating scattering problems, one may benefit from computing only the scattered field. To this end, the total field is decomposed into an incident field, $\{\boldsymbol{E}^{\mathrm{i}}, \boldsymbol{H}^{\mathrm{i}}\}$, and a scattered field, $\{\boldsymbol{E}^{\mathrm{s}}, \boldsymbol{H}^{\mathrm{s}}\}$:

$$\{\boldsymbol{E}, \boldsymbol{H}\} = \{\boldsymbol{E}^{\mathrm{i}}, \boldsymbol{H}^{\mathrm{i}}\} + \{\boldsymbol{E}^{\mathrm{s}}, \boldsymbol{H}^{\mathrm{s}}\}. \tag{9.1.4}$$

The incident field will be taken, as usual, to be the field generated by all the sources in the system radiating in free space. This implies that the incident field satisfies the following system of Maxwell's equations:

$$\begin{cases} \nabla \times \boldsymbol{H}^{\mathrm{i}} - \varepsilon_0 \frac{\partial \boldsymbol{E}^{\mathrm{i}}}{\partial t} = \boldsymbol{J} \\ \nabla \times \boldsymbol{E}^{\mathrm{i}} + \mu_0 \frac{\partial \boldsymbol{H}^{\mathrm{i}}}{\partial t} = \boldsymbol{K}, \end{cases} \tag{9.1.5}$$

where ε_0 and μ_0 are the electric permittivity and the magnetic permeability of vacuum, respectively. The $\boldsymbol{J}$ and $\boldsymbol{K}$ are the independent sources of the original problem. On the boundary of the domain, we do not put special conditions on the incident field because we consider the incident field to be defined in free space.

With thin obstacles, modeled by continuity conditions, we have

$$\begin{cases} n \times \left[\boldsymbol{H}_{\mathrm{tg}}^{\mathrm{i}}\right]_{-}^{+} = \boldsymbol{J}_{\Sigma} \\ n \times \left[\boldsymbol{E}_{\mathrm{tg}}^{\mathrm{i}}\right]_{-}^{+} = \boldsymbol{K}_{\Sigma}. \end{cases} \tag{9.1.6}$$

The equations satisfied by the scattered field follow by subtracting those for the incident field from the equations for the total field. The Maxwell's equations for the scattered field are then

$$\begin{cases} \nabla \times \boldsymbol{H}^{\mathrm{s}} - \varepsilon \frac{\partial \boldsymbol{E}^{\mathrm{s}}}{\partial t} - \sigma \boldsymbol{E}^{\mathrm{s}} = \boldsymbol{J}^{\mathrm{s}} \\ \nabla \times \boldsymbol{E}^{\mathrm{s}} + \mu \frac{\partial \boldsymbol{H}^{\mathrm{s}}}{\partial t} = \boldsymbol{K}^{\mathrm{s}}, \end{cases} \tag{9.1.7}$$

where the source terms are now dependent on the incident field, $\boldsymbol{J}^{\mathrm{s}} = (\varepsilon - \varepsilon_0)(\partial \boldsymbol{E}^{\mathrm{i}}/\partial t) + \sigma \boldsymbol{E}^{\mathrm{i}}$ and $\boldsymbol{K}^{\mathrm{s}} = -(\mu - \mu_0)(\partial \boldsymbol{H}^{\mathrm{i}}/\partial t)$.

On the boundary of the domain, we have one of the conditions,

$$\begin{cases} \boldsymbol{E}_{\mathrm{tg}}^{\mathrm{s}}|_{\partial\Omega} = \boldsymbol{e}^{\mathrm{s}} \\ \boldsymbol{H}_{\mathrm{tg}}^{\mathrm{s}}|_{\partial\Omega} = \boldsymbol{h}^{\mathrm{s}}, \end{cases} \tag{9.1.8}$$

where $\boldsymbol{e}^{\mathrm{s}} = \boldsymbol{e} - \boldsymbol{E}_{\mathrm{tg}}^{\mathrm{i}}|_{\partial\Omega}$ and $\boldsymbol{h}^{\mathrm{s}} = \boldsymbol{h} - \boldsymbol{H}_{\mathrm{tg}}^{\mathrm{i}}|_{\partial\Omega}$, in which, for the special cases of perfect conductors or perfectly magnetizable media, one should substitute $\boldsymbol{e} \equiv 0$ or $\boldsymbol{h} \equiv 0$, respectively.

The thin conducting obstacle equations are as follows:

$$\begin{cases} \boldsymbol{a}_{\mathrm{n}} \times \left[\boldsymbol{H}^{\mathrm{s}}_{\mathrm{tg}}\right]^{+}_{-} - \sigma_{\Sigma}\left\{\boldsymbol{E}^{\mathrm{s}}_{\mathrm{tg}}\right\}^{+}_{-} = \boldsymbol{J}^{\mathrm{s}}_{\Sigma} \\ \boldsymbol{a}_{\mathrm{n}} \times \left[\boldsymbol{E}^{\mathrm{s}}_{\mathrm{tg}}\right]^{+}_{-} = \boldsymbol{K}^{\mathrm{s}}_{\Sigma}, \end{cases} \tag{9.1.9}$$

where $\boldsymbol{J}^{\mathrm{s}}_{\Sigma} = \sigma_{\Sigma}\boldsymbol{E}^{\mathrm{i}}_{\mathrm{tg}}$ and $\boldsymbol{K}^{\mathrm{s}}_{\Sigma} \equiv 0$. Furthermore, the continuity conditions on surfaces of discontinuity in the constitutive parameters are the same as those for the total field because such discontinuities are absent in the incident field. Therefore,

$$\left[\boldsymbol{E}^{\mathrm{s}}_{\mathrm{tg}}, \boldsymbol{H}^{\mathrm{s}}_{\mathrm{tg}}\right]^{+}_{-} = 0$$

$$[\varepsilon\boldsymbol{a}_{\mathrm{n}} \cdot \boldsymbol{E}^{\mathrm{s}}, \mu\boldsymbol{a}_{\mathrm{n}} \cdot \boldsymbol{H}^{\mathrm{s}}]^{+}_{-} = 0.$$

One observes that the equations for the scattered field, constituting the so-called scattered-field formulation, are formally identical to those for the total field. In fact, the only difference is in the definition of the source terms. In the so-called total-field formulation, the source terms can be chosen completely independently, but in the scattered-field formulation the source terms are interrelated and determined by the definition of some incident field.

In the rest of this section, we shall only work with the total-field formulation because this shows all the essential properties and we do not need the mutual dependencies in the source terms imposed by an incident field. The scattered-field formulation will be discussed later.

We conclude our introduction with some generalities on the solutions of Maxwell's equations in the time domain. Mathematical study of the problems of existence and uniqueness is mostly performed in spaces of generalized functions. One takes into consideration classes of square integrable vector-valued functions of the space coordinates, for which the curl operator again gives square integrable functions which are considered as continuous distribution-valued functions of time, $E, H \in C((0, T), L^2(\text{curl}, \Omega))$. It is known that the initial value problem has a unique solution for any $J, K \in C((0, T), L^2(\Omega)^3)$ and, moreover, that the solutions depend continuously on the sources. These results are not the optimal ones, but here we only want to emphasize the continuity properties, which underlie the possibility to apply the finite-volume method in principle.

The remaining part of this section is divided into three subsections. In Section 9.1.1, we cast the Maxwell's equations into the form which is usual in the study of linear conservation laws. In Section 9.1.2, we present some results concerning the propagation of discontinuities in the electromagnetic field. In Section 9.1.3, we present an elementary finite-volume method for Maxwell's equations in free space and study the numerical problems in a formal way.

9.1.1 The Conservative Form of Maxwell's Equations

To facilitate the comparison of the FVTD method as applied to Maxwell's equations with applications to other systems of equations, we shall now cast the system of Maxwell's equations into a form which is usual in the study of general (linear) conservative systems. Note that this is only a formal exercise which does not change in any way the fundamental properties of the theory. Also note that vectors in R^3 can be mapped to antisymmetric 3×3 matrices $\Lambda^2(R^3)$. We denote this mapping by s:

$$s : R^3 \longrightarrow \Lambda^2(R^3),$$

by

$$s : (v_1, v_2, v_3) \longmapsto \begin{bmatrix} 0 & -v_3 & v_2 \\ v_3 & 0 & -v_1 \\ -v_2 & v_1 & 0 \end{bmatrix}.$$

Next, we define a divergence operator on functions with values in the space of antisymmetric matrices [note that any $m \in \Lambda^2(R^3)$ is of the form $s(\boldsymbol{v})$ with $\boldsymbol{v} = s^{-1}(m)$]:

$$\mathrm{Div}(s(\boldsymbol{v})) = \nabla \times \boldsymbol{v}$$

or, componentwise,

$$[\mathrm{Div}(s(\boldsymbol{v}))]_q = \sum_{p=1}^{3} \frac{\partial}{\partial x^p}[s(\boldsymbol{v})]_{pq}.$$

With these definitions, we obtain the following for Maxwell's equations:

$$\begin{cases} \varepsilon \frac{\partial \boldsymbol{E}}{\partial t} - \mathrm{Div}(s(\boldsymbol{H})) + \sigma \boldsymbol{E} = -\boldsymbol{J} \\ \mu \frac{\partial \boldsymbol{H}}{\partial t} + \mathrm{Div}(s(\boldsymbol{E})) = \boldsymbol{K}, \end{cases} \tag{9.1.10}$$

or, grouping the electromagnetic field components into a single vector $\boldsymbol{U} \in R^3 \oplus R^3$,

$$\alpha \frac{\partial \boldsymbol{U}}{\partial t} + \mathrm{Div}(\mathcal{A}\boldsymbol{U}) + \mathcal{B}\boldsymbol{U} = \boldsymbol{G}, \tag{9.1.11}$$

where

$$\begin{aligned} \alpha &: R^3 \oplus R^3 \longrightarrow R^3 \oplus R^3, \\ \mathcal{A} &: R^3 \oplus R^3 \longrightarrow \Lambda^2(R^3) \oplus \Lambda^2(R^3), \\ \mathcal{B} &: R^3 \oplus R^3 \longrightarrow R^3 \oplus R^3, \end{aligned}$$

with matrix representations

$$[\alpha] = \begin{bmatrix} \varepsilon & 0 \\ 0 & \mu \end{bmatrix} \quad [\mathcal{A}] = \begin{bmatrix} 0 & -s \\ s & 0 \end{bmatrix} \quad [\mathcal{B}] = \begin{bmatrix} \sigma & 0 \\ 0 & 0 \end{bmatrix}$$

and

$$\boldsymbol{G} = \begin{bmatrix} -\boldsymbol{J} \\ \boldsymbol{K} \end{bmatrix},$$

and the definition of the operator Div is extended to the direct sum space in the obvious way.

Because we have only regrouped the various field quantities in a new way, we can recover the uniqueness and existence results in a simple way from those for the original Maxwell's equations. We have a unique $\boldsymbol{U} \in C((0,T), L^2(\mathrm{Div} \circ \mathcal{A}, \Omega))$ for any $\boldsymbol{G} \in C((0,T), L^2(\Omega)^6)$ satisfying $\boldsymbol{U}|_{t=0} = 0$, where $L^2(\mathrm{Div} \circ \mathcal{A}, \Omega)$ is the domain of the operator $\mathrm{Div} \circ \mathcal{A}$ in $L^2(\Omega)^6$, i.e., classes for which the image of the distributionally defined composite operator $\mathrm{Div} \circ \mathcal{A}$ is again equivalent to a distribution in $L^2(\Omega)^6$.

9.1.2 Characteristics and Wavefront Propagation

In this section, we discuss two phenomena related to the hyperbolic nature of Maxwell's equations which are fundamental for the finite-volume method. First, the nature of the partial differential equations allows for discontinuities in the field solutions, but the space–time behavior of these discontinuities is subject to certain constraints. To take advance of the results we will show that the discontinuities propagate along well-identified paths, called characteristics, and the structure of the discontinuities is locally that of a so-called plane wave. This constraint has important consequences because the possibility of propagating discontinuities is important in the decision of which numerical approximations we will use. In general, we have to take care not to smooth our solutions too much, implicitly or explicitly, because "ripples" can contain important information about the field structure, especially when the temporal frequency spectrum is of interest. The second phenomenon we wish to discuss is the fact that the evolution of the field in a certain point is dependent only on the initial data on a finite sphere around this point. This is in fact a consequence of the first phenomenon. The properties of wavefront propagation dictate that the elementary solutions of Maxwell's equations are singularities which expand with finite speed, and hence the field in a certain space–time point can be influenced only by initial data on a restricted domain. We shall now study the evolution of a wavefront in free space, i.e., a smooth hypersurface of which the spatial sections at fixed times, $S \subset R^3$, are smooth and depend smoothly on time. We assume that on either side of this hypersurface there exists an electromagnetic field solution of the vacuum Maxwell's equations, which is a smooth function of space–time, but on S there is a discontinuity in the field components. Without loss of generality, we may assume that on one side (e.g., the positive side with respect to the normal) the field is in fact identically zero, such that the discontinuity is identical to the field limit from the negative side on S. We want to derive the relations between the structure of the hypersurface and the structure of the discontinuity of the field components as imposed by Maxwell's equations.

If we substitute the discontinuous field described previously into the Maxwell's equations, taken in the distributional sense, we obtain the following equations for the discontinuity in the fields $\{\boldsymbol{E}, \boldsymbol{H}\}$ on S:

$$\begin{cases} \boldsymbol{a}_{\mathrm{n}} \cdot \varepsilon \boldsymbol{E} = 0 \\ \boldsymbol{a}_{\mathrm{n}} \cdot \mu \boldsymbol{H} = 0 \end{cases} \qquad \begin{cases} \boldsymbol{E}_{\mathrm{tg}} + \mu \boldsymbol{v} \times \boldsymbol{H} = 0 \\ \boldsymbol{H}_{\mathrm{tg}} - \varepsilon \boldsymbol{v} \times \boldsymbol{E} = 0, \end{cases}$$

or, equivalently,

$$\begin{cases} (\boldsymbol{a}_{\mathrm{n}} \cdot \boldsymbol{v}) \mu \boldsymbol{H}_{\mathrm{tg}} - \boldsymbol{a}_{\mathrm{n}} \times \boldsymbol{E}_{\mathrm{tg}} = 0 \\ (\boldsymbol{a}_{\mathrm{n}} \cdot \boldsymbol{v}) \varepsilon \boldsymbol{E}_{\mathrm{tg}} + \boldsymbol{a}_{\mathrm{n}} \times \boldsymbol{H}_{\mathrm{tg}} = 0, \end{cases} \tag{9.1.12}$$

where $\boldsymbol{a}_{\mathrm{n}}$ is at any time the unit normal to the instantaneous surface, and $\boldsymbol{v}$ is at any time the instantaneous velocity vector of the spatial surface.

These equations show that the discontinuities can only be tangential to a moving surface. Furthermore, we observe that nonzero solutions to these equations can only exist for $(\boldsymbol{a}_{\mathrm{n}} \cdot \boldsymbol{v}, \boldsymbol{a}_{\mathrm{n}})$ satisfying (substitute the second equation in the first to obtain $[(\boldsymbol{a}_{\mathrm{n}} \cdot \boldsymbol{v})^2 \varepsilon\mu - \boldsymbol{a}_{\mathrm{n}} \cdot \boldsymbol{a}_{\mathrm{n}}] \boldsymbol{E}_{\mathrm{tg}} = 0$)

$$\frac{(\boldsymbol{a}_{\mathrm{n}} \cdot \boldsymbol{v})^2}{c^2} - \boldsymbol{a}_{\mathrm{n}} \cdot \boldsymbol{a}_{\mathrm{n}} = 0, \tag{9.1.13}$$

where $c = (\varepsilon\mu)^{-1/2}$. This equation is in fact identical to the characteristic equation for the Maxwell system because it is the characteristic equation of the principal matrix symbol obtained by replacing the partial derivative operators in Maxwell's equations by corresponding imaginary multipliers $\partial/\partial x \rightarrow i\xi$. [This also explains the qualification "hyperbolic," used for the time domain Maxwell's equations, because this characteristic equation describes (degenerate) hyperboloids in R^4.] Substitution of $(\boldsymbol{a}_\mathrm{n} \cdot \boldsymbol{v}) = c$ into the equations satisfied by the discontinuities shows that they locally have the structure of a plane wave, propagating in the direction $\pm\boldsymbol{a}_\mathrm{n}$:

$$\begin{cases} \boldsymbol{E} = \pm Z\boldsymbol{a}_\mathrm{n} \times \boldsymbol{H} \\ \boldsymbol{H} = \mp Y\boldsymbol{a}_\mathrm{n} \times \boldsymbol{E}, \end{cases} \tag{9.1.14}$$

where $Z = \sqrt{\mu/\varepsilon} = Y^{-1}$. Our conclusion is that smooth surfaces of discontinuities can appear in solutions of Maxwell's equations, which necessarily propagate with the speed of light in directions orthogonal to the surface. Moreover, the discontinuities are tangential to the surface and have the structure of a plane wave. The most obvious discontinuous solution to Maxwell's equations is the point source solution or the Green distribution. This distribution can be derived from the simpler one for the scalar d'Alembert equation, $(\frac{\partial^2}{\partial(ct)^2} - \Delta)G(x,t) = \delta$:

$$G(x,t) = \mathrm{Pv}\frac{1}{4\pi t} \otimes \delta(t - \|x\|/c), \tag{9.1.15}$$

i.e., the Cauchy principal value distribution of $1/(4\pi t)$, on the time axis multiplied by the Dirac measure of a sphere expanding with the speed of light. The Green distributions, or elementary solutions, for Maxwell's equations can be constructed from vector potentials of which the Cartesian components satisfy the scalar wave equation. Let $\boldsymbol{\pi}^\mathrm{e}$, $\boldsymbol{\pi}^\mathrm{m}$ denote the elementary electric and the magnetic vector potential, respectively, satisfying

$$\begin{cases} \left(\Delta - \frac{\partial^2}{\partial(ct)^2}\right)\boldsymbol{\pi}^\mathrm{e} = \boldsymbol{J}\delta \\ \left(\Delta - \frac{\partial^2}{\partial(ct)^2}\right)\boldsymbol{\pi}^\mathrm{m} = -\boldsymbol{K}\delta, \end{cases} \tag{9.1.16}$$

with $\boldsymbol{J}$ and $\boldsymbol{K}$ arbitrary, fixed vectors in R^3. Then, we can verify, without too much difficulty, that the four vector fields,

$$\begin{cases} \boldsymbol{G}^{ej} = \mu\frac{\partial\boldsymbol{\pi}^\mathrm{e}}{\partial t} - \frac{1}{\varepsilon}\int_0^t \nabla\nabla\cdot\boldsymbol{\pi}^\mathrm{e} \\ \boldsymbol{G}^{hj} = -\nabla\times\boldsymbol{\pi}^\mathrm{e} \\ \boldsymbol{G}^{ek} = \nabla\times\boldsymbol{\pi}^\mathrm{m} \\ \boldsymbol{G}^{hk} = \varepsilon\frac{\partial\boldsymbol{\pi}^\mathrm{m}}{\partial t} - \frac{1}{\mu}\int_0^t \nabla\nabla\cdot\boldsymbol{\pi}^\mathrm{m} \end{cases} \tag{9.1.17}$$

define two pairs, $\{\boldsymbol{G}^\mathrm{ej}, \boldsymbol{G}^\mathrm{hj}\}$ and $\{\boldsymbol{G}^\mathrm{ek}, \boldsymbol{G}^\mathrm{hk}\}$, constituting elementary solutions of Maxwell's equations with the two independent space–time Dirac sources $\boldsymbol{J}\delta$ and $\boldsymbol{K}\delta$, respectively.

In the next section, we use the evaluation of the Green distribution on initial data to compute the evolution of an electromagnetic field in time. Again, it is simpler to proceed using vector potentials. If we elaborate the integral representation for a vector potential in terms of its initial data, being a zero value and a prescribed time derivative $\boldsymbol{E}_0$, at time $t = 0$ say, we obtain

$$\boldsymbol{\pi}^\mathrm{e}(x,t) = \frac{1}{4\pi}\int_{\theta\in S^2} \boldsymbol{E}_0(x + t\theta)\, t\, d\theta. \tag{9.1.18}$$

One verifies quite easily that $\lim_{t\to 0}\{\pi^{\mathrm{e}}(x,t), \partial_t \pi^{\mathrm{e}}(x,t)\} = \{0, \boldsymbol{E}_0(x)\}$ indeed holds. Then, we can solve an initial value problem for the Maxwell's equations in vacuum, with prescribed $\boldsymbol{E}(x,0) = \boldsymbol{E}_0$ and $\boldsymbol{H}(x,0) = \boldsymbol{H}_0$ which we assume to be smooth enough, by taking

$$\begin{cases} \boldsymbol{E} = \mu \frac{\partial \pi^{\mathrm{e}}}{\partial t} - \frac{1}{\varepsilon}\int_0^t \nabla\nabla\cdot\pi^{\mathrm{e}} + \nabla\times\pi^{\mathrm{m}} \\ \boldsymbol{H} = \varepsilon \frac{\partial \pi^{\mathrm{m}}}{\partial t} - \frac{1}{\mu}\int_0^t \nabla\nabla\cdot\pi^{\mathrm{m}} - \nabla\times\pi^{\mathrm{e}}, \end{cases} \tag{9.1.19}$$

with

$$\begin{cases} \pi^{\mathrm{e}}(x,t) = \frac{1}{4\pi\mu}\int_{\theta\in S^2} \boldsymbol{E}_0(x+ct\theta)\,t\,d\theta \\ \pi^{\mathrm{m}}(x,t) = \frac{1}{4\pi\varepsilon}\int_{\theta\in S^2} \boldsymbol{H}_0(x+ct\theta)\,t\,d\theta. \end{cases} \tag{9.1.20}$$

What we can learn from these integral representations is that the evolution of the electromagnetic field is a local phenomenon. This means that the evolution of the field at a point x, up to a certain time t, is completely determined by the initial data in a ball of finite radius equal to ct centered at x.

9.1.3 An Elementary Form of the Finite-Volume Method

In this section, we discuss the application of the finite-volume method on a semiformal level. We introduce a space discretization in a generic way and demonstrate how the original problem is translated into a numerical update scheme for a vector of finite-volume integrals over the electromagnetic field. We discuss the essential problems of numerical analysis related to the convergence of the method; however, we do not present any mathematical proofs.

We consider the simplest application possible—a pure initial value problem for Maxwell's equations in free space on the domain $\Omega = R^3$. One might consider this the problem of the evolution of a free-space electromagnetic field after all the sources have been switched off. Let Ω be triangulated by an infinite number of elementary volumes and surfaces described by

$$C^3 \xrightarrow{\partial} C^2, \tag{9.1.21}$$

where C^p, $(p = 2, 3)$ are vector spaces generated by formal linear combinations, with coefficients in R, of the cells of dimension p, called p-cells and denoted by $\gamma^p \in C^p$, and ∂ is the boundary homomorphism which associates the formal linear combination of boundary surface cells to any volume in the triangulation. We suppose that all cells have finite, nonvanishing size such that balls of finite radius are covered by a finite number of 3-cells. Due to the hyperbolic nature of our system, we can work in spaces of bounded functions having spatial support in only a finite number of cells.

We now apply a complete discretization of the partial differential equations by integrating them over space–time cells. The time interval $\mathcal{T} = (0, T) \subset R_+$ is covered by a finite set of intervals, $I_q = (t_{q-1}, t_q]$, where $q = 1, \cdots, N$ $(t_q - t_{q-1} = \Delta t)$, and the integral of $\boldsymbol{U}$ over a space–time cell $\gamma_p^3 \times I_q$ is denoted by $\langle \boldsymbol{U}, \gamma_p^3 \times I_q\rangle$. Integration of the partial differential equation, with $\mathcal{B} = 0$ and $\boldsymbol{G} = 0$, gives

$$\forall\gamma_p^3\,\forall I_q \quad \alpha\big\langle \boldsymbol{U}, \gamma_p^3 \times \partial I_q\big\rangle = -\big\langle \boldsymbol{a}_{\mathrm{n}}\cdot\mathcal{A}\boldsymbol{U}, \partial\gamma_p^3 \times I_q\big\rangle, \tag{9.1.22}$$

where

$$[\boldsymbol{a}_{\mathrm{n}}\cdot\mathcal{A}] = \begin{bmatrix} 0 & -s(\boldsymbol{n}) \\ s(\boldsymbol{n}) & 0 \end{bmatrix}.$$

Using $\partial I_q = \{t_q\} - \{t_{q-1}\}$, we get

$$\langle \boldsymbol{U}, \gamma_p^3 \times \{t_q\} \rangle = \langle \boldsymbol{U}, \gamma_p^3 \times \{t_{q-1}\} \rangle - \alpha^{-1} \langle \boldsymbol{a_n} \cdot \mathcal{A}\boldsymbol{U}, \partial \gamma_p^3 \times I_q \rangle. \tag{9.1.23}$$

Now, if we had a relation such as

$$\alpha^{-1} \langle \boldsymbol{a}_{\mathrm{n}} \cdot \mathcal{A}\boldsymbol{U}, \partial \gamma_p^3 \times I_q \rangle = \sum_k E_{pk} \langle \boldsymbol{U}, \gamma_k^3 \times \{t_{q-1}\} \rangle \Delta t \tag{9.1.24}$$

(the factor Δt has been added for normalization purposes), then we could compute the volume integrals at time t_q from those at time t_{q-1} and hence compute the evolution in an explicit iteration scheme:

$$\langle \boldsymbol{U}, \gamma_p^3 \times \{t_q\} \rangle = \sum_k (\mathrm{Id} - \Delta t E)_{pk} \langle \boldsymbol{U}, \gamma_k^3 \times \{t_{q-1}\} \rangle, \tag{9.1.25}$$

where Id is the identity operator.

The problem then is to estimate the fluxes through the cell faces, integrated over a time interval, from the knowledge of the spatial cell integrals computed earlier. Before we discuss the details of this flux estimation, we introduce some general ideas concerning the numerical approximation.

What we wish to do is to construct our finite-volume integrals, over all space and time, such that we can retrieve the space–time field distributions from them. This brings forward a number of other problems which, in fact, are hard problems of applied mathematics. Some work is available in this area, but the proofs are very technical and we cannot discuss these within the context of this book. Nevertheless, we discuss the general ideas. We believe that even a formal discussion will make the numerical approximations (used later) more intuitively clear.

The first problem to address is how we retrieve a function from a set of integrals over this function. This is the reconstruction problem, and its solution depends on the a priori regularity information concerning the function and hence on advanced mathematical results concerning the smoothness properties of the solutions of the initial value problems with which we are dealing.

However, we are immediately confronted with the question of whether the finite-volume integrals, which we have constructed with our numerical scheme, are good approximations to the integrals over the exact solution. This is a complicated problem of the convergence of the numerical method, i.e., can we approximate the exact solution with arbitrary precision if we sufficiently increase our efforts? This convergence problem has two subproblems.

The first subproblem of the convergence problem is related to the spatial and temporal discretization. The finite set of spatial integrals over the field quantities at a certain time do not give us control over all aspects of the field distributions. Moreover, we use finite time integrals, so we do not control all aspects of the time evolution. This inevitably leads to differences between our approximations through sampled functions and the actual solutions of Maxwell's equations. We require from our approximation that, if we refine the spatial/temporal discretization, the field samples correspond to samples of exact solutions of Maxwell's equations and the error diminishes with the mesh size. We can measure the error by substituting the approximation $\{\boldsymbol{E}_h, \boldsymbol{H}_h\}$, corresponding to a mesh size h, in Maxwell's equations

$$\begin{cases} \nabla \times \boldsymbol{H}_h - \varepsilon \frac{\partial \boldsymbol{E}_h}{\partial t} = \boldsymbol{J}_h \\ \nabla \times \boldsymbol{E}_h + \mu \frac{\partial \boldsymbol{H}_h}{\partial t} = \boldsymbol{K}_h. \end{cases} \tag{9.1.26}$$

A consistent discretization would give an estimate such as $\boldsymbol{J} - \boldsymbol{J}_h = o(h^k)$, for some k, and similarly for $\boldsymbol{K} - \boldsymbol{K}_h$, in which case we would speak of a consistent numerical scheme of order

k. The continuity results for solutions of the initial/boundary value problems, $\boldsymbol{J}, \boldsymbol{K} \to \{\boldsymbol{E}, \boldsymbol{H}\}$, then imply that the errors in the fields tend to zero when h diminishes.

The second subproblem of the convergence problem has to do with the construction of the evolution of the solution, which is the problem of stability. When any error (measured in some natural norm) introduced in the solution at a certain time t_q does not grow when we advance the solution in time, the numerical scheme is called stable.

In general, the convergence problem can be analyzed in the following way. One first shows that the original problem is well posed, i.e., for a well-identified class of initial data there exists a unique solution which depends continuously on the initial data in the following sense:

$$\boldsymbol{U}(t) = S(t)\boldsymbol{U}(0), \tag{9.1.27}$$

where $\boldsymbol{U} : t \mapsto \boldsymbol{U}(t) \in L^2(\mathrm{Div} \circ \mathcal{A}, \Omega)$ is the distribution valued function solution of the initial value problem, and $S(t)$ is a bounded linear operator in $L^2(\mathrm{Div} \circ \mathcal{A}, \Omega)$ for any $t \in R_+$, which depends continuously on t. The unique solvability of the initial value problem implies $S(t_1+t_2) = S(t_2) \circ S(t_1)$ and $\lim_{t\to 0} S(t) = S(0) = \mathrm{Id}$; these operators define a semigroup representation of R_+ in $L^2(\mathrm{Div} \circ \mathcal{A}, \Omega)$. Using these relations we can formally obtain

$$\partial_t S(t) = -ES(t),$$

with $E = \lim_{\Delta t\to 0}(1 - S(\Delta t))/\Delta t$ and hence $S(t) = \exp(-tE)$.

Then, one shows consistency and stability of the numerical scheme to arrive at the conclusion that the error as a function of space and time can always be kept arbitrarily small by making h small enough (at the cost, of course, of ever-growing computational effort). Although the pertaining mathematics is beyond the scope of this book, the arguments may be understood intuitively. Indeed, consistency of the approximation guarantees that the local deviations between the exact and approximate solutions can be made arbitrarily small, whereas stability of the numerical scheme implies that errors, once introduced into the approximate solution, do not grow such that we can obtain uniform approximations over any finite time interval.

Now, return to the discussion of discretization because the operator we were looking for is exactly the one that allows us to approximate $S(\Delta t)$. Observe that the matrix operator $(\mathrm{Id} - \Delta t E)$ can be iterated r times on the initial data to find the solution at time t_r:

$$\langle \boldsymbol{U}, \gamma_p^3 \times \{t_r\}\rangle = \sum_k [(1 - \Delta t E)^r]_{pk} \langle \boldsymbol{U}, \gamma_k^3 \times \{t_0\}\rangle. \tag{9.1.28}$$

This means that, when $\Delta t \to 0$, for a fixed t we should have $t = r\Delta t$ and $r \to \infty$:

$$\lim_{r\to\infty} \left(\mathrm{Id} - \frac{t}{r}E\right)^r = \exp(-tE) = S(t). \tag{9.1.29}$$

Therefore, with growing spatial and temporal precision our operator E should approximate the so-called infinitesimal generator of the semigroup $S(t)$, $t \in R_+$. In fact, the operators, $S(t)$, have a kernel distribution which we know explicitly (as Green functions) in many canonical problems such as free-space problems and problems with planar interfaces between media. Actually, the free-space case was encountered in the previous section.

To satisfy the consistency requirement, we have to relate the construction of the time evolution to the spatial, finite-volume integral discretization in a special way such that we obtain an approximation to a Maxwellian field. A natural conclusion, therefore, is to base our numerical scheme on flux estimations using the Green distributions. We now elaborate such estimations in the current simplified case. The more elaborate cases will be handled in subsequent sections with different techniques but which lead to essentially the same type of results.

We consider the field at a point x near the middle of some interface surface element at a time $t = \Delta t/2$, supposing that we have at our disposal the complete distribution, $\boldsymbol{E}_0$, $\boldsymbol{H}_0$, over space at time $t = 0$. We can use the integral representations for the evolution of the electromagnetic field as derived previously. First, consider the electric field

$$\boldsymbol{E}(x,t) = \mu\frac{\partial}{\partial t}\int_{\theta\in S^2}\frac{\boldsymbol{E}_0(x+c_0t\theta)}{4\pi\mu}t\,d\theta - \frac{1}{\varepsilon}\int_0^t\nabla\nabla\cdot\int_{\theta\in S^2}\frac{\boldsymbol{E}_0(x+c_0t\theta)}{4\pi\mu}t\,d\theta$$
$$+\nabla\times\int_{\theta\in S^2}\frac{\boldsymbol{H}_0(x+ct\theta)}{4\pi\varepsilon}t\,d\theta.$$

To simplify the analysis, we assume that $\nabla\cdot\boldsymbol{E}_0 = 0$, i.e., that the computations up to $t = 0$ have produced divergenceless fields such as is the case for the true solutions to the vacuum Maxwell's equations. The actual representations we have to elaborate are

$$\boldsymbol{E}(x,t) = \mu\frac{\partial}{\partial t}\left(\int_{\theta\in S^2}\frac{\boldsymbol{E}_0(x+ct\theta)}{4\pi\mu}t\,d\theta\right) + \nabla\times\int_{\theta\in S^2}\frac{\boldsymbol{H}_0(x+ct\theta)}{4\pi\varepsilon}t\,d\theta. \qquad (9.1.30)$$

We are searching for a relation between cell fields, so we consider separately the contributions from the volume cells on the positive and those on the negative sides of the interface surface. This amounts to computing certain half-spherical means of the field distributions. Let $S^2_{n+} = \{x\in R^3 : \|x\| = 1 \wedge \boldsymbol{x}\cdot\boldsymbol{n} > 0\}$ be the "northern hemisphere" of $S^2 \in R^3$ with axis along $\boldsymbol{a}_\mathrm{n}$ and radius ct; then

$$\frac{1}{4\pi}\frac{\partial}{\partial t}\int_{S^2_{n+}}(\boldsymbol{E}t^{-1}) = \frac{1}{2}\boldsymbol{E}(0) + \frac{1}{2}\frac{\partial\boldsymbol{E}(0)}{\partial n}ct + o(t^2), \qquad (9.1.31)$$

which means that first-order variations count only in the $\boldsymbol{a}_\mathrm{n}$ direction. If we are interested in $\boldsymbol{a}_\mathrm{n}\times\boldsymbol{E}$, the first-order term contributes half the normal derivative of $\boldsymbol{a}_\mathrm{n}\times\boldsymbol{E}_0$.

The lowest order contribution of the curl term is determined by the normal variation of the tangential components of $\boldsymbol{H}_0$ if we neglect the tangential variations of the normal component ($\boldsymbol{a}_\mathrm{n}\times(\nabla\times\boldsymbol{H}) = -\boldsymbol{a}_\mathrm{n}\times(\boldsymbol{a}_\mathrm{n}\times\nabla)(\boldsymbol{a}_\mathrm{n}\cdot\boldsymbol{H}) + (\boldsymbol{a}_\mathrm{n}\cdot\nabla)[\boldsymbol{a}_\mathrm{n}\times(\boldsymbol{a}_\mathrm{n}\times\boldsymbol{H})]$):

$$\boldsymbol{a}_\mathrm{n}\times\left[\nabla\times\int_{\theta\in S^2_{n+}}\frac{\boldsymbol{H}_0(x+ct\theta)}{4\pi\varepsilon}t\,d\theta\right] = \frac{1}{2\varepsilon}\boldsymbol{a}_\mathrm{n}\times\left[\frac{\partial(\boldsymbol{a}_\mathrm{n}\times\boldsymbol{H}_0)}{\partial n}\right]t + o(t^2). \qquad (9.1.32)$$

The same kind of computations can be done with the integral representation for the evolution of the magnetic field. Collecting the results, we obtain for the contributions from the 3-cell on the positive side

$$\begin{cases}\boldsymbol{a}_\mathrm{n}\times\boldsymbol{E}(x,\Delta t/2) = \frac{1}{2}\boldsymbol{a}_\mathrm{n}\times\boldsymbol{E}_0(x) + \frac{1}{2}\frac{\partial}{\partial n}[\boldsymbol{a}_\mathrm{n}\times\boldsymbol{E}_0(x) + Z\boldsymbol{a}_\mathrm{n}\times(\boldsymbol{a}_\mathrm{n}\times\boldsymbol{H}_0(x))]\frac{c\Delta t}{2}\\ \boldsymbol{a}_\mathrm{n}\times\boldsymbol{H}(x,\Delta t/2) = \frac{1}{2}\boldsymbol{a}_\mathrm{n}\times\boldsymbol{H}_0(x) + \frac{1}{2}\frac{\partial}{\partial n}[\boldsymbol{a}_\mathrm{n}\times\boldsymbol{H}_0(x) - Y\boldsymbol{a}_\mathrm{n}\times(\boldsymbol{a}_\mathrm{n}\times\boldsymbol{E}_0(x))]\frac{c\Delta t}{2}.\end{cases} \qquad (9.1.33)$$

For the contributions from the 3-cell on the negative side, we obtain similar results relative to $-\boldsymbol{a}_\mathrm{n}$. In terms of the fluxes, $\boldsymbol{a}_\mathrm{n}\cdot\mathcal{A}\boldsymbol{U}$, the complete expression can be written as

$$\boldsymbol{a}_\mathrm{n}\cdot\mathcal{A}\boldsymbol{U}\left(x,\frac{\Delta t}{2}\right) = \boldsymbol{a}_\mathrm{n}\cdot\mathcal{A}\boldsymbol{U}_0(x) + \pi^+\frac{\partial}{\partial(-n)}(\boldsymbol{a}_\mathrm{n}\cdot\mathcal{A}\boldsymbol{U}_0)(x)c\frac{\Delta t}{2} + \pi^-\frac{\partial}{\partial n}(\boldsymbol{a}_\mathrm{n}\cdot\mathcal{A}\boldsymbol{U}_0)(x)c\frac{\Delta t}{2},$$

where we have written $\partial/\partial(-n)$ to emphasize that the normal derivative with respect to $-n$ on the negative side of the surface has to be taken. The $\pi^{\pm}$ in this expression are projection operators on current pairs corresponding to a plane wave propagating in the $\pm\boldsymbol{a}_{\rm n}$ direction. These projection operators are defined as

$$\pi^{\pm} : R^3 \oplus R^3 \to R^3 \oplus R^3$$

by

$$\pi^{\pm} : (j, k) \mapsto \frac{1}{2}(j \mp Yn \times k, k \pm Zn \times j),$$

or, in matrix notation,

$$[\pi^{\pm}] = \frac{1}{2}\begin{bmatrix} \mathrm{Id} & \mp Ys(n) \\ \pm Zs(n) & \mathrm{Id} \end{bmatrix}.$$

One can verify, without difficulty,

$$\begin{aligned} &\text{(i)}\ \pi^{\pm} \circ \pi^{\pm} = \pi^{\pm} \\ &\text{(ii)}\ \pi^{+} \circ \pi^{-} = 0 = \pi^{-} \circ \pi^{+} \\ &\text{(iii)}\ \pi^{-} + \pi^{+} = \mathrm{Id} \end{aligned}$$

Using (iii) we can write the flux as

$$\begin{aligned} \boldsymbol{a}_{\rm n} \cdot \mathcal{A}\boldsymbol{U}\left(x, \frac{\Delta t}{2}\right) = \pi^{+}\left(\boldsymbol{a}_{\rm n} \cdot \mathcal{A}\boldsymbol{U}_0(x) + \frac{\partial}{\partial(-n)}(\boldsymbol{a}_{\rm n} \cdot \mathcal{A}\boldsymbol{U}_0(x))c\frac{\Delta t}{2}\right) \\ + \pi^{-}\left(\boldsymbol{a}_{\rm n} \cdot \mathcal{A}\boldsymbol{U}_0(x) + \frac{\partial}{\partial n}(\boldsymbol{a}_{\rm n} \cdot \mathcal{A}\boldsymbol{U}_0(x))c\frac{\Delta t}{2}\right), \end{aligned}$$

or, in a first-order Taylor expansion,

$$\boldsymbol{a}_{\rm n} \cdot \mathcal{A}\boldsymbol{U}\left(x, \frac{\Delta t}{2}\right) = \pi^{+}\left(\boldsymbol{a}_{\rm n} \cdot \mathcal{A}\boldsymbol{U}_0\left(x - nc\frac{\Delta t}{2}\right)\right) + \pi^{-}\left(\boldsymbol{a}_{\rm n} \cdot \mathcal{A}\boldsymbol{U}_0\left(x + nc\frac{\Delta t}{2}\right)\right).$$

Using (ii), we obtain two equations from the last expression:

$$\pi^{+}\left(\boldsymbol{a}_{\rm n} \cdot \mathcal{A}\boldsymbol{U}\left(x, \frac{\Delta t}{2}\right) - \boldsymbol{a}_{\rm n} \cdot \mathcal{A}\boldsymbol{U}_0\left(x - nc\frac{\Delta t}{2}\right)\right) = 0$$

and

$$\pi^{-}\left(\boldsymbol{a}_{\rm n} \cdot \mathcal{A}\boldsymbol{U}\left(x, \frac{\Delta t}{2}\right) - \boldsymbol{a}_{\rm n} \cdot \mathcal{A}\boldsymbol{U}_0\left(x + nc\frac{\Delta t}{2}\right)\right) = 0,$$

which gives a system of two relations between the required quantity, $\boldsymbol{a}_{\rm n} \cdot \mathcal{A}\boldsymbol{U}(x, \Delta t/2)$, and the given field distributions at $t = 0$ in the adjacent 3-cells. These two equations state that the two constituent plane wave fluxes in the unique decomposition of the flux on an interface cell are in fact plane wave components in the initial data on the 3-cells at the two sides of the interface and, indeed, those which correspond to plane waves propagating toward the interface from a distance $(c\Delta t/2)$ away.

This concludes our discussion of an elementary finite-volume method for Maxwell's equations. We have discussed the basic ingredients of the construction of the solution of Maxwell's equations in the time domain through a simple explicit numerical scheme. In the following sections, more general constructions will be presented. In those sections, one will encounter the basic ingredients again, but in slightly different form and in the context of more elaborate numerical procedures.

9.2 FINITE-VOLUME DISCRETIZATION OF MAXWELL'S EQUATIONS

In this section, we present the numerical implementation of the finite-volume method for the conservative form of Maxwell's equations. In contrast to the previous discussion, the discussion in this section relies only on the analysis of the partial differential equations. This more conventional approach can be seen to lead to the same formulas as the one of the preceding section, in which use has been made of the Green distributions.

We start with the Maxwell's equations in conservative form, such as they were presented in the previous section. In contrast to the complete discretization of the previous section, we now first semidiscretize the equations by integrating them over only a set of elementary spatial cells—the finite-volume integrals. Relations will be derived for the field values in the cell centers and the limiting values on the cell boundary. By using the appropriate continuity conditions at each interface, the fluxes can be computed explicitly from the field values on the adjacent finite volumes. In this way, an ordinary differential equation is obtained for the time evolution of the vector of finite-volume integrals. The truncation of the possibly unbounded domain of the fields to a limited computational domain is discussed. Then, the discretization with respect to the time dependence is discussed. The section concludes with a detailed study of the consistency and stability of the method.

9.2.1 Spatial Discretizations

In this section, we present the spatial, finite-volume discretization of Maxwell's equations. We study the Maxwell's equations, given by

$$\begin{aligned} \varepsilon\frac{\partial \boldsymbol{E}}{\partial t} - \nabla\times\boldsymbol{H} &= 0 \\ \mu\frac{\partial \boldsymbol{H}}{\partial t} + \nabla\times\boldsymbol{E} &= 0, \end{aligned} \tag{9.2.1}$$

where ε and μ are the electric permittivity and magnetic permeability of the medium, $\boldsymbol{E} = (E_x, E_y, E_z)^t$ (the electric field vector), and $\boldsymbol{H} = (H_x, H_y, H_z)^t$ (the magnetic field vector).

As discussed in the previous section, these equations can be written in conservative form as follows:

$$\alpha\frac{\partial \boldsymbol{U}}{\partial t} + \mathrm{Div}\,\mathcal{A}\boldsymbol{U} = 0, \tag{9.2.2}$$

with $U = (E_x, E_y, E_z, H_x, H_y, H_z)^t$ and $\mathcal{A}$ as defined in the previous section and we recall the definition:

$$\alpha = \begin{bmatrix} \varepsilon\mathrm{Id} & 0 \\ 0 & \mu\mathrm{Id} \end{bmatrix}.$$

We now introduce a slightly different notation, which is the usual one in the theory of (nonlinear) conservative systems. We write

$$\mathcal{A}U = F(U) = (F_1(U); F_2(U); F_3(U)),$$

and $F_p(U)$, where $p = 1, \cdots, 3$, are the columns of $\mathcal{A}U$:

$$F_1(U) = \begin{bmatrix} 0 \\ H_z \\ -H_y \\ 0 \\ -E_z \\ E_y \end{bmatrix} \quad F_2(U) = \begin{bmatrix} -H_z \\ 0 \\ H_x \\ E_z \\ 0 \\ -E_x \end{bmatrix} \quad F_3(U) = \begin{bmatrix} H_y \\ -H_x \\ 0 \\ -E_y \\ E_x \\ 0 \end{bmatrix}.$$

Note that, in the case of the Maxwell's equations, F is defined by a set of three linear functions on R^6. We consider $F_p(U)$, where $p = 1, \cdots, 3$, as an R^3 vector function on space–time, with its three components taking values in R^6 (we use Cartesian coordinates only). The partial differential equations are then written as

$$\alpha \frac{\partial U}{\partial t} + \frac{\partial F_1(U)}{\partial x} + \frac{\partial F_2(U)}{\partial y} + \frac{\partial F_3(U)}{\partial z} = 0 \qquad (9.2.3)$$

or, in condensed form, as

$$\alpha \frac{\partial U}{\partial t} + \operatorname{div} F(U) = 0. \qquad (9.2.4)$$

To obtain a spatial discretization of this equation, we must integrate it over a set of subdomains $\{V_i\}$, the "finite volumes." There exist essentially two techniques to construct such subdomains for a given triangulation. These techniques are named after the location of the degrees of freedom of the unknown field components: They are called the cell vertex and the cell-centered formulations [14]. In the first case, the degrees of freedom are associated to the nodes of the cells in the triangulation and in the second the degrees of freedom are associated to the barycenters of those cells. With the cell-centered formulation, a finite volume coincides with a cell of the triangulation itself. Figure 9.1 shows an example of the finite volumes in a two-dimensional mesh.

With the cell vertex formulation, the finite volumes are defined by particular subdomains surrounding the nodes of the triangulation. This particular subdomain does not coincide with cells of the triangulation, it is defined by connecting the barycenters of the cells to which the node belongs. In two dimensions, the finite volumes are polygons centered around the nodes as shown in Fig. 9.2.

These two different approaches do not present the same advantages or inconveniences. The extension to higher spatial orders is easier for the cell vertex formulation; on the other hand, the boundary conditions are taken into account more naturally using a cell-centered formulation. For example, in a cell vertex formulation, specific constructions of truncated finite volumes around a face S on the boundary of the domain are needed (Fig. 9.3).

In the cell-centered formulation, this face S corresponds already to a boundary of a finite volume (Fig. 9.4), and consequently boundary conditions can be taken into account without any difficulty.

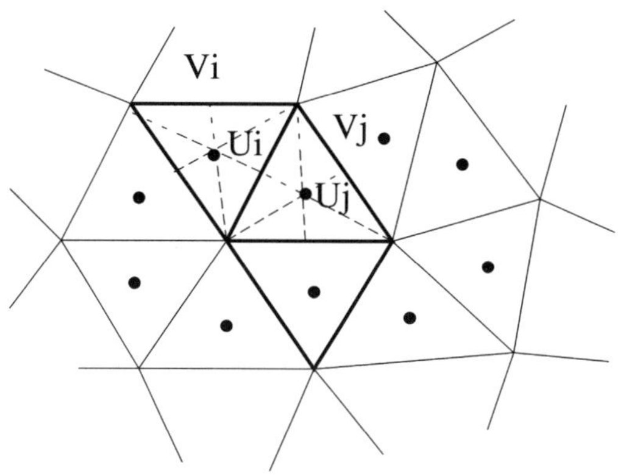

FIGURE 9.1
Definition of finite volumes for the cell-centered formulation.

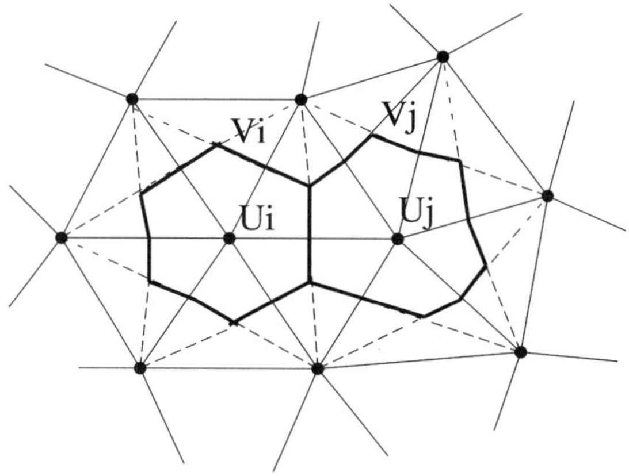

FIGURE 9.2
Definition of finite volumes for the cell vertex formulation.

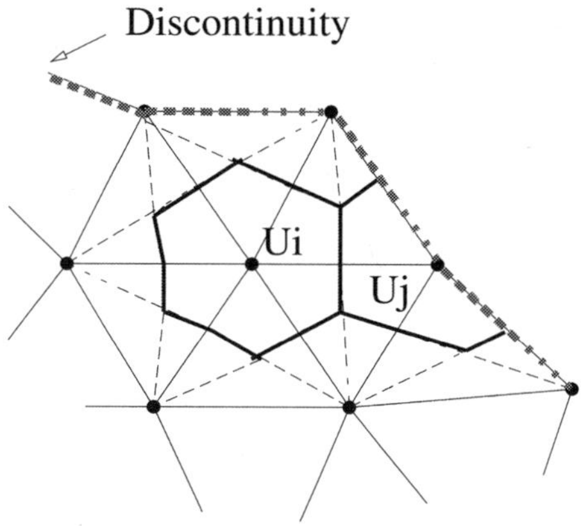

FIGURE 9.3
Definition of finite volumes for the cell vertex formulation near a boundary.

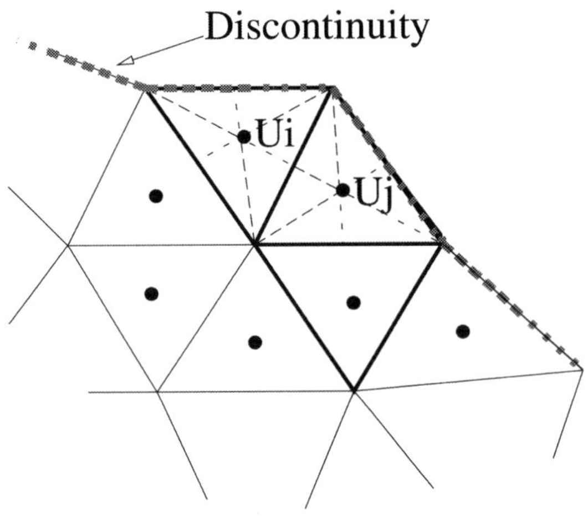

FIGURE 9.4
Definition of finite volumes for a cell-centered formulation near a boundary.

Regarding the spatial discretization, the only distinction between the two methods resides in the construction and the general shape of the finite volumes. The essential point for our purposes is that we have a computational domain, $\Omega = \cup_i V_i$, subdivided into finite volumes V_i, each of which has a boundary surface, S, which consists of a number, m_i, of planar faces $\partial V_i = S = \cup_{k=1}^{m_i} S_k$.

Now, let Eq. (9.2.4) be integrated over a volume V, with boundary surface S:

$$\alpha \int_V \frac{\partial \boldsymbol{U}}{\partial t} dV + \int_V \operatorname{div} F(\boldsymbol{U})\, dV = 0. \tag{9.2.5}$$

Using the divergence theorem, we obtain

$$\alpha \frac{\partial}{\partial t} \int_V \boldsymbol{U} dV + \int_S F(\boldsymbol{U}^*) \cdot \boldsymbol{a}_{\mathrm{n}}\, dS = 0, \tag{9.2.6}$$

where $\boldsymbol{U}^* = (\boldsymbol{E}^*, \boldsymbol{H}^*)$, $\boldsymbol{E}^*$ and $\boldsymbol{H}^*$ denote the boundary limits of $\boldsymbol{E}$ and $\boldsymbol{H}$, respectively, on the surface S, and $\boldsymbol{a}_{\mathrm{n}}$ denotes the unit normal vector to the surface S pointing outward from volume V. The quantity $\boldsymbol{a}_{\mathrm{n}} \cdot F(\boldsymbol{U}^*)$ is called the flux.

For an elementary volume, V_i, in a given triangulation, the volume integral in Eq. (9.2.6) will be approximated by the value of $\boldsymbol{U}$ at the center multiplied by the volume of the cell, $|V_i|$. The surface integral in Eq. (9.2.6) will be approximated by the sum of the fluxes on the centers of the faces, S_k, multiplied by their area, $|S_k|$. For the cell i (volume V_i), we obtain

$$\frac{\partial \boldsymbol{U}_i}{\partial t} = -\frac{1}{|V_i|} \sum_{k=1}^{m_i} |S_k|\, \alpha^{-1} F(\boldsymbol{U}_k^*) \cdot \boldsymbol{a}_{\mathrm{n}k}, \tag{9.2.7}$$

where $\boldsymbol{U}_i$ denotes the value of $\boldsymbol{U}$ in the center of volume V_i, $F(\boldsymbol{U}_k^*) \cdot \boldsymbol{a}_{\mathrm{n}k} = \left[\begin{smallmatrix} -\boldsymbol{a}_{\mathrm{n}k} \times \boldsymbol{H}_k^* \\ \boldsymbol{a}_{\mathrm{n}k} \times \boldsymbol{E}_k^* \end{smallmatrix}\right]$ is the flux in the center of S_k, m_i is the number of faces S_k in the boundary of volume V_i, $\boldsymbol{a}_{\mathrm{n}k}$ the unit normal vector to the surface S_k pointing outward from V_i. The major difficulty now is the evaluation of the numerical fluxes. On each face of the triangulation, the flux has to be related to the values of $\boldsymbol{U}$ on the two volumes on both sides.

Consider a cell i bounded by m_i surfaces S_k, with $\boldsymbol{a}_{\mathrm{n}k}$, $k = 1, \cdots, m_i$, the m_i unit normal vectors to these surfaces. For a nondegenerate cell, this set of unit normal vectors can serve to parameterize R^3, i.e., any point $x \in R^3$ can be expressed as a linear combination of the $\boldsymbol{a}_{\mathrm{n}k}$ (though not in a unique way). This means that we parameterize R^3 by coordinates in R^{m_i}:

$$R^{m_i} \ni (\xi_1, \cdots, \xi_{m_i}) \mapsto x = \sum_{k=1}^{m_i} \xi_k \boldsymbol{a}_{\mathrm{n}k} \in R^3.$$

The symbol ξ_k will also be used for those (nonunique) linear functions on R^3 which satisfy $x = \sum_{k=1}^{m_i} \xi_k(x)\boldsymbol{a}_{\mathrm{n}k}$. Then, we can write Eq. (9.2.3) on the cell i as follows:

$$\frac{\partial \boldsymbol{U}}{\partial t} = -\alpha^{-1} \sum_{k=1}^{m_i} \left(\frac{\partial F_1}{\partial \boldsymbol{U}} \frac{\partial \xi_k}{\partial x} + \frac{\partial F_2}{\partial \boldsymbol{U}} \frac{\partial \xi_k}{\partial y} + \frac{\partial F_3}{\partial \boldsymbol{U}} \frac{\partial \xi_k}{\partial z} \right) \frac{\partial \boldsymbol{U}}{\partial \xi_k}. \tag{9.2.8}$$

This equation can also be written as

$$\frac{\partial \boldsymbol{U}}{\partial t} = -\alpha^{-1} \sum_{k=1}^{m_i} A(\boldsymbol{a}_{\mathrm{n}k}) \frac{\partial \boldsymbol{U}}{\partial n_k}, \tag{9.2.9}$$

where, $\partial \boldsymbol{U}/\partial \boldsymbol{a}_{\mathrm{n}k} \equiv \partial \boldsymbol{U}/\partial \xi_k$ and the 6×6 matrix $A(\boldsymbol{a}_{\mathrm{n}k})$ is defined by

$$A(\boldsymbol{a}_{\mathrm{n}k}) = \frac{\partial F_1}{\partial \boldsymbol{U}} \frac{\partial \xi_k}{\partial x} + \frac{\partial F_2}{\partial \boldsymbol{U}} \frac{\partial \xi_k}{\partial y} + \frac{\partial F_3}{\partial \boldsymbol{U}} \frac{\partial \xi_k}{\partial z}.$$

The matrix $A(\boldsymbol{a}_{\mathrm{n}})$ is given explicitly by

$$A(\boldsymbol{a}_{\mathrm{n}}) = \begin{bmatrix} 0 & 0 & 0 & 0 & n_z & -n_y \\ 0 & 0 & 0 & -n_z & 0 & n_x \\ 0 & 0 & 0 & n_y & -n_x & 0 \\ 0 & -n_z & n_y & 0 & 0 & 0 \\ n_z & 0 & -n_x & 0 & 0 & 0 \\ -n_y & n_x & 0 & 0 & 0 & 0 \end{bmatrix},$$

and the product $A(\boldsymbol{a}_{\mathrm{n}})\boldsymbol{U}^*$ can be seen to define the flux $\left[\begin{smallmatrix} -\boldsymbol{a}_{\mathrm{n}} \times \boldsymbol{H}^* \\ \boldsymbol{a}_{\mathrm{n}} \times \boldsymbol{E}^* \end{smallmatrix}\right]$ on the face S associated with the unit normal vector $\boldsymbol{a}_{\mathrm{n}} = (n_x, n_y, n_z)$.

We introduce the shorthand notation

$$\tilde{A}(\boldsymbol{a}_{\mathrm{n}}) = \alpha^{-1} A(\boldsymbol{a}_{\mathrm{n}}).$$

Each matrix $\tilde{A}(\boldsymbol{a}_{\mathrm{n}k})$ has six real eigenvalues λ_j, given by

$$\left(0, 0, \frac{1}{\sqrt{\mu\varepsilon}}, \frac{1}{\sqrt{\mu\varepsilon}}, -\frac{1}{\sqrt{\mu\varepsilon}}, -\frac{1}{\sqrt{\mu\varepsilon}}\right),$$

and six independent eigenvectors. Thus, Eq. (9.2.9) defines a hyperbolic system [15].

We can write the matrix $\tilde{A}(\boldsymbol{a}_{\mathrm{n}})$ in the form $\tilde{A} = P \, \Lambda \, P^{-1}$, where $\Lambda = (\mathrm{diag}(\lambda_j))_{j=1,\cdots,6}$ is the diagonal matrix of eigenvalues, and P is given by

$$P = \begin{bmatrix} n_x & 0 & \frac{n_x n_z}{v\varepsilon} & -\frac{n_x n_y}{v\varepsilon} & -\frac{n_x n_z}{v\varepsilon} & \frac{n_x n_y}{v\varepsilon} \\ n_y & 0 & \frac{n_y n_z}{v\varepsilon} & \frac{n_x^2+n_z^2}{v\varepsilon} & -\frac{n_y n_z}{v\varepsilon} & -\frac{n_x^2+n_z^2}{v\varepsilon} \\ n_z & 0 & -\frac{n_x^2+n_y^2}{v\varepsilon} & -\frac{n_y n_z}{v\varepsilon} & \frac{n_x^2+n_y^2}{v\varepsilon} & \frac{n_y n_z}{v\varepsilon} \\ 0 & n_x & -n_y & -n_z & -n_y & -n_z \\ 0 & n_y & n_x & 0 & n_x & 0 \\ 0 & n_z & 0 & n_x & 0 & n_x \end{bmatrix},$$

where $v = 1/\sqrt{\varepsilon\mu}$ is the speed of light in the medium.

If we consider the variation of $\boldsymbol{U}$ to be only with respect to the $\boldsymbol{a}_{\mathrm{n}}$ direction, Eq. (9.2.9) gives

$$\frac{\partial \boldsymbol{U}}{\partial t} = -\tilde{A}(\boldsymbol{a}_{\mathrm{n}}) \frac{\partial \boldsymbol{U}}{\partial n}. \tag{9.2.10}$$

Using the previous decomposition of $\tilde{A}(\boldsymbol{a}_{\mathrm{n}})$ and introducing the vector $\boldsymbol{V} = P^{-1}\boldsymbol{U}$, Eq. (9.2.10) can be written as a system of six independent scalar equations:

$$\frac{\partial V_j}{\partial t} + \lambda_j \frac{\partial V_j}{\partial n} = 0 \quad j = 1, \cdots, 6,$$

with solutions $V_j(\xi, t) = f(\lambda_j t - \xi n)$, where f is any differentiable function on R and the λ_j can be seen to play the role of propagation velocities. For each of these equations, the solution is constant on the space–time curves $t = \xi n/\lambda_j +$ const., the characteristics of the equation. In particular, this means that the value of V_j at the cell's boundary face, at $\xi = \xi^*$, corresponding to the chosen direction $\boldsymbol{a}_{\mathrm{n}}$ at time t is equal to the value of V_j at the center of the cell, $\xi = \xi_0$, at time $t - d/\lambda_j$, where $d = |\xi^* - \xi_0|$ is the distance between the boundary face and the center:

$$V_j\left(\xi_0, t - \frac{d}{\lambda_j}\right) = V_j(\xi^*, t). \tag{9.2.11}$$

We now introduce an additional approximation by neglecting the propagation time between the center of the cell and the boundary cells. This is justified when the cell size is very small compared to the shortest relevant wavelength in the excitation. Therefore, we shall work with the following relation:

$$V_j(t) = V_j^*(t), \tag{9.2.12}$$

where V_j is again the value at the center of the volume cell and V_j^* the boundary limit on the face corresponding to the normal $\boldsymbol{a}_{\mathrm{n}}$. For preserving the stability of the numerical schemes, which will be made clear in Section 9.2.3, we consider only waves propagating in the positive $\boldsymbol{a}_{\mathrm{n}}$ direction, i.e., from the center toward the boundary faces. Because we consider normals in multiple directions, this does not imply a lack of generality in principle. The point is that certain variations of the fields in a finite volume are used to derive relations between fields at the cell center and only certain boundary faces. Hence, for each of the normals, we consider only the positive eigenvalues.

The matrix $\tilde{A}(n)$ can, in fact, be split into a positive part and a negative part in the following way:

$$\tilde{A}(\boldsymbol{a}_{\mathrm{n}}) = P\Lambda P^{-1} = P(\Lambda^{+} + \Lambda^{-})P^{-1} = P\Lambda^{+}P^{-1} + P\Lambda^{-}P^{-1} = \tilde{A}(\boldsymbol{a}_{\mathrm{n}})^{+} + \tilde{A}(\boldsymbol{a}_{\mathrm{n}})^{-},$$

where Λ^{+} is the diagonal matrix of the positive eigenvalues of $\tilde{A}(\boldsymbol{a}_{\mathrm{n}})$ and Λ^{-} the diagonal matrix of the negative eigenvalues.

The explicit matrix representation of $\tilde{A}(\boldsymbol{a}_{\mathrm{n}})^{+}$ is given by

$$\tilde{A}(\boldsymbol{a}_{\mathrm{n}})^{+} = \frac{1}{2}\begin{bmatrix} (n_z^2+n_y^2)\,v & -n_x n_y v & -n_x n_z v & 0 & n_z\frac{1}{\varepsilon} & -n_y\frac{1}{\varepsilon} \\ -n_x n_y v & (n_x^2+n_z^2)\,v & -n_y n_z v & -n_z\frac{1}{\varepsilon} & 0 & n_x\frac{1}{\varepsilon} \\ -n_x n_z v & -n_y n_z v & (n_x^2+n_y^2)\,v & n_y\frac{1}{\varepsilon} & -n_x\frac{1}{\varepsilon} & 0 \\ 0 & -n_z\frac{1}{\mu} & n_y\frac{1}{\mu} & (n_z^2+n_y^2)\,v & -n_x n_y v & -n_x n_z v \\ n_z\frac{1}{\mu} & 0 & -n_x\frac{1}{\mu} & -n_x n_y v & (n_x^2+n_z^2)\,v & -n_y n_z v \\ -n_y\frac{1}{\mu} & n_x\frac{1}{\mu} & 0 & -n_x n_z v & -n_y n_z v & (n_x^2+n_y^2)\,v \end{bmatrix},$$

where $v = 1/\sqrt{\varepsilon\mu}$ is the speed of light in the medium. From this formula, the expression for $\tilde{A}(\boldsymbol{a}_{\mathrm{n}})^{-}$ can be derived by replacing v by $-v$. Observe that we have the following relation between the two parts:

$$\tilde{A}(\boldsymbol{a}_{\mathrm{n}})^{-} = -\tilde{A}(-\boldsymbol{a}_{\mathrm{n}})^{+}. \tag{9.2.13}$$

Now, we obtain the desired relation between $\boldsymbol{U}$ at the cell center and $\boldsymbol{U}^{*}$ on the boundary face corresponding to the normal $\boldsymbol{a}_{\mathrm{n}}$ by multiplying Eq. (9.2.12) by $P\Lambda^{+}$ on both sides:

$$\tilde{A}(\boldsymbol{a}_{\mathrm{n}})^{+}\boldsymbol{U} = \tilde{A}(\boldsymbol{a}_{\mathrm{n}})^{+}\boldsymbol{U}^{*}$$

or, in terms of the $\boldsymbol{E}$ and $\boldsymbol{H}$ fields,

$$Y\boldsymbol{a}_{\mathrm{n}} \times \boldsymbol{E} - \boldsymbol{a}_{\mathrm{n}} \times (\boldsymbol{a}_{\mathrm{n}} \times \boldsymbol{H}) = Y\boldsymbol{a}_{\mathrm{n}} \times \boldsymbol{E}^{*} - \boldsymbol{a}_{\mathrm{n}} \times (\boldsymbol{a}_{\mathrm{n}} \times \boldsymbol{H}^{*}), \tag{9.2.14}$$

with $Y = \sqrt{\frac{\varepsilon}{\mu}}$.

A similar relation can be found for the volume cell on the other side of S. Therefore, for each interface we have the following set of equations ($\boldsymbol{a}_{\mathrm{n}}$ pointing from left to right):

$$\begin{cases} Y^{\mathrm{L}}\boldsymbol{a}_{\mathrm{n}} \times \boldsymbol{E}^{*} - \boldsymbol{a}_{\mathrm{n}} \times (\boldsymbol{a}_{\mathrm{n}} \times \boldsymbol{H}^{*}) = Y^{\mathrm{L}}\boldsymbol{a}_{\mathrm{n}} \times \boldsymbol{E}^{\mathrm{L}} - \boldsymbol{a}_{\mathrm{n}} \times (\boldsymbol{a}_{\mathrm{n}} \times \boldsymbol{H}^{\mathrm{L}}) \\ Y^{\mathrm{R}}\boldsymbol{a}_{\mathrm{n}} \times \boldsymbol{E}^{**} + \boldsymbol{a}_{\mathrm{n}} \times (\boldsymbol{a}_{\mathrm{n}} \times \boldsymbol{H}^{**}) = Y^{\mathrm{R}}\boldsymbol{a}_{\mathrm{n}} \times \boldsymbol{E}^{\mathrm{R}} + \boldsymbol{a}_{\mathrm{n}} \times (\boldsymbol{a}_{\mathrm{n}} \times \boldsymbol{H}^{\mathrm{R}}), \end{cases} \tag{9.2.15}$$

where Y^{L} and Y^{R} are the characteristic admittances on the left- and right-hand sides of the interface and $(\boldsymbol{E}^{\mathrm{L}}, \boldsymbol{H}^{\mathrm{L}})$, $(\boldsymbol{E}^{\mathrm{R}}, \boldsymbol{H}^{\mathrm{R}})$, $(\boldsymbol{E}^{*}, \boldsymbol{H}^{*})$, and $(\boldsymbol{E}^{**}, \boldsymbol{H}^{**})$ are as depicted in Fig. 9.5.

The Eqs. (9.2.15) give, for each interface S_k, a relation between the $\boldsymbol{U}$, taken at the centers of the two adjacent volume cells, and the two boundary limits, $\boldsymbol{U}^{*}$ and $\boldsymbol{U}^{**}$, taken from either side of the surface. By supplementing these equations with the appropriate continuity conditions for the interface S_k, we can compute the fluxes. We shall elaborate the case of dielectric contrast and a perfect conductor in detail in the following section. In both cases, we obtain a system of ordinary differential equations for the vector of field samples at the centers of the finite volumes.

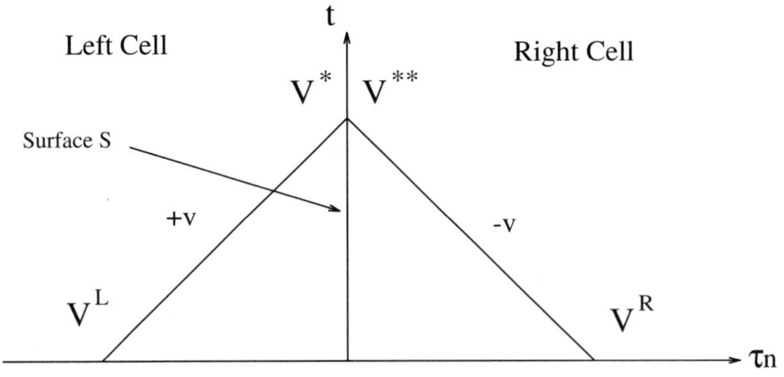

FIGURE 9.5
Characteristic curves defined by $v^2 = 1/\mu\varepsilon$ and by $-v$.

Dielectric Contrast When the interface coincides with a surface of discontinuity in the constitutive parameters, we impose the continuity of the tangential electric and magnetic field vectors:

$$\begin{cases} \boldsymbol{a}_{\mathrm{n}} \times \boldsymbol{E}^{*} = \boldsymbol{a}_{\mathrm{n}} \times \boldsymbol{E}^{**} \\ \boldsymbol{a}_{\mathrm{n}} \times \boldsymbol{H}^{*} = \boldsymbol{a}_{\mathrm{n}} \times \boldsymbol{H}^{**}. \end{cases}$$

The solution of Eq. (9.2.15) together with these interface conditions gives

$$\boldsymbol{a}_{\mathrm{n}} \times \boldsymbol{E}^{*} = \boldsymbol{a}_{\mathrm{n}} \times \boldsymbol{E}^{**} = \frac{1}{Y^{\mathrm{L}} + Y^{\mathrm{R}}}(\boldsymbol{a}_{\mathrm{n}} \times (Y^{\mathrm{L}} E^{\mathrm{L}} + Y^{\mathrm{R}} \boldsymbol{E}^{\mathrm{R}}) - \boldsymbol{a}_{\mathrm{n}} \times (\boldsymbol{a}_{\mathrm{n}} \times (\boldsymbol{H}^{\mathrm{L}} - \boldsymbol{H}^{\mathrm{R}}))) \quad (9.2.16)$$

$$\boldsymbol{a}_{\mathrm{n}} \times \boldsymbol{H}^{*} = \boldsymbol{a}_{\mathrm{n}} \times \boldsymbol{H}^{**} = \frac{1}{Z^{\mathrm{L}} + Z^{\mathrm{R}}}(\boldsymbol{a}_{\mathrm{n}} \times (Z^{\mathrm{R}} \boldsymbol{H}^{\mathrm{R}} + Z^{\mathrm{L}} \boldsymbol{H}^{\mathrm{L}}) - \boldsymbol{a}_{\mathrm{n}} \times (\boldsymbol{a}_{\mathrm{n}} \times (\boldsymbol{E}^{\mathrm{R}} - \boldsymbol{E}^{\mathrm{L}}))), \quad (9.2.17)$$

where $Y^{L,R}$ are the characteristic admittances of the media in the volumes on the left- and right-hand sides, relative to the interface, and $Z^{L,R} = (Y^{L,R})^{-1}$ are the corresponding characteristic impedances.

We rewrite these expressions in a more formal way:

$$F(\boldsymbol{U}^{*}) \cdot \boldsymbol{a}_{\mathrm{n}} = F(\boldsymbol{U}^{**}) \cdot \boldsymbol{a}_{\mathrm{n}} = \alpha^{\mathrm{L}} T^{\mathrm{L}} \tilde{A}^{\mathrm{L}}(\boldsymbol{a}_{\mathrm{n}})^{+} U^{\mathrm{L}} + \alpha^{\mathrm{R}} T^{\mathrm{R}} \tilde{A}^{\mathrm{R}}(\boldsymbol{a}_{\mathrm{n}})^{-} U^{\mathrm{R}},$$

where the superscripts L and R on α and $\tilde{A}(\boldsymbol{a}_{\mathrm{n}})$ indicate that the ε and μ of the media on the left- and right-hand side of S are to be substituted and where we introduced the transmission matrices

$$T^{\mathrm{L,R}} = \begin{bmatrix} \frac{2Z^{\mathrm{L,R}}}{Z^{\mathrm{L}}+Z^{\mathrm{R}}} & 0 \\ 0 & \frac{2Y^{L,R}}{Y^{\mathrm{L}}+Y^{\mathrm{R}}} \end{bmatrix}.$$

Now that the fluxes appearing in Eq. (9.2.7) have been related to the fields in the finite volumes, we have obtained a system of ordinary differential equations, with respect to time, for the vector of field samples at the centers of the volume cells

$$\alpha \frac{\partial U_i}{\partial t} = \frac{-1}{|V_i|} \sum_{k=1}^{m_i} |S_k| [\alpha^{\mathrm{L}} T^{\mathrm{L}} \tilde{A}^{\mathrm{L}} (\boldsymbol{a}_{\mathrm{n}k})^+ U_{k^-} + \alpha^{\mathrm{R}} T^{\mathrm{R}} \tilde{A}^{\mathrm{R}} (\boldsymbol{a}_{\mathrm{n}k})^- U_{k^+}],$$

where $k^{\pm}$ is the index of the volume cell on the $\pm$ side of the surface S_k.

Observe that in the special case in which the media on both sides are identical, $T^{\mathrm{L,R}} = \mathrm{Id}$ and $\alpha = \alpha^{\mathrm{L}} = \alpha^{\mathrm{R}}$, and hence the differential equation simplifies to

$$\frac{\partial U_i}{\partial t} = \frac{-1}{|V_i|} \sum_{k=1}^{m_i} |S_k| [\tilde{A}(\boldsymbol{a}_{\mathrm{n}k})^+ U_{k^-} + \tilde{A}(\boldsymbol{a}_{\mathrm{n}k})^- U_{k^+}]. \tag{9.2.18}$$

Perfect Conductors In the case in which the interface coincides with a perfectly conducting surface, the fluxes are evaluated by supplementing Eq. (9.2.14) with the condition

$$\boldsymbol{a}_{\mathrm{n}} \times \boldsymbol{E}^* = 0$$

in the total field formulation. This gives

$$\boldsymbol{a}_{\mathrm{n}} \times \boldsymbol{H}^* = Y \boldsymbol{a}_{\mathrm{n}} \times (\boldsymbol{a}_{\mathrm{n}} \times \boldsymbol{E}) + \boldsymbol{a}_{\mathrm{n}} \times \boldsymbol{H}.$$

This leads to the following expression for the flux:

$$F(\boldsymbol{U}^*) \cdot \boldsymbol{a}_{\mathrm{n}} = \alpha^{\mathrm{L}} T_{pc}^{\mathrm{L}} \tilde{A}^{\mathrm{L}} (\boldsymbol{a}_{\mathrm{n}})^+ \boldsymbol{U}^{\mathrm{L}},$$

where

$$T_{pc}^{\mathrm{L}} = \lim_{Z^{\mathrm{R}} \to 0} T^{\mathrm{L}}$$

or, in matrix form,

$$[T_{pc}^{\mathrm{L}}] = \begin{bmatrix} 2\mathrm{Id} & 0 \\ 0 & 0 \end{bmatrix}.$$

In the scattered-field formulation, we have

$$\boldsymbol{a}_{\mathrm{n}} \times \boldsymbol{E}^{\mathrm{s}*} = -\boldsymbol{a}_{\mathrm{n}} \times \boldsymbol{E}^{\mathrm{i}},$$

which leads to the flux

$$F(\boldsymbol{U}^*) \cdot \boldsymbol{a}_{\mathrm{n}} = \alpha^{\mathrm{L}} T_{pc}^{\mathrm{L}} \tilde{A}^{\mathrm{L}} (\boldsymbol{a}_{\mathrm{n}})^+ \boldsymbol{U}^{\mathrm{L}} + 2\tilde{A}^{\mathrm{L}} (\boldsymbol{a}_{\mathrm{n}})^+ G^{\mathrm{L}},$$

with the source term

$$\boldsymbol{G}^{\mathrm{L}} = \begin{bmatrix} \boldsymbol{E}^{\mathrm{L;i}} \\ 0 \end{bmatrix}$$

being the value of the incident electric field on the negative side of the interface surface.

Of course, the case of an infinitely thin perfectly conducting screen can be handled by using the previous relations on both sides.

Radiation Boundary Conditions Since computer storage capacity is limited, the computational domain of the unknown fields has to be limited as well. If the electromagnetic fields of a radiating antenna, for example, are to be analyzed, they must in principle be calculated in an unbounded open space and the support of the radiated fields, although finite for all finite times, increases rapidly with time. Therefore, a certain type of boundary condition needs to be applied on the outer boundaries of the computational domain in order to simulate the unbounded physical space. The difficulty of this special boundary condition is that quantities beyond the outer boundary of the computational domain do not appear in the numerical scheme. Therefore, conditions similar to the continuity conditions of the preceding subsection which use the fields in the volume cells on both sides of the interface cannot be used. The special boundary conditions we must use appear in one of the two following families: (i) conditions of nonreflection for waves going out of the computational domain and (ii) absorbing layers located at the outer boundaries of the computational domain.

In the finite-volume method, the most frequently used and the most natural boundary condition applied on the outer boundary is the "Silver–Muller" condition [18]. It belongs to the family of boundary conditions referred to in the literature as local nonreflecting or radiation boundary conditions. These local conditions are less accurate than exact or global outer boundary conditions based on boundary integral equations. However, they are very easy to implement and require less computer resources. The computational volume is truncated by introducing a surface S_0 enclosing the scatterer. Consider a cell V with a face S which belongs to the surface S_0 and the electromagnetic fields $\boldsymbol{E}^*$ and $\boldsymbol{H}^*$ defined on S, with unit normal $\boldsymbol{a}_\mathrm{n}$ pointing outward, as illustrated in Fig. 9.6.

Then, the Silver–Muller condition is simply that the flux on the outer boundary corresponds to outgoing waves only. This leads to the following relation for the fields on S:

$$\sqrt{\frac{\varepsilon_0}{\mu_0}}\boldsymbol{a}_\mathrm{n} \times \boldsymbol{E}^* + \boldsymbol{a}_\mathrm{n} \times (\boldsymbol{a}_\mathrm{n} \times \boldsymbol{H}^*) = 0. \tag{9.2.19}$$

This condition ensures that waves which come normal to the artificial boundary are not reflected, whereas other waves are only partially reflected. The condition Eq. (9.2.19) together with Eq. (9.2.14) can be used to express the numerical fluxes on the surface S_0 in terms of the cell-center values without difficulties. In order to avoid reflections on the outer boundary, we still have to place it sufficiently far away from the sources to ensure that the waves can be assumed to be locally plane waves upon arrival at this artificial boundary.

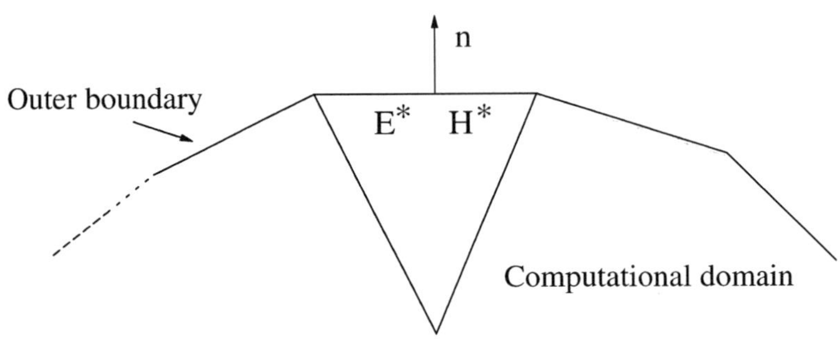

FIGURE 9.6
Triangulation and orientation near the outer boundary.

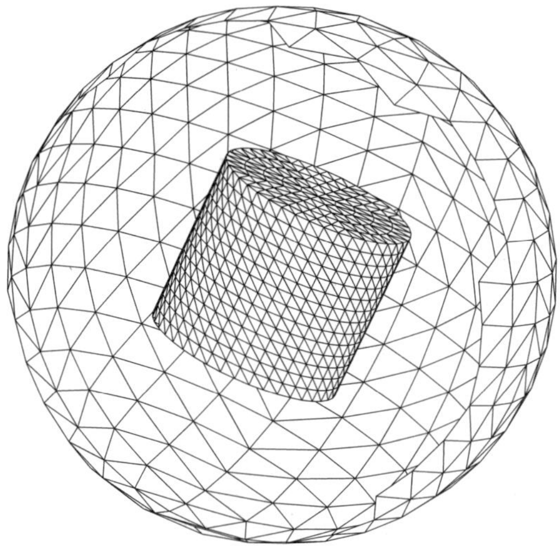

FIGURE 9.7
Mesh sizes near a scatterer and near the outer boundary.

This constraint generates a relatively large computational volume. However, from numerical experience, we conclude that the mesh size of the triangulation near the artificial boundary does not have to be as small as it must be near scattering obstacles while preserving the accuracy of the computation. On the average, scattering obstacles will be discretized using about 10 cell diameters per wavelength, and less than five grid lines per wavelength seem to be sufficient for the outer boundary of the computational domain (an example is shown in Fig. 9.7). Thus, the unstructured grid generated from the surface of the object toward the outer boundary is altogether not very large. However, higher order conditions on the outer boundary allow for outer boundaries closer to the sources and hence decrease the number of required elementary cells and, consequently, improve computation times.

Another possibility to truncate the computational domain is based on the concept of absorbing layer. This concept has been employed, and greatly improved, by J. P. Berenger, who proposed the perfectly matched layer (PML) [19]. In the FDTD community, this method has been studied and used by many authors [20–24]. The idea is to surround the computational domain by some layers with electric and magnetic conductivities, σ and σ^*, satisfying

$$\frac{\sigma}{\varepsilon_0} = \frac{\sigma^*}{\mu_0} \tag{9.2.20}$$

and to split each component of the fields into two components in the absorbing layers. For example, the E_x component satisfying Maxwell's equation

$$\varepsilon_0 \frac{\partial E_x}{\partial t} + \sigma E_x = \frac{\partial H_z}{\partial y} - \frac{\partial H_y}{\partial z} \tag{9.2.21}$$

is split into $E_x = E_{xy} + E_{xz}$, satisfying

$$\begin{cases} \varepsilon_0 \frac{\partial E_{xy}}{\partial t} + \sigma_y E_{xy} = \frac{\partial H_z}{\partial y} = \frac{\partial (H_{zx} + H_{zy})}{\partial y} \\ \varepsilon_0 \frac{\partial E_{xz}}{\partial t} + \sigma_z E_{xz} = -\frac{\partial H_y}{\partial t} = -\frac{\partial (H_{yx} + H_{yz})}{\partial t}. \end{cases} \tag{9.2.22}$$

With these equations in the absorbing layer, the fields are absorbed, without reflections on the different interfaces vacuum–PML and PML–PML. At the boundary, after the PML, we still have to impose a boundary condition to limit the computational domain. One can use the condition, $\boldsymbol{a}_{\mathrm{n}} \times \boldsymbol{E} = 0$, which is equivalent to taking a perfectly conducting surface or, as an improvement, the Silver–Muller condition, for example.

If we attempt to use the same PML formalism in the FVTD method, there is a problem in obtaining the values of the fluxes in the absorbing layers. Indeed, we can write, without any difficulties, the previous equations in the absorbing layers into a conservative form as it has been done for Maxwell's equations. However, we cannot diagonalize the matrix of the system and, hence, we cannot obtain a system of characteristic equations to determine the fluxes on the faces of the cells in the absorbing layers.

However, one can still benefit from the PML formalism by combining the FVTD scheme with the FDTD scheme. Indeed, as discussed in a following section, we can use an FVTD method, with an unstructured mesh, close to the scattering objects and an FDTD method elsewhere, with a structured mesh and a PML formalism on the outer boundary. This strategy permits the number of cells in the mesh to be decreased considerably and so the CPU time and the memory requirements are lessened.

Higher Order Spatial Discretizations In order to generate higher order spatial approximations and consequently to improve the spatial precision of the numerical scheme, a MUSCL (monotonic upwind schemes for conservation laws) approach is often used in finite-volume techniques [1]. The MUSCL approach requires an approximation of the gradient of the field values at the center of each finite volume. With these gradients, we compute the values at the boundary faces of the cell from the field values at the center of the cell. For example, in a finite-volume V_i, where the gradient at the center is given by ∇U_i and the field value by U_i, we define two values at the boundary S_k between the cell V_i and its neighbor V_{i+1} by

$$U^{\mathrm{L}} = U_i + \nabla U_i \cdot dl_i$$

and

$$U^{\mathrm{R}} = U_{i+1} + \nabla U_{i+1} \cdot dl_{i+1},$$

where dl_i and dl_{i+1} are respectively the distance between the center of cells V_i and V_{i+1} and the center of surface S_k. The values U^{L} and U^{R} are then used in Eq. (9.2.15) to evaluate the numerical fluxes.

Of course, how the gradients are defined in a cell-centered formulation is different from how they are used in a cell vertex formulation. For a cell vertex formulation, the most common way is to use an average of the Galerkin gradient evaluated at the nodes of the mesh [25]. For a cell-centered formulation, we can use the same approach [25] by computing first the values of the fields at the nodes of the mesh from the cell-centered values. However, with this gradient evaluation, we keep the difficulty to evaluate the field values at the nodes of the mesh on a boundary, as in the cell vertex formulation. Thus, using a cell-centered formulation, it is more convenient to approximate the gradients by the values at the center of the finite volumes. To evaluate these gradients, one can write in each cell V_i

$$\int_{V_i} \nabla U \, dv = \sum_{k=1}^{m_i} \int_{S_k} U^* \boldsymbol{a}_{\mathrm{n}k} \, ds, \tag{9.2.23}$$

where m_i is the number of faces S_k bounding the cell V_i. Using the same approximations to the integrals as before, we obtain

$$|V_i|\,\nabla U_i = \sum_{k=1}^{m_i} |S_k|\, U_k^*\, \boldsymbol{a}_{nk}, \tag{9.2.24}$$

where $|V_i|$ is the volume of V_i, $|S_k|$ is the area of S_k, and $\boldsymbol{a}_{nk}$ is the unit outer normal on S_k. For the value U_k^*, we use the value U_i in cases in which S_k may not be defined or may be near a surface of discontinuity in the fields (e.g., the boundary of the computational domain or a perfectly conductive face). In all other cases, we take the average value between the centered values U_i and U_{i+1} of the cells V_i and V_{i+1}, weighted by the distances between the centers of the cells and the center of the face. In this case,

$$U^* = \frac{dl_{i+1} \cdot U_i + dl_i \cdot U_{i+1}}{dl_{i+1} + dl_i},$$

where dl_i and dl_{i+1} define the distance between the center of the face and the centers of V_i and V_{i+1}, respectively.

9.2.2 Temporal Discretization

In this section, we discuss the most popular numerical schemes employed for the discretization of the time derivative in the finite-volume method for Maxwell's equations. First, we present the different numerical schemes for a generic differential equation of the form $\partial u/\partial t = f(u)$ and we give for each of these an estimate of the order of the error.

Let the generic differential equation be defined as

$$\begin{cases} \frac{\partial u}{\partial t} = f(u) \\ u(t=0) = u_0. \end{cases} \tag{9.2.25}$$

The most well-known numerical scheme is the explicit Euler method [8], given by

$$\begin{cases} \frac{u^{n+1}-u^n}{dt} = F(u^n) \\ u^0 = u_0, \end{cases} \tag{9.2.26}$$

where dt defines the time step, u^n is the value of u taken at the time $t = n\,dt$, and $F(u)$ is the discretization of the function $f(u)$.

To analyze the properties of this scheme we use a Taylor expansion of $\partial u/\partial t$ at the time $n\,dt$. We obtain

$$u^{n+1} = u^n + dt\left(\frac{\partial u}{\partial t}\right)^n + \frac{dt^2}{2}\left(\frac{\partial^2 u}{\partial t^2}\right)^n + o(dt^3), \tag{9.2.27}$$

where the superscript n denotes evaluation at $t = n\,dt$. This gives the following estimate:

$$\left(\frac{\partial u}{\partial t}\right)^n = \frac{u^{n+1} - u^n}{dt} - \frac{dt}{2}\left(\frac{\partial^2 u}{\partial t^2}\right)^n + o(dt^2). \tag{9.2.28}$$

When we substitute Eq. (9.2.28) in the differential equation written at the time $t = n\,dt$, we obtain Eq. (9.2.26). Substitution of the solution of the discretized equation into the original differential equation gives an error term of order dt. When the time step dt tends to zero, or the number of discretization intervals in the time domain tends to infinity, the difference between the original equation and the discretized form of the equation tends to zero. Then, we say that the numerical scheme is consistent in time with the equation and, in addition, that it is of order 1 in time.

The second numerical scheme presented here is the Lax–Wendroff scheme [26]. This scheme improves the precision in time of the explicit Euler scheme by a best approximation of the time derivative. In the Taylor expansion (Eq. 9.2.27) given previously, we approximate the term $\partial^2 u/\partial t^2$ by a derivative in time of the differential equation

$$\frac{\partial^2 u}{\partial t^2} = \frac{\partial (f(u))}{\partial t} = \frac{\partial f}{\partial u}(u)\frac{\partial u}{\partial t} = \frac{\partial f}{\partial u}(u) f(u). \tag{9.2.29}$$

Using this equation and Eq. (9.2.28), we obtain the following discretized form of Eq. (9.2.25):

$$\frac{u^{n+1} - u^n}{dt} = F(u^n) + \frac{dt}{2}\left(\frac{\partial F(u)}{\partial u} F(u^n)\right) + o(dt^2),$$

so we have a scheme of order 2 in time and hence a more precise one than the explicit Euler scheme.

In the same way, we also have a so-called "predictor–corrector" scheme [8] with two steps:

1. Predictor step:

$$\frac{u^{n+1/2} - u^n}{dt/2} = F(u^n) + o(dt/2)$$

2. Corrector step:

$$\frac{u^{n+1} - u^n}{dt} = F(u^{n+1/2}) + o(dt^2)$$

In the linear case in wh.ch $f(u) = Au$, where A is a matrix independent of u and t as in Maxwell's equations, the Lax–Wendroff scheme and this predictor–corrector scheme are, in fact, identical, [13, 26]. We obtain the following scheme:

$$\frac{u^{n+1} - u^n}{dt} = Au^n + \frac{dt}{2}A^2 u^n + o(dt^2).$$

Finally, we discuss the Runge–Kutta method [9]. This method can be found in the literature in the following general form:

$$\begin{cases} u^{(1)} = u^n \\ u^{(2)} = u^n + \alpha_2\, dt\, F\big(u^{(1)}\big) \\ u^{(3)} = u^n + \alpha_3\, dt\, F\big(u^{(2)}\big) \\ \dots \\ u^{(k)} = u^n + \alpha_k\, dt\, F\big(u^{(k-1)}\big) \\ \dots \\ u^{n+1} = u^n + dt \sum_{j=1}^{q} \beta_j\, F\big(u^{(j)}\big), \end{cases} \tag{9.2.30}$$

with the constraint $\sum_{j=1}^{q} \beta_j = 1$ in order to derive a consistent numerical scheme. With this constraint the scheme is of order 1 in time, but by choosing the coefficients α_j and β_j appropriately we can obtain greater accuracy. The frequently used Runge–Kutta scheme is the fourth-order Runge–Kutta method in which the coefficients are given by

$$\begin{aligned}
&q = 4\\
&\alpha_2 = \tfrac{1}{2} \quad \alpha_3 = \tfrac{1}{2} \quad \alpha_4 = 1\\
&\beta_1 = \tfrac{1}{6} \quad \beta_2 = \beta_3 = \tfrac{1}{3} \quad \beta_4 = \tfrac{1}{6}.
\end{aligned}$$

9.2.3 Consistency and Stability

As indicated in Section 9.1, a numerical scheme must be consistent and stable to ensure convergence of the solution of the discretized equations to the solution of the original problem. To study these two properties of a numerical scheme, we must consider the discretizations in all independent variables, i.e., for Maxwell's equations the discretizations with respect to space and time. In the case of Maxwell's equations, the spatial discretization can be performed using a structured or unstructured grid.

For a spatially structured grid, the conservative form of Maxwell's equations, in a cell (i, j, k) of the structured grid, is given by

$$\begin{aligned}
&\alpha\frac{\partial \boldsymbol{U}(i,j,k)}{\partial t} + \frac{\partial F_1(\boldsymbol{U}(i,j,k))}{\partial x} + \frac{\partial F_2(\boldsymbol{U}(i,j,k))}{\partial y} + \frac{\partial F_3(\boldsymbol{U}(i,j,k))}{\partial z}\\
&\quad = \alpha\frac{\partial \boldsymbol{U}(i,j,k)}{\partial t} + A_1(n_x)\frac{\partial \boldsymbol{U}(i,j,k)}{\partial x} + A_2(n_y)\frac{\partial \boldsymbol{U}(i,j,k)}{\partial y} + A_3(n_z)\frac{\partial \boldsymbol{U}(i,j,k)}{\partial z} = 0
\end{aligned}$$

With $\partial F_1/\partial \boldsymbol{U} = A_1(n_x)$, $\partial F_2/\partial \boldsymbol{U} = A_2(n_y)$, and $\partial F_3/\partial \boldsymbol{U} = A_3(n_z)$. This last equation can also be written as

$$\begin{aligned}
&\alpha\frac{\partial \boldsymbol{U}(i,j,k)}{\partial t} + \left(A_1^+(n_x) + A_1^-(n_x)\right)\frac{\partial \boldsymbol{U}(i,j,k)}{\partial x}\\
&\quad + \left(A_2^+(n_y) + A_2^-(n_y)\right)\frac{\partial \boldsymbol{U}(i,j,k)}{\partial y} + \left(A_3^+(n_z) + A_3^-(n_z)\right)\frac{\partial \boldsymbol{U}(i,j,k)}{\partial z} = 0.
\end{aligned}$$

By using a Taylor expansion for the spatial and time derivatives, one obtains

$$\begin{aligned}
&\alpha\frac{\boldsymbol{U}^{n+1}(i,j,k) - \boldsymbol{U}^n(i,j,k)}{dt} + A_1^+(n_x)\frac{\boldsymbol{U}^n(i,j,k) - \boldsymbol{U}^n(i-1,j,k)}{dx}\\
&\quad + A_1^-(n_x)\frac{\boldsymbol{U}^n(i+1,j,k) - \boldsymbol{U}^n(i,j,k)}{dx} + A_2^+(n_y)\frac{\boldsymbol{U}^n(i,j,k) - \boldsymbol{U}^n(i,j-1,k)}{dy}\\
&\quad + A_2^-(n_y)\frac{\boldsymbol{U}^n(i,j+1,k) - \boldsymbol{U}^n(i,j,k)}{dy} + A_3^+(n_z)\frac{\boldsymbol{U}^n(i,j,k) - \boldsymbol{U}^n(i,j,k-1)}{dz}\\
&\quad + A_3^-(n_z)\frac{\boldsymbol{U}^n(i,j,k+1) - \boldsymbol{U}^n(i,j,k)}{dz} = 0.
\end{aligned}$$

This equation can be also written as

$$\alpha \frac{U^{n+1}(i,j,k) - U^n(i,j,k)}{dt} + \frac{A_1^+(n_x)U^n(i,j,k) + A_1^-(n_x)U^n(i+1,j,k)}{dx}$$
$$+ \frac{A_1^+(-n_x)U^n(i,j,k) + A_1^-(-n_x)U^n(i-1,j,k)}{dx}$$
$$+ \frac{A_2^+(n_y)U^n(i,j,k) + A_2^-(n_y)U^n(i,j+1,k)}{dy}$$
$$+ \frac{A_2^+(-n_y)U^n(i,j,k) + A_2^-(-n_y)U^n(i,j+1,k)}{dy}$$
$$+ \frac{A_3^+(n_z)U^n(i,j,k) + A_3^-(n_z)U^n(i,j,k+1)}{dz}$$
$$+ \frac{A_3^+(-n_z)U^n(i,j,k) + A_3^-(-n_z)U^n(i,j,k+1)}{dz} = 0.$$

This form can be interpreted as the integrated form of the Maxwell's equation given in Section 9.2.1. Then, because the matrices $A_1(n_x)$, $A_2(n_y)$, and $A_3(n_z)$ are independent of dx, dy, dz, and dt, the numerical scheme used for the integrated form of Maxwell's equations in a structured grid is consistent. For numerical schemes based on spatially unstructured grids, consistency is more difficult to analyze.

To show that a numerical scheme is stable generally requires more work. For a numerical scheme based on a spatially structured grid, generally we can use the well-known Von Neumann method. The principle of this method is to study the evolution of the error ϵ in time and to adjust the parameters, if possible, such that this error decreases or remains constant. One considers an error $\epsilon(t,x)$, represented by a sum of harmonic functions in space $\sum_{i=1}^{m} \bar{\epsilon}(t)\ e^{-jk\,idx}$, and one studies the evolution of a general term $\bar{\epsilon}(t)e^{-jk\,idx}$ of this sum between two time steps.

To show the application of the method, we use as an example the generic equation introduced previously, where we suppose that u is now a function of one spatial variable and $f(u) = a(\partial u)/(\partial x)$. Then, we obtain the following equation:

$$\frac{\partial u}{\partial t} = a\frac{\partial u}{\partial x}. \tag{9.2.31}$$

If we approximate the term $\partial u/\partial x$ by $(u_{i+1} - u_i)/dx$, where dx is the step in the spatial discretization and u_i the value of u taken at the position $x_i = (i-1)\,dx$, we obtain, in the case of the explicit Euler method, the numerical scheme

$$\frac{u_i^{n+1} - u_i^n}{dt} = a\frac{u_{i+1}^n - u_i^n}{dx}, \tag{9.2.32}$$

where the superscript n denotes the sample at time $t_n = (n-1)\,dt$ and dt is the time step.

Let a harmonic component of the error be given by $\epsilon^n e^{-jk\,idx}$; if we substitute this component for u_i^n in Eq. (9.2.32), we obtain

$$\epsilon^{n+1}e^{-jk\,idx} = \epsilon^n e^{-jk\,idx}\left(1 + a\frac{dt}{dx}(e^{-jkdx} - 1)\right).$$

Then, we have the following expression for the rate of change in the error:

$$\frac{\epsilon^{n+1}}{\epsilon^n} = 1 - a\frac{dt}{dx} + a\frac{dt}{dx}\cos(kdx) + ja\frac{dt}{dx}\sin(kdx). \tag{9.2.33}$$

To guarantee the stability of the scheme we must have

$$\frac{\|\epsilon^{n+1}\|}{\|\epsilon^n\|} \leq 1. \tag{9.2.34}$$

Equation (9.2.33) shows that the quantity $\epsilon^{n+1}/\epsilon^n$ lies on a circle of radius $a(dt/dx)$ and origin $(1 - a(dt/dx), 0)$ in the complex plane. Hence, stability is only possible, in this scheme, when a is positive and, furthermore, we must require

$$a\frac{dt}{dx} \leq 1.$$

That a has to be positive is the reason why we based our flux estimations in Section 9.2.1 on waves propagating toward the interfaces only.

We now consider Maxwell's equations in three dimensions, on a structured spatial grid with steps $dx = dy = dz = h$, using the following numerical scheme:

$$\begin{aligned}
&\alpha\frac{\boldsymbol{U}^{n+1}(i,j,k) - \boldsymbol{U}^n(i,j,k)}{dt} + \frac{A_1^+(n_x)\boldsymbol{U}^n(i,j,k) + A_1^-(n_x)\boldsymbol{U}^n(i+1,j,k)}{dx}\\
&\quad + \frac{A_1^+(-n_x)\boldsymbol{U}^n(i,j,k) + A_1^-(-n_x)\boldsymbol{U}^n(i-1,j,k)}{dx}\\
&\quad + \frac{A_2^+(n_y)\boldsymbol{U}^n(i,j,k) + A_2^-(n_y)\boldsymbol{U}^n(i,j+1,k)}{dy}\\
&\quad + \frac{A_2^+(-n_y)\boldsymbol{U}^n(i,j,k) + A_2^-(-n_y)\boldsymbol{U}^n(i,j+1,k)}{dy}\\
&\quad + \frac{A_3^+(n_z)\boldsymbol{U}^n(i,j,k) + A_3^-(n_z)\boldsymbol{U}^n(i,j,k+1)}{dz}\\
&\quad + \frac{A_3^+(-n_z)\boldsymbol{U}^n(i,j,k) + A_3^-(-n_z)\boldsymbol{U}^n(i,j,k+1)}{dz} = 0.
\end{aligned}$$

Using the Von Neumann method, the stability condition is

$$dt \leq \frac{1}{c}\frac{h}{2}.$$

In the case of unstructured spatial grids and an explicit Euler scheme of spatial order 1, where the field values are constant in each cell, we can write the numerical scheme for a homogeneous medium as follows:

$$|V_i|\frac{\boldsymbol{U}_i^{n+1} - \boldsymbol{U}_i^n}{dt} + \sum_{k=1}^{m_i}|S_k|\big(\tilde{A}^+(\boldsymbol{a}_{nk})U_i^n + \tilde{A}^-(\boldsymbol{a}_{nk})U_k^n\big) = 0, \tag{9.2.35}$$

where $\boldsymbol{U}$ defines the field vector $(E_x, E_y, E_z, H_x, H_y, H_z)^t$, $|V_i|$ is the volume of the cell V_i, m_i is the number of boundary faces of the cell, and $|S_k|$, for $k = 1, \cdots, m_i$, are the surfaces of these faces. The field value inside the cell V_i is given by $\boldsymbol{U}_i$, and in a neighboring cell V_k of cell V_i, the field value is denoted by $\boldsymbol{U}_k$. If the field value $\boldsymbol{U}_i^n$ is constant inside the cell i, the value $F(\boldsymbol{U}_i^n)$ is also constant inside the cell i and $\operatorname{div}(F(\boldsymbol{U}_i^n)) = 0$.

Therefore,

$$\int_{V_i} \operatorname{div}\big(F\big(\boldsymbol{U}_i^n\big)\big)\, dV = \int_{\partial V_i} \tilde{A}(\boldsymbol{a}_{\mathrm{nk}})\boldsymbol{U}_i^n\, dS = \sum_{k=1}^{m_i} |S_k|\tilde{A}(\boldsymbol{a}_{\mathrm{nk}})\boldsymbol{U}_i^n = 0. \tag{9.2.36}$$

By subtracting this last equation from Eq. (9.2.35), we obtain

$$|V_i|\frac{\boldsymbol{U}_i^{n+1} - \boldsymbol{U}_i^n}{dt} + \sum_{k=1}^{m_i} |S_k|A^-(\boldsymbol{a}_{\mathrm{nk}})\big(\boldsymbol{U}_k^n - \boldsymbol{U}_i^n\big) = 0. \tag{9.2.37}$$

Now, by defining the vector $\boldsymbol{W}_i^n = MU_i^n$, with the matrix M given by

$$M = \begin{bmatrix} \frac{1}{v}\mathrm{Id} & 0 \\ 0 & \mu\mathrm{Id} \end{bmatrix}, \tag{9.2.38}$$

where Id denotes the identity matrix in R^3 and $v = 1/\sqrt{\mu\varepsilon}$ the speed of the light in the medium, the numerical scheme (Eq. 9.2.37) becomes

$$|V_i|\frac{\boldsymbol{W}_i^{n+1} - \boldsymbol{W}_i^n}{dt} + \sum_{k=1}^{m_i} |S_k|M\tilde{A}^-(\boldsymbol{a}_{\mathrm{nk}})M^{-1}\big(\boldsymbol{W}_k^n - W_i^n\big) = 0. \tag{9.2.39}$$

In this scheme, the matrix $B^-(\boldsymbol{a}_{\mathrm{nk}}) = M\tilde{A}^-(\boldsymbol{a}_{\mathrm{nk}})M^{-1}$ is a symmetric negative matrix given by

$$B^-(\boldsymbol{a}_{\mathrm{nk}}) = v\begin{bmatrix} -n_y^2 - n_z^2 & n_x n_y & n_x n_z & 0 & n_z & -n_y \\ n_x n_y & -n_x^2 - n_z^2 & n_y n_z & -n_z & 0 & n_x \\ n_x n_z & n_y n_z & -n_x^2 - n_y^2 & n_y & -n_x & 0 \\ 0 & -n_z & n_y & -ny^2 - n_z^2 & n_x n_y & n_x n_z \\ n_z & 0 & -n_x & n_x n_y & -n_x^2 - n_y^2 & n_y n_z \\ -n_y & n_x & 0 & n_x n_z & n_y n_z & -n_x^2 - n_y^2 \end{bmatrix}, \tag{9.2.40}$$

and the eigenvalues of $B^-(\boldsymbol{a}_{\mathrm{nk}})$ are $\{0, 0, 0, 0, -v, -v\}$.

The Eq. (9.2.39) scalar multiplied by $\boldsymbol{W}_i^n$ gives

$$\big\langle \boldsymbol{W}_i^{n+1}, \boldsymbol{W}_i^n\big\rangle - \big\langle \boldsymbol{W}_i^n, \boldsymbol{W}_i^n\big\rangle + \sum_{k=1}^{m_i} \frac{|S_k|}{|V_i|}\, dt\big(\big\langle B^-(\boldsymbol{a}_{\mathrm{nk}})\boldsymbol{W}_k^n, \boldsymbol{W}_i^n\big\rangle - \big\langle B^-(\boldsymbol{a}_{\mathrm{nk}})W_i^n, W_i^n\big\rangle\big) = 0 \tag{9.2.41}$$

where $\langle\cdot,\cdot\rangle$ denotes the scalar product in R^6, but

$$\big\langle \boldsymbol{W}_i^{n+1}, \boldsymbol{W}_i^n\big\rangle = -\frac{1}{2}\big\langle \boldsymbol{W}_i^{n+1} - \boldsymbol{W}_i^n, \boldsymbol{W}_i^{n+1} - \boldsymbol{W}_i^n\big\rangle + \frac{1}{2}\big\langle \boldsymbol{W}_i^{n+1}, \boldsymbol{W}_i^{n+1}\big\rangle + \frac{1}{2}\big\langle \boldsymbol{W}_i^n, \boldsymbol{W}_i^n\big\rangle \tag{9.2.42}$$

and

$$\langle B^-(\boldsymbol{a}_{\mathrm{nk}})(\boldsymbol{W}_k^n - \boldsymbol{W}_i^n), \boldsymbol{W}_k^n - \boldsymbol{W}_i^n\rangle = \langle B^-(\boldsymbol{a}_{\mathrm{nk}})\boldsymbol{W}_k^n, \boldsymbol{W}_k^n\rangle \tag{9.2.43}$$

$$+ \langle B^-(\boldsymbol{a}_{\mathrm{nk}})\boldsymbol{W}_i^n, \boldsymbol{W}_i^n\rangle \tag{9.2.44}$$

$$- 2\langle B^-(\boldsymbol{a}_{\mathrm{nk}})\boldsymbol{W}_k^n, \boldsymbol{W}_i^n\rangle. \tag{9.2.45}$$

Using the last two relations together with Eq. (9.2.41) yields

$$-\frac{1}{2}\langle \boldsymbol{W}_i^{n+1} - \boldsymbol{W}_i^n, \boldsymbol{W}_i^{n+1} - \boldsymbol{W}_i^n\rangle + \frac{1}{2}\langle \boldsymbol{W}_i^{n+1}, \boldsymbol{W}_i^{n+1}\rangle - \frac{1}{2}\langle \boldsymbol{W}_i^n, \boldsymbol{W}_i^n\rangle \tag{9.2.46}$$

$$-\sum_{k=1}^{m_i} \frac{|S_k|}{2|V_i|} dt\big(\langle B^-(\boldsymbol{a}_{\mathrm{nk}})(\boldsymbol{W}_k^n - \boldsymbol{W}_i^n), \boldsymbol{W}_k^n - \boldsymbol{W}_i^n\rangle \tag{9.2.47}$$

$$-\langle B^-(\boldsymbol{a}_{\mathrm{nk}})\boldsymbol{W}_k^n, \boldsymbol{W}_k^n\rangle + \langle B^-(\boldsymbol{a}_{\mathrm{nk}})\boldsymbol{W}_i^n, \boldsymbol{W}_i^n\rangle\big) = 0. \tag{9.2.48}$$

Substituting Eq. (9.2.39), written as

$$\boldsymbol{W}_i^{n+1} - \boldsymbol{W}_i^n = -\sum_{k=1}^{m_i} \frac{|S_k|}{|V_i|} dt\, B^-(\boldsymbol{a}_{\mathrm{nk}})(\boldsymbol{W}_k^n - \boldsymbol{W}_i^n), \tag{9.2.49}$$

into Eq. (9.2.48) yields

$$\langle \boldsymbol{W}_i^{n+1}, \boldsymbol{W}_i^{n+1}\rangle - \langle \boldsymbol{W}_i^n, \boldsymbol{W}_i^n\rangle = r_{1i} + r_{2i}, \tag{9.2.50}$$

with

$$r_{1i} = -\frac{dt}{|V_i|}\sum_{k=1}^{m_i} |S_k|\big(\langle B^-(\boldsymbol{a}_{\mathrm{nk}})\boldsymbol{W}_k^n, \boldsymbol{W}_k^n\rangle + \langle B^+(\boldsymbol{a}_{\mathrm{nk}})\boldsymbol{W}_i^n, \boldsymbol{W}_i^n\rangle\big) \tag{9.2.51}$$

(where we use Eq. 9.2.36 again) and

$$r_{2i} = \frac{dt^2}{|V_i|^2}\left\langle \sum_{k=1}^{m_i} |S_k| B^-(\boldsymbol{a}_{\mathrm{nk}})(\boldsymbol{W}_k^n - \boldsymbol{W}_i^n), \sum_{k=1}^{m_i} |S_k| B^-(\boldsymbol{a}_{\mathrm{nk}})(\boldsymbol{W}_k^n - \boldsymbol{W}_i^n)\right\rangle$$

$$+ \frac{dt}{|V_i|}\sum_{k=1}^{m_i} |S_k|\langle B^-(\boldsymbol{a}_{\mathrm{nk}})(\boldsymbol{W}_k^n - \boldsymbol{W}_i^n), \boldsymbol{W}_k^n - \boldsymbol{W}_i^n\rangle.$$

If we sum Eq. (9.2.50) over all the N_v cells of the computational domain, we get

$$\frac{\sum_{i=1}^{N_v} \langle \boldsymbol{W}_i^{n+1}, \boldsymbol{W}_i^{n+1}\rangle}{\sum_{i=1}^{N_v} \langle \boldsymbol{W}_i^n, \boldsymbol{W}_i^n\rangle} = 1 + \frac{\sum_{i=1}^{N_v} (r_{1i} + r_{2i})}{\sum_{i=1}^{N_v} \langle \boldsymbol{W}_i^n, \boldsymbol{W}_i^n\rangle}. \tag{9.2.52}$$

This equation shows that the numerical scheme is globally stable if the term $\sum_{i=1}^{N_v}(r_{1i} + r_{2i})$ is negative. From Eq. (9.2.51), we derive

$$\sum_{i=1}^{N_v} |V_i| r_{1i} = -\sum_{k=1}^{N_{s\infty}} dt|S_k|\{\langle B^-(\boldsymbol{a}_{\mathrm{nk}})\boldsymbol{W}_k^n, \boldsymbol{W}_k^n\rangle + \langle B^+(\boldsymbol{a}_{\mathrm{nk}})\boldsymbol{W}_i^n, \boldsymbol{W}_i^n\rangle\}.$$

In this equation, the summation is only over the $N_{s\infty}$ faces on the outer boundary of the computational domain because the contributions of all the internal faces can be seen to vanish [use Eq. (9.2.13) and the fact that any internal face appears two times in the sum, but with a permutation of the indices]. Because the first term in the previous sum refers to cells exterior to the computational domain, it is in fact equal to zero. The second term in the sum is positive; hence, $\sum_{i=1}^{N_v} |V_i|\, r_{1i}$ is negative. Because all values of $|V_i|$ are positive and bounded, one concludes that $\sum_{i=1}^{N_v} r_{1i}$ is negative.

Now, to ensure that $\sum_{i=1}^{N_v} r_{2i}$ be negative as well, one needs to take care that each term r_{2i} is negative. This leads to a condition on the time step dt:

$$dt \le \min_i \left(-|V_i| \frac{\sum_{k=1}^{m_i} |S_k| \langle B^-(\boldsymbol{a}_{\text{nk}})(\boldsymbol{W}_k^n - \boldsymbol{W}_i^n),\, \boldsymbol{W}_k^n - \boldsymbol{W}_i^n \rangle}{\langle \sum_{k=1}^{m_i} |S_k| B^-(\boldsymbol{a}_{\text{nk}})(\boldsymbol{W}_k^n - \boldsymbol{W}_i^n),\, \sum_{k=1}^{m_i} |S_k| B^-(\boldsymbol{a}_{\text{nk}})(\boldsymbol{W}_k^n - \boldsymbol{W}_i^n) \rangle} \right). \tag{9.2.53}$$

In the numerator on the right-hand side, we use the eigenvector decomposition to obtain

$$\sum_{k=1}^{m_i} |S_k| \langle B^-(\boldsymbol{a}_{\text{nk}})(\boldsymbol{W}_k^n - \boldsymbol{W}_i^n),\, \boldsymbol{W}_k^n - \boldsymbol{W}_i^n \rangle = v \sum_{k=1}^{m_i} |S_k| \sum_{l=1}^{6} \lambda_l (w_{kl}^n - w_{il}^n)^2,$$

where $\lambda_l = 0$ or -1 and $(w_{kl}^n)_{l=1,\cdots,6}$ is the eigenvector decomposition of $\boldsymbol{W}_k^n$. In the denominator, we make the following estimate:

$$\left\langle \sum_{k=1}^{m_i} |S_k| B^-(\boldsymbol{a}_{\text{nk}})(\boldsymbol{W}_k^n - \boldsymbol{W}_i^n),\, \sum_{k=1}^{m_i} |S_k| B^-(\boldsymbol{a}_{\text{nk}})(\boldsymbol{W}_k^n - \boldsymbol{W}_i^n) \right\rangle$$

$$= v^2 \left(\sum_{k=1}^{m_i} |S_k| \sum_{l=1}^{6} \lambda_l (w_{kl}^n - w_{il}^n) \right)^2 \le v^2 \left(\sum_{k=1}^{m_i} |S_k| \right) \left(\sum_{k=1}^{m_i} |S_k| \sum_{l=1}^{6} \left(\lambda_l (w_{kl}^n - w_{il}^n)\right)^2 \right).$$

The condition (Eq. 9.2.53), then, is satisfied if the following condition is satisfied:

$$dt \le \frac{1}{v} \min_i \left(\frac{|V_i|}{\sum_{k=1}^{m_i} |S_k|} \right),$$

which is the stability criterion we were looking for. Satisfying this criterion leads to a globally stable numerical scheme (Eq. 9.2.39) and, because there is a fixed transformation $\boldsymbol{U} = M\boldsymbol{W}$, also to a globally stable scheme (Eq. 9.2.35).

9.3 HYBRIDIZATION OF THE FVTD METHOD WITH OTHER MODELS AND METHODS

In this section, we discuss some techniques which extend the applicability of the FVTD method as discussed in the previous sections. To this point, we have only discussed the FVTD method as a stand-alone technique to compute electromagnetic fields. In fact, the method can be extended in various ways. A very important extension is the coupling of the field computations to models for wire structures. This will be discussed in Section 9.3.1. Another extension consists of a hybridization of the FVTD method with the FDTD method. As will be discussed in Section 9.3.2, the combination of the two methods improves on the performances of both. Finally, in Section 9.3.3, we present an alternative finite-volume method which can also be seen as a mix between the finite-volume method as we presented it previously and a finite-surface method or conformal finite-difference method.

9.3.1 Thin-Wire Models in the FVTD Method

In electromagnetic compatibility, as well as in studies of antennas, a thin-wire model is often required in numerical codes. In the FDTD method, a "thin-wire" formalism was introduced by Holland and Simpson [38]. This formalism is suitable for wires with diameters which are very small compared to the mesh size. In the FDTD method, wires are generally kept parallel to a cell axis and, when this is not the case, the wire is approximated by a staircase wire. In some cases, this approximation is not very good because it modifies the effective length of the wire and hence its resonance frequencies. In the FVTD approach, the advantage is that the wires can have any direction in the computional domain and, hence, there is no need to approximate them by a staircase model.

The Thin-wire Model The formalism introduced in FVTD for thin wires is similar to the one used for the FDTD method. As with the FDTD method, a set of differential equations for the net current and net charge on the wire are coupled to the field quantities appearing in the finite-volume scheme.

Because these subjects are treated in detail elsewhere, we present a rapid derivation of the essential formulas from Maxwell's equations. We begin with the first Maxwell equation,

$$\nabla \times \boldsymbol{H} - \varepsilon_0 \frac{\partial \boldsymbol{E}}{\partial t} = 0. \tag{9.3.1}$$

We integrate the normal component to the surface of the wire of this equation along the boundary of the cross section, $\mathcal{C}$, of the wire. Using $\boldsymbol{a}_{\mathrm{n}} \cdot (\nabla_T \times \boldsymbol{H}_T) = 0$, $\boldsymbol{a}_{\mathrm{n}} \times \boldsymbol{a}_l = \boldsymbol{a}_\tau$, $\int_{\mathcal{C}} \boldsymbol{a}_{\mathrm{n}} \cdot (\nabla_T H_l \times \boldsymbol{a}_l) = \int_{\mathcal{C}} \boldsymbol{a}_\tau \cdot \nabla_T H_l = 0$, where $\boldsymbol{a}_l$ is a unit vector along the wire axis, $\nabla = \nabla_T + \partial_l \boldsymbol{a}_l$, and the subscripts T and l refer to the transversal and longitudinal coordinates and components, respectively, we obtain

$$\frac{\partial}{\partial l} \int_{\mathcal{C}} \boldsymbol{a}_\tau \cdot \boldsymbol{H}_T + \frac{\partial}{\partial t} \int_{\mathcal{C}} \varepsilon_0 \boldsymbol{a}_{\mathrm{n}} \cdot \boldsymbol{E}_T = 0.$$

We define the net current, I, and the net charge per unit length, q, as functions of the longitudinal coordinate on the wire by

$$I = \int_{\mathcal{C}} \boldsymbol{a}_\tau \cdot \boldsymbol{H}_T$$

$$q = \int_{\mathcal{C}} \varepsilon_0 \boldsymbol{a}_{\mathrm{n}} \cdot \boldsymbol{E}_T.$$

In this way, we obtain the first differential equation for the wire quantities:

$$\frac{\partial}{\partial l} I + \frac{\partial}{\partial t} q = 0. \tag{9.3.2}$$

This is the charge conservation equation.

A second equation is obtained from

$$\nabla \times \boldsymbol{E} + \mu_0 \frac{\partial \boldsymbol{H}}{\partial t} = 0. \tag{9.3.3}$$

We integrate "$\boldsymbol{a}_l \times$" times this equation over a curve, $\mathcal{L}$, starting on a point $w(l)$ on the wire surface and arriving at some point $x(l)$ in a plane transversal to the wire at longitudinal coordinate l. Using $\boldsymbol{a}_l \times (\nabla \times \boldsymbol{E}) = \nabla_T E_l - \partial_l \boldsymbol{E}_T$, we obtain

$$\int_{\mathcal{L}} \boldsymbol{a}_\tau \cdot \left(\nabla_T E_l - \frac{\partial}{\partial l} \boldsymbol{E}_T \right) = -\mu_0 \frac{\partial}{\partial t} \int_{\mathcal{L}} \boldsymbol{a}_\tau \cdot (\boldsymbol{a}_l \times \boldsymbol{H}_T) \tag{9.3.4}$$

or

$$E_l(x(l)) - E_l(w(l)) - \frac{\partial}{\partial l} \int_{\mathcal{L}} \boldsymbol{a}_\tau \cdot \boldsymbol{E}_T = -\mu_0 \frac{\partial}{\partial t} \int_{\mathcal{L}} \boldsymbol{a}_\nu \cdot \boldsymbol{H}_T, \tag{9.3.5}$$

where $\boldsymbol{a}_\nu = \boldsymbol{a}_l \times \boldsymbol{a}_\tau$ is a unit normal to the curve in the transversal plane.

We can relate the distribution of $\boldsymbol{E}_T$ and $\boldsymbol{H}_T$ to the net charge and the net current on the wire with a two-dimensional static model. In the general case, i.e., general cross sections and inhomogeneous media, this may require a numerical solution of two two-dimensional potential problems, but in the case of a circular cylindrical wire in free space, this can be done analytically as follows.

We consider a wire of radius a and a cylindrical coordinate system (l, r, θ). The l axis of this system is chosen parallel to the wire and its origin located on the wire. Due to the circular symmetry, the transversal electric field has only a radial component and the transversal magnetic field has only an azimuthal component, given by

$$E_r = \frac{q}{2\pi \varepsilon_0 r} \tag{9.3.6}$$

$$H_\theta = \frac{I}{2\pi r}, \tag{9.3.7}$$

where $r > a$ and I and q are as defined previously.

The second equation for the wire quantities is then

$$L(x(l)) \left\{ \frac{\partial}{\partial t} I + c^2 \frac{\partial}{\partial l} q \right\} = E_l(x(l)) - E_l(w(l)), \tag{9.3.8}$$

where $c^2 = 1/\varepsilon_0 \mu_0$ and $L(x)$ has the dimension of inductance per unit length and is given by

$$L(x) = \frac{\mu_0}{2\pi} \ln \left(\frac{r(x)}{a} \right), \tag{9.3.9}$$

where $r(x)$ is the distance from x to the wire axis. In this equation, the coupling to the fields around the wire is represented for by the term $E_l(x(l))$. We next discuss the discretization of the thin-wire model equations and give the details of the coupling of this model with the finite-volume scheme for the fields.

Coupling the Wire Model and the FVTD Method In the FDTD discretization for the wire model, we use two dual space–time grids. The wire is subdivided in segments labeled with an integer i and time is discretized in equal time intervals, of length dt, labeled with an integer n. The center of segment i will be denoted by the longitudinal coordinate l_i and the two endpoints by $l_{i\pm 1/2}$, respectively, and its length by dl_i. The charges will be sampled on the space–time grid with the labels $(i \pm 1/2, n)$ for integer i and n. We write $q^n_{i\pm 1/2} = q(l_{i\pm 1/2}, n\,dt)$ for the limit of the charge density on approaching the corresponding endpoint of segment i (through increasing or decreasing longitudinal coordinate) at time $t^n = n\,dt$, but we distinguish $q^n_{i+1/2}$ from $q^n_{(i+1)-1/2}$ so our index notation must be interpreted as a formal one. The currents will be sampled on the space–time grid with labels $(i, n \pm 1/2)$ for integer i and n. We write $I_i^{n\pm 1/2} = I(l_i, (n \pm 1/2)\,dt)$ and, because we consider currents continuous in space and time, in this case computations on the indices are allowed.

We begin by discretizing Eq. (9.3.8) using central differences for the space and time differentiations:

$$L(x_i)\left\{\frac{I_i^{n+1/2} - I_i^{n-1/2}}{dt} + c^2\frac{q^n_{i+1/2} - q^n_{i-1/2}}{dl_i}\right\} = E^n_l(x_i) + \frac{V^n_i}{dl_i}, \tag{9.3.10}$$

where $x_i = x(l_i)$ is a point in the transversal plane of segment i and $V^n_i = -dl_i\,E^n_l(w(l_i))$ is the voltage over segment i.

We can now define precisely how this finite-difference discretization of the wire model can be coupled to the finite-volume scheme for the electromagnetic field. In the finite-volume discretization, wires run along the edges of the elementary cells; in practice, we always use prism-shaped cells around wires so there are several cells adjacent to each wire segment, as is illustrated in Fig. 9.8. Now, we could couple the wire equations to the finite-volume scheme by taking $x(l_i)$ in the center of one of the adjacent cells. However, it is more accurate to take the average of the coupled equations to all adjacent cells. Consider a wire segment i; averaging Eq. (9.3.10) with $x(l_i)$ at the center of each of the volumes bordering this segment gives the following equation:

$$\langle L_i\rangle\left\{\frac{I_i^{n+1/2} - I_i^{n-1/2}}{dt} + c^2\frac{q^n_{i+1/2} - q^n_{i-1/2}}{dl_i}\right\} = \left\langle E^n_{l;i}\right\rangle + \frac{V^n_i}{dl_i}, \tag{9.3.11}$$

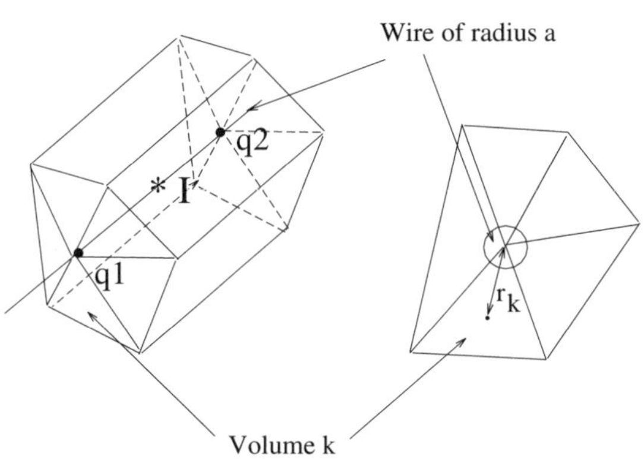

FIGURE 9.8
The mesh around a wire segment.

where $\langle f_i \rangle = \sum_{k=1}^{N_v} f(x_k)/N_v$, with N_v being the number of prisms adjacent to the segment i and x_k the center point of the kth prism adjacent to this segment.

We next discretize the charge conservation equation finite difference approximation to obtain

$$\frac{q_{i+1/2}^{n+1} - q_{i+1/2}^{n}}{dt} = -\frac{I_{i+1/2}^{n+1/2} - I_i^{n+1/2}}{dl_i/2} \tag{9.3.12}$$

$$\frac{q_{(i+1)-1/2}^{n+1} - q_{(i+1)-1/2}^{n}}{dt} = -\frac{I_{i+1}^{n+1/2} - I_{(i+1)-1/2}^{n+1/2}}{(dl_{i+1})/2}. \tag{9.3.13}$$

We can eliminate the quantity $I_{i+1/2}^{n+1/2} = I_{(i+1)-1/2}^{n+1/2}$ from these equations by using the continuity of the transversal electric field, expressed by the following equation:

$$c^2 \langle L_i \rangle q_{i+1/2}^{n} = c^2 \langle L_{i+1} \rangle q_{(i+1)-1/2}^{n} \tag{9.3.14}$$

(we assume that the speed of light is the same around all wire segments). We obtain in a straightforward way

$$\frac{q_{i+1/2}^{n+1} - q_{i+1/2}^{n}}{dt} = -\frac{I_{i+1}^{n+1/2} - I_i^{n+1/2}}{\frac{dl_i}{2} + \frac{\langle L_i \rangle}{\langle L_{i+1} \rangle} \frac{dl_{i+1}}{2}} \tag{9.3.15}$$

$$\frac{q_{(i+1)-1/2}^{n+1} - q_{(i+1)-1/2}^{n}}{dt} = -\frac{I_{i+1}^{n+1/2} - I_i^{n+1/2}}{\frac{dl_{i+1}}{2} + \frac{\langle L_{i+1} \rangle}{\langle L_i \rangle} \frac{dl_i}{2}}. \tag{9.3.16}$$

This completes the finite-difference discretization of the wire model. We have to supplement the equations with the following boundary conditions:

- On an endpoint of a wire located on a perfect conductor, we take $q = 0$.
- On a segment of a wire having an endpoint in free space, we take $I = 0$.
- The voltage V_i on a segment can be related to the current by an equation, which we can think of as representing a lumped circuit element replacing the wire. For example, we can impose a Thevenin-type model

$$V_i = \mathcal{E}_i + \mathcal{R}_i I_i + \mathcal{L}_i \frac{\partial}{\partial t} I, \tag{9.3.17}$$

where $\mathcal{E}$, $\mathcal{R}$, and $\mathcal{L}$ are the Thevenin source, internal resistance, and internal inductance of the wire segment, respectively. Here, we only consider the case $V_i = 0$.

We can now construct the solution of the combined system in two possible ways:

1. Using an explicit Euler scheme of order 1 in time and 2 in space for the field evaluation and a finite-difference "leapfrog" scheme in time and space for the evaluation of current and charge on wires:

- For any wire segment i;
  - Compute I_i^n from known values using

$$\frac{I_i^{n+1/2} - I_i^{n-1/2}}{dt} + c^2 \frac{q_{i+1/2}^n - q_{i-1/2}^n}{dl_i} = \frac{E^n}{\langle L_i \rangle}.$$

  - Compute $q_{i+1/2}^{n+1}$ from known values using

$$\frac{q_{i+1/2}^{n+1} - q_{i+1/2}^n}{dt} = -\frac{I_{i+1}^{n+1/2} - I_i^{n+1/2}}{\frac{dl_i}{2} + \frac{\langle L_i \rangle}{\langle L_{i+1} \rangle} \frac{dl_{i+1}}{2}}.$$

  - Compute $q_{i-1/2}^{n+1}$ from known values using

$$\frac{q_{i-1/2}^{n+1} - q_{i-1/2}^n}{dt} = -\frac{I_i^{n+1/2} - I_i^{n+1/2}}{\frac{dl_i}{2} + \frac{\langle L_i \rangle}{\langle L_{i-1} \rangle} \frac{dl_{i-1}}{2}}.$$

  - For any finite volume p, compute U_p^{n+1} using

$$\alpha |V_i| \frac{U_p^{n+1} - U_p^n}{dt} + \frac{1}{2}\left(J_p^{n+1/2} + J_p^{n-1/2}\right) = \sum_k^{m_i} |S_k| F(U_p^*)^n \cdot \boldsymbol{a}_{\mathrm{nk}},$$

  where we used the notations of Section 9.2.1 and

$$[J_p^n] = \begin{bmatrix} \theta_p^{\mathrm{i}} I_i^n \boldsymbol{a}_l \\ 0 \end{bmatrix}$$

  is the fraction of the current on wire segment i flowing through volume p. The coefficient θ_p^{i} is zero when volume p is not adjacent to wire segment i and if it is θ_p^{i} gives the solid angle, normalized to 2π, spanned by the faces of the prism along the segment.

2. Using a predictor–corrector scheme for the fields and a leapfrog scheme in time and space for the current and charge on wires:
   - For any wire segment i,
     - Compute I_i^n from known values using

$$\frac{I_i^{n+1/2} - I_i^{n-1/2}}{dt} + c^2 \frac{q_{i+1/2}^n - q_{i-1/2}^n}{dl_i} = \frac{E^n}{\langle L_i \rangle}.$$

     - Compute $q_{i+1/2}^{n+1}$ from known values using

$$\frac{q_{i+1/2}^{n+1} - q_{i+1/2}^n}{dt} = -\frac{I_{i+1}^{n+1/2} - I_i^{n+1/2}}{\frac{dl_i}{2} + \frac{\langle L_i \rangle}{\langle L_{i+1} \rangle} \frac{dl_{i+1}}{2}}.$$

- Compute $q_{i-1/2}^{n+1}$ from known values using

$$\frac{q_{i-1/2}^{n+1} - q_{i-1/2}^{n}}{dt} = -\frac{I_i^{n+1/2} - I_i^{n+1/2}}{\frac{dl_i}{2} + \frac{\langle L_i \rangle}{\langle L_{i-1} \rangle}\frac{dl_{i-1}}{2}}.$$

- Predictor step: For any finite volume p, compute $U_p^{n+1/2}$ using

$$\alpha|V_i|\frac{U_p^{n+1/2} - U_p^n}{1/2\,dt} + \frac{1}{2}\left(J_p^{n+1/2} + J_p^{n-1/2}\right) = \sum_k^{m_i} |S_k| F(U_p^*)^n \cdot \boldsymbol{a}_{\mathrm{n}k}.$$

- Corrector step: For any finite volume p, compute U_p^{n+1} using

$$\alpha|V_i|\frac{U_p^{n+1} - U_p^n}{dt} + J_p^{n+1/2} = \sum_k^{m_i} |S_k| F(U_p^*)^{n+1/2} \cdot \boldsymbol{a}_{\mathrm{n}k}.$$

This concludes the definition of the coupling of a finite-difference model for wires with the finite-volume scheme for the fields.

9.3.2 Hybridization of the FVTD and the FDTD Methods

In order to improve the CPU time and memory requirements, a good approach [10, 27] is to combine the finite-volume time domain method described in the previous section with the standard FDTD method [4]. Indeed, with the finite-volume method, we can use an unstructured conformal mesh to represent the object and its immediate neighborhood with relatively few cells. With the finite-difference method, we can use a structured mesh for the remaining part of the computational domain and apply an efficient boundary condition, for example, the PML formalism [28, 29] which reduces considerably the number of unknowns. The gain in CPU time can also be improved for both methods by using a local time step. For the finite-volume method, we have shown in the previous section that the stability condition on the time step leading to a convergent numerical scheme is

$$dt \leq \min_i \left(\frac{1}{v} \frac{|V_i|}{\sum_{k=1}^{m_i} |S_k|} \right), \tag{9.3.18}$$

where $|V_i|$ is the volume of the cell V_i, m_i is the number of faces enclosing the cell V_i, S_k is the surface of the face k, and v is the speed of light in the medium. For the finite-difference scheme of Yee, this condition is given by

$$dt \leq \min_i \left(\frac{1}{v} \frac{1}{\sqrt{\frac{1}{dx_i^2} + \frac{1}{dy_i^2} + \frac{1}{dz_i^2}}} \right), \tag{9.3.19}$$

where dx_i, dy_i, and dz_i are the lengths of the edges of a rectangular cell in the grid.

We observe that the stability criterion for the finite-volume method is generally more restrictive than that for the finite-difference method. This implies that if we can use the two different time steps for each numerical scheme, we can expect a reduction in CPU time. In this section, we discuss

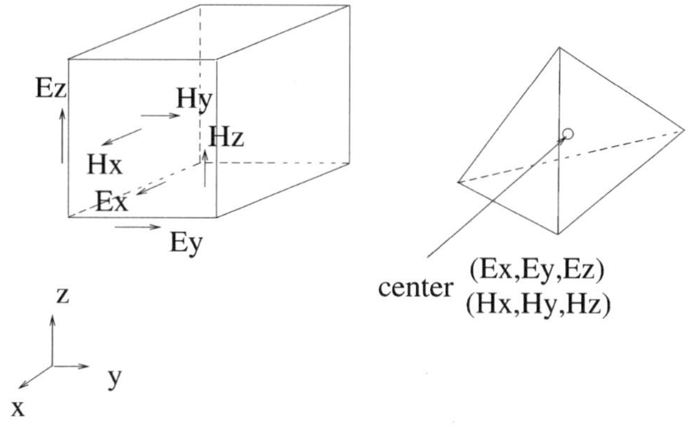

FIGURE 9.9
Locations of the field components in the two schemes.

the interface between the unstructured and the structured part of the mesh, the computation of the field at the interface, and the use of a local time step in the resulting hybrid method.

This hybrid method combines a finite-volume time domain method, in which the unknown electric and magnetic fields are evaluated at the centers of the cells (we shall consider only hexahedra), and a finite-difference time domain method, in which the unknowns are computed on the edges of orthogonal parallelepiped cells according to Yee's scheme. Figure 9.9 depicts the locations of the various field components in the cells for both schemes. The evaluation of the fields in the unstructured and structured areas by using respectively the finite-volume and the finite-difference methods has been explained in other chapters. The major difficulty of the hybridization of the two methods is caused by the interface between the structured and the unstructured grids. Indeed, we have to combine two schemes that use fields computed at different locations. With this purpose in mind, we create an "electric wall" on the interface for the two schemes. Therefore, the two numerical schemes in each area are correctly defined. The electric wall condition consists of imposing the tangential electric field values on the boundary surface of a volume.

Figure 9.10 presents the transition between an unstructured cell and a structured cell. In this transition, an intermediate area defined by two hexahedral cells (A and B) can be seen in which the fields are computed with finite-volume and finite-difference schemes as we shall now explain.

In cells A and B, the fields are computed with the FVTD scheme at the center of the cells. For cell A, there is no difficulty in applying the FVTD scheme presented in the previous section. Indeed, the fields are computed at the center of all the neighboring cells of the current cell A. For cell B, it is more difficult. The numerical flux on the surface S_B cannot be evaluated because the fields at the center of the cell C are not accessible. However, we do have available the electric fields on the edges of the face S_B, E_p, where $p = 1, \cdots, 4$, and we can compute, using an interpolation between these fields, an estimate of the value at the center of this face:

$$E^* = \sum_{p=1}^{4} \lambda_p E_p,$$

where the interpolation coefficients λ_p is dependent on the geometry of the particular surface. Then, the values $[\begin{smallmatrix} \boldsymbol{a}_n \times \boldsymbol{E}^* \\ \boldsymbol{a}_n \times \boldsymbol{H}^* \end{smallmatrix}]$ on the face are obtained by using the characteristic equation of the FVTD scheme. The component $\boldsymbol{a}_n \times \boldsymbol{E}^*$ follows immediately, and the other component is computed

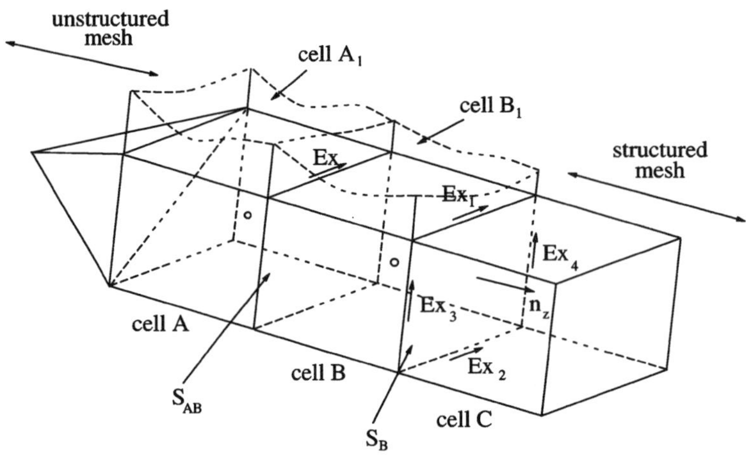

FIGURE 9.10
Transition between the unstructured area and the structured area.

from

$$Y\boldsymbol{a}_\mathrm{n} \times \boldsymbol{E}^* - \boldsymbol{a}_\mathrm{n} \times \boldsymbol{a}_\mathrm{n} \times \boldsymbol{H}^* = Y\boldsymbol{a}_\mathrm{n} \times \boldsymbol{E} - \boldsymbol{a}_\mathrm{n} \times \boldsymbol{a}_\mathrm{n} \times \boldsymbol{H}, \tag{9.3.20}$$

where Y is the characteristic admittance of the medium. This gives

$$\boldsymbol{a}_\mathrm{n} \times \boldsymbol{H}^* = Y\boldsymbol{a}_\mathrm{n} \times \boldsymbol{a}_\mathrm{n} \times (\boldsymbol{E} - \boldsymbol{E}^*) + \boldsymbol{a}_\mathrm{n} \times \boldsymbol{H}.$$

This defines the flux at the face S_B in terms of the field inside cell B and the FVTD scheme can be applied to the unstructured part of the mesh.

Now, for the FDTD method, we want to evaluate the electric fields at the edges of the surface S_B. To do this, we have to compute the magnetic fields inside cell B and then the electric fields on the edges of the surface S_AB. To obtain these fields, we use the electrical fields inside cells B and A, computed by the FVTD method and we interpolate these fields to obtain the desired values. For example, for edge No. 1 of the face S_AB, if we suppose that all the cells have the same size,

$$E_x = \frac{E_{x\mathrm{A}} + E_{x\mathrm{B}} + E_{x\mathrm{A1}} + E_{x\mathrm{B1}}}{4}.$$

In the case of cells with different sizes, we multiply each component field of the previous expression with weights dependent on the size of the surrounding cells. As with the FVTD scheme, we have obtained an electric wall condition at the surface S_AB terminating the FDTD scheme.

For both schemes, the electric fields at the electric walls will be computed for the same time sample. Therefore, it is not necessary to make a particular interpolation or extrapolation in time to combine the schemes. We only have to do spatial interpolations on the electric fields at different surfaces. We can resume the procedure of the hybridization of the two schemes through the following simplified algorithm:

Repeat

- Interpolate the electric fields given by the FDTD scheme at the surface S_B to obtain a boundary condition for the FVTD scheme.

- Interpolate the electric fields given by the FVTD scheme at the surface S_{AB} to obtain a boundary condition for the FDTD scheme;
- Compute the electric and magnetic fields with the FDTD scheme $(E^{n+1}, H^{n+1/2}) = f_1(E^n, H^{n-1/2})$.
- Compute the electric and magnetic fields with the FVTD scheme $(E^{n+1}, H^{n+1}) = f_2(E^n, H^n)$.

  Go_to repeat

The difficulty with the hybrid FVTD–FDTD method is to obtain a mesh for the structure. A possible method is as follows:

- Obtain a mesh composed with triangular patches of the skin of the structure.
- Surround the object by a surface S defined by a set of rectangular patches, each one split into two triangular patches.
- Generate a mesh between the skin of the structure and the latter surface S to obtain a domain with an unstructured mesh.
- Mesh outside the surface S until an orthogonal parallellepid boundary surface S_0 to merge with a structured mesh.

The last step can easily be done as follows: We can create a regular grid for the totality of the interior of S_0 and then we eliminate the part of this mesh corresponding to the unstructured area by putting on each cubic cell a flag, for example. In this way, we can identify the different parts of the mesh. To illustrate these different operations, Figs. 9.11–9.13 show the different steps of the mesh generation around an aeroplane.

As discussed previously, an interesting method to obtain a reduction in CPU time is to use a local time step in the different parts of the mesh. We can define in the presented hybrid FVTD–FDTD method three time steps:

- One in the unstructured grid given by $dt \leq \frac{1}{c}\frac{h\sqrt{3}}{24}$ for the FVTD scheme
- One in the area between unstructured grid and structured grid for the FVTD sheme given by $dt \leq \frac{1}{c}\frac{h}{2}$
- One in the structured grid for the FDTD scheme given by $dt \leq \frac{1}{c}\frac{h}{\sqrt{3}}$

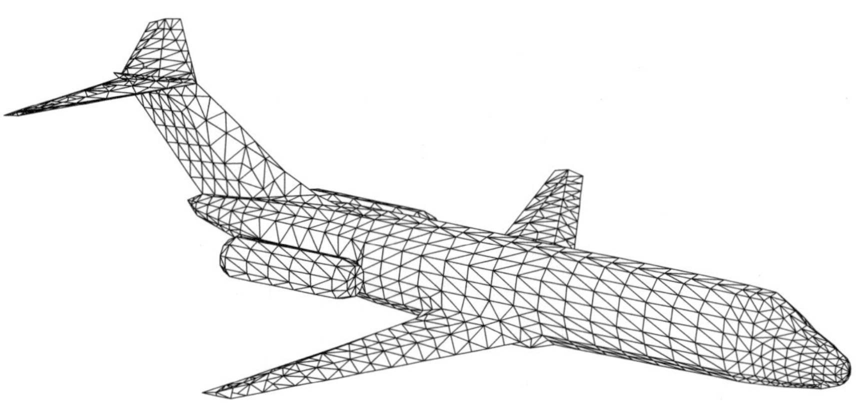

FIGURE 9.11
Different steps in a mesh generation: mesh of the structure.

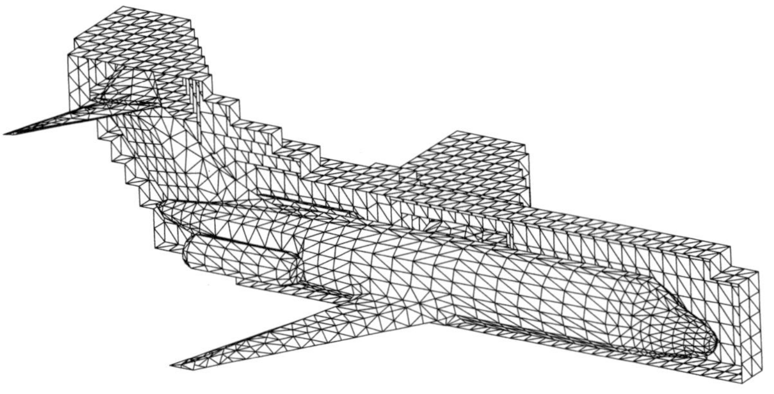

FIGURE 9.12
Different steps in a mesh generation: boundary of the area containing the unstructured mesh.

For these values, we have assumed that the unstructured grid is composed only of equilateral tetrahedral cells of edge h and that the structured grid is composed only of cubic cells of edge h.

To use a numerical scheme with a local time step, the determination of the time steps to be used can be broken up into several steps:

- Find the lowest value $dt_{\min}$ for the time step considering all the cells of the mesh and all the schemes applied.
- For each cell V_i compute a local time step dt_i with the given stability criterion, a ratio $r_i = dt_i/dt_{\min}$ and an integer n_i corresponding to the integer part of r_i.
- Compute the field values in cell V_i with the iterative process at each $n_i\, dt_{\min}$ time step.

With this method, not all field values in every cell are computed, at every step. In case the fields are not computed, they are kept equal to the value from the previous step. Although, good results

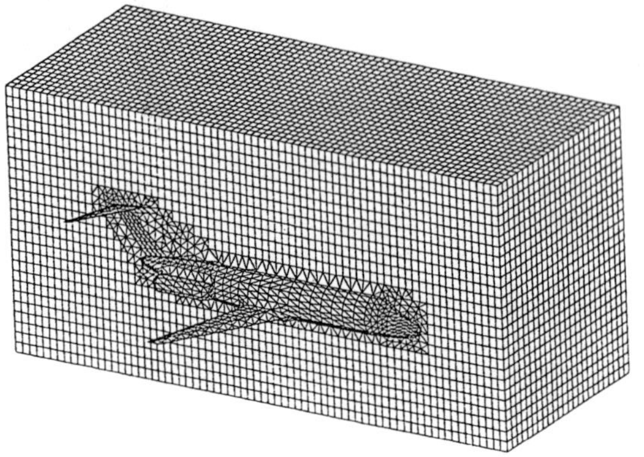

FIGURE 9.13
Different steps in a mesh generation: final mesh.

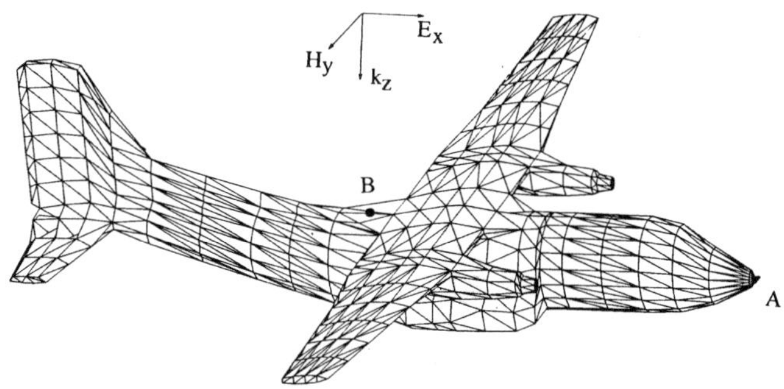

FIGURE 9.14
Mesh of the aircraft.

have been obtained with this method, numerical experience shows that there cannot be too large of a difference in the time steps on two neighboring cells.

To show the efficiency of the hybrid FVTD–FDTD method, we use a local time step to compare the CPU time and the memory requirements obtained with this method and with the FVTD method alone. When we used the FVTD method alone, we did not use a local time step and the mesh is unstructured everywhere. The example chosen concerns the scattered field, evaluated at two points A and B located on the skin of an aircraft illuminated by an incident plane wave (Fig. 9.14). In this example, a maximum ratio of eight between time steps on neighboring cells has been chosen.

Figures 9.15 and 9.16 present the results of the scattered fields obtained with the hybrid FVTD–FDTD method, the FVTD method alone, and the FDTD method alone. In these figures, the solid curves, the dotted curves, and the dashed curves correspond respectively to the results of the FVTD method alone, the hybrid FVTD–FDTD method, and the FDTD method alone. Table 9.1 gives the CPU time and the memory requirements for each method. These results were obtained on a Cray computer.

In the hybrid FVTD–FDTD method, regarding the actual mesh sizes used, we should be able to allow for the following three time steps: $v\,dt = 7.4\ 10^{-3}$ s in the unstructured area, $v\,dt = 1.02$ s in the intermediate area for the FVTD scheme, and $v\,dt = 1.18$ s in the structured area for the FDTD scheme, where v is the speed of light. The quantities which have actually been used, respect with to a not too large ratio between the time steps on neighboring cells, are

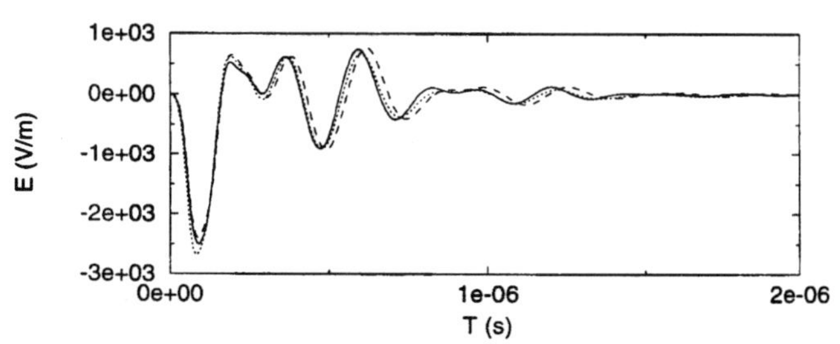

FIGURE 9.15
Comparison of the component E_x at point A computed with different methods.

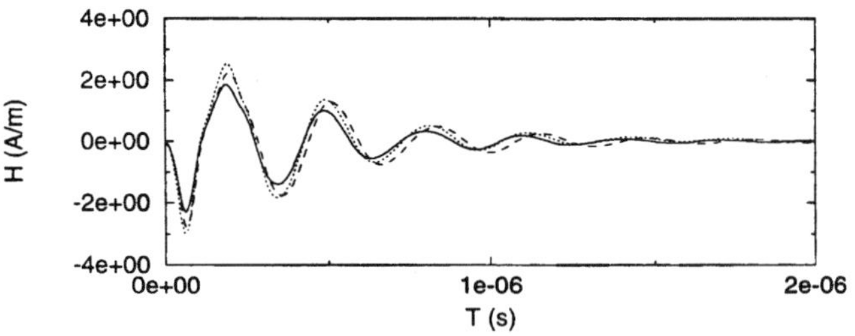

FIGURE 9.16
Comparison of the component H_y at point B computed with different methods.

- $v\,dt = 7.4\ 10^{-3}$ s in the unstructured area for the FVTD scheme
- $v\,dt = 5.92\ 10^{-2}$ s in the intermediate area for the FVTD scheme
- $v\,dt = 0.2368$ s in the structured area for the FDTD scheme

In this example, the numbers of cells in the various parts of the mesh are as follows: 13, 425 tetrahedra for the unstructured area, 6119 hexahedra for the intermediate area, and 27, 020 cubes for the structured area. In the FVTD method alone, there were 79, 361 tetrahedra in the computational domain and a time step was equal to $v\,dt = 7.4\ 10^{-3}$ s. Figures 9.17 and 9.18 show the meshes used for each method.

9.3.3 Another Approach of the Finite-Volume Method

Another approach to discretize Maxwell's equations with a finite-volume numerical scheme is based on a generalization of the well-known finite-difference method introduced by Yee [4]. The aim of this approach is different from that of the one described previously. Now, one tries to improve on the possibilities of the standard FDTD method to take into account any polyhedral cells such as tetrahedra, prisms, and distorted cubes. As in the FDTD method, a primal and a dual grid are defined, where respectively electric and magnetic fields are computed. In this approach, when the cells of the mesh have an orthogonal cubic form, Yee's FDTD numerical scheme is retrieved.

The basic idea of this method is to apply a conformal mesh on the structure and sufficiently far away to use a rectangular mesh and the classical FDTD method. Therefore, the boundary conditions on the outer boundary of the computational domain are the same as those in the FDTD method. For example, the PML formalism of Berenger can be applied without additional difficulties.

In this section, we present a general overview of this approach. More details and information on this method can be found in the Bibliography. As indicated previously, a mesh with two dual grids is necessary to utilize this method. The primal grid of the mesh is generally given by a set of cells obtained by a triangulation of the computational domain. In this mesh, there are structured

Table 9.1 Efficiency of the Methods

Approach	CPU time (s)	Memory required (Mw)
Hybrid FVTD–FDTD	13,205	2.7
FVTD alone	100,115	8

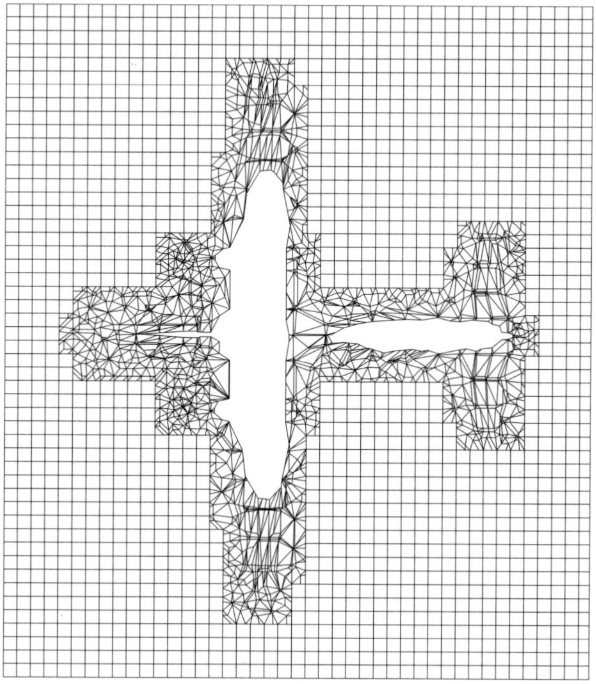

FIGURE 9.17
Cross section of the mesh used in the hybrid FVTD–FDTD method.

and unstructured parts. The dual grid is then obtained by computing the centroids of all the cells of the primal mesh and by connecting the centroids of neighboring cells to form the edges of the dual grid. These dual grids are generally composed of very complicated cells with many faces, edges, and nodes.

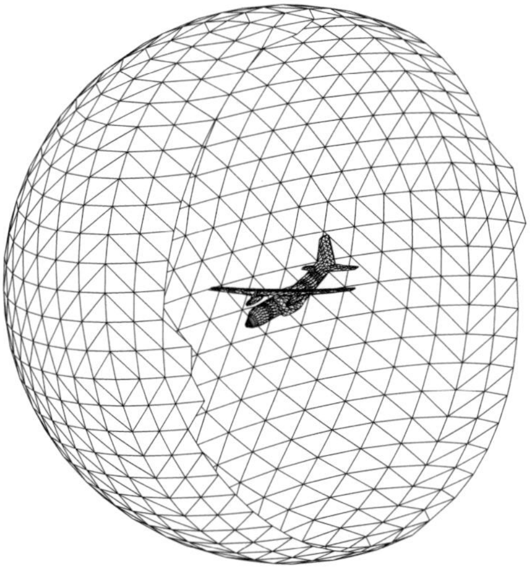

FIGURE 9.18
Mesh used in the FVTD method alone.

After the mesh has been obtained, the formalism used to compute the fields is based on two integral forms of Maxwell's equations. The first form is given by

$$\begin{cases} -\frac{\partial}{\partial t}\int_A \boldsymbol{B}\cdot d\boldsymbol{s} = \int_{\partial A} \boldsymbol{E}\cdot d\boldsymbol{l} \\ \frac{\partial}{\partial t}\int_{A^*} \boldsymbol{D}\cdot d\boldsymbol{s} = \int_{\partial A^*} \boldsymbol{H}\cdot d\boldsymbol{l}, \end{cases} \tag{9.3.21}$$

and the second form is given by

$$\begin{cases} -\frac{\partial}{\partial t}\int_V \boldsymbol{B}\, dv = \int_{\partial V}(\boldsymbol{a}_\mathrm{n}\times \boldsymbol{E})\, ds \\ \int_{V^*} \boldsymbol{D}\, dv = \int_{\partial V^*}(\boldsymbol{a}_\mathrm{n}\times \boldsymbol{H})\, ds. \end{cases} \tag{9.3.22}$$

Equation (9.3.21) was obtained by integrating Maxwell's equations on the faces A and A^* taken respectively from the primal grid and the dual grid. This form of the Maxwell's equations is a generalization of the FDTD scheme, and when the cells of the mesh are orthogonal this form corresponds to the form used with the well-known Yee scheme.

Equation (9.3.22) was obtained by integrating the Maxwell's equations on the volume of a primal cell V or a dual cell V^*. This formalism corresponds to a finite-volume formalism in which $\int_{\partial V}(\boldsymbol{a}_\mathrm{n}\times \boldsymbol{E})\, ds$ and $\int_{\partial V^*}(\boldsymbol{a}_\mathrm{n}\times \boldsymbol{H})\, ds$ define the numerical fluxes on the different faces ∂V or ∂V^* surrounding the primal volume cell V or the dual volume cell V^*.

For both systems of equations, the localization of the unknown field components in the grid can be chosen at different points. Those most commonly used in the literature are the following:

- Components of the electric fields are evaluated along the edges of the cells of the primal grid and components of the magnetic fields along the edges of the cells of the dual grid. This is an exact generalization of the FDTD method.
- Components of the electric fields are evaluated at the nodes of the cells of the primal grid and components of the magnetic fields at the nodes of the cells of the dual grid. In this case, we can also use a structured mesh far from the object, and for the evaluation of the fields in this area only the first form of Maxwell's equations is applied. Then, in this part of the mesh, the components of the fields have the same location as that in the FDTD method.

The first approach was proposed by Madsen and Ziolkowsky [30], and it is considered as the reference by many authors. This technique was presented as a modified finite-volume method. It was first presented exclusively for distorted cubes, but we can apply it equally well to different cell geometries such as tetrahedra. Le Martret and Le Potier [31] give an example of the use of this method on a Delaunay–Voronoi mesh not composed exclusively of hexahedra to evaluate the radar cross section of perfectly conducting obstacles. In this method, the principle is to project the electric or magnetic fields in each cell in three vector directions: the first one is normal to the primal or dual face where the field is computed and two other independent vector directions on the surface. Figure 9.19 shows the vector directions used to compute the field components in a primal and a dual cell. The quantities computed by this method are given by $\boldsymbol{H}\cdot\boldsymbol{\tau}^*$ and $\boldsymbol{E}\cdot\boldsymbol{\tau}$, where $\boldsymbol{\tau}^*$ and $\boldsymbol{\tau}$ define the vector direction of an edge on a primal or dual cell as depicted in Fig. 9.19.

If we consider a dual-face A^*, we can write for the electric field $\boldsymbol{E}$ the following equation:

$$\boldsymbol{E} = (\boldsymbol{E}\cdot\boldsymbol{a}_\mathrm{n}^*)\boldsymbol{a}_\mathrm{n}^* + (\boldsymbol{E}\cdot\boldsymbol{\nu}_1^*)\boldsymbol{\nu}_1^* + (\boldsymbol{E}\cdot\boldsymbol{\nu}_2^*)\boldsymbol{\nu}_2^*,$$

where $\boldsymbol{a}_\mathrm{n}^*$ is the unit normal on the face A^* and the vectors $\boldsymbol{\nu}_1^*$ and $\boldsymbol{\nu}_2^*$ are two independent unit vectors tangential to the face. If we project $\boldsymbol{E}$ on the direction $\boldsymbol{\tau}$ which defines the unit vector on

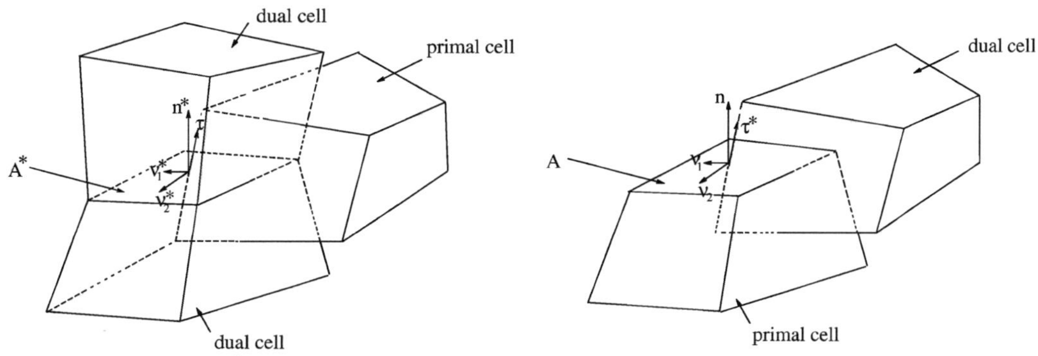

FIGURE 9.19
Vector directions in a primal cell and a dual cell.

the edge of the primal cell which intersects the face A^*, we obtain

$$\boldsymbol{E} \cdot \boldsymbol{\tau} = (\boldsymbol{E} \cdot \boldsymbol{a}_\mathrm{n}^*)(\boldsymbol{a}_\mathrm{n}^* \cdot \boldsymbol{\tau}) + (\boldsymbol{E} \cdot \boldsymbol{\nu}_1^*)(\boldsymbol{\nu}_1^* \cdot \boldsymbol{\tau}) + (\boldsymbol{E} \cdot \boldsymbol{\nu}_2^*)(\boldsymbol{\nu}_2^* \cdot \boldsymbol{\tau})$$

and then

$$\frac{\partial(\boldsymbol{E} \cdot \boldsymbol{\tau})}{\partial t} = \frac{\partial(\boldsymbol{E} \cdot \boldsymbol{a}_\mathrm{n}^*)}{\partial t}(\boldsymbol{a}_\mathrm{n}^* \cdot \boldsymbol{\tau}) + \frac{\partial(\boldsymbol{E} \cdot \boldsymbol{\nu}_1^*)}{\partial t}(\boldsymbol{\nu}_1^* \cdot \boldsymbol{\tau}) + \frac{\partial(\boldsymbol{E} \cdot \boldsymbol{\nu}_2^*)}{\partial t}(\boldsymbol{\nu}_2^* \cdot \boldsymbol{\tau}). \tag{9.3.23}$$

In this equation the first term is evaluated with the first form of the Maxwell's equations given previously. Indeed, on the dual-surface A^*,

$$\varepsilon|A^*|\frac{\partial(\boldsymbol{E} \cdot \boldsymbol{a}_\mathrm{n}^*)}{\partial t} = \int_{A^*} (\nabla \times \boldsymbol{H}) \cdot \boldsymbol{a}_\mathrm{n}^* \, ds = \oint_{\partial A^*} \boldsymbol{H} \cdot \boldsymbol{dl},$$

where ∂A^* defines the boundary edges of the face A^*, $|A^*|$ is its surface, and $\oint_{\partial A^*}$ is the line integral over the boundary of this face. The different components of the magnetic fields in this expression are known because they correspond to the field values computed by the method. Then, the line integral can be written as follows:

$$\oint_{\partial A^*} \boldsymbol{H} \cdot \boldsymbol{dl} = \sum_{i=1}^{4} (\boldsymbol{h} \cdot \boldsymbol{\tau}_i^*) l_i,$$

where $\boldsymbol{\tau}_i^*$ is the unit vector of the ith edge of the dual-face A^* and l_i is its length.

The evaluation of the other terms of Eq. (9.3.23) is done using the second form of Maxwell's equations (given previously). In this form, the volume of integration corresponds to the two dual cells that have the face A^* in common. By projecting the integrated form on the two unit vectors $\boldsymbol{\nu}_1^*$ and $\boldsymbol{\nu}_2^*$, the following equations are obtained:

$$\varepsilon|V^*|\frac{\partial(\boldsymbol{E} \cdot \boldsymbol{\nu}_1^*)}{\partial t} = \int_{\partial V^*} (\boldsymbol{a}_\mathrm{n}^* \times \boldsymbol{H}) \cdot \boldsymbol{\nu}_1^* \, ds \tag{9.3.24}$$

$$\varepsilon|V^*|\frac{\partial(\boldsymbol{E} \cdot \boldsymbol{\nu}_2^*)}{\partial t} = \int_{\partial V^*} (\boldsymbol{a}_\mathrm{n}^* \times \boldsymbol{H}) \cdot \boldsymbol{\nu}_2^* \, ds, \tag{9.3.25}$$

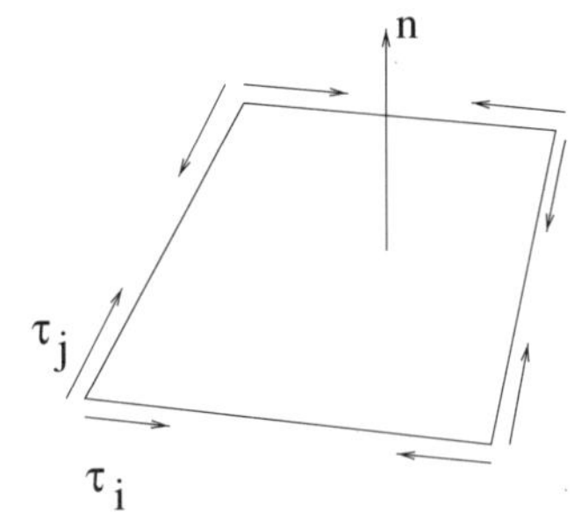

FIGURE 9.20
Magnetic or dual face in the evaluation of the electric field.

where $\boldsymbol{a}_{\text{n}}^{*}$ is the unit outgoing normal on the boundary face of the volume cell V^{*} and $|V^{*}|$ is its volume. As in the first form, in these equations we know the magnetic components on the edges of the volume cell V^{*} and we must compute the $\boldsymbol{a}_{\text{n}}^{*} \times \boldsymbol{H}$ on each face. To do this, take for example a face S on the boundary of the volume V^{*}; we then have four unit vector edges as shown in Fig. 9.20.

We compute for each corner a value of $\boldsymbol{a}_{\text{n}}^{*} \times \boldsymbol{H}$ by

$$\boldsymbol{a}_{\text{n}}^{*} \times \boldsymbol{H} = -\boldsymbol{H} \times \boldsymbol{a}_{\text{n}}^{*} = -\boldsymbol{H} \times (\boldsymbol{\tau}_i \times \boldsymbol{\tau}_j) = -(\boldsymbol{H} \cdot \boldsymbol{\tau}_i)\boldsymbol{\tau}_j + (\boldsymbol{H} \cdot \boldsymbol{\tau}_j)\boldsymbol{\tau}_i,$$

where $\boldsymbol{\tau}_i$ and $\boldsymbol{\tau}_j$ are the two unit vectors along the edges incident on the specific corner. Then we obtain the value of $\boldsymbol{a}_{\text{n}}^{*} \times \boldsymbol{H}$ on the face S as an interpolation of the values computed at each corner.

By combining the two forms of Maxwell's equations, we obtain the desired value of the electric field component in the direction $\boldsymbol{\tau}$. In the same way, we compute the magnetic field component in the direction $\boldsymbol{\tau}^{*}$ of the dual grid by taking an electric or primal surface A and an electrical or primal volume V corresponding to the two electric cells which have the surface A in common as shown in Fig. 9.19.

When the mesh is a rectangular mesh we retrieve the well-known Yee FDTD scheme with only the first form of the Maxwell's equation; the second form disappears because the edge directions are normal at the faces and the projections are identical to zero.

To complete the definition of this method, we must provide a numerical scheme for the time derivative. As in the FDTD Yee method, the numerical scheme chosen here is the leapfrog scheme, defined by

$$\begin{aligned} H^{n+1/2} &= H^{n-1/2} + dt\, g(E^{n}) \\ E^{n+1} &= E^{n} + dt f(H^{n+1/2}). \end{aligned}$$

In the second approach, proposed by Riley and Turner [11] under the name finite-volume hybrid grid [11] and by Yee and Chen under the name hybrid FDTD–FVTD method [12, 32], the magnetic and the electric fields are not evaluated on the edges of the mesh in the FVTD scheme but on the nodes of the cells. The electric fields are located at the nodes of the primal or electric cells and the magnetic fields at the nodes of the dual or magnetic cells, as shown in Fig. 9.21.

In this case, the dual grid is obtained in the same way as with the modified finite-volume method. However, with this method, we compute the electric and magnetic fields inside the cells by using the second integral form of Maxwell's equations respectively on a dual and a primal cell.

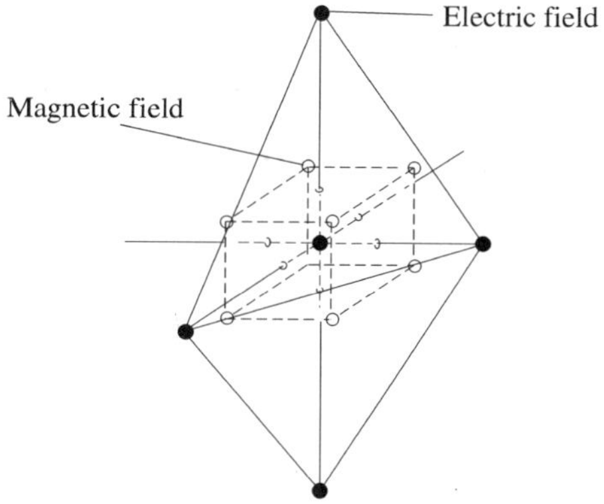

FIGURE 9.21
Electric and magnetic field locations on the primal and dual grids.

Then, we obtain

$$\int_{V^*} \varepsilon \frac{\partial \boldsymbol{E}}{\partial t}\, dv = \int_{\partial V^*} \boldsymbol{a}_{\mathrm{n}}^* \times \boldsymbol{H}\, ds \tag{9.3.26}$$

and

$$\int_{V} \mu \frac{\partial \boldsymbol{H}}{\partial t}\, dv = -\int_{\partial V} \boldsymbol{a}_{\mathrm{n}} \times \boldsymbol{E}\, ds, \tag{9.3.27}$$

where V^* and ∂V^* define a dual cell and its boundary and V and ∂V a primal cell with its boundary, and $\boldsymbol{a}_{\mathrm{n}}$ and $\boldsymbol{a}_{\mathrm{n}}^*$ denote the outgoing normal on the surface ∂V and ∂V^*, respectively.

The values $\boldsymbol{a}_{\mathrm{n}}^* \times \boldsymbol{H}$ and $\boldsymbol{a}_{\mathrm{n}} \times \boldsymbol{E}$ can be evaluated on each face, S, of the boundaries ∂V^* and ∂V with an interpolation of their values on each vertex of this face as in the method of Madsen presented previously. At this step, Riley and Turner [11] propose to improve the computation of the flux by using the first integrated form of Maxwell's equations. If we suppose that we compute the magnetic field inside a primal or electric cell, the value $\boldsymbol{a}_{\mathrm{n}} \times \boldsymbol{E}$ on a face of the boundary of the primal cell is given, as in the method of Madsen, by

$$\int_S \boldsymbol{a}_{\mathrm{n}} \times \boldsymbol{E} = -\frac{1}{2n_c}\sum_{k=1}^{n_c} \boldsymbol{E}_k \times (\boldsymbol{a}_k \times \boldsymbol{b}_k) = -\frac{1}{2n_c}\sum_{k=1}^{n_c}((\boldsymbol{E}_k \cdot \boldsymbol{a}_k)\boldsymbol{b}_k - (\boldsymbol{E}_k \cdot \boldsymbol{b}_k)\boldsymbol{a}_k),$$

where n_c is the number of the vertices of the face and $\boldsymbol{a}_k$ and $\boldsymbol{b}_k$ are the unit vectors on the edges incident on the kth vertex. Instead of expressing the integral in terms of the electric field values at the vertices, Riley and Turner prefer to work with the average values along the edges, which allows for a comparison with the electric fields appearing on the surfaces of the dual grid.

Then, on each edge we have to evaluate in this formalism a value $(\boldsymbol{E}_1 + \boldsymbol{E}_2)/2 \cdot \boldsymbol{a}$, where $\boldsymbol{E}_1$ and $\boldsymbol{E}_2$ are the values of the fields at the extremal points of the edge. The value $(\boldsymbol{E}_1 + \boldsymbol{E}_2)/2$ can be viewed as the electric field taken at the middle of the edge. If we project this value on the unit vector $\boldsymbol{\tau}$, i.e., the unit normal of the dual face intersected by the edge, we should find that this projection is equal to the computed electric field obtained by the first integrated form of Maxwell's equations. This, however, is generally not exactly true and the idea proposed by

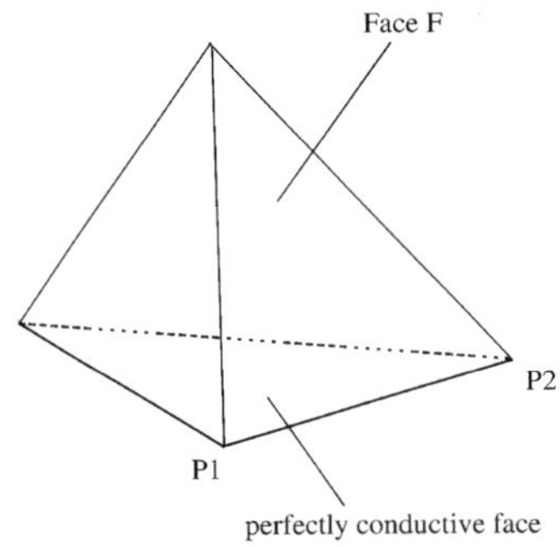

FIGURE 9.22
Cell with a boundary face on a perfect conductor.

Riley and Turner [11] is to improve the computation of the fluxes so as to obtain this equality. Therefore, we replace the component of the electrical field $(\boldsymbol{E}_1 + \boldsymbol{E}_2)/2$ along the unit vector $\boldsymbol{\tau}$ by the value computed with the first integrated form of Maxwell's equations. Then, the corrected projection of the electric field on the edge unit vector $\boldsymbol{a}$ is given by

$$\frac{(\boldsymbol{E}_1 + \boldsymbol{E}_2)}{2} \cdot \boldsymbol{a} + \left(\boldsymbol{E} \cdot \boldsymbol{\tau} - \frac{(\boldsymbol{E}_1 + \boldsymbol{E}_2)}{2} \cdot \boldsymbol{\tau} \right) \boldsymbol{\tau} \cdot \boldsymbol{a},$$

where $\boldsymbol{E} \cdot \boldsymbol{\tau}$ is the value computed by the first integrated form of the Maxwell equations.

However, with this approach, where the fields are computed at the nodes in the FVTD scheme there is a problem in evaluating electric fields on a node located on a perfectly conductive surface. Indeed, we know only the tangential component of the field on such faces but not its normal component. This problem also exists in the FDTD method, but it is not necessary to know the normal component of the electric field on a perfectly conductive surface in order to be able to evaluate the projection of the magnetic fields on the edges. If we want to evaluate the magnetic field inside a cell, with a boundary face on a perfect conductor, using the new approach, all the components of the electric field at points P_1 and P_2 are needed (Fig. 9.22).

To avoid this difficulty, we approximate these values by a field which has its tangential component equal to zero and the normal component equal to the average of the electric field taken at the other end of the edges incident on the node P (Fig. 9.23).

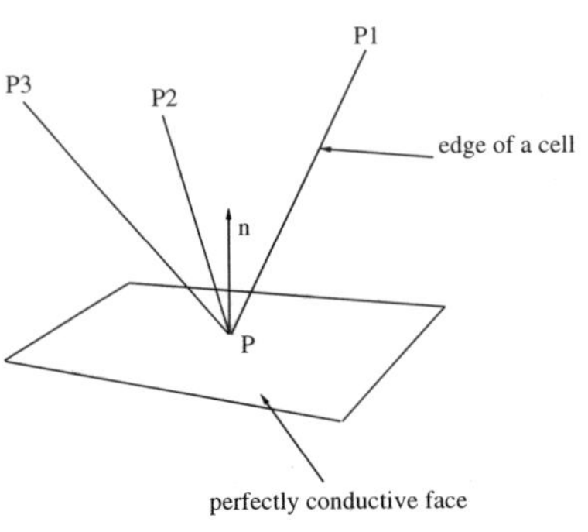

FIGURE 9.23
Evaluation of the electric field on a perfect conductor.

Then, a first approximation of the value $\boldsymbol{E}$ at point P is given by

$$\boldsymbol{E} = (\boldsymbol{e} \cdot \boldsymbol{a}_{\rm n})\boldsymbol{a}_{\rm n} = \frac{1}{nb}\sum_{k=1}^{nb}(\boldsymbol{E}_k \cdot \boldsymbol{a}_{\rm n})\boldsymbol{a}_{\rm n},$$

where nb is the number of edges incident on the point P, $\boldsymbol{a}_{\rm n}$ is the unit normal of face at the point P, and $\boldsymbol{E}_k$ are the fields at the points $P_1, P_2, \cdots, P_k$, as shown in Fig. 9.23.

A second approximation of the electric field $\boldsymbol{E}$ at the point P is given by an average of the values on each edge given by an extrapolation between a central edge value E_l and the other value taken at the end opposite P. The central edge value is computed by the first integrated form of Maxwell's equations on the dual face intersected by the edge under consideration. In this way, we obtain on an edge i incident on the point P the value of the electric field given by

$$\boldsymbol{E} = (\boldsymbol{E} \cdot \boldsymbol{a}_{\rm n})\boldsymbol{a}_{\rm n} = 2\,\boldsymbol{E}_{li} - (\boldsymbol{E}_i \cdot \boldsymbol{a}_{\rm n})\boldsymbol{a}_{\rm n},$$

where $\boldsymbol{a}_{\rm n}$ denotes the unit normal of the face.

By summing over all the nb edges which have P as an endpoint, we obtain for the electric field at point P the following value [12]:

$$\boldsymbol{E} = (\boldsymbol{E} \cdot \boldsymbol{a}_{\rm n})\boldsymbol{a}_{\rm n} = \frac{1}{nb}\sum_{i=1}^{nb}(2\boldsymbol{E}_{li} - (\boldsymbol{E}_i \cdot \boldsymbol{a}_{\rm n})\boldsymbol{a}_{\rm n}),$$

where $\boldsymbol{a}_{\rm n}$ is defined as it was previously.

9.4 NUMERICAL EXAMPLES

In this section, we discuss many examples computed with the methods and techniques introduced previously. We begin with examples of dielectric structures in Section 9.4.1. Then, we provide comparisons of analytical and other numerical models on configurations with thin conductive screens in Section 9.4.2. We conclude with results computed with the thin-wire model.

9.4.1 Dielectric Structures

In order to introduce dielectric regions with a varying real relative permittivity in the finite-volume formalism, we assume that these dielectric regions coincide exactly with a set of elementary volumes of the mesh and that the value of the permittivity on each elementary volume is constant. The permittivity is then included in the numerical fluxes by substituting the pertinent local values of ε and μ in Eq. (9.2.15).

Now, to study objects composed of only dielectric structures and illuminated by an incident plane wave, we introduce an artificial surface S_0 around these objects, as in the FDTD method [33]. Then, we compute the scattered fields in the area between this surface S_0 and the outer boundary and the total field in the area between S_0 and the objects. In the examples presented here, we choose a sphere for the surface S_0 because it is easy to create a mesh for a sphere.

To validate the modeling of dielectric structures in the finite-volume formalism, two examples are presented in which FVTD results are compared with the results obtained by Rao [34] using time and frequency domain integral equation methods. Figure 9.24 shows the far field scattered by a dielectric sphere illuminated by a Gaussian plane wave. In this example, the sphere has relative permittivity $\varepsilon_r = 10$, radius $r = 1$ m, and the time dependence is given by $f(t) = Ae^{-((t-t_0)/\sigma)^2}$ with $A = 0.5641895$, $\sigma = 0.333\,10^{-8}$, and $t_0 = 2\,10^{-8}$. In Fig. 9.24, curves a and b represent the backscattered far field and curves c and d the far field scattered in a direction perpendicular to

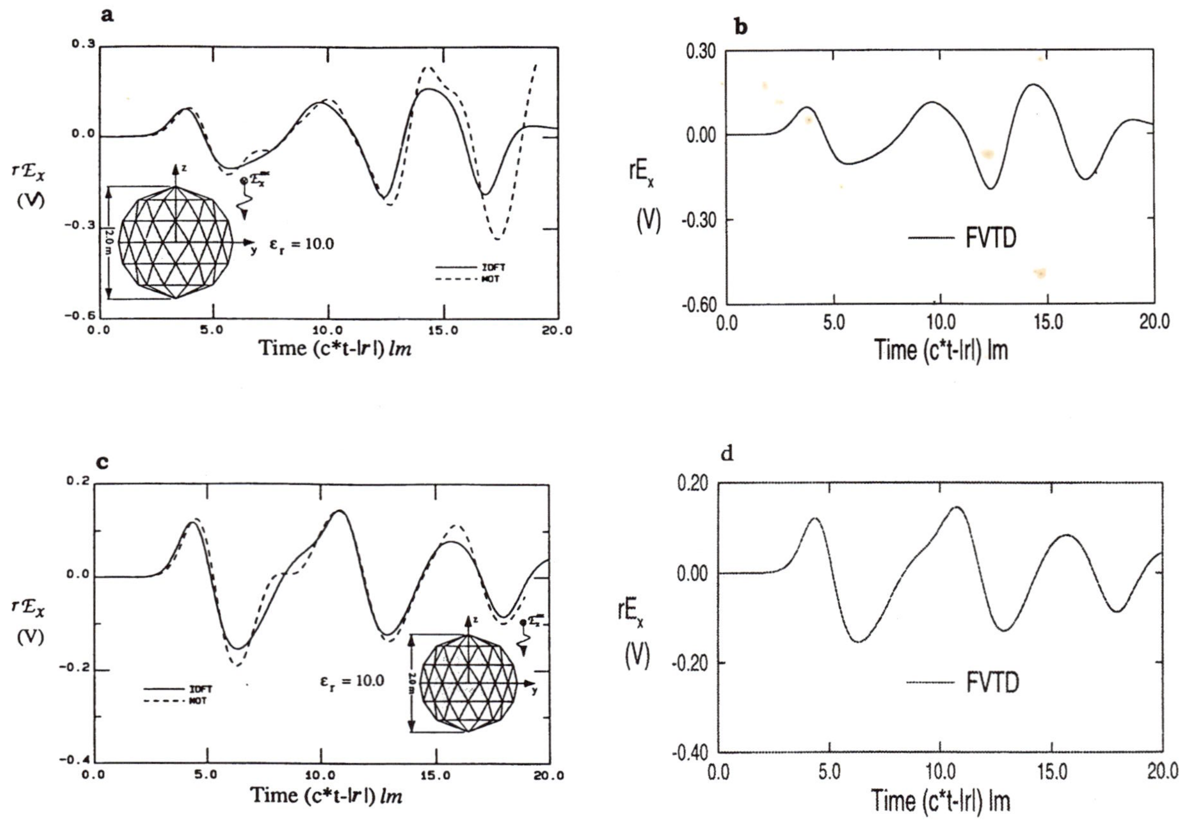

FIGURE 9.24
Comparison of FVTD–integral equation method for a dielectric sphere with a radius $r = 1$ m and a relative permittivity $\varepsilon_r = 10$.

the incident wave's propagation direction. The results obtained with the FVTD method are consistent with those obtained by the frequency domain integral equation method.

9.4.2 Thin Screens with Finite Conductivity

Thin screens with finite conductivity, such as some composite materials, are commonly used in the construction of modern airplanes and helicopters. Hence, it is very important to be able to model them.

A model for thin material with finite conductivity can be given by means of a surface impedance defined by $Z_s = 1/\sigma d$, where σ is the bulk conductivity of the material and d is its thickness [35,36]. In the FVTD method, we can introduce this model through a relation between the electric fields, $\boldsymbol{E}^*$ and $\boldsymbol{E}^{**}$, and the magnetic fields, $\boldsymbol{H}^*$ and $\boldsymbol{H}^{**}$, taken on either side of a surface S in the mesh, representing the thin screen [13]. This relation is given by

$$\boldsymbol{a}_\mathrm{n} \times \boldsymbol{E}^* = \boldsymbol{a}_\mathrm{n} \times \boldsymbol{E}^{**}$$
$$\boldsymbol{a}_\mathrm{n} \times \boldsymbol{H}^{**} - \boldsymbol{a}_\mathrm{n} \times \boldsymbol{H}^* = -\frac{1}{Z_\mathrm{s}}\boldsymbol{a}_\mathrm{n} \times (\boldsymbol{a}_\mathrm{n} \times \boldsymbol{E}^*),$$

where the quantities $\boldsymbol{a}_\mathrm{n}$, $\boldsymbol{E}^*$, $\boldsymbol{E}^{**}$, $\boldsymbol{H}^*$, and $\boldsymbol{H}^{**}$ are located on S as shown in Fig. 9.25.

This relation is to be combined with Eq. (9.2.15), given in Section 9.2, for evaluating the fluxes on the cell faces. Thus, the tangential fields on the interface can be expressed as follows:

$$\begin{aligned}\boldsymbol{a}_\mathrm{n} \times \boldsymbol{E}^* &= \boldsymbol{a}_\mathrm{n} \times \boldsymbol{E}^{**}\\ &= \frac{1}{Y^\mathrm{L} + Y^\mathrm{R} + \sigma d}[\boldsymbol{a}_\mathrm{n} \times (Y^\mathrm{L}\boldsymbol{E}^\mathrm{L} + Y^\mathrm{R}\boldsymbol{E}^\mathrm{R}) - \boldsymbol{a}_\mathrm{n} \times (\boldsymbol{a}_\mathrm{n} \times (\boldsymbol{H}^\mathrm{L} - \boldsymbol{H}^\mathrm{R}))] \qquad (9.4.1)\end{aligned}$$

$$\begin{aligned}\boldsymbol{a}_\mathrm{n} \times \boldsymbol{H}^* = {}& \frac{1}{Z^\mathrm{L} + Z^\mathrm{R} + \sigma d Z^\mathrm{R} Z^\mathrm{L}}[\boldsymbol{a}_\mathrm{n} \times (Z^\mathrm{R}\boldsymbol{H}^\mathrm{R} + (1 + \sigma d Z^\mathrm{R})Z^\mathrm{L}\boldsymbol{H}^\mathrm{L})\\ & - \boldsymbol{a}_\mathrm{n} \times (\boldsymbol{a}_\mathrm{n} \times (\boldsymbol{E}^\mathrm{R} - (1 + \sigma d Z^\mathrm{R})\boldsymbol{E}^\mathrm{L}))] \qquad (9.4.2)\end{aligned}$$

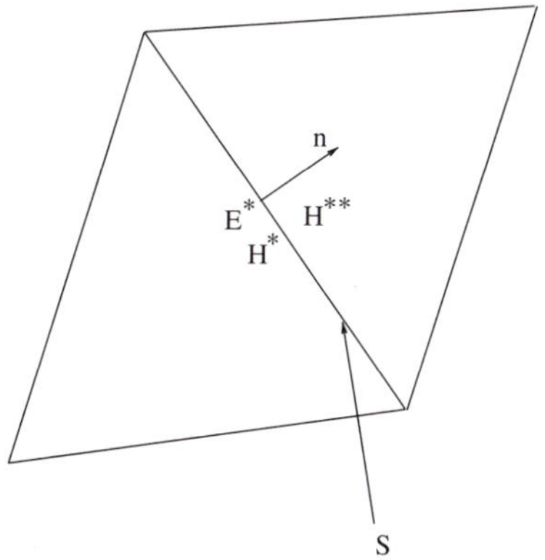

FIGURE 9.25
Field quantities relative to a composite surface S.

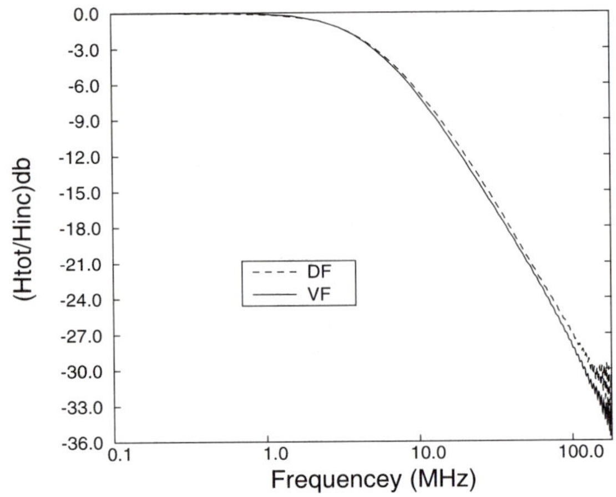

FIGURE 9.26
Comparison between FDTD and FVTD methods on a cubic box of thin material with $Z_s = 5\Omega$.

$$\begin{aligned}\boldsymbol{a}_n \times \boldsymbol{H}^{**} = \frac{1}{Z^L + Z^R + \sigma d Z^R Z^L} [&\boldsymbol{a}_n \times (Z^L \boldsymbol{H}^L + (1 + \sigma d Z^L) Z^R \boldsymbol{H}^R) \\ &- \boldsymbol{a}_n \times (\boldsymbol{a}_n \times (\boldsymbol{E}^L - (1 + \sigma d Z^L) \boldsymbol{E}^R))], \end{aligned} \tag{9.4.3}$$

where Z^L, Z^R, Y^L, and Y^R are as defined in Section 9.2.

Figure 9.26 presents the FDTD and FVTD results on a cubic box, made from thin material with a finite conductivity. The length of the cube's edge is 1 m, and the surface impedance of the material is $Z_s = 5\,\Omega$. The object is illuminated by a plane wave. The curves in the figure give the frequency domain magnetic field, computed at the center of the cube, normalized to the incident magnetic field at the same point. There is good agreement between the results of the two methods.

Figure 9.27 presents the frequency domain magnetic field computed in the center of a sphere, illuminated by a plane wave, normalized to the incident magnetic field at the same point. The sphere in this example has a radius $r = 12$ cm and is made from imperfectly conducting thin material with $Z_s = 50\,\Omega$. The solid curve in the figure represents the theoretical attenuation of the magnetic field at the center of the sphere [37]. The dotted curve is obtained with the

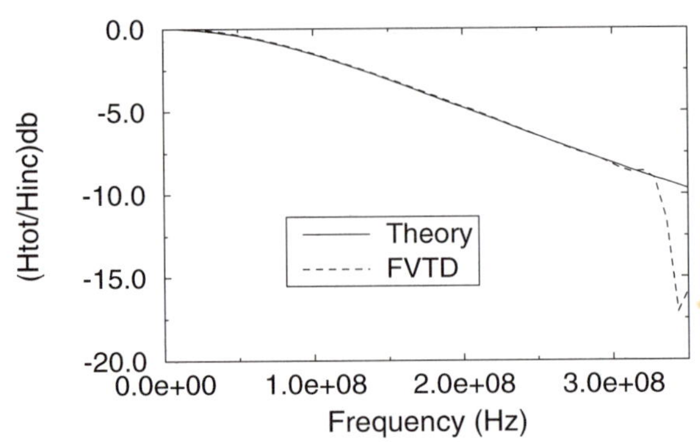

FIGURE 9.27
Comparison of the FVTD method with a theorical curve for a sphere with $Z_s = 50\,\Omega$.

FVTD method. The results are in agreement with the frequency domain data regarding the 3 dB attenuation frequency and the slope of the curves. For the FVTD method, this result was obtained with a mesh and an incident field frequency spectrum correct only up to 300 MHz, which explains the difference between the two curves for higher frequencies.

9.4.3 Thin Wires

To illustrate the thin-wire formalism introduced in the FVTD method, we make some comparisons with a finite-difference method, with a geometrical theory of diffraction (GTD) method, and with measurements. The source strength as a function of time, in all these examples, is a Gaussian pulse $f(t) = e^{-((t-t_0)/\sigma)^2}$ with $\sigma = 5 \times 10^{-9}$ and $t_0 = 20 \times 10^{-9}$.

The first example concerns the radiation pattern of a $\lambda/4$ dipole antenna at the frequency $f = 1$ GHz, located on a perfectly conducting circular plate of radius $r = 0.61$ m (Fig. 9.28).

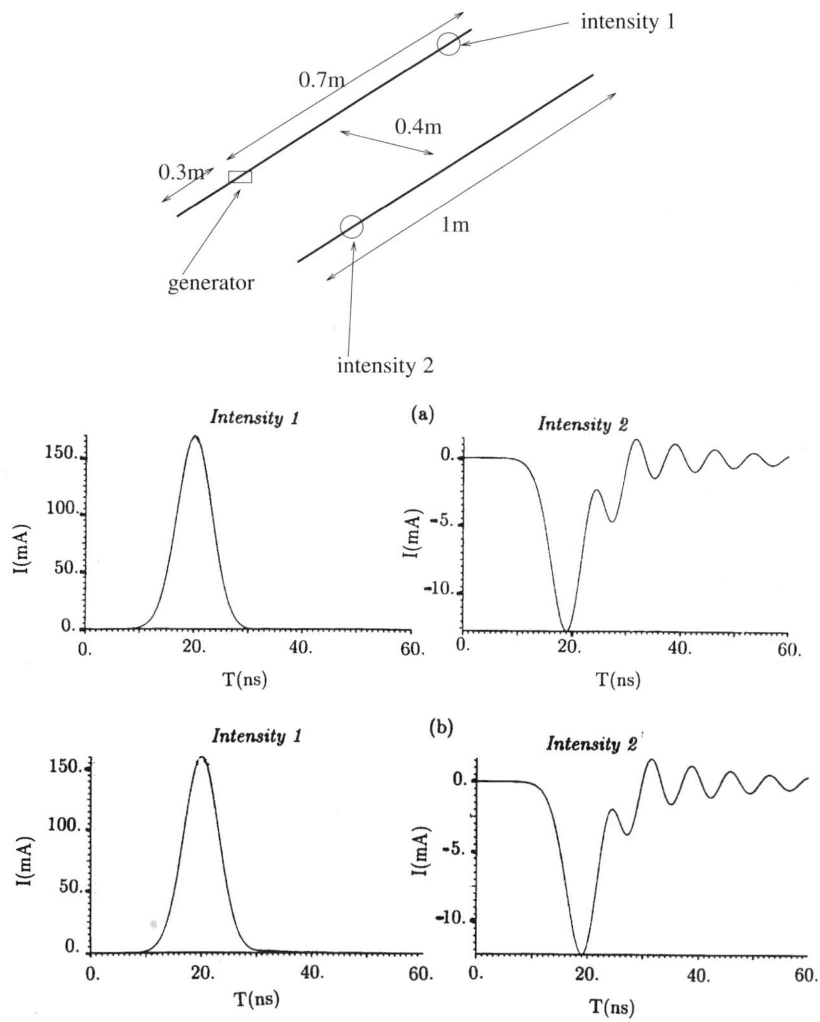

FIGURE 9.28
$\lambda/4$ dipole antenna on a circular, perfectly conducting plate.

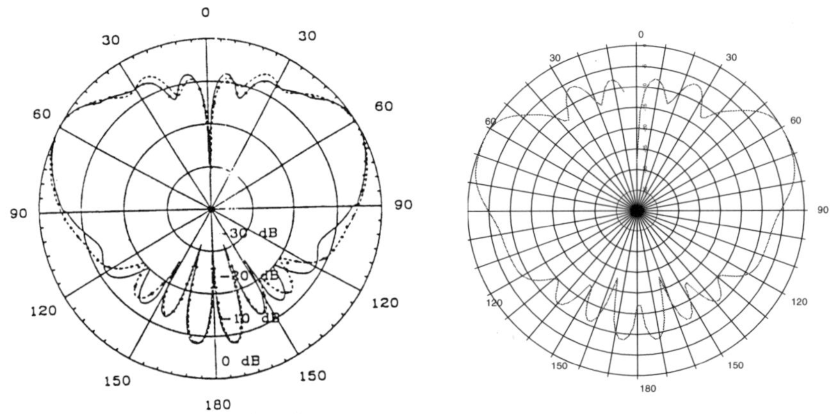

FIGURE 9.29
Radiation patterns of a $\lambda/4$ dipole antenna obtained by GTD and measurements (left) and FVTD (right).

The results obtained by GTD and measurements [39] are compared to the FVTD code results (Fig. 9.29). In this example, there is good agreement between results obtained by both methods (FVTD and FDTD).

The second example shows a comparison between FVTD and an experimental study. The scatterer is a cylindrical empty box with an open side. The open side of the box is placed on a perfectly conducting plate (Fig. 9.30). The thick walls of the box are made with polycarbonate material. The relative permittivity of this material is assumed to be frequency independent and has a value of $\varepsilon_r = 3$. The dielectric walls are externally covered by a layer of conductive paint.

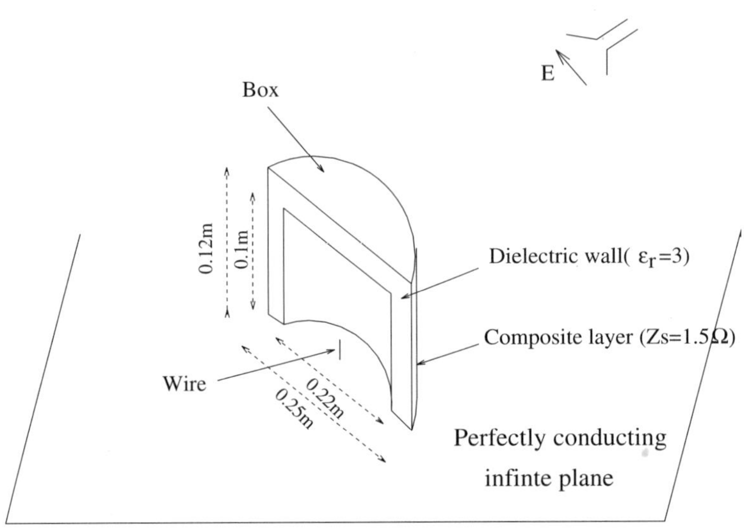

FIGURE 9.30
Section of the studied structure lying on an infinite perfectly conducting plane.

The value of the corresponding surface impedance ($Z_s = 1.5\ \Omega$) has been measured. For the experimental tests, carried out in an anechoic chamber, the box was illuminated by a plane wave and the source was moved in the xy plane in order to obtain angles of incidence from 0 to 180° in steps of 2°. The incident wave has a constant amplitude, in the frequency domain, up to 3 GHz. A thin wire (length, 7 mm; radius, 0.4 mm) inside the box (Fig. 9.30) was used as a receiving antenna to measure the voltage on the wire due to the transmitted field through the walls of the box. The experimental value obtained is the effective area of the receiving structure is given by $A_{\text{eff}} = (V^2/E^2) \cdot (377/50)$, where V is the measured voltage and E is the amplitude of the incident electric field.

For practical reasons, the experimental and numerical conditions are different. First, in the numerical simulation, in accordance with image theory the plane is removed and a symmetrized box is considered by adding the mirror image of the structure. Then, we computed the electric far field $\boldsymbol{E}_{\text{F}}$ generated by a voltage source V feeding the dipole located inside the box.

The reciprocity theorem of antenna theory [40] shows that the experimental value of the current I and the computed value of the electric far field E_F should be proportional. The mesh used for the numerical simulation consists of 139,150 surfaces and 69,363 volumes. At 2 GHz, the box is triangulated with a mesh size of $\lambda/10$, but near the sphere on which the far field is computed the mesh size is only $\lambda/5$. The experimental and numerical results for the normalized radiation pattern in the plane containing the wire, at three frequencies $f = 0.86$ GHz, $f = 1.342$ GHz, and $f = 2.188$ GHz, are presented in Figs. 9.31 and 9.32. In all cases, the main lobe directions, found numerically and experimentally, are in good agreement and the 3 dB beam widths are comparable. For $f = 1.342$ GHz and $f = 2.188$ GHz, the first nulls are slightly different because measurements were taken only every 2° for the experimental curve as opposed to every 1° for the simulation.

9.5 CONCLUDING REMARKS

In this chapter, we presented an overview of the finite-volume method for the solution of Maxwell's equations in the time domain. The following are essential features of the method:

- The possibility of obtaining a conformal mesh of the structure by using unstructured meshes
- The facility to account for different materials on each side of a surface in the mesh by a special flux-splitting formulation
- The possibility to implement models for particular components, such as thin wires and thin screens. There exist in the literature some FVTD methods coupled with the Vlasov equations to consider particle models [41, 42]
- The possibility to combine this method with other methods, such as the FDTD, which can significantly improve the performances with respect to computation time and memory requirements
- The possibility of local subgridding without excessive modifications on the mesh and on the computational procedure

For these reasons, we can expect that the FVTD method for Maxwell's equations will play an ever more important role in the modeling of real-world electromagnetic interaction problems.

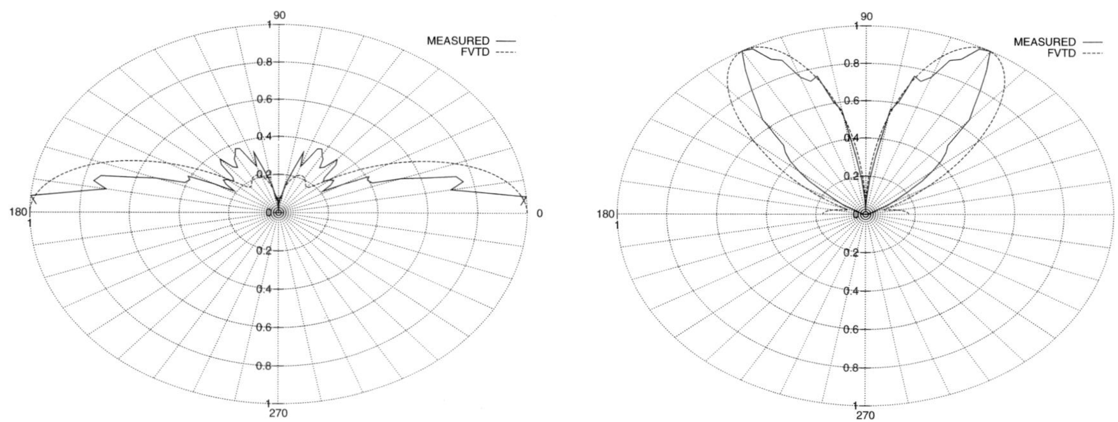

FIGURE 9.31
Radiation patterns obtained by measurements and by FVTD at $f = 860$ MHz (left) and at $f = 1.384$ GHz (right).

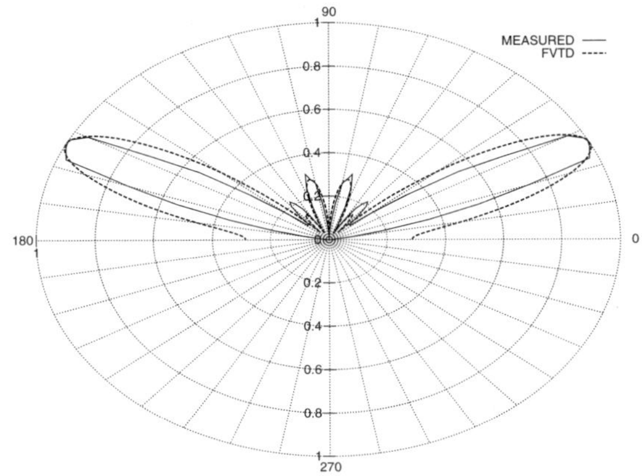

FIGURE 9.32
Radiation patterns obtained by measurements and by FVTD at $f = 2.188$ GHz.

ACKNOWLEDGMENTS

The authors thank Prof. Fontaine and Dr. Paladian of LASMEA (Clermont-Ferrand) and Dr. J. C. Alliot and Dr. J. L. Boulay of ONERA for the interest they have shown in our work on the finite-volume method. The support of Prof. Mazet, Prof. Kalfon, and Dr. Paintandre of ONERA and of the University Paul Sabatier of Toulouse, with respect to the result on the stability of the finite-volume method, was of great value. Finally, we thank the direction of ONERA, ElectroMagnetism and Radar Department, for allowing us to write this chapter.

BIBLIOGRAPHY

[1] C. Hirsch, *Numerical Computation of Internal and External Flows*, Wiley, New York, 1988.

[2] V. Shankar and W. F. Hall, "A Time Domain Differential Solver for Elecromagnetic Scattering," *URSI Meeting*, Univ. of Colorado, Boulder, January 1988.

[3] N. K. Madsen and R. W. Ziolkowski, "Numerical Solutions of Maxwell's Equations in the Time Domain Using Irregular Nonorthogonal Grids," *Waves Motion*, vol. 10, pp. 538–596, 1988.

[4] K. S. Yee, "Numerical Solution of Initial Boundary Value Problems Involving Maxwell's Equations in Isotropic Media," *IEEE Trans. Antennas Propagat.*, vol. 14, pp. 302–307, May 1966.

[5] J. S. Shang, "A Fractional Step Method for Solving 3D Time Domain Maxwell Equations," *J. Comput. Phys.*, vol. 118, pp. 109–219, 1995.

[6] J. S. Shang and R. H. Fitten, "A Comparative Study of Characteristic-Based Algorithms for the Maxwell Equations," *J. Comput. Phys.*, vol. 125, pp. 378–394, 1996.

[7] J. S. Shang, "Characteristic-Based Algorithms for Solving the Maxwell Equations in the Time Domain," *IEEE Antennas Propagat. Mag.*, vol. 37, pp. 15–25, June 1995.

[8] P. Bonnet, X. Ferrieres, F. Issac, F. Paladian, J. Grando, J. C. Alliot, and J. Fontaine, "Numerical Modeling of Scattering Problems Using a Time Domain Finite Volume Method," *J. Electromag. Waves Appl.*, vol. 11, pp. 1165–1189, 1997.

[9] J. P. Cioni, L. Fezoui, and D. Issautier, "High Order Upwind Schemes for Solving Time Domain Maxwell Equations," *La Recherche Aérospatiale*, No. 5, 1994.

[10] D. Kalfon and P. Harran-Klotz, "An Hybrid Finite Difference–Finite Volume Method for Solving Maxwell Equations in Time Domain," *URSI Meeting*, Newport Beach, CA, June 1995.

[11] D. J. Riley and C. D. Turner, "Local Tetrahedron Modeling of Microelectronics Using the Finite-Volume Hybrid-Grid Technique," SANDIA Report No. SAND95-2790, UC-706, December 1995.

[12] K. S. Yee and J. S. Chen, "The Finite-Difference Time-Domain (FDTD) and the Finite-Volume Time-Domain (FVTD) Methods in Solving Maxwell's Equations," *IEEE Trans. Antennas Propagat.*, vol. 45, pp. 354–363, 1997.

[13] A. H. Mohammadian, V. Shankar, and W. F. Hall, "Computation of Electromagnetic Scattering and Radiation Using a Time-Domain Finite-Volume Discretization Procedure," *Comput. Phys. Commun.*, vol. 68, pp. 175–196, 1991.

[14] J. P. Cioni and M. Remaki, *Comparaison de Deux Méthodes de Volumes Finis en Électromagnétisme*, Rapport de recherche INRIA, No. 3166, May 1997.

[15] S. K. Godunov, *Equations de la Physique Mathématique*, Editions MIR, Moscow, Russia, 1973.

[16] J. L. Steger and R. F. Warming, "Flux Vector Splitting for the Inviscid Gas Dynamic Equations with Applications to Finite Difference Methods," *Comp. Phys.*, vol. 40, pp. 263–293, 1981.

[17] R. Courant, E. Isaacson, and M. Reeves, "On the Solution of Nonlinear Hyperbolic Differential Equations by Finite Differences," *Commun. Pure Appl. Mathematics*, vol. 5, pp. 243–255, 1952.

[18] C. Müller, *Foundations of the Mathematical Theory of Electromagnetic Waves*, Springer-Verlag, Berlin, 1969.

[19] J. P. Berenger, "Perfectly Matched Layer for the FDTD Solution of Wave–Structure Interaction Problems," *IEEE Trans. Antennas Propagat.*, vol. 44, pp. 110–117, January 1996.

[20] J. R. Wait, "On the PML Concept: A View from the Outside," *IEEE Antennas Propagat. Mag.*, vol. 38, pp. 48–51, April 1996.

[21] Z. Wu and J. Fang, "Numerical Implementation and Performance of Perfectly Matched Layer Boundary Condition for Waveguide Structures," *IEEE Trans. Microwave Theory Tech.*, vol. 43, pp. 2676–2683, December 1995.

[22] W. C. Chew and J. M. Jin, "Perfectly Matched Layers in the Discretized Space: An Analysis and Optimization," *Electromagnetics*, vol. 16, pp. 325–340, 1996.

[23] J. W. Nehrbass, J. F. Lee, and R. Lee, "Stability Analysis for Perfectly Matched Layered Absorbers," *Electromagnetics*, vol. 16, pp. 385–397, 1996.

[24] D. C. Wittwer and R. W. Ziolkowski, "How to Design the Imperfect Berenger PML," *Electromagnetics*, vol. 16, pp. 465–485, 1996.

[25] M. Remaki, L. Fezoui, and F. Poupaud, *Un Nouveau Schéma de Type Volumes Finis Appliqué aux Équations de Maxwell en Milieu Hétérogène*, Rapport de recherche INRIA, No. 3351, 1998.

[26] M. A. Jensen and Y. Rahmat-Samii, "Finite Difference and Finite Volume Time Domain Techniques in Electromagnetics: A Comparative Study," *Radio Sci.*, vol. 31, pp. 1823–1836, 1996.

[27] D. Kalfon, P. Harran-Klotz, and D. Volpert, "Mixed Methods for Solving Maxwell Equations," *Workshop on Approximations and Numerical Methods for the Solution of the Maxwell Equations*, Oxford University, Oxford, March 1995.

[28] J. P. Berenger, "A Perfectly Matched Layer for the Absorption of Electromagnetic Waves," *J. Comput. Phys.*, vol. 114, pp. 185–200, 1994.

[29] J. P. Berenger, "Three-Dimensional Perfectly Matched Layer for the Absorption of Electromagnetic Waves," *J. Comput. Phys.*, vol. 127, pp. 363–379, 1996.

[30] N. K. Madsen and R. W. Ziolkowski, "A Three-Dimensionnal Modified Finite Volume Technique for Maxwell's Equations," *Electromagnetics*, vol. 10, pp. 147–161, 1990.

[31] R. Le Martret and C. Le Potier, "Finite-Volume Solution of Maxwell's Equations in Nonsteady Mode," *La Recherche Aerospatiale*, No. 5, pp. 329–342, 1994.

[32] K. S. Yee and J. S. Chen, "Impedance Boundary Condition Simulation in the FDTD/FVTD Hybrid," *IEEE Trans. Antennas Propagat.*, vol. 45, pp. 921–925, 1997.

[33] A. Taflove and K. R. Umashankar, "Finite Difference Time Domain Method," *Electromagnetics*, vol. 10, pp. 105–126, 1990.

[34] S. M. Rao, *Séminaire ACOSTE*, vol. 1, Paris, April 1–3, 1992.

[35] V. Gobin, J. P. Aparicio, J. Grando, and J. C. Alliot, "The Surface Impedance: A Pertinent Parameter to Describe Finite Conductivity Materials in Numerical Codes," *EMC Symposium*, Zurich, March 12–14, 1991.

[36] V. Gobin, F. Issac, and F. Jaillot, "Modeling of Electrically Thick Materials Theoretical and Experimental Aspects," *ICOLSE 91*, Cocoa Beach, FL, April 16–19, 1991.

[37] G. Franceschetti, "Fundamentals of Steady-State and Transient Electromagnetic Fields in Shielding Enclosures," *IEEE Trans. Electromagnetic Compatibility*, vol. 21, 1979.
[38] R. Holland and L. Simpson, "Finite Difference Analysis of EMP Coupling to Thin Struts and Wires," *IEEE Trans. Electromagnetic Compatibility*, vol. 23, 1981.
[39] A. Panayiotis, A. Tirkas, and C. A. Balanis, "Finite-Difference Time-Domain Method for Antenna Radiation," *IEEE Trans. Antennas Propagat.*, vol. 40, pp. 334–340, 1992.
[40] A. T. De Hoop, "The N-Port Receiving Antenna and Its Equivalent Electrical Network," *Philips Res. Rep.*, vol. 30, pp. 302–315, 1975.
[41] F. Hermeline, "Two Coupled Particle–Finite Volume Methods Using Delaunay–Voronoi Meshes for the Approximation of Vlasov–Poisson and Vlasov–Maxwell Equations," *J. Comput. Phys.*, vol. 106, pp. 1–18, 1993.
[42] D. Issautier, F. Poupaud, J. P. Cioni, and L. Fezoui, "A 2D Vlasov–Maxwell Solver on Unstructured Meshes," *Waves 95, Mathematical and Numerical Aspects of Wave Propagation*, SIAM/INRIA, Mandelieu-La Napoule, pp. 355–371, 1995.

Index